A Companion to Anthropological Genetics

The *Blackwell Companions to Anthropology* offers a series of comprehensive syntheses of the traditional subdisciplines, primary subjects, and geographic areas of inquiry for the field. Taken together, the series represents both a contemporary survey of anthropology and a cutting-edge guide to the emerging research and intellectual trends in the field as a whole.

1. *A Companion to Linguistic Anthropology*, edited by Alessandro Duranti
2. *A Companion to the Anthropology of Politics*, edited by David Nugent and Joan Vincent
3. *A Companion to the Anthropology of American Indians*, edited by Thomas Biolsi
4. *A Companion to Psychological Anthropology*, edited by Conerly Casey and Robert B. Edgerton
5. *A Companion to the Anthropology of Japan*, edited by Jennifer Robertson
6. *A Companion to Latin American Anthropology*, edited by Deborah Poole
7. *A Companion to Biological Anthropology*, edited by Clark Larsen
8. *A Companion to the Anthropology of India*, edited by Isabelle Clark-Decès
9. *A Companion to Medical Anthropology*, edited by Merrill Singer and Pamela I. Erickson
10. *A Companion to Cognitive Anthropology*, edited by David B. Kronenfeld, Giovanni Bennardo, Victor C. de Munck, and Michael D. Fischer
11. *A Companion to Cultural Resource Management*, edited by Thomas King
12. *A Companion to the Anthropology of Education*, edited by Bradley A. Levinson and Mica Pollock
13. *A Companion to the Anthropology of the Body and Embodiment*, edited by Frances E. Mascia-Lees
14. *A Companion to Paleopathology*, edited by Anne L. Grauer
15. *A Companion to Folklore*, edited by Regina F. Bendix and Galit Hasan-Rokem
16. *A Companion to Forensic Anthropology*, edited by Dennis Dirkmaat
17. *A Companion to the Anthropology of Europe*, edited by Ullrich Kockel, Máiréad Nic Craith, and Jonas Frykman
18. *A Companion to Border Studies*, edited by Thomas M. Wilson and Hastings Donnan
19. *A Companion to Rock Art*, edited by Jo McDonald and Peter Veth
20. *A Companion to Moral Anthropology*, edited by Didier Fassin
21. *A Companion to Gender Prehistory*, edited by Diane Bolger
22. *A Companion to Organizational Anthropology*, edited by D. Douglas Caulkins and Ann T. Jordan
23. *A Companion to Paleoanthropology*, edited by David R. Begun
24. *A Companion to Chinese Archaeology*, edited by Anne P. Underhill
25. *A Companion to the Anthropology of Religion*, edited by Janice Boddy and Michael Lambek
26. *A Companion to Urban Anthropology*, edited by Donald M. Nonini
27. *A Companion to the Anthropology of the Middle East*, edited by Soraya Altorki
28. *A Companion to Heritage Studies*, edited by William Logan, Máiréad Nic Craith, and Ullrich Kockel
29. *A Companion to Dental Anthropology*, edited by Joel D. Irish and G. Richard Scott
30. *A Companion to Anthropology of Environmental Health*, edited by Merrill Singer
31. *A Companion to South Asia in the Past*, edited by Gwen Robbins Schug and Subhash R. Walimbe
32. *A Companion to the Anthropology of Africa*, edited by Roy Richard Grinker, Euclides Gonçalves, Christopher B. Steiner, and Stephen C. Lubkemann
33. *A Companion to Anthropological Genetics*, edited by Dennis H. O'Rourke

Forthcoming

A Companion to Witchcraft and Sorcery, edited by Bruce Kapferer
A Companion to Oral History, edited by Mark Tebeau

A Companion to Anthropological Genetics

Edited by
Dennis H. O'Rourke

WILEY Blackwell

This edition first published 2019

Registered Office
John Wiley & Sons, Inc., 111 River Street, Hoboken, NJ 07030, USA

Editorial Office
9600 Garsington Road, Oxford, OX4 2DQ, UK

For details of our global editorial offices, customer services, and more information about Wiley products visit us at www.wiley.com.

Wiley also publishes its books in a variety of electronic formats and by print-on-demand. Some content that appears in standard print versions of this book may not be available in other formats.

Library of Congress Cataloging-in-Publication data applied for

9781118768990 (hardback)

Cover Design: Wiley
Cover Images: © Mina De La O/Getty Images; © Matej Ograjensek/Alamy Stock Photo; © MarcelClemens/Shutterstock

Set in 10.5/13pt Galliard by SPi Global, Pondicherry, India

Printed in the United States of America

10 9 8 7 6 5 4 3 2 1

Derek F. Roberts (1925–2016)
Foundational scholar, mentor to many, and friend

Contents

About the Editor

Dennis H. O'Rourke, PhD, is Foundation Distinguished Professor of Anthropology at the University of Kansas, USA, and Professor Emeritus at the University of Utah, USA. His current research focuses on the analysis of genetic and genomic data to reconstruct regional population histories, particularly in the North American Arctic, and to elucidate the earliest arrival and dispersal of people in the Western Hemisphere by merging the genetic, archaeological, and paleo-ecological records. He is also particularly engaged in aiding the development and promotion of ethical practices in community based genetic research. He is a co-editor of *Human Biology: An Evolutionary and Biocultural Perspective, 2nd Edition* (Wiley Blackwell, 2012), an associate editor of *The International Encyclopedia of Biological Anthropology* (Wiley Blackwell, 2018), and has authored numerous articles and book chapters on topics in anthropological genetics.

Notes on Contributors

Jessica D. Bardill is an assistant professor in the Department of English at Concordia University in Montreal, Quebec. Alongside a range of literatures from Indigenous communities, she engages with the possibilities of biological sciences and policy, with particular emphasis on genetics and genomics, by, for, and with Indigenous peoples. Since 2011, she has helped to lead the decolonial bioethics training for the Summer internship for INdigenous peoples in Genomics (SING) workshops.

Abigail W. Bigham is an assistant professor of anthropology at the University of Michigan. Abby received her BA from the University of Arizona and her PhD from The Pennsylvania State University. She completed a postdoctoral fellowship at the University of Washington. Her current research focuses on understanding human genetic adaptation to environmental pressures and how these adaptations affect the range of modern human phenotypic diversity.

Rafael Bisso-Machado graduated in biology at the Federal University of Rio Grande do Sul (UFRGS) in Porto Alegre, Brazil, holding master's (2010) and PhD (2014) degrees in Genetics and Molecular Biology (UFRGS). He spent one year as a postdoctoral researcher at UFRGS, and was associate professor at Universidad de la República, Uruguay, until 2017. His main topics of research are historical and anthropological genetics, especially concerning Y-chromosome, mitochondrial DNA; intra- and interpopulational diversity in Native and neo-American populations; and gene–culture coevolution based on an interdisciplinary approach.

Graciela S. Cabana is associate professor of anthropology at the University of Tennessee, Knoxville. Her expertise is in anthropological genetics, and she focused her early research efforts on developing core analytical and simulation methods for the analysis of genetic data, particularly ancient DNA data. She is the co-editor of the 2011 edited volume *Rethinking Anthropological Perspectives on Migration*. Her current efforts are directed toward actively engaging social and ethical concerns into anthropological genetic research.

Murray P. Cox is a computational biologist specializing in human genomics, computer modeling, and statistical inference. Dr. Cox is currently an associate professor in the Institute of Fundamental Sciences at Massey University, New Zealand, and an inaugural Rutherford Fellow of the Royal Society of New Zealand.

Michael H. Crawford, professor of anthropological genetics and director of the Laboratory of Biological Anthropology, University of Kansas, has conducted extensive field investigations in Siberia, the Aleutian Islands, Central America, and the Caribbean, and religious and geographic genetic isolates in the Midwestern United States and Europe. Professor Crawford served as editor-in-chief of the journal *Human Biology* from 1989 to 2000. He has published widely with over 400 articles, chapters, and books on various aspects of anthropological genetics.

Stefan A. Czerwinski received his BA in anthropology from Binghamton University and his PhD in biological anthropology from the University at Albany. He completed a postdoc in genetic epidemiology at the Southwest Foundation for Biomedical Research in San Antonio, Texas. From there, he moved to Wright State University, where he eventually became director of the Fels Longitudinal Study at the Lifespan Health Research Center. He is currently professor of epidemiology at the University of Texas School of Public Health, Brownsville Campus. Dr. Czerwinski's research interests include growth and development, obesity, and chronic disease risk.

Ellen W. Demerath received her AB from Harvard University and her PhD from the University of Pennsylvania in biological/physical anthropology and human biology. In 1997, she joined the Lifespan Health Research Center at Wright State University School of Medicine in Dayton, Ohio, where she worked with the Fels Longitudinal Study on the developmental origins of obesity. In 2007, she joined the faculty of the University of Minnesota School of Public Health, where she has continued her federally funded research program on the interplay of genetic and early-life environmental factors on child development and cardiometabolic disease risk, including an interdisciplinary cohort study on maternal obesity, breast milk composition, and infant outcomes. Professor Demerath is the director of the Maternal and Child Health MPH program and co-directs the university's Driven to Discover Research Facility.

Nelson Jurandi Rosa Fagundes graduated in biology at the Federal University of Rio Grande do Sul (UFRGS) in Porto Alegre, Brazil, holding master's (2001) and PhD (2007) degrees in genetics and molecular biology (UFRGS). After a stage as a postdoctoral researcher at UFRGS, he acted as adjunct professor at the Federal University of Minas Gerais, Brazil. He has been an adjunct professor at

UFRGS since 2011, working with evolutionary biology and population genetics of several different organisms, and advising students in both the genetics and molecular biology and animal biology postgraduation programs.

Nanibaa' A. Garrison (Navajo), PhD, is a faculty member in the Treuman Katz Center of Pediatric Bioethics at Seattle Children's Research Institute and assistant professor of pediatrics in the Division of Bioethics at the University of Washington. Her research focuses on the ethical, social, and cultural implications of genetic and genomic research for Indigenous communities. Using community-based research approaches, she engages with tribal communities to develop policies and guidance for tribes.

Omer Gokcumen graduated from Bogazici University in Istanbul, Turkey, with a degree in molecular biology and genetics. He defended his thesis, "Ethnohistorical and Genetic Survey of Four Central Anatolian Settlements," and was awarded a PhD in anthropology from the University of Pennsylvania. He worked on human and primate genomic variation as a postdoctoral fellow at Harvard Medical School. He is currently an assistant professor in the Department of Biological Sciences at the University at Buffalo. The research in his laboratory uses anthropological genomics tools to address the broad question "What makes us human?"

Henry C. Harpending was an anthropological geneticist whose work focused on patterns of genetic diversity among human populations. Following a PhD from Harvard, he held faculty positions at Yale, the University of New Mexico, The Pennsylvania State University, and the University of Utah. At the latter institution, he was the Thomas Distinguished Professor of Anthropology. He was also a fellow of the American Association for the Advancement of Science and a member of the National Academy of Sciences.

Toomas Kivisild graduated with a PhD in genetics from the University of Tartu, Estonia, in 2000. After a one-year postdoctoral fellowship at Stanford University (2002–2003), he worked at the Estonian Biocentre (2003–2005) and the Department of Evolutionary Biology, University of Tartu (2005–2006). Since 2006, he has led the Human Evolutionary Genetics research group at the University of Cambridge. His research interests have included human evolution and evolutionary population genetics, with a particular focus on questions relating global genetic population structure with evolutionary processes such as selection, drift, migration, and admixture.

Leslie A. Knapp received her degrees in anthropology from UCLA, and she was a postdoc at the University of Wisconsin. From 1997 to 2013, she was a faculty member in biological anthropology at the University of Cambridge, where her PrIME (Primate Immunogenetics and Molecular Ecology) laboratory hosted 8 postdoctoral researchers and 16 PhD students, including international scholars from Japan, France, Switzerland, Chile, Brazil, and Mexico. Currently, she is professor and chair of anthropology at the University of Utah. Her research concentrates on the evolution of immune response genes and the relationship between genes, disease, and behavior in primates.

Ripan S. Malhi is a Richard and Margaret Romano Professor in the Department of Anthropology, School of Integrative Biology, and the Carl R. Woese Institute for Genomic Biology at the University of Illinois Urbana-Champaign. He earned his PhD in anthropology at the University of California, Davis, and completed a postdoc in human genetics at the University of Michigan Medical School. Prior to starting at the University of Illinois, he cofounded a biotechnology company, Trace Genetics, Inc. Ripan is a molecular anthropologist who partners with Indigenous communities to study the evolutionary histories of Indigenous peoples of North America. He organizes the Summer internship for INdigenous peoples in Genomics (SING), an NSF and NIH-funded program for Indigenous community members to learn innovative concepts and methods at the cross section of indigeneity and genomics. He is executive editor of *Human Biology* and an associate editor of the *American Journal of Physical Anthropology*.

Aida T. Miró-Herrans is a postdoctoral researcher at the University of Texas at Austin. Her previous work studied how the combined use of simulated genetic data and nongenetic data enhanced demographic reconstructions of the migration of modern humans out of Africa. Currently, she studies genome-wide patterns of variation in Native Americans from the southern United States to identify the effect of European contact on Native American evolutionary history.

Amy L. Non is an associate professor of anthropology at the University of California, San Diego. She is a molecular anthropologist with an interest in genetic and sociocultural contributors to racial and social inequalities in health. Her research explores how social experiences can become biologically embedded early in life to affect health, through epigenetics and other mechanisms. Her current focus is on epigenetic mechanisms that can link early adverse environmental exposures with altered gene expression, potentially resulting in long-term health consequences.

Heather L. Norton is an assistant professor of anthropology at the University of Cincinnati. Her research focuses on two primary lines of investigation relating to the human pigmentary phenotype. In the first, she identifies genetic variants associated with quantitatively assessed variation in skin, hair, and eye pigmentation across a wide range of populations. In the second, she characterizes the evolutionary forces, including natural selection, that have shaped global variation in the human pigmentary phenotype.

Jennifer A. Raff is an assistant professor of anthropology at the University of Kansas, Lawrence. Her research focuses on the initial peopling of the Americas and regional population histories of the North American Arctic, the Aleutian

Islands, and Mid-Continental North America. She is also interested in public science literacy and science communication. She is currently serving as the vice president of the American Association of Anthropological Genetics.

John H. Relethford is a biological anthropologist specializing in human evolutionary biology, with major foci on modern human origins and anthropological genetics (specifically quantitative traits). He received his PhD in anthropology at the State University of New York at Albany in 1980. He is currently a SUNY Distinguished Teaching Professor. He is past president and vice president of both the American Association of Physical Anthropologists and the American Association of Anthropological Genetics, and is a fellow (and former chair of Section H) of the American Association for the Advancement of Science.

Christopher A. Schmitt is assistant professor of anthropology and biology at Boston University, where he is also affiliated faculty in the Women's, Gender, and Sexuality Program. His research interests include juvenile growth and development, using quantitative genetics and genomics techniques to better understand the interactions between climatic adaptions, energetics, human impacts, and obesity. He also works with several professional societies to advocate for the increased visibility and success of fellow LGBTQIA scholars.

Anne C. Stone is a professor in the School of Human Evolution and Social Change at Arizona State University. Her research examines population history and how humans and the great apes have adapted to their environments, including their disease and dietary environments. This has three foci: Native American population history, the evolutionary history of the great apes, and the co-evolutionary history of pathogens (particularly those causing tuberculosis and leprosy) with humans and other primates.

Zaneta M. Thayer is an assistant professor of anthropology at Dartmouth College. Her research is focused on understanding how social inequalities create health inequalities, and how these effects can be passed on across generations. She conducts research with a multiethnic birth cohort in New Zealand and with Native American populations in the United States.

Nicole S. Torosin received her bachelor's degree in anthropology and biology at Northwestern University. She obtained her MS in biological anthropology at the University of Utah and is pursuing her PhD studies in Dr. Leslie Knapp's laboratory. Her research is on the evolution of the primate immune system, focusing on howler monkeys and humans in Central and South America.

Bradford Towne received undergraduate degrees in psychology and anthropology from the University of Northern Colorado, and MA and PhD degrees in physical anthropology from the University of Utah. His postdoctoral training was in psychiatric genetics at Washington University and genetic epidemiology at the Southwest Foundation for Biomedical Research (now Texas Biomedical Research

Institute). Dr. Towne joined the faculty of Wright State University School of Medicine in 1991 to develop the Fels Longitudinal Study as a resource for genetic epidemiologic studies of growth and development, body composition, and complex disease risks over the lifespan. Dr. Towne is professor and curator of the Fels Longitudinal Study.

Amanda VanSteelandt is a visiting researcher at Arizona State University. She conducts interdisciplinary research on infectious diseases and the social, cultural, and biological factors that promote or prevent their spread in human populations. Her research draws on methods from biocultural anthropology, epidemiology, microbiology, and genetics. Her previous research projects have varied from examinations of tuberculosis, intestinal parasitosis, and Chagas disease in indigenous communities of Paraguay to hospital-acquired, antibiotic resistant infections in a Canadian hospital.

Carrie C. Veilleux is currently a postdoctoral research affiliate in the Anthropology Department at the University of Texas at Austin. Her research focuses on the ecology and evolution of sensory systems in humans and nonhuman primates. She is particularly interested in effects of foraging strategy and habitat preference on intraspecific and interspecific variation in color vision and taste discrimination.

Noreen von Cramon-Taubadel is an associate professor of anthropology at the University at Buffalo and director of the Buffalo Human Evolutionary Morphology Lab. Her research focusses on the evolution of the primate skeleton, with a particular emphasis on the microevolution of the human cranium. She is especially interested in assessing the relative effects of past evolutionary processes, such as drift and selection, in shaping contemporary patterns of human and nonhuman primate skeletal shape variation.

Michael D. Weight is currently a research assistant in the Anthropology and Mathematics Departments at the University of Utah. His previous work emphasized the phenotypic consequences of recent selection. His current work focuses on developing a rigorous study of adaptation following the Formal Darwinism Project and employs costly signaling theory to account for the evolution of collective action.

Sloan R. Williams is an associate professor and head of anthropology at the University of Illinois at Chicago. She is an anthropological geneticist and bioarchaeologist whose work centers on the use of both modern and ancient DNA to study migration patterns and trade networks among living and archeological populations in the Andes and East Africa. She is also interested in burial practices and mortuary archeology, and has conducted fieldwork in Peru and Kenya.

Preface

Like most academic disciplines, anthropological genetics experienced a long period of development and maturation, characterized by "hybridization" between physical (biological) anthropology, human genetics, and molecular biology. The first acknowledgment, of which I'm aware, of anthropological genetics as a distinct academic field is in an article title ("Assumption and Fact in Anthropological Genetics") by Derek Roberts in 1965. Thus, we have just recently passed the 50-year anniversary of the "emergence" of the discipline. The real foundation of the field, however, is usually pinned to Crawford and Workman's seminal volume, *Method and Theory in Anthropological Genetics*, published 45 years ago this spring.

Initially, the field was primarily focused on assessing gene frequency differences between populations, identifying evolutionary forces that resulted in population differences as measured by genetic distance statistics, and linking patterns of genetic diversity and population structure to underlying demographic processes – what today might be characterized as genetic histories of regional populations. The scope of anthropological genetics has expanded rapidly in recent decades, encompassing studies in the genetics of disease and health, primate genetics and behavior, phylogenetics and molecular evolution, the genetics of quantitative variation, as well as the evolution and genetics of human adaptation to diverse environments. Anthropological genetics has also made significant contributions to archeology and the study of prehistory via the developing field of ancient DNA (aDNA) analysis. The dramatic growth of anthropological genetics is evidenced by the large scientific literature that now characterizes the field, spanning several disciplinary journals as well as numerous books focused on specific issues and/or analytical approaches. Given this growth, and the notable advances in genetic, molecular, and statistical methods that have dramatically influenced the field in recent years, a new synthetic treatment of the scope of anthropological genetics is appropriate and timely.

The following chapters are designed to cover the scope of research now routinely conducted under the rubric of anthropological genetics. **Part 1** of the volume sets the stage for what follows by providing context for the development of the field and current best practices. Chapter 1 provides the background for the development of the discipline, from its earliest efforts at gene frequency descriptions in select populations to the modern genomic approaches that characterize human evolutionary genetics, by one of the founders and promoters of the field, M.H. Crawford. As genetic and genomic analyses have become more powerful, risks to personal and community identities have become of greater concern should there be breaches of confidentiality or misuse of individual genomic profiles. In addition, consultation with identifiable communities that are often the subject of such study has taken on primary importance. Increasingly, investigators partner with participant communities as research collaborators, attempt to carefully identify both community and individual risks and benefits of the proposed research, and solicit input and advice from the community on acceptable questions and approaches. Accordingly, Garrison and Bardill (in Chapter 2) explore the ethical issues attendant to the rapidly expanding use of genomics in ancestry testing, at both the individual and population levels, while Malhi (Chapter 3) draws on his own experience in human population genetics and community consultation to explicate ethical issues in community-based research programs.

Part 2 of the volume explores topics in macroevolution and phylogenetics, topics that were not really emphasized in the early years of the field, but that have become increasingly important and prominent with the development of newer genomic and bioinformatic tools. A fundamental tenet of molecular evolution is that genomes evolve at a constant rate. But not all genomes, or components of genomes, evolve at the same rate. Knowing how to calibrate the various molecular clocks is essential for using molecular data to date evolutionary events, a topic summarized by Cox in Chapter 4. Use of the "molecular clock" is central to the review of primate genomics and phylogenetics by Schmitt and Gokcumen (in Chapter 5), who discuss the variation in mutation rates and evolutionary histories of different primate lineages. In addition, this chapter highlights the enhanced resolution afforded by next-generation sequencing approaches, from new insights into the relationship between functional genomic and phenotypic variation to the role of new genomic methods in field primatology for study of behavioral and ecological variation. As in so many cases, the newer genomic approaches utilized in primate studies permit an examination of the demographic history of both species and local populations.

Use of molecular dating methods is also central to Williams' (Chapter 6) summary of hominin phylogenetic diversity. The role of molecular genomic approaches in addressing questions of human evolution and diversity is a recent adjunct to the traditional methods of paleoanthropology, and such approaches have been made possible by advances in molecular methods via aDNA analyses. Integrating separate streams of knowledge from different fields to focus on common questions is always a major challenge in science. Von Cramon-Taubadel takes on the task by assessing the correspondence between molecular diversity, geography, ecology, and quantitative phenotypic variation in modern humans, illustrating how the methods are both powerful and broadly applicable to questions of subspecific diversity and history (Chapter 7). Completing this section of the book, Miró-Herrans (Chapter 8) demonstrates the importance of demographic processes in structuring genetic variability in time and space. She illustrates how simulation is used to reconstruct and test alternative demographic scenarios and, thus, human evolutionary history using approximate Bayesian computational methods.

The five chapters in **Part 3** treat topics that are nearer to the core of anthropological genetics at its inception, with one possible exception. Relethford provides a review of traditional population structure and population history approaches, and clarifies their differences (Chapter 9). He brings those traditions into the present with updated methods and examples from the study of mate choice, population and cultural subdivision, and isolation by distance. Importantly, he shows how such local analyses are related to broader questions and analyses of both regional and global human population history. Similarly, Cabana examines the evolutionary roles of migration (gene flow) and genetic drift in structuring the patterns of genetic variation observed in human populations (Chapter 10). She discusses clearly the importance of population size in evaluating such effects in the context of classical population genetic models. This chapter also introduces analyses of aDNA from archeological samples and the concomitant use of simulation to test hypotheses of ancestry in regional population studies. Taking an explicitly geographic perspective, Kivisild (Chapter 11) reviews the patterns of genetic and genomic variation observed in Eurasian populations and explores what they tell us about how areas of the region were colonized. In particular, this chapter emphasizes the role of environmental change in structuring genomic variation over time, in this case the importance of climate change during the Last Glacial Maximum, and the shift to a more settled agrarian lifeway during the Neolithic. In a similar vein, Bisso-Machado and Fagundes (Chapter 12) examine patterns of genetic and genomic diversity in the Western Hemisphere and describe alternative models of human dispersal into the hemisphere that can be tested by both archeology and genetics. Finally, in a contribution highlighting recent innovations in methods, Raff (Chapter 13) explores the utility of aDNA analyses in contributing to bioarchaeological studies. In so doing, the chapter links to many other contributions in the volume by touching on the importance of community consultation and ethical precepts in aDNA research, the role of these methods in assessing identity, contributions to paleopathology, and reconstructing regional population histories.

Part 4 of the volume contains chapters that survey how we examine patterns in human variation, with particular reference to the role of natural selection. Many of the topics covered here relate to the evolution of specific traits or phenotypes that are of long-standing interest in anthropology and human evolutionary biology, and they are brought up-to-date with modern analytical methods in the chapters included. Weight and Harpending (Chapter 14) provide a historical perspective on the debate between the biometrician and Mendelian schools in accounting for patterned variation, and argue that this continues in some ways in molecular versus organismal biological approaches. Having set this context, these authors continue to explore recent research on how we think about tempo and mode in human evolution. Taking a finer grained look at human adaptability, Bigham (Chapter 15) reviews the current evidence for the role of natural selection in permitting humans to adapt to extreme environments, permitting our species to successfully disperse to all continents except Antarctica. Evidence for natural selection is amassed from populations that occupy environments from high-altitude locales in the Andes, the Himalayas, and Ethiopia to the extreme cold of the Arctic. In each of these environmental contexts, the physiological responses and health effects are examined. In Chapter 16, Veilleux reviews the evidence for selection on sensory polymorphisms and, therefore, its role in human dietary adaptation. In all mammals, including primates, sensory systems mediate the interactions of an organism with the external environment. Among primates, particularly humans, the evolution of color vision and taste sensitivities affects diet and general foraging strategies. Variation in opsin genes (responsible for color vision) and taste receptor genes in humans is reviewed in this chapter, along with an examination of how adaptive dietary shifts may have influenced the evolution of these sensory polymorphisms.

Human skin color has a long, and sometimes notorious, history in anthropology. In Chapter 17, Norton examines the evidence for the operation of natural selection on the geographically structured variation in human skin pigmentation. The correlation between ultraviolet radiation and degree of skin pigmentation strongly implicates natural selection in the evolution of skin color. As noted by Norton in this chapter, most studies to date have been conducted on European populations, with much less attention given to the genetic architecture and evolution of pigmentation in other geographic areas. She encourages future research to address this deficiency in our understanding of the evolution of human skin color. The role of genetics in growth trajectories that produce such common phenotypes as stature and body mass index across the life course is the subject of Chapter 18 by Czerwinski, Demerath, and Towne. Historically, genetic studies in growth and development have been few, but recent developments in analytical techniques in statistical genetics have reversed this trend. Accordingly, the authors of this chapter provide an up-to-date review of the genetic epidemiology of normal human growth and development. Concluding this section, Non and Thayer (Chapter 19) remind us that not all variation derives from genetic differences among individuals. Epigenetic markers in the genome affect gene function and expression and serve as one link between the environment and the genome. Epigenetic markers can enhance human plasticity across environments without alteration in the underlying DNA sequence. Although conceptually epigenetics as an area of inquiry is decades old, recent methods of analysis and the recognition that both the external environment and the genome itself affect epigenetic variation have stimulated an explosion of research. To address this relatively new area of analysis, this chapter illustrates the rapid pace of discovery and evolution of molecular methods, their effect on the discipline, and our constantly evolving knowledge base.

The final section of the volume, **Part 5**, contains two chapters relating to anthropological genetic research on the evolution of health and disease. Knapp and Torosin (in Chapter 20) review the evolution of immune response in the primate order. Summarizing work on the evolution of both the innate and adaptive immune systems, the chapter clearly describes the role of different cell types in mounting immune responses in each system. The chapter particularly focuses on the major histocompatibility complex (MHC) and its role in immune response. Finally, VanSteelandt and Stone (Chapter 21) treat the related area of the evolution of human pathogens and how such studies inform a newer perspective in evolutionary medicine. These authors explore human–pathogen co-evolution and summarize the ways in which pathogens have evolved strategies for avoiding a host's immune response. Through examples and case studies, this chapter describes research on how evolution has shaped patterns of human health and disease and how such work links anthropological genetics with evolutionary medicine.

The chapters contained herein survey a remarkably diverse array of modern methods and results in contemporary research on nonhuman primate and human evolutionary genetics. Almost all chapters highlight, either explicitly or implicitly, the importance of the continual evolution of genetic analytical methods and the importance of population demography in providing critical context for genetic and genomic studies of human evolution and adaptation. Similarly, many of the chapters relate core topics to historical, or continuing, debates in anthropology and genetics, for example the relative importance of selection versus drift in structuring variation in population, or quantitative versus Mendelian approaches to the study of phenotypic diversity. Collectively, I hope these chapters, by top researchers in the relevant specialties, provide a snapshot of the continuing growth of anthropological genetics and the field's role in explicating the evolution of our species.

Dennis H. O'Rourke
Lawrence, Kansas, May 2018

PART I Anthropological Genetics in Context

CHAPTER 1

History and Evolution of Anthropological Genetics

Michael H. Crawford

Laboratory of Biological Anthropology, University of Kansas, Lawrence, KS, USA

Introduction

"Anthropological genetics is a synthetic discipline that applies the methods and theories of genetics to evolutionary questions posed by anthropologists" (Crawford 2007a, p. 1). These questions involve the processes of human evolution, reconstruction of the human diaspora (e.g. out-of-Africa migrations), causes and patterns of human variation, and biocultural interactions in complex, multifactorial traits and diseases such as coronary and vascular disease, autoimmune conditions, diabetes, obesity, cancer, and biological aging. Anthropological genetics not only shares with the field of human genetics traditional methods of analyses but also adds methods of field investigation in small, non-Western populations with limited numbers of founders. A contrast of the methodologies employed in human genetics versus anthropological genetics is contained in Table 1.1 (updated from Crawford 2000b).

A Companion to Anthropological Genetics, First Edition. Edited by Dennis H. O'Rourke.

Table 1.1 Comparison of tendencies in human genetics versus anthropological genetics.

Anthropological genetics	*Human genetics*
1. Broader biocultural perspective on genetic–environmental interactions	1. Mechanisms and processes – particularly disease
2. Population focus; pedigrees utilized to measure familial resemblance	2. Families of probands, twins, and twin families
3. Small, reproductively isolated populations – often non-Western	3. Larger, urban, and clinical samples
4. Culturally homogeneous populations	4. Populations may be heterogeneous by race, socioeconomic factors, occupation, and lifestyle
5. Sampling representative of normal variation in populations and between populations	5. Sampling based on clinical ascertainment
6. Attempts made to characterize and measure the effects of the environment	6. Environmental variation rarely assessed. It is often assumed that $e^2 = 1 - h^2$.
7. Focus on normal variation in complex traits	7. Dichotomy of disease versus normality

Source: Crawford (2000a).

Roots of Anthropological Genetics

The ancestral roots of anthropological genetics are intertwined with early developments in evolutionary biology, population genetics, and biological anthropology. O'Rourke (2003) noted that the field of anthropological genetics was further cross-fertilized by the developments in molecular genetics and bioinformatics. However, the roots of anthropological genetics were preceded by more than a century of discovery in evolutionary theory and Mendelian genetics. The concept of natural selection, traced to Charles Darwin's publication of *Origin of Species* in 1859, differed slightly from the contemporary version. Darwin stressed differential survivorship (survival of the fittest) instead of stressing fertility, was unaware of Mendel's research, and explained genetic variation using the Lamarckian concept of inheritance of acquired characteristics. More recent research in epigenetics has revealed that the environment does influence the expression of the genome through methylation. Table 1.1 provides a timeline of significant developments in genetics and anthropological genetics.

The field of anthropological genetics resulted from the convergence of two academic streams: (i) biological or physical anthropology, and (ii) human genetics. These disciplines initially flowed together in a symposium entitled "Methods in Anthropological Genetics," convened by me on February 24–28, 1971, at the School of American Research in Santa Fe, New Mexico. Twelve researchers

Figure 1.1 The participants in the School of American Research symposium on anthropological genetics, held in 1970 in Santa Fe, New Mexico. Back row (standing from left): Steven Vandenberg, Jean Benoist, Frank Livingstone, Gabriel Lasker, Peter Workman, Eugene Giles, Christy Turner, Francis Johnston, and James Spuhler. Front row (seated): Michael Crawford, Derek Roberts, and William Howells. *Photo*: Michael H. Crawford. *Source*: Crawford (2007a). Reproduced with permission of Cambridge University Press.

from the United States, France, and the United Kingdom participated in this symposium (see Figure 1.1). With the addition of several key contributions, primarily covering evolutionary and epidemiological issues, a volume entitled *Methods and Theories of Anthropological Genetics* was published by the University of New Mexico Press (Crawford and Workman 1973). An introduction to this volume by Derek F. Roberts, entitled "Anthropological Genetics: Problems and Pitfalls," pointed out that this field focused on the

> "terrain shared by human genetics and physical anthropology, the exploration of whose fertile soil will continue to be both adventurous and profitable" (Roberts 1973, p. 2).

How prescient Roberts turned out to be about the development of the field of anthropological genetics throughout the 1980s and 1990s and well into the new millennium with the methodological stimulation from the field of molecular genetics. This was the first of five volumes that were devoted entirely to anthropological genetics and defined the state of the art in the 1970s.

During the 1980s, three volumes, published in the Plenum Press series *Current Developments in Anthropological Genetics*, provided key updates in theory and methodology for this new synthetic field. Volumes 1 and 2 were based on a distinguished lecture program developed at the University of Kansas (KU). The most eminent specialists in biological anthropology and genetics each spent one week providing public lectures and training graduate students and faculty. This program was financially made possible through the administrative release of my salary, after I was awarded a five-year National Institutes of Health Career Development Award. The first volume focused on theory and methods, and included the application of quantitative genetics to complex diseases as visualized by the research of Theodore Reich, R.C. Elston, D.C. Rao, Newton Morton, and C.C. Li (Mielke and Crawford 1980). The second volume emphasized ecology and population structure with geographically widely cast nets to include population structures of: nonhuman primates; the !Kung of Africa; circumpolar populations of Siberia, Alaska, and Greenland; the pastoral Tuaregs of the Sahara; Romany Gypsies; South American Swidden agriculturalists; the Malaysian Semai; Bougainville Islanders: New Guinea populations; the Aland islands, Finland; the Hutterite isolates of Alberta; and Irish anthropometrics. This was the most extensive worldwide analysis of variation in the genetic structure of human populations (Crawford and Mielke 1982). The third volume in this series, entitled *Black Caribs: A Case Study in Biocultural Adaptation*, was devoted to the genetic structure, origins, admixture, morphology, and social organization of the Black Carib (Garifuna) people of Central America. This research was primarily built around a series of field expeditions that I organized to Belize, Guatemala, and St. Vincent Island (Crawford 1984). Additional research on Black Caribs of St. Vincent and their origins and migrations appeared in a series of articles (Crawford 1983).

In 1989, when I became editor-in-chief of the journal *Human Biology*, I narrowed the journal's focus from general human biology to anthropological genetics – in the broadest sense. This was in part necessitated by the disassociation of the Human Biology Council from Wayne State University Press and the classic journal *Human Biology* (founded in 1929 by Raymond Pearl) and the creation of a new journal (*American Journal of Human Biology*). In 1994, the American Association of Anthropological Genetics (AAAG) was founded (after a series of workshops and meetings in Lawrence, Denver, and San Antonio) to promote the field of anthropological genetics and to provide guidance and support for the implementation of the new direction of *Human Biology*. In the new millennium, AAAG became a rapidly growing association with an assortment of activities and scientific platforms, usually nestled within the annual programs of the American Association of Physical Anthropology (AAPA) and the American Society of

Human Genetics (ASHG). A number of highly successful workshops on topics dealing with genetic epidemiology, molecular genetics, histocompatibility (HLA), and ethics were organized by AAAG members and generated considerable interest in the millennial anthropological geneticists. Beginning in 1989, a series of special issues of *Human Biology*, devoted to anthropological genetics, were published on a yearly basis. The first of the series, "Foundations of Anthropological Genetics," was built around the selection and reprinting of the most significant articles to appear in *Human Biology* from 1929 to 1980. Each article was updated by its author, or if the original author was deceased, an eminent, contemporary specialist in the same discipline wrote an introduction (Crawford and Lasker 1989). During the 60-year history of *Human Biology*, many contributors to genetics and theory published key articles. They included J.B.S. Haldane, Theodosius Dobzhansky, J.V. Neel, Arno Motulsky, Frank Livingstone, J.F. Crow, L.L. Cavalli-Sforza, and D.F. Roberts.

Table 1.2 provides a timeline of the significant developments in Mendelian, population, molecular, and anthropological genetics. It traces the first glimmers of evolutionary theory from the pages of Charles Darwin's tome on the *Origin of Species* to August Weismann's demonstration of the separation of germ plasm from somatic cells, thus challenging the underlying basis of Lamarck's theory of inheritance of acquired characteristics. This timeline notes Mendel's extensive research on the inheritance of discrete phenotypes of peas as well as Watson and Crick's pioneering deciphering of the genetic code. Key methodologies are traced from the development of electrophoresis by Orville Smithies to Kary Mullis' highly innovative polymerase chain reaction (PCR) system to sequencing methodology by Sanger to high-throughput sequencing and the characterization of the human genome. These innovative methodologies provided anthropological geneticists with the tools to explain human adaptation through chance mutations and selection in hemoglobin, glucose-6-phosphate dehydrogenase (G6PD), and Duffy null genes to malaria; the high-throughput sequencing of a 38 000-year-old Neanderthal genome; reconstruction of the recent human diaspora out of Africa; and the peopling of the Americas by populations from Siberia. During my 50 years of research, my earliest experiments focused on genetic variation in hemoglobin molecules based on electrophoresis and protein chemistry in *Macaca nemestrina* (Crawford 1966). This was the *pre-DNA era*, and the only available genetic tools at that time were blood group typing, starch gel electrophoresis for phenotyping proteins, and an assortment of immunological techniques (such as microcomplement fixation [MCF] and immunodiffusion) for quantifying genetic similarity and phylogeny. At this time, computation and analyses involved mainframe computers with the necessary key-punching of programs and data on cards that had to be hand-carried to a computer center for analysis. Currently, laptops have the equivalent storage capacity of the mainframes of the 1960s, while the field of bioinformatics requires supercomputers to handle billions of bytes of data. With the molecular and informational revolutions, we are now able to sequence

Table 1.2 Timeline of significant developments in genetics and anthropological genetics.

1859	Publication of Charles Darwin's *On the Origin of Species*
1860	Friedrich Meischer first isolated DNA from the cell nucleus in the form of nuclein
1880	August Weismann demonstrated separation of germ plasm from somatic cells
1900	Rediscovery of Mendel's laws of inheritance
1901	First Mendelian polymorphism ABO blood group system described by Karl Landsteiner
1902	Archibald Garrod demonstrated the mode of inheritance of inborn errors of metabolism – alkaptonuria
1905	Edmund Wilson and Nettie Stevens independently discovered that XY chromosomes determined gender in humans
1908	Formulation of the principle of genetic equilibrium, initially attributed to G.H. Hardy (Wilhelm Weinberg was added in the 1940s), followed by the discovery that W.W. Castle preceded both in 1903
1919	Population variability in the frequency of blood group genes described by Ludwik Hirschfeld and Hanka Hirschfeld in troops during World War I
1930–1932	R.A. Fisher, J.B.S. Haldane, and S. Wright published the mathematical bases of the modern population genetic theory of evolution
1937	Theodosius Dobzhansky published *Genetics and the Origin of Species*, which further fleshed out the modern synthesis by reconciling the evidence of naturalists with that of theoretical geneticists
1949	A molecular basis for sickle cell disease is demonstrated by Linus Pauling et al., followed by the research of J.V. Neel (1957)
1952	A.D. Hershey and M. Chase experiments demonstrated that DNA, not protein, is the genetic material
1953	James Watson and Francis Crick break the genetic code
1954	Anthony C. Allison reveals relationship between the sickle cell trait and malaria
1955	Oliver Smithies developed a medium (starch gel) for separating classes of proteins based on the charge and size of molecules through electrophoresis
1956	Human chromosomal numbers correctly characterized by J. Hin Tjio and A. Levan
1972	Richard Lewontin apportioned human genetic diversity through analysis of variance and demonstrated that 85% of the variation is within populations
1973	Publication of the first major synthesis of anthropological genetics by M.H. Crawford and P.L. Workman
1977	DNA-sequencing dideoxy method described by Fred Sanger
1978	Restriction fragment length polymorphism (RFLP) first described
1981	Human mitochondrial DNA (mtDNA) genome sequenced at Sanger's laboratory at Cambridge
1984	Methods of DNA fingerprinting developed by Alex Jeffries and applied to forensic cases
1985	Development of polymerase chain reaction (PCR) methods by Kary Mullis
1988	Commencement of the Human Genome Project
1997	First Neanderthal mtDNA sequenced by Max Planck, Leipzig
1998	Completion of sequencing of first human chromosome
1998	Drafts of human genome sequence published in *Nature* and *Science*
2006	Neanderthal genome sequenced by R. Green et al. and J.P. Noonan et al.
2010	Development by GeoGenetics, University of Copenhagen: ancient DNA (aDNA) extraction and whole-genomic sequencing from hair of 4000-year-old Paleo-Eskimo nicknamed Inuk

Source: Crawford (2007b).

more than 80% of the total human genome based on a 4000-year-old clump of hair and, through association in modern humans, obtain a glimpse of the evolution of disease predispositions in an east Greenland Paleo-Eskimo nicknamed Inuk (Rasmussen et al. 2010).

Early Anthropological Genetics

Until the 1950s, few anthropologists were trained in the methodologies and theories of genetics. Most anthropologists were preoccupied with the application of typology to studies of race and its classification (Washburn 1951). In the 1940s and early 1950s, several of Albert Hooton's doctoral students at Harvard University developed interests and obtained some background in genetics. Most prominent among these students were:

1. Gabriel W. Lasker (Department of Anatomy, Wayne State University), who applied Sewall Wright's formulations to Mexican and Peruvian populations (Lasker and Kaplan 1964). He also applied James Crow's isonomy statistics to assess the intrapopulation structure of various human populations (Lasker 1985). For 34 years, he influenced the fields of human biology and physical anthropology through his editorship of the journal *Human Biology* – he encouraged and guided young anthropologists with interests in human genetics to publish in his journal. In 1989, Gabriel Lasker reviewed the significant publications on genetics that graced the pages of *Human Biology* from 1929 to the 1980s.
2. Frederick Hulse (Universities of Washington and Arizona) focused on the effects of heterosis on Swiss Alpine villages and conducted extensive research on Native American populations, particularly on the Northwest coast populations by examining blood group variation (Hulse 1955, 1964).
3. James N. Spuhler (University of Michigan), together with cultural anthropologist Clyde Kluckhohn, first applied Sewall Wright's pathway measures of inbreeding among the Ramah Navajo (Spuhler and Kluckhohn 1953). Spuhler went on to examine migration, mutation rates, and selection in human populations (Spuhler 1956, 1963; Spuhler and Clark 1961).
4. William Laughlin (Universities of Wisconsin and Connecticut) was best known for his archeological excavations at the Anangula Blade Site in the eastern Aleutian Islands and documentation of the presence of Aleuts at approximately 9000 years BP (Laughlin 1963). In addition, he published a few blood group frequencies for the Aleutian Islands (Laughlin 1951). Laughlin also provided blood samples from St. Paul in the Pribilof and Kodiak Islands for a series of genetic markers (Majumder et al. 1988; Merriwether et al. 1995). His interest in genetics mainly concerned the origins of Aleuts and Eskimos (Laughlin 1963).

5. Alice Brues (University of Colorado), who received her PhD under Hooton's mentorship, wrote a dissertation on the genetics of morphological traits based on sib resemblance (Brues 1940). Later, she applied genetics, the concept of populations, and their discontinuities in her classical treatise on race, *People and Races* (Brues 1977).

Although Derek F. Roberts (University of Newcastle upon Tyne) was trained in anatomy at Oxford University, he led a generation of genetically oriented scientists from the United Kingdom. In 1969, he coined the field of *anthropological genetics* in a lecture for the Royal Anthropological Institute (Roberts 1968). His application of Sewall Wright's concepts to a small population from Tristan da Cuhna and the measurement of genetic drift and bottleneck effects are the finest examples in the field of anthropological genetics. He also contributed to the development of genetic demography through his innovative field investigations of the Dinka population of southern Sudan (Roberts 1956). Roberts and Hiorns (1962) developed multivariate methods for estimating gene flow for triracial hybrid groups. Other British scholars who conducted some genetic or genetic-related projects included: Anthony Boyce, Geoffrey Ainsworth Harrison, Nigel Barnicot, and Malcom Smith. The beneficiaries of Falconer's quantitative genetic research on animal models and formulations included genetic epidemiologists, who played a prominent role in gene mapping and disease predispositions analyses.

Francisco Mauro Salzano (Federal University of Rio Grande do Sul, Brazil) started his academic career focusing on the genetics of *Drosophila*, but after a one-year postdoctoral fellowship in human genetics at the University of Michigan, he shifted to what can now be called anthropological genetics. Salzano conducted more than 50 years of research on the genetic structures of indigenous populations of South America (Salzano and Callegari-Jacques 1988). He developed a highly productive research center in Porto Alegre with a focus on populations and the molecular genetics of indigenous populations of South America.

The second generation of anthropological geneticists included Frank Livingstone, a former student of biological anthropologist James Spuhler and J.V. Neel, a geneticist at the University of Michigan. Livingstone examined the effects of culture (i.e. slash-and-burn agriculture) on the spread of falciparum malaria in Sub-Saharan Africa. This cultural phenomenon resulted in a shift from an epidemic to an endemic form of malaria. Livingstone trained a number of anthropological geneticists at Michigan, including Kenneth Weiss, Alan Fix, and Richard Ward. Jack Kelso received his doctoral degree at the University of Michigan with a dissertation on assortative mating. He found associations between blood types, body morphology, and disease. Charlotte Otten was a graduate student with Jack Kelso at Michigan and collaborated with Fred Theme (who became provost at the University of Washington) on blood-typing ancient bone – an early attempt at using genetics for reconstruction of population history. The University of Michigan had a rich tradition of combining genetics and biological anthropology.

At Harvard, the successor of Albert Hooton was W.W. Howells, who trained several researchers (Eugene Giles, Jonathan Friedlander, and Henry Harpending) with research in anthropological genetics. Giles conducted field investigations on the genetics of small populations in New Guinea and Mexico, but focused his later academic career on forensics (Giles et al. 1966). Jonathan Friedlander (1971) followed in the steps of Albert Damon at Harvard, and conducted anthropological genetic studies in Oceania, primarily in Bougainville. Henry Harpending was mentored by Howells but conducted his dissertation research on the genetic structure of !Kung Southern African populations. He developed R-matrix methods for characterizing populations, and his chapter with Trefor Jenkins in the volume *Methods and Theories of Anthropological Genetics* is the most highly cited chapter from this classical volume (Harpending and Jenkins 1973). Harpending also made significant contributions to the analyses of population structure and evolutionary history.

The subsequent generations of anthropological geneticists received equal training in genetics and biological anthropology. Ken Weiss (University of Michigan) was trained by Livingstone and Spuhler in anthropology and J.V. Neel in genetics. Richard (Ryk) Ward accompanied James Neel in the classic research on the genetic structure of the Yanomami (Ward 1972; Ward and Neel 1976). David Glenn Smith received a PhD at the University of Colorado in biological anthropology, and then further honed his skills in human genetics at Michigan Medical School. He developed a highly productive program in anthropological genetics at the University of California, Davis, by developing ancient DNA research and examining the biogeography of the widespread genus *Macaca*. At Davis, Smith trained the next generation of specialists working on ancient DNA: Ripan Malhi, Brian Kemp, Debra Bolnick, and Fredrika Kaestle all matriculated from the program at Davis and further applied molecular genetics to key questions concerning the peopling of the Americas.

During the 1960s, I received training in biochemical genetics from Arno Motulsky and Stanley Gartler at the University of Washington–Seattle and blood group typing from Eloise Giblett at the King County Blood Bank. I served as an "apprentice" to Derek F. Roberts, a visiting professor for Oxford University who was analyzing genetic and demographic data from Tristan da Cuhna islanders. However, in my final year of graduate school, Marshall T. Newman was appointed professor of anthropology and stimulated my interest in the peopling of the Americas (Crawford 1998). I developed interests in the genetic–environmental interactions in complex diseases, such as baboon models for lymphoma, admixture and risk of hypertension, genetic adaptation to malaria, and biological aging in Mennonites (Crawford and O'Rourke 1978; Crawford 1980a; Crawford and Rogers 1982). The Laboratory of Biological Anthropology (University of Kansas) initiated a longitudinal study of the genetics of biological aging (1979–2015) and became a center for anthropological genetics and genetic epidemiology while training a total of 36 PhDs (Crawford 2000b). Some of these graduates applied anthropological genetics and molecular genetics to complex diseases: Ravindrinath Duggirala (diabetes mellitus), Anthony Comuzzie (primate models of disease), Kari North (obesity), M.J. Mosher (epigenetics), Sobha Puppala (disease susceptibility gene mapping), Janis Hutchinson (admixture and risk of hypertension), Geetha Chittoor (complex disease – obesity, diabetes, hypertension), Rohina Rubicz (prostate cancer, serological measures of infection), Kris Young (obesity, cancer), Anne Justice (body mass index [BMI], cardiometabolic risk), Laurine (Rogers) Oberdieck and Meredith Uttley (biological aging in Mennonite populations), Philip E. Melton (cardiovascular disease), and Lisa Martin (application of statistical measures to disease). Most of these students completed further training in genetic epidemiology in the Department of Genetics, Southwest Foundation for Biomedical Research, and the Department of Epidemiology, University of North Carolina.

Field Investigations

Physical (biological) anthropology adds a rich tradition of fieldwork to anthropological genetics. Fieldwork provides a comparative dimension plus unique environmental and cultural context for the study of evolution and genetic–environmental interactions for polygenic traits. Physical anthropologists, such as William Laughlin (Aleutian archipelago), James Spuhler (Navajo), Gabriel W. Lasker (Mexico and Peru), and Derek F. Roberts (Dinka of Sudan), conducted field investigations worldwide. These pioneers were followed by the next generation of field researchers, such as Frank Livingstone (Africa on sickle cell), Francisco Salzano and Ryk Ward (genetic structure of Amazonian tribes – Xavante, Makiritare, and Yanomama), and Jean Benoist (St. Barthelemy; Benoist 1964). My field research included the following: (i) in 1968, my first fieldwork, in collaboration with the University of Torino, commenced in the Alpine valleys of northern Italy (Crawford 1980b), followed by: (ii) 1969–1974: on the genetic–environmental effects of relocation of Tlaxcaltecans of Mexico, from the altiplano of central Mexico to the arid regions of the north (Crawford 1976); (iii) 1970: origins of the Irish itinerants (Tinkers): were the Irish Travelers of Romany Gypsy origins, or were they Irish displaced by the potato blight and the dissolution of the monastic settlements (Crawford and Gmelch 1975)? (iv) 1975–1985: The successful colonization of the coast of Central America by the Black Caribs, who were forcibly deported out of St. Vincent Island to the Bay Islands and eventually Honduras (Crawford 1983, 1984); (v) 1977–1995: Siberian origins of Native Americans: where in Siberia did Native Americans originate? My first project in 1976, funded by the National Science Foundation (NSF), was to compare the genetics of Siberians from the Chukchi Peninsula with Eskimo populations on the Alaskan side of the Bering Straits (Crawford et al. 1981). Unfortunately, the Cold War limited this comparison to St. Lawrence and King Islands and Wales,

Alaska. Only in 1989 was I able to access the indigenous populations of Siberia (Crawford et al. 1994). (vi) 1999–2005: Genetic structure of the Aleutian archipelago (Crawford et al. 2010). A summary and more detailed explanation of my field research projects is included in Chapter 4, "The Importance of Field Research in Anthropological Genetics: Methods, Experiences and Results," of the volume *Anthropological Genetics* (Crawford 2007b).

Molecular Genetic Revolution

From 1973 to the 1990s, the molecular revolution was precipitated by a series of methodological developments and diffused into the field of anthropological genetics. During the 1970s and 1980s, DNA extraction was a time-consuming and tissue-requiring methodology. In the 1980s, with the development of phenol chloroform techniques followed by salting procedures, high-yield commercial kits became available and the process of extraction was simplified and considerably sped up. In 1978, Werner Arber, Hamilton O. Smith, and Daniel Nathan won Nobel Prizes for the discovery of the specificity of bacterial restriction enzymes permitting the characterization of DNA variation through restriction fragment length polymorphisms (RFLPs) in human populations. This breakthrough was followed by Kary Mullis' development of the PCR, which permitted DNA analyses based on minute quantities of DNA (Mullis et al. 1986). As a result, blood and tissue previously used for DNA extraction would be replaced by buccal swabs and/or saliva, thus facilitating DNA collection in the field. The combination of DNA extraction, RFLP analyses, and the use of Southern blots resulted in the characterization of DNA fingerprints and their application to forensic cases (Jeffreys 1987). This methodology was applied to celebrated forensic cases, such as the identification of Josef Mengele's remains in Brazil (Jeffreys et al. 1992). Dental analyses proved to be ambiguous in Mengele's case, but a comparison of DNA fingerprints of Mengele's son with those of the drowning victim from Brazil proved to be conclusive. Similarly, the skeletal remains of Tsar Nicholas II, his family, and their attendants were identified using mtDNA and portions of the genome. In addition, the reliability of the original analysis was improved by obtaining an archived 117-year-old blood sample from the tsar, resulting from an assassination attempt (Gill et al. 1994; Rogaev et al. 2009).

With the breakup of the Soviet Union in 1989, I was able to conduct field research via collaboration with the Soviet Academy of Sciences on native Siberians to determine the origins of Native Americans. The DNA collected from the Evenki reindeer herders, Kets from the shores of the Yenesey River, and Kizhi pastoral populations of the Altai were compared with other Siberian and Native American populations using DNA fingerprints, as well as mtDNA (in collaboration with Douglas Wallace, and NRY with Fabricio Santos). The Siberian DNA fingerprints clustered tightly with Native American populations and differed significantly from European and African populations (McComb et al. 1995). Similar results were shown for mtDNA and Y-chromosome markers (Torroni et al. 1993; Santos et al. 1999). Thus, it appeared that the Southern blot–based DNA fingerprints could be used not only for forensic cases but also for population comparisons (Crawford and Beaty 2013).

Future of Anthropological Genetics

Advances in anthropological genetics have primarily been generated by technological and methodological breakthroughs in molecular genetics. Key innovations in PCR, DNA extraction, and sequencing methodologies of mtDNA resulted in the replacement of the multiregional model (based on paleontological evidence) by a recent-out-of-Africa (ROA) model (Cann et al. 1987). Instead of an early 1.8 million years ago migration of *Homo erectus*, the mtDNA data indicated that all humans alive today had more recent common ancestry (ROA). Molecular genetics has been applied to the reconstruction of the human diaspora, focusing on a number of key questions: (i) what were the chronology and number of migrations by anatomically modern *Homo sapiens* out of Africa, and the routes that were taken to populate the world? (ii) Peopling of the Americas – how many migrations came out of Siberia, when, and were they coastal, inland, or both? (iii) Was Polynesia settled by Asians directly from Taiwan, or did the early settlers intermix with Melanesian populations? (iv) Was *Homo floresiensis* an example of island dwarfing or a pathological *H. sapiens* (microcephalic or cretin)? (v) Based on genomic signatures, is there evidence of selection operating on migrants living under extreme environments such as hypoxic stress, heat and cold, and various forms of zoonoses? (vi) Are the Basques remnants of Paleolithic settlers of Europe, or are they descendants of early farming communities? (vii) Does epigenetic inheritance provide an archive of environmental influences on the genome? (viii) Are populations with small numbers of founders more informative in the mapping of genes than using genome-wide association studies (GWAS)?

The future for anthropological genetics looks promising with a number of new, significant methodological developments that will affect the future of the field:

1. DNA sequencing breakthroughs permitted the characterization of the human genome in 2001 and the relatively inexpensive whole-genome sequencing currently employed. Until the use of dideoxyribose chain terminators and synthetic oligonucleotides as primers (Sanger et al. 1977), DNA sequencing was cumbersome and time-consuming. Primer-linked fluorescent reporters and fluorescent wavelength detection led to an automated base identification and high-throughput, parallel, next-generation sequencing. With the addition of real-time PCR and stringent controls, fragmented DNA can be clustered and contamination sorted out. This millennium has seen the whole-genomic

sequencing of numerous paleontological remains, including Neanderthals; the identification of introgression of Neanderthal gene complexes in contemporary human populations; and their involvement in human innate immunity against bacteria, parasite components, and fungal infections (Dannemann et al. 2016).

The DNA sequencing of 4000 years BP Paleo-Eskimo hair from east Greenland has also contributed to the evolutionary examination of disease predispositions. By searching for single nucleotide polymorphisms (SNPs) that are associated with specific diseases in contemporary populations, it is possible to detect evolutionary antecedents to metabolic conditions such as glucose metabolism (Rasmussen et al. 2010). Thus, a future fusion of GWAS-based genetic epidemiology and ancient DNA sequencing offers a unique glimpse into the evolution of complex diseases. Similarly, methylation patterns can be detected in ancient genomes, exposing population adaptation and environmental influences that took place during the human out-of-Africa diaspora (Orlando and Willerslev 2014). Pedersen et al. (2014) followed up on the original DNA sequencing of the 4000-year-old Saqqaq individual by measuring levels of methylation. Using ancient methylation, they were able to estimate the age at death of the Saqqaq man and illustrate how epigenetic information can be used to infer ancient gene expression. Non and Thayer (2015, p. 302) conclude their overview of epigenetics for anthropologists as follows:

> Epigenetic data have tremendous potential to transform the field of anthropology by adding a new layer of complexity to our understanding of human biological variation, and a new way to explore human adaptation to changing environments.

2. Development of the CRISPR (clustered regularly interspersed short palindromic repeats) technology provides a new tool for gene editing. Besides the obvious importance of this methodology in repairing proteins that cause an assortment of diseases (such as Duchenne muscular dystrophy), this gene-editing tool can be effective in identifying genomic causes of diseases, enhancer elements that regulate gene expression, and the further mapping of the genome. The editor of *Science*, Marcia McNutt, has recognized the development of the CRISPR gene-editing technology as "The 2015 Breakthrough of the Year." She states: "Because CRISPR is poised to revolutionize research, the international community gathered together earlier this month to address the implications of this technique for modifying human germ cells and embryos, articulating guidelines that clarify the ethical bounds for researchers, funders and publishers" (McNutt 2015, p. 1445).

Future technological innovations, particularly in molecular genetics, will further facilitate the exploration of phenomena relevant to anthropological genetics. This synthetic field, through fieldwork and an evolutionary perspective, should also enhance the development of new methodologies and perspectives. The future is bright for anthropological genetics, with new ethical issues arising with our newfound abilities to edit and control the human genome.

CHAPTER 2

The Ethics of Genetic Ancestry Testing

Nanibaa' A. Garrison[1] *and Jessica D. Bardill*[2]

[1] *Treuman Katz Center for Pediatric Bioethics, Seattle Children's Research Institute, Seattle, WA, USA*
[2] *Department of Pediatrics, University of Washington School of Medicine, Seattle, WA, USA*
Department of English, Concordia University, Montreal, QC, Canada

Introduction

Genetic ancestry testing has grown in popularity among the general public, in part due to the increased availability and affordability of ancestry tests offered by direct-to-consumer (DTC) genetic companies. Curious about their ancestral past, consumers may submit a saliva or buccal swab sample, from which DNA is extracted and examined for particular genetic markers used to infer ancestral origins.

This chapter describes several types of genetic ancestry tests, the history of major genetic ancestry research projects, and companies that offer these tests, and further discusses the limitations and ethical considerations that arise from such tests. The scope and the interrelations of genetic ancestry research and consumer interest necessitate an evaluation of the ethics of this testing, including the basics

A Companion to Anthropological Genetics, First Edition. Edited by Dennis H. O'Rourke.

and limitations of the technology, understanding of its results and implications, as well as how ancient and present-day populations contribute to growing databases that house ancestry information for many populations. We recognize that genetic ancestry tests are conducted in other animal and plant species, but in this chapter, we focus on human ancestry testing, although some of the ethical considerations discussed are applicable to testing on other species. We focus as well on the USA and regulations therein, principally because it has been the largest market for DTC genetic testing and the home to the largest genetic research projects historically. Regardless, laboratories in other countries and collaborative efforts with peoples in other regions are robustly growing.

Background on Ancestry Tests

Genealogy is the study of family histories and tracing ancestral lineages, traditionally done by tracing through historical documentation and uncovering family stories. Genetic genealogy is a newer tool that examines these questions of family history and lineage in tandem with genetic testing, where the introduction of such testing may provide additional clues, and has been used in both research and recreational (such as DTC) settings. Genetic ancestry tests may reveal relationships between close biological relatives or may suggest more distant population affiliations. Additional tools, such as the use of social media to draw connections, are made available through some companies as well as coordinated by the genetic genealogy community. Several companies host DNA surnames projects where consumers are able to trace their last names with the Y-chromosomal DNA pattern inherited by male relatives in their families (Pomery 2009). DTC testing companies have stimulated consumers' interest in genetic ancestry through the marketing of these tests directly to the general public.

In addition to using genetic ancestry tests to learn more about human biological relationships, researchers use these tests to control for the ancestral contributions to one's genome in studies of disease susceptibility across human populations (Burgner et al. 2006; Manolio et al. 2008). By separating out the ancestral component, researchers seek to hone in on the genetic variants that are associated with diseases or other phenotypes.

While genetic ancestry testing has become popular in scientific practice and in recreational settings, scholars have noted the ethical implications of the concept of genetic ancestry testing, including the interpretation of results, and have raised concerns about the general lack of standards and uniformity across DTC companies (Lee et al. 2009). Some researchers are dedicated to identifying additional genetic markers to increase the precision and specificity of these tests (Nassir et al. 2009), whereas others are examining consumer attitudes and knowledge about genetic ancestry testing (Bloss et al. 2011; Wagner et al. 2012).

Estimating ancestry through genetic tests

Genetic ancestry testing relies on both biological examination and subsequent statistical analyses to make inferences about a person's past. Individuals who are interested in researching their family history through genealogical information, surname searches, or ancestral histories may turn to genetic ancestry testing for additional clues (Royal et al. 2010). The broad designation of genetic ancestry testing encompasses a variety of tests in the laboratory setting and the consumer marketplace. These tests include analyses of mitochondrial DNA (mtDNA), Y-chromosomal DNA, and autosomal (non–sex chromosome) DNA. Each type of test has its strengths and limitations: they may glean information from a single ancestral line (mtDNA and Y chromosome) or uncover broader information regarding multiple potential ancestral backgrounds (autosomal DNA).

Examinations of single ancestral lineages are conducted with mtDNA or Y-chromosome DNA by tracing haplotypes, which are large blocks of DNA that are passed from one generation to the next. Because mtDNA and Y-chromosome DNA are not subject to recombination, large sections of DNA are largely unchanged from one generation to the next. Scientists are able to track the genetic variations of haplotypes to reconstruct population histories of human populations. For a scientific study, haplotypes are organized into haplogroups to determine the degree of similarity from one haplotype to the next. Generally, haplogroups are associated with geographical regions. For example, mtDNA haplogroup L is found most frequently in people of African descent and thus is thought to have originated in Africa. Similarly, Y chromosomes are sex-linked male chromosomes that are inherited in a paternal fashion. The Y chromosome has a non-recombining section (NRY) that contains genetic variants across populations, but it does not change much from a biological father to son, allowing scientists to trace paternal lineages. Females do not have Y chromosomes, but they may learn about their paternal lineage through the genetic testing of biologically close male relatives.

Autosomal DNA analysis can be used in concert with mtDNA or Y-chromosome testing, and it involves the use of Ancestry Informative Markers (AIMs) and bioinformatic tools to estimate biogeographical ancestry. AIMs are a collection of genetic markers with significant allelic frequency variation across populations (such as 30–50%). Researchers use probabilities based on the frequency of AIMs to infer the genetic ancestry of individuals or groups. AIMs are typically used to estimate ancestry because they are common in the human genome, representing 90% of human variation, and are generally quicker and easier to genotype and analyze than other types of polymorphisms. Most AIMs are single nucleotide polymorphisms (SNPs), which are allelic variants that occur on the level of one nucleotide change (A, T, C, or G) and must be present in at least 1% of a given population in order to be useful for analyses of differences. AIMs are used in aggregate, and they have involved as few as 34 (Shriver et al. 2003) to several thousand markers across the genome that are correlated with continental origins as well as race/ethnicity categories (Yang et al. 2005).

Genetic recombination, a process that shuffles genetic contributions from each parent, takes place at every generation. Each chromosomal segment containing genetic variants such as AIMs originates from one of the parents and is passed down from one generation to the next; thus, each segment is tied to larger population ancestral histories. Geneticists are able to analyze the genetic variation across chromosomal segments to estimate the geographic origin of one's ancestors. For example, one chromosomal segment containing AIMs at high frequencies in one population would indicate a shared ancestor from another population in the same geographic region, and thus it is an indicator of shared or similar ancestral background. Some chromosomal segments might be derived from populations with common ancestors originating in Africa, whereas other segments might be traced to populations in Europe or Asia. Genetic ancestry tests use and make predictions about a person's ancestry based on comparisons of their genetic variation with other present-day populations. Researchers use these tests to theorize about human migration patterns, as well as to examine how human populations have changed over time and how they differ at the molecular level.

Variation in genomic databases

Multiple databases containing data derived from AIMs are useful for capturing the variation for analysis to make ancestry estimates. Genetic ancestry testing companies rely on these databases to infer ancestral origins for their consumer base. Current databases are limited by the composition of individuals and populations sampled in a given geographic area. The individuals who are sampled do not necessarily reflect the entire genetic diversity within a geographic area, as they may actually possess rare alleles that are not found in the rest of the population or may lack common alleles. This limitation is a particular concern because lack of representation of genetic diversity within the databases may lead to inaccurate conclusions. A consumer who sends their sample to two different companies that utilize separate proprietary databases, with reference samples representing different global populations, may receive differing or conflicting ancestry estimates based on the AIMs used by each company and the reference populations within their separate databases (Kutz 2010).

Statistical methods and ancestry inferences

Inferences that are made about ancestry from a DNA sample are based on probability estimates and thus are not determinative. A common computational tool called *structure* is used to infer ancestral contributions to a person's genome under the assumption that individuals in a study group represent mixtures from different genetic populations (Rosenberg et al. 2002). *Structure* requires distinct reference populations, usually from distinct continental populations, to anchor the analyses of the study group. The AIMs represented in the study group are then compared to the reference populations, and the program estimates ancestral contributions coupled with a measurement of the likelihood that an individual's ancestry is derived from the different ancestral groups. In some populations, such as Native American tribes, there is a lack of specificity to distinguish the differences between tribes or geographic regions because there is typically a paucity of data representing tribes to anchor the results. Furthermore, the analysis of reference samples across databases, such as the level of accuracy and completeness at which they are sequenced, can influence the results and interpretation.

Statistical methods carry an amount of uncertainty, often not clearly conveyed to consumers, which may lead to incorrect assumptions about one's ancestral past. Rather, results may appear in the form of percentages of ancestral composition or DNA portraits complete with maps and pictorial descriptions based on the proprietary and public databases referenced for the given sample (Scott 2012). Definitive results beyond close biological relatives and recent ancestry are quite rare, as expansion of reference samples can alter the analysis and probability of the results for a tested sample. Genetic variants occur within multiple populations around the world and can occur more frequently with a range of other variants, providing clues to a population connection. Certainty would only emerge for ancestral testing from an allelic variant unique to a particular population, and while genetic variants occur in different proportions in different populations, unique variants are more likely a result of limits of databases and sampling practices than of the population. This point also holds true in research, where the tested samples and the inferences from them depend on the laboratory and public databases of information.

Continental ancestry and race

Ancestry tests use statistical estimates that infer the likelihood of a person having ancestors originating from a certain geographical region, based on the genetic profiles of people who live there today. Biogeographical ancestry combines estimates of genetic ancestry, based on AIMs, with continental ancestries that are tied to geographical locations around the world. *Continental ancestry* typically refers to geographic origins represented by five major parental populations, originating from Africa, Europe, Asia, the Americas, or Oceania. A geographic origin is then inferred starting from the recent common ancestor based on the genetic differentiation seen in present-day populations that are considered Indigenous to these geographic areas (Nassir et al. 2009; Royal et al. 2010). Inferences are made about a person's presumed ancestral past based on certain DNA markers that are found at higher frequencies in one population compared to other populations. Researchers may use statistical estimates of data derived from multiple AIMs to infer deep ancestral relationships. Certain population groups who are relatively homogeneous might harbor DNA markers that are found at higher frequencies compared to other populations. Some of these populations are classified as reference populations and are then used as comparison groups to determine degrees of similarities and differences with other populations. Ancestry estimates for a person's genome may be presented as multiple proportions of ancestral contributions, adding up to 100%, a reflection of contributions of different ancestral pasts based on the reference populations. For example, a person's genome may be composed of estimated proportions of ancestry from three continental regions, such as 60% African, 35% European, and 5% Native American, but may not necessarily provide precise locations or affiliation with any particular group.

Critics of genetic ancestry testing have cautioned against conflating social constructs of race with biogeographical ancestry (Lee et al. 2009; Royal et al. 2010; TallBear 2013). Historically, biological differences have undergirded constructs of race, such as the use of phenotypes (including skin color, hair texture, and eye shape) to differentiate between populations, and have associated those phenotypes with geographic origins. Given this history of racial typology and larger histories, continental origins have become loaded with additional meaning such that the social construction of geography, including the continents named by that discipline, means that receipt of ancestral links to a particular continent correlates to how race and cultural differences have been used to construct that place and its peoples (Beinin 2010). In part, this conflation relies on the use of present-day human populations to make inferences about the past (Bolnick et al. 2007), but those inferences are not themselves separable from the past or the present. In discussions regarding the return of genetic ancestry or medical results, it is important to remember that race (and its constructions) along with ancestry estimates may influence the identity constructs of those receiving the results, even in the face of contrasting information from cultures, communities, and environments.

Ancestry Estimations of Present-Day Populations

While genetic ancestry testing is the technological backbone for many different companies, the technology began and is consistently advanced in institutional research, within both the academy and the private industry sector. This research then informs the products, but the major projects of genetic ancestry do not exist within a vacuum, and therefore their aims, progress, and problems are useful to review here. Many of these projects emerged from the success of the Human Genome Project, for which a draft genome was completed in 2001, and the projects attempt to address the need for larger and more diverse information about human genomes, including the AIMs of various populations. Subsequently, the more fully developed technology has been used for forensic applications, recreational purposes, and political disputes, and now as a part of new initiatives on precision medicine.

Large-scale studies of human history

Beginning in the 1990s, scientists began to share and distribute DNA samples that were collected for previous research studies to carry out new studies to explore human genetic variation. From a practical and funding standpoint, this production of a genetic research commons makes sense; however, from various Indigenous and community-based standpoints, this work is viewed as extractive at best and harmful at worst, with little if any benefits accruing to their communities particularly. These pooled samples were often used to examine human genetic variation not only in the present but also through a model of how the present may represent the past, with a smaller grouping of ancient remains contributing to the knowledge of genetic variation in the ancestral past.

To estimate and model how present populations may be related and come from one another in the past, scientists use genetic ancestry testing along with statistics and geography, all of which include particular assumptions. Even as the testing results may impact a whole group or community, an individual is often used as a representative of that group; however, conclusions from a sample can impact more than just an individual, revealing genetic information that may affect the whole group. This potential harm can be especially concerning if the migration and ancestry research is done as secondary research on samples. Relatedly, using present-day peoples and their samples as proxies for ancestral populations has possibilities for models but can also come with harms, but explicit and more embedded within the research. For example, assumptions about the continuity and lack of admixture of populations track to older notions of differentiations in "races" but can be more submerged in the analysis than reporting about the likelihood of inbreeding or stigmatized health conditions within a population, which are more explicit harms. This assumption also has to take into account that genetic development has occurred between the present and the ancestral past, so there is not a perfect correlation even in conserved regions of DNA. Attention to the amount of certainty and uncertainty of these inferences can help mitigate this concern. While the availability and quality of ancestral genetic samples have expanded through obtaining DNA from sources such as coprolites, hair locks, and tooth enamel, the amount of this resource is still remarkably small and ethically complicated by concerns over how these remains are treated and used for purposes that their owner could not and did not consent to, and how they are used to represent present-day peoples. Therefore, inferences from present-day populations into human ancestral past will continue to be made, but they can be made with attention to larger impacts and the limitations of the inferences.

The Human Genome Diversity Project In 1991, the Human Genome Diversity Project (HGDP) sought to increase the number and the ethnic and geographical diversity of individuals included in genomic research (Cavalli-Sforza 2005).

In seeking diversity, the researchers identified populations throughout the world who were considered to be Indigenous to a given geographical area, particularly as these individuals were imagined to have less genetic admixture (based on the idea of ancestral and present-day population distinction instead of interaction) due to their isolation. However, many groups who refused to participate had previous experiences in which multiple forms of intellectual property were taken from them, such as knowledge from ceremonies, medicinal plants, art, and stories. The idea that undergirded the HGDP research of purity versus admixture was catalogued by Jenny Reardon as one of its fatal flaws, along with the targeting of communities instead of cultivating partners through engagement and thereby successfully recruiting participants (Reardon 2005). Indigenous peoples who did not perceive benefits, or who felt that the harms (such as social, psychological, political, and epistemological) outweighed any perceived benefit, from this research called for its halting. In the end, the HGDP-CEPH (Centre d'Etude du Polymorphisme Humain, or Human Polymorphism Study Center) database included samples from 51 distinct populations from 1064 individuals, and it is still widely used for a range of secondary research, including human genetic ancestry research. None of the populations that are included are living in the USA despite many of the researchers being based in the USA, in large part due to pushback by tribal communities who joined together to voice their concerns. Communities who are represented in the database have not been engaged in the research process since the inception of the database. In the end, the Indigenous peoples did find a small benefit – building a global coalition to have their voices heard instead of only being viewed as a natural resource from which to extract samples and data.

National Genographic Project The National Genographic Project (NGP) emerged to conduct genetic ancestry research on a global scale. This project aimed to draw on the lessons learned from the HGDP by working to build partnerships with target populations, including having Indigenous peoples on their advisory board. The NGP engages with some of their target populations by creating a range of products, such as research papers, news stories, documentary films, and other educational media. In some cases, the NGP has invited advisory board members to coauthor publications (Zhadanov et al. 2010). However, the narratives that are perpetuated in these materials have been critiqued from many sources in how they explicitly state their purpose to be antiracist while re-inscribing racial difference at the molecular level and repeating differentiations between populations (Wald 2006; TallBear 2007). It is worth noting here that *population* often refers to a statistically bounded group, while *people* refers to the social community that may or may not be reflected in a population (Barker 2004).

Furthermore, in an effort to expand their research datasets, to engage a largely first-world public, and to garner additional funding, the NGP has a DTC arm that offers collection tests to paying consumers much like other DTC companies but with the explicit information that the consumer will be a part of a larger research project. Whether DTC information is used for research, and therefore includes proper consent to such, has been a contentious issue that has moved forward the discussion on tiered and secondary-use consents in general. While other DTC companies do engage in research with their databases, the NGP explicitly notes this work and has created a hybrid research participant/consumer scenario.

International HapMap Project The International HapMap Project, begun in 2002, uses a deliberate focus on engagement with partner communities, including consultation. Like many of the projects preceding it, and unlike the business model of DTC companies, the information from the HapMap Project is released to the public as it is available. Overall, the HapMap Project intends the genetic ancestry information to be used in health fields, particularly for the identification of disease genes and pharmaceutical response genes. Four populations began the HapMap, with multiple members of the partner communities contributing to the population information through a robust informed consent process: Yoruban-descent persons in Nigeria; Han Chinese–descent persons in Beijing, China; Japanese-descent persons in Tokyo, Japan; and European-descent persons in Utah, USA (International HapMap Consortium 2003). These four reference populations have helped in producing an understanding of ancestral similarities and distinctions, and now seven additional populations have begun partnerships with HapMap. Like the first four communities, these seven new populations are also seeking benefits of biomedical research, and they have seen good practices involving the initial partners.

Public genomics projects Given the popularity of personal genomics, research intersections have emerged that bring together the genomes of people willing to have them publicly presented, such as the Personal Genomes Project and the 1000 Genomes Project. The first project, founded in 2005 and that spanned four countries (USA, Canada, UK, and Austria), sought to present the data from a range of genomes publicly to better support research into health and traits derived from genomic information (Church 2005). The individuals who have submitted their information to this project have been vulnerable to re-identification, emphasizing the importance of protecting the privacy and identities of those who have contributed. The 1000 Genomes Project (2010) sought to use the reduced cost of genomic sequencing to similarly create a large, publicly accessible database of genomic information, and to further share that information through networks of public databases.

Linking ancient DNA with present-day populations

While research on present-day humans has a vast array of possibilities, including ethical considerations and limitations, work conducted with ancient DNA (aDNA) samples is both more finite and more acute in its ethical considerations. Many aDNA studies help to address the often-problematic conflation of present-day populations with ancestral populations in genetic ancestry testing (Royal et al. 2010), providing evidence of the DNA of peoples found in a given region dated to a particular time. However, these studies are conducted to address questions that not all peoples believe are unaddressed in their epistemologies, such as: where did we come from, and how do we relate to one another? Furthermore, the studies venture into a particularly vexed area of ethics and law, as well as considerations around what kinds of inferences are made and what those say about contemporary populations. For example, the remains that have been used for aDNA studies, while not covered under the Belmont Principles for informed consent, are often disputed between scientists and communities who claim them in order to protect them.

The Native American Graves Protection and Repatriation Act In response to historic injustices including looting of gravesites, the USA enacted legislation to repatriate remains of Native Americans to the present-day tribal nation that claims or traces responsibility and lineage to those remains. The Native American Graves Protection and Repatriation Act (NAGPRA) of 1990 also came on the heels of the developing genetic and genomic technologies, and there have been multiple disputes about newly discovered remains as well as illumination of loopholes in the original law, such as inapplicability of the law to remains found on private lands.

The Kennewick Man, also known as the Ancient One, has been one such dispute. Multiple tribes laid claim to protect the remains discovered by the Columbia River in 1996, but efforts from scientists including James Chatters allowed the remains to be examined and experimented on for a short period of time in the midst of the nine-year court battle regarding custodianship of the remains. At the time these examinations were conducted, the remains were being held by the US Army Corps of Engineers, and at that time no genetic evidence was successfully found. A recent publication showed that DNA was extracted from a sliver of bone tissue, a sample that was entirely destroyed in the process, and the resulting information confirmed the assertion of the tribal groups in the area that the Ancient One is Indigenous to the Americas (Rasmussen et al. 2015).

Use of ancient DNA to reconstruct history Two recent and large-impact studies of aDNA and ancestry involve remains found in places that are not protected under US federal regulations – private lands and in an underwater cave in Mexico. The remains of a young child were found on the private lands of the Anzick family in Montana in 1968 and later were analyzed by Sarah Anzick, the daughter of this family, who trained and worked at the National Institutes of Health on genomic technologies. This analysis demonstrated the Clovis child to have a genetic ancestor in common with 80% of all Native American peoples (Rasmussen et al. 2014), and that the remains were 12 600 years old. While quite literally the child had no direct descendants, this ancestry information has been touted as valuable information gained from the remains, and after such genome sequencing was successful and complete, the Montana tribes who had petitioned for the remains to be returned to the ground were able to hold such a ceremony in June 2014. Importantly, these tribes were not necessarily shown to be connected genetically to the remains but based their claim on ties to place, to land, given the vast majority of Native American samples in databases are from South American peoples, not US tribes, many of whom refused participation in such genetic sampling and databases.

Naia, the "Hoyo Negro" woman, named after the "Black Hole" cave in which her well-preserved and quite complete remains were found near the Yucatan Peninsula in 2007, was dated to be between 12 000 and 13 000 years old, and was also found to be genetically related to present-day Native American populations (Chatters et al. 2014). Given the completeness of her remains and further development in age and facial structure at the time of her death in comparison to the Clovis child remains, Naia provided scientists with another way to think about how morphology and genetics have changed over time (Chatters et al. 2014).

Overall, aDNA has been used to look at the connections between remains and present-day peoples, to give different points along the genetic history of migrations and evolutions that have led to our differences today. However, with all of these remains, there exist concerns around the treatment of the remains, their destruction in processing to obtain genetic information, and their liminal positionality as not quite human and not quite object or resource, bringing forth the question of who owns or who speaks consent for the remains.

Contradictory results

Results may not only be contradictory to one another across databases, but also be contradictory to the knowledge of a community or a family. Given the limitations and proprietary nature of many genetic databases, samples may be tested against different sets of reference AIMs, and therefore the results can easily be distinct from one another. Statistical estimates of ancestry are based in probabilities with room for error. In 2003, the General Accounting Office report submitted a standard DNA sample to four different DTC testing companies for 10 different tests and found the range of results to be contradictory and inconsistent with one another (Kutz 2010).

Issues may arise when consumers of this information seek personal connections to a community based on estimations of shared ancestry or when findings contradict one's beliefs about their ancestry. Two well-publicized examples include the media-circulated and produced stories of the ancestry of Oprah Winfrey and of Henry Louis Gates, Jr., both of whom initially were told they had particular South African and European ancestry connections, respectively. Subsequent testing from other companies indicated their origins were more likely West African and included African ancestry, respectively. Currently, there is a push for standard samples to be used to test the accuracy of the companies' and researchers' analyses, but those standards would be incapable of covering the full range of human genetic diversity. These contradictions, though, can lead to confusion around identity, as well as concerns for laboratory practices that utilize these results for other applications and analyses, such as migration studies and pharmacogenomic studies. This last concern points to the concern that ancestry information, or a proxy of race, may be used, but the specific allelic variations in the genome are the best markers for determining not ancestry but the possible effects of drugs or commonly shared genetics.

Even when results are consistent with each other, they may be contradictory to the ancestral beliefs of the individual or community. Many tribes are concerned about use of their DNA samples in human migration studies because the scientific evidence may contradict cultural and deeply held beliefs about tribal origins. Furthermore, studies that suggest that a tribe migrated across the Bering Strait from Asia may have political implications and challenge tribal sovereignty and land rights. Many Indigenous and other knowledge systems, such as those found across and throughout religious texts, include information about the origins and movements of peoples, particularly how they are connected to place and in time. Contradictions between these two sets of knowledge can have harmful implications for the nonscientists, as their knowledge undergirds claims to homelands or territory, and responsibilities to other groups as well. For example, both the Havasupai and the Nuu-chah-nulth peoples donated their DNA samples for biomedical research, but samples were in scientific publications on human migration studies in which the tribes were identified without their permission (Rubin 2004; Wiwchar 2004; Harmon 2010). The scientific results indicated that the tribes originated from Asia and crossed the Bering Strait. These studies contradicted tribal origin stories and beliefs about where their ancestors originated from, causing stigma and psychological harms to their identities and knowledge systems. In both cases, the researchers did not partner with the tribes for richer analysis, such as regarding elaborate clan relational systems that were trivialized by the use of inbreeding coefficients, thus suggesting that the tribes had engaged in taboo-related activities. For both of these peoples who make a claim to homeland in their current and historic geographic locations, such migration studies contradict their knowledge and their concordant responsibilities to that geographic location. As opposed to achieving additional benefits through consultation and partnership, such as culturally informed analysis and additional data to support scientific analysis, the research caused harms to the community and to many further communities in their stigma against participating in research.

Recreational and Forensic Uses of DNA Ancestry

Data derived from DNA can reveal information about a person's ancestry, make predictions about a person's phenotype, or influence a person's perception of who they are related to. Recreational use of genetic testing often refers to nonmedical applications of genetic technology for the purposes of learning something "fun" about oneself. Recreational use of genetic tests may include seeking additional information about one's ancestral background to identify close or distant relatives or to reconstruct family trees. Recently, genetic ancestry tests have been used in the forensic setting, either to re-identify a suspect based on genetic markers, or to find close relatives of a suspect based on similarities between a DNA sample and data contained in databases of ancestry companies.

Satisfying familial curiosity

Many consumers of genetic ancestry tests seek out results to satisfy their curiosity regarding ancestral connections to groups of people and places, and to find potential connections to extended family members. DTC genetic-testing companies, such as 23andMe, Ancestry DNA, and African Ancestry, offer genetic-testing kits to consumers who are interested in learning more about their genetic ancestry by mapping connections through close relatives or distant kinship. Consumers may purchase a kit with which they collect a saliva sample and send it back to the company with payment. These ancestry companies utilize their own proprietary databases of AIMs and reference samples, which may produce different results across companies. Given the resource of their own databases, these companies are also venturing into research by comparing the prevalence of phenotypic traits in relation to one another and connecting participant surveys with data based on their genetic markers.

Following conversations within the American Society of Human Genetics (ASHG), an Ancestry Task Force was formed, and in 2008 they published a set of recommendations to increase the amount of research to improve the accuracy of the testing results, to develop guidelines to facilitate the return of and interpretations of results, and to inform the public of the limitations of ancestry testing, particularly the testing offered by DTC companies, calling for greater accountability of these firms as well (American Society of Human Genetics 2008). In a subsequent 2010 paper, the ASHG Ancestry Task Force further recommended "thoughtful and rigorous use of ancestry estimation" by academic researchers. Their recommendation for a national roundtable to discuss the issues of concern and work to develop solutions between DTC companies, researchers, consumers, and other key stakeholders has occurred in two iterations to date (Royal et al. 2010).

Furthermore, presentations made to the US Congress have described the interrelations between consumer and researcher attitudes, as well as the lack of standards, and have argued for increased policy and legislation for the protection of individuals and groups (Sarata 2008). This policy briefing and empirical research on consumer motivations have demonstrated variable degrees of knowledge about genetic science and statistics, and argue for increased regulation of the tests and guidelines for standardizing the ways in which results are returned to the consumer.

Genetic nonrelatedness

Nonrelatedness refers to the ethical concern of uncovering through genetic ancestry testing, including maternity and paternity testing, that someone thought to be a biological parent of a given person is not their biological parent. It is commonly noted that nonpaternity, where the person who is believed to be the biological father is not such, likely occurs in up to 10% of the global population. Unexpected discovery of nonmaternity, on the other hand, is much less common and may occur through a form of relatedness, such as a sister, mother, or aunt raising the child born from their sister, daughter, or niece. Nonrelatedness therefore occurs when children are raised by other families entirely, through adoption or mistaken identity for example. As technology has advanced, and the use of *in vitro* fertilization has increased, the possibility of nonmaternity through implantation of a fertilized egg that originated in another woman increases, and newly approved three-parent embryos also alter our social and biological understandings of motherhood (Gallagher 2015). Often, information about nonrelatedness may not trouble the family unit, which has socialized and developed schemas that do not limit kinship to genetics, but it does have the possibility to alter these familial dynamics.

The general concern about genetic ancestry testing of participants and consumers not fully understanding the results and their implications is further complicated by nonrelatedness events that either are revealed by or remain obscured within the data. Particularly with genetic ancestry testing, including SNP analysis and Y-chromosome testing, the deep ancestry that a person and family thought they knew about themselves may differ, as well as the near ancestry of relatedness. This information, once received by a participant or consumer and contrary to the previous understanding of self, may alter their identification and have wider implications for their families and their communities. Given the commonality of nonrelatedness, and the very social and not only biological ways in which we create families and communities, genetic ancestry testing could be vastly complicated by relatively common nonrelatedness events, including issues around self-identification of samples and family who may correlate genetic markers to a geographic location in Europe while a nonrelated event, of which they do not know and the genetics may not show, could have brought those genetic markers from Asia. Furthermore, given the greater participation of individuals in Europe in research studies and of peoples of European descent in the consumer community, the results might make better predications for some groups of peoples compared to others who do not participate so fully or benefit from the sciences (Bustamante and De La Vega 2011).

Beyond nonrelatedness, there are also potential secondary uses of samples that may impact individuals and their families, such as the recent revelation about 23andMe selling data from their proprietary database (Regalado 2015). The partnership created with Genentech appears to put a value on the consent of participants and to allow a sharing of proprietary information through a partnership in which Genentech paid $60 million for access to Parkinson's patient information – a scenario that appears to entice participants to consent by offering money in exchange for data, rather than a business partnership.

Ancestry estimates in health and medicine

Several efforts have aimed to incorporate genetic ancestry estimates into medical studies to promote integrated clinical care. These efforts, called *precision medicine*, may utilize ancestry information to control for ancestry in order to disentangle clinical phenotypes that are more common within a certain population. Precision medicine aims to utilize patients' genomic information to determine the most optimal therapeutic options for each patient. A potential application that integrates genomic information with drug treatments is pharmacogenomics, which tailors drug and dosing prescriptions to individuals with certain genotypes, thus enhancing clinical treatments to individuals.

The use of racial categories in clinical practice has created challenges around the interpretation of genetic ancestry information, both for physicians and for the public. First, racial categories do not overlap neatly with genetic variation. Genetic variants that occur more frequently in a continental population are not proxies for racial categories. Furthermore, given the varying degrees of admixture within a population, physicians may incorrectly conclude a person's continental ancestry, demonstrating a limited utility of the information. Even self-identification has limited utility, as the genetic variants in question could occur in other populations or a patient may have no knowledge of a particular continental ancestry. For example, BiDil, a drug that combines two medications for congestive heart failure in a fixed dosage, was tested in clinical trials but initially did not gain approval from the US Food and Drug Administration (FDA). Subsequently, tests were done in African Americans, and data given to the FDA the second time proved more convincing about efficacy, so it was approved to be marketed to African Americans. The company noted that the African American population is genetically distinct, thereby leading to differential effectiveness of the drug in that population compared to others. However, those claims were later shown to be scientifically inaccurate.

In 2015, US President Obama announced the Precision Medicine Initiative in his presidential address (White House Office of the Press Secretary 2015). The NIH began a series of workshops that summer, with much of the discussion focused on the promise of delivering tailored treatments using genomics. However, few studies to date have demonstrated a clear pathway forward. Furthermore, only a few genes have been directly connected to specific disease states, such as Huntington's disease, and even fewer genes have been directly connected to better understanding effective drug interactions.

Forensic applications of ancestry tests

Forensic applications of ancestry tests have been in use for several years and are being applied more frequently. Genetic ancestry testing is emerging in forensic sciences, primarily for re-identification purposes.

Reconstruction and facial predictions New research, called *forensic DNA phenotyping*, uses genetic polymorphisms of certain genes to predict a person's appearance, from skin color to facial structures and features, and tries to determine a person's ancestry based on AIMs. However, this approach has raised concerns about the scientific accuracy of the technology and the potential to exacerbate racial profiling among law enforcement, and it may infringe on the privacy of many people who fit certain profiles (Pollack 2015). In China, using the same technology, artists and environmentalists have partnered with researchers to collect trace amounts of DNA from litter on public streets to re-identify or to predict the appearance of the person who disposed of the trash and posted pictorial depictions of the presumed suspect in attempts to publicly shame litterbugs (Cope 2015).

Familial Identification In Idaho, police obtained a court order to access Ancestry.com's database of genetic data and names, locations, and birthdays in an attempt to crack a nearly 20-year-old cold case (https://www.genomeweb.com/scan/when-cops-met-ancestrycom). The police identified a suspect based on close DNA matches and information contained in the company's database of social networks. The suspect was eventually cleared in the case. This approach prompted concerns about customer privacy. These secondary uses of ancestry information to identify family members of crime suspects are troubling because, as well as conducting research with their consumers, the forensic possibilities demonstrated by the release of information to police by Ancestry.com have been viewed as a privacy violation and an unintended possibility by many participants, consumers, and their families.

Many of the people who have donated a sample for one purpose (such as learning about their ancestral past or finding close relatives) were not anticipating uses such as access by law enforcement. Such use of database access may limit participation in ancestry and other research, and limit the ability of ancestry companies to increase coverage among certain populations who already have a history of being racially profiled.

Politics and belonging

Genetic genealogy incorporates genetic ancestry testing information, usually from DTC companies, to augment the stories of families. For example, on Ancestry.com, a consumer can search through historical records and build their family tree with that information, but if there are gaps or if one wishes to know more about additional branches on the tree, then they can send in a cheek swab to AncestryDNA.com and find extended genetic relatives as well. This information, however, is not without larger impacts than the fleshing out of a family tree, as genetic ancestry information may be viewed singularly as "the" information about identity, or as "more accurate" information about identity in the face of conflicting familial histories or documentation. This centralizing of genetic information is being questioned from many angles, including how it is consumptive and extractive of resources from marginalized populations by imperialist populations (Reardon and TallBear 2012; Kolopenuk 2014). Furthermore, the tests may contradict not only known genealogical information but also one another, as the proprietary databases at different companies or with different researchers may lead to distinct results about lineage, such as percentages of admixture or more precise locations on continental ancestry estimates. This information and its conflicts can also inform belonging, not only for self or group identification, but also for citizenship and entitlements. The risks to psychological well-being should be considered along with the assumptions, accuracy, and limitations of any given genetic ancestry testing.

Native American Tribal Membership The Freedmen of the Five Civilized Tribes have utilized genetic ancestry testing to attempt to prove their relationship to the tribes, information that those sovereign nations have in general rejected for citizenship applications (Koerner 2005). Other Native American tribes have utilized maternity and paternity testing to disenroll tribal citizens (Taylor 2011) or to limit tribal enrollment eligibility (Eastern Band of Cherokee Indians 2010). These policies use scientific ideas of relatedness through genetic ancestry testing to harm both individuals and communities, privileging particular biological forms of relatedness over social forms, and thereby limiting the possible kinds of family structures. With disenrollment, harms include loss of access to resources and support, denial of identity and psychological harm from such, as well as social stigma.

National Citizenship Recently, the US Immigration and Naturalization Service has turned to genetic ancestry testing to prove the stated relationship of a sponsor to an applicant for citizenship (US Department of State Bureau of Consular Affairs 2015). Yet, some nations use this testing not to exclude or to idealize scientific evidence as the penultimate kind, but instead to allow those who have found roots in their country to become citizens and thereby contribute to the responsibilities and share in the rights of such a role. For example, Isaiah Washington successfully applied for citizenship in Sierra Leone after an ancestry test noted the people of that present nation as his most likely African ancestors, and Oprah Winfrey used information from one ancestry test to commit to serving the young women in South Africa, where she was told her ancestors were likely to have come from (Watanabe 2009; 23andMe 2014).

Like so many other stories we tell of our families and our past, the stories of genetic ancestry are not without their constructions and their extensions, such as those seen around political belonging as discussed in this chapter. Genetic genealogy, like that used by Winfrey and Washington, is catching on as a reality television trend, and celebrities do not always consent to the genetic testing, nor are they immune from the desire to shape the narrative that emerges from the testing and research. When research showed actor Ben Affleck's ancestor was a slave owner, he attempted to keep that information out of the PBS show *Finding Your Roots* (PBS 2015). News stories erupted over this effort, and the stories of the multiple slaveholding ancestors were later told, but this effort demonstrates the importance we place on our ancestry within this country, not just regarding the general geographic ancestry but also in regard to the impact of our ancestors' specific actions as a reflection of ourselves, such that this major celebrity felt it important to distance himself from the history and to not engage with larger questions of privilege.

Conclusion and Emerging Concerns

The technology of genetic ancestry testing continues to develop, and so must attention to and awareness of the ethical issues involved in the technology, the research, and the uses of this testing for a variety of purposes, some of which could not be imagined even at this writing. The most recent progress in human migration research has further shown that continental populations are more likely to be diverse than singular (Raghavan et al. 2015; Skoglund et al. 2015), so continental ancestry would still refer to a wide variety of peoples. These contrasting studies also highlight the concerns about taking such testing as concrete, unequivocal evidence of ancestry, as the findings may differ, or be at odds with each other, and may not represent the whole picture. Concerns around the use of present-day populations as stand-ins for ancient populations should be taken together with concerns about testing of ancient remains, the ethical concerns of which are made clear in the cases of Kennewick Man, who was recently "proven" to be Native American (Rasmussen et al. 2015); the remains found in the Hoyo Negro cave; and the remains found on the Anzick property – all of which are connected to living populations through claims, analyses, and/or geography.

Furthermore, the coordination of technologies is an emerging issue in the ethical evaluation of ancestry testing, as concerns about identity have largely been tested and theorized in relation to person-to-person interactions. However, recently a computer programmer showed that such information could be used to discriminate on the Internet, limiting sites to those with certain demonstrated percentages of continental ancestry through a program called Genetic Access Control (Lee 2015). Even the "possible uses," which are distinguished from the "flaws," of this program using the database information include ones that essentialize gender, religion, and ethnicity, which, like social ancestry, are fluid constructions with biological components. While here the consent to how the data are stored and shared returns as an important concern, how the data can be requested and required for access brings up new and not fully explored concerns, which hearken to older exclusions seen in anti-miscegenation and Jim Crow laws, as well as the separation of groups that contributed eventually to the Nuremberg Principles.

Even as curiosity leads more individuals each day to seek out their biological ancestry through genetic testing, the limits of that information and the ethical concerns inherent to it and to its further applications should be further understood. For those researchers engaged with the testing, communicating and understanding those limitations as well as contexts of appropriation and resource extraction are important for facilitating and gaining truly informed consent, to create research relationships and not extractive gathering of data from human bodies.

CHAPTER 3

Community-Oriented Research and the Future of Anthropological Genetics

Ripan S. Malhi

Department of Anthropology, School of Integrative Biology, and the Carl R. Woese Institute for Genomic Biology, University of Illinois Urbana-Champaign, Urbana, IL, USA

Anthropological genetics is a discipline that seeks to use genetic and biological variation of humans and nonhuman primates to better understand the evolutionary history of humans. A major goal of anthropological genetics includes using genetic data to infer the evolutionary history of Indigenous peoples (present-day and past). During the time that I have been an active researcher in the field of anthropological genetics (since 1996), views and ethical considerations regarding working with Indigenous peoples have changed dramatically. For example, when I was in graduate school, an article entitled "Freezer Anthropology: New Uses for Old Blood" (Merriwether 1999) was published in a widely respected, peer-reviewed journal. At that time, it was generally acceptable in the field to extract and sequence DNA from biological tissues stored in freezers for over 30 years, with no contact with the individuals who provided the samples, to study genetic diversity and the evolutionary history of humans. Since then, many of the samples stored in freezers for decades have been

A Companion to Anthropological Genetics, First Edition. Edited by Dennis H. O'Rourke.

returned to the Indigenous communities from which they were originally sampled (Couzin-Frankel 2010), and tissue samples that remain in freezers are considered "orphan samples" and generally not ethical to use for research purposes (Kowal 2013). In this chapter, I begin by discussing the turbulent relationships between scientists and Indigenous peoples in North America. Then, I examine reasons why Indigenous communities are interested in working with anthropological geneticists on DNA projects and what steps anthropological geneticists have taken to redirect these relationships toward respectful, mutually beneficial partnerships. I end with considerations and comments on DNA analysis of ancient human remains.

The Legacy of Mistrust of Scientists by Indigenous Community Members

Human genetic research requires that the research studies and protocols used have approval from the appropriate Institutional Review Boards (IRBs). However, IRB approval only provides the minimum assurance that a research study is adhering to the guidelines of respect for persons, beneficence, and justice, yet still may not meet the necessities of ethical research. For example, in the well-known study where a team of researchers from Arizona State University (ASU) began collecting blood samples from the Havasupai tribe to identify genetic components to diabetes in the early 1990s, the research team had an IRB-approved informed consent (albeit a broad consent) document and approval for a study on diabetes from the Tribal Council (Hart 2003). However, additional requirements in the protocol were needed to ensure a research study that followed the ethical guidelines of the Belmont Report, a document created by the National Commission for the Protection of Human Subjects of Biomedical and Behavioral Research that summarizes ethical principles and guidelines and is a primary resource for human genetics researchers. Specifically, genuine community engagement and consistent communication with the Tribal Council (or the appropriate governing body) and participants of the study would have likely greatly assisted the potential for success of this project. Instead, the research team did not actively communicate and engage with the Havasupai, and as a result, the Havasupai filed a lawsuit against the Arizona Board of Regents and ASU researchers for misuse of their DNA samples.

Anthropological geneticists are still feeling the downstream effects of this lawsuit. I have personally engaged with an Indigenous community whose cultural heritage council was interested in partnering on a DNA study to infer demographic history. However, the legal team of the community recommended not engaging with such research, citing the Havasupai case. Unfortunately, the negative experience of the Havasupai with scientists is a shared experience of many Indigenous communities. The scientific community and other colonial powers have a long history of exploitation and marginalization of Indigenous communities that has often resulted in a palpable resentment toward scientists by Indigenous community members. I have also witnessed the cumulative results of this exploitation firsthand during encounters with Native American community members. Elders relayed stories of scientists taking blood samples, never to return or be heard from again. One community member asked me, given these experiences, why he should trust me. Many others have been vocally opposed to researchers telling them where they came from, government officials making them "jump through hoops" to repatriate their ancestors from museums, and outsiders in general telling them how to live their lives (Malhi, 2009). The cumulative effect of this exploitation by scientists likely contributes to the large gaps of data from Indigenous communities of North America in genomic studies (Reich et al., 2012; Figure 3.1).

Figure 3.1 Map of geographic sampling for genomic study conducted by Reich et al. (2012) that displays the gaps in sampling in North America. © Springer Nature. Reproduced with permission.

Why Are Native American Communities Interested in Anthropological Genetics Research?

There has been a transition in the time that I've been engaging with Native American communities on DNA research. Early on in my career, I approached Indigenous communities about partnering on a DNA study and discussed questions that I wanted to address using genetic variation obtained from individuals in the community. A number of communities would thank me for my time but then decline involvement in such a project. For the communities that were interested in partnering, I would then ask what questions they might like to address in a DNA study of their community members. In the past few years, this has transitioned to Indigenous communities contacting me to initiate a partnership on a project involving DNA analysis. The DNA projects vary among communities, and the reasons for pursuing such a project can also be different for the Indigenous government entities and the volunteer participants from the community. A consistent underlying reason for tribal governments wanting to pursue DNA research is likely due to the observation that settler colonial governments tend to privilege genetic results over Indigenous forms of knowledge (TallBear 2013; Bardill et al. 2018). Some of the communities that have approached me to pursue a DNA project are in treaty negotiations or are pursuing access and resources promised to them years ago. The tribal governments wish to use DNA studies along with traditional knowledge in an applied way to support their legal cases. Many participants in DNA studies of Indigenous communities have expressed interest in learning about genetic affinities with other communities who speak a related language in a distant geographic region. Others want to see if components of oral histories they've heard from their Elders, such as long-distance travel and settlements along the coastline, might be visible in patterns visualized for genetic data. Lastly, Indigenous communities that are open to ancient DNA analysis of ancestral remains are interested in using archeology and DNA along with Indigenous knowledge to help reconstruct information that was lost following European contact.

Mutually Beneficial Partnerships with Indigenous Communities

Community-oriented research that takes into account participant perspectives and concerns provides a path to ethically engage Indigenous communities as partners in genetics research. Along with providing general safeguards to protect participant confidentiality and avoid stigmatization of the community, a primary goal of community-oriented research is to learn about the specific concerns of the group to tailor a research protocol that addresses these concerns. Anthropological geneticists considering forming partnerships with an Indigenous community on a DNA research project should be informed about the difficult histories of

Indigenous encounters with science, and associated issues relevant to the culture of the Indigenous community (Deloria, 1988). As mentioned in the previous section, the bias toward science and scientific results (that are rooted in Western thought) over Indigenous forms of knowledge by settler colonial governments needs to be considered in the risk–benefit assessment of a DNA-focused research project.

In a partnership on a DNA research project, Indigenous community members should be involved in numerous aspects of the project – from developing protocols for sample collection and storage, to developing questions and hypotheses to address, interpreting results, and writing and reporting the results of a study in articles for the scientific community and the general public (Bardill et al., 2018). Formalized research agreements as well as consistent timely meetings between Indigenous community members and scientists allow the expectations for stakeholders to be defined and, if necessary, modified. The National Congress of American Indians (NCAI) has an online Genetics Resource Center with templates and models of research agreements between scientists and Indigenous communities that can be used to help initiate partnerships and can be tailored to the specific needs of the community. At multiple times during the research project, visits (electronic and in person) to the community are vital. Such visits allow the research team to provide the most recent updates on the project, allow for participants and community members to provide input on results, and also allow the research team and community to revisit the goals of the project and update them if necessary. I've found that partnerships with Indigenous communities are most productive when the community members of the research team are enthusiastic and invested in the project. However, it is important to keep in mind the time burden on community research team members. Members who are the community contacts and organizers for the nonlocal part of the research team often have multiple additional responsibilities in the tribes, and so efforts need to be made to acknowledge and appropriately reward this part of the research team.

Diversifying the Field of Anthropological Genetics

As first stated by Roderick McInnes (2011) during his 2010 presidential address of the American Society of Human Genetics, scientists of non-Indigenous descent likely lack detailed knowledge of Indigenous cultures and their unique sociopolitical concerns. This social and political ignorance can lead to cultural harm in scientific research, mistrust, and weakened political authority for communities whose members participate in such studies. Indigenous scientists are more likely to successfully recognize potentially delicate social and political issues. In addition, having diverse scholars in a field provides different historical experiences and views that result in better science because diverse groups bring multiple perspectives to a problem and have lower levels of confirmation bias (Paige 2008). However, if the American Association of Anthropological Genetics (AAAG) membership is similar in demographics to its parent organization, the American Association of Physical Anthropologists (AAPA), then the membership is remarkably nondiverse (Antón et al., 2018). This means that anthropological genetics researchers are mainly of European descent, while the subjects and participants of anthropological genetics research are much more diverse in ancestry and ethnicity, a composition that has not changed much from early colonial times.

This lack of diversity of researchers with different experiences and perspectives is recognized, and programs have been created to increase diversity in the field. For example, the AAPA Committee on Diversity (COD) launched the Increasing Diversity in Evolutionary Anthropological Sciences (IDEAS) program at the annual AAPA meeting in Atlanta in 2016. The IDEAS program, funded by the National Science Foundation (2016–2019), works to induce culture change in biological anthropology and to promote the training and inclusion of ethnically diverse scholars. The IDEAS program complements other programs in the sciences, with an aim of specifically increasing representation of Indigenous scholars among anthropological genetics researchers. Genetic Education for Native Americans (GENA) workshops pioneered efforts to diversify science by holding workshops to increase genetic literacy for Native American communities. Since 2011, the SING (Summer internship for INdigenous peoples in Genomics) program offers a one-week workshop to facilitate discussions on how genomics can be used as a tool for Native American interests. The SING program is helping to dissolve barriers between Indigenous communities and scientists, and it provides indigenous community members access to the latest advances in genomic science. Along with hands-on training in bioinformatics and molecular biology, the workshop highlights Indigenous values and how scientific methods can be beneficially incorporated with these values. Combining ethical, legal, and social discussions surrounding historical Native American encounters with science with hands-on training, the SING workshop helps prepare participants for future leadership positions in science, research, and teaching careers. Programs like SING also help to foster support networks for Indigenous scientists and scholars (Malhi and Bader 2015).

Ancient DNA Analysis of Human Remains in Colonized Regions

Recent conflicts between Indigenous community members and scientists centered on the treatment of ancestral human remains, such as the Ancient One or "Kennewick Man" and the ancient individual from Spirit Cave. Scientists (including anthropological geneticists) desired to study the genomes of these ancient individuals, as they may provide information to test hypotheses about the evolutionary history of Native Americans. As many Native Americans feel ancestral remains to be sacred – and, if disturbed, should be reburied and left in peace – the study and possible destruction of skeletal remains through DNA analysis are offensive.

In 1990, the Native American Graves Protection and Repatriation Act (NAGPRA) was passed that requires Native American skeletal remains to be repatriated to Native American communities who demonstrate an affiliation with the skeletal remains. In the mid-1990s, the skeletal remains of the Ancient One were requested for repatriation under NAGPRA by a consortium of five Native American tribes who live in the geographic region where the remains were found. The Secretary of the Department of the Interior complied and ordered the repatriation of the skeletal remains. However, a group of scientists sued the government to block the repatriation to be able to scientifically study the remains. The lawsuit ultimately resulted in a judicial ruling allowing scientific study of the skeletal remains (Owsley and Jantz 2014). DNA analyses conducted by three labs were unsuccessful, and almost 20 years after the initial NAGPRA request of the Ancient One, DNA analysis of a sample from the ancient individual using more recent techniques was successful (Rasmussen et al. 2015). Unlike the morphological analyses that suggested that the Ancient One was more closely related to Polynesians or the Ainu (Rasmussen et al. 2015), the DNA analysis demonstrated that the Ancient One is more closely related to Indigenous peoples of the Americas than to any other population on other continents in the world. Also, the analysis suggests that the Ancient One is most closely related to Indigenous populations in the same geographic region as where the remains were found in the interior of the Pacific Northwest, implying genetic continuity of populations over the past 8000 years. This pattern of temporal genetic continuity of populations in a geographic region in the Americas is not unique. Other populations have also shown the same pattern of temporal continuity in the same geographic region over thousands of years, in multiple geographic regions (Raff et al. 2011; Cui et al. 2013; Raghavan et al. 2015; Lindo et al. 2017). Even though the case of the Ancient One is portrayed as one of conflict between scientists and Indigenous peoples, and even though many Native Americans see the judicial rulings as an infringement on Native American sovereignty, the DNA study published in 2015 was one where scientists and Indigenous community members engaged in a non-antagonistic manner, and the consortium of five Native American tribes who originally requested repatriation of the remains in the mid-1990s used the DNA analysis as evidence in their renewed and ultimately successful repatriation case (Green 2017).

Should ancient human remains in colonized regions (e.g. Americas, Australia, and New Zealand) be subject to scientific study and specifically to DNA analysis? This question cannot be answered generally, as it depends on the communities, scientists, and government agencies involved. However, the process to obtain the answer is clear for both affiliated and unaffiliated human remains on public or private lands and for museums that store ancestral remains. Community engagement with the affiliated or geographically proximate Indigenous communities needs to occur. By engaging with the communities, researchers may be able to demonstrate how DNA analysis may serve community interests. If a community is opposed to DNA analysis of ancient human remains, other avenues, such as DNA analysis of nonsacred ancient human commensals (Witt et al. 2015), non-destructive (or minimally destructive) DNA analysis (Bolnick et al. 2012), or the analysis of ancient human byproducts that may not be considered sacred, such as dental calculus (Warinner et al. 2014) or feces (Poinar et al. 2001), might provide the needed information to serve the same purpose. What cannot happen is the continued disrespect and marginalization of Indigenous views and community members in the scientific and research process.

Conclusion

Some anthropological geneticists have been engaging with Indigenous communities since before the 1960s (Friedlander 2004), and the trend is growing toward involving Indigenous communities in research in a way that is respectful and mutually beneficial. In addition, recent efforts to increase the diversity of scientists in the anthropological genetics field are growing. As the field of anthropological genetics becomes more inclusive and has lead researchers who are similar in ancestry, ethnicity, and experience to the research participants studied around the globe, the field will more effectively meet the needs of research participants by providing a more nuanced understanding of research and related sociopolitical situations and provide knowledge and experiences that will improve the scientific endeavor as a whole.

ACKNOWLEDGMENTS

I'd like to thank the Summer internship for INdigenous peoples in Genomics (SING) community, where many of the ideas in this chapter were initially discussed. I also thank Dennis O'Rourke for inviting me to write this chapter.

PART II Macroevolution and Phylogenetics

CHAPTER 4

Calibrating the Clock

Murray P. Cox

Institute of Fundamental Sciences, Massey University, Palmerston North, New Zealand

Dating past events using genetic evidence remains one of the holy grails of molecular anthropology (Ho et al. 2014b). Accurate estimation of the relative and absolute timing of past events and processes is crucial if they are to be placed in their appropriate anthropological context, as framed by archeology and linguistics. However, dating is more than just pinning times on events (Ho and Duchêne 2014). Uncertainty in genetic dating can lead to real contention about how we interpret human history. As just one key example, it is still surprisingly unclear when our species first left Africa. The out-of-Africa expansion is generally considered to have occurred around 50 kya (kya = thousand years ago), but mitochondrial DNA (mtDNA) evidence has variously been used to support this date (Underhill and Kivisild 2007) or the much earlier time of 75 kya (Petraglia et al. 2007). This has important implications for how lithic assemblies outside Africa are interpreted. For instance, the Jwalapuram lithics of southern India span the ash layer laid down by the Toba eruption and are therefore securely dated to 74 kya. But who made them? Depending on the dating of the out-of-Africa expansion, this lithic assemblage might reflect (i) the first movements of modern humans out of Africa, (ii) an early expansion of modern humans that turned out to be an evolutionary dead end, or (iii) one of the many archaic hominin groups that are now known to have populated Eurasia over much of the last million years. Dates matter, and uncertainty in dating is preventing us from answering fundamental questions about our history.

A Companion to Anthropological Genetics, First Edition. Edited by Dennis H. O'Rourke.

The role that molecular evidence can play in assigning dates to past events was first identified in the early 1960s with the discovery of the molecular clock. Zuckerkandl and Pauling (1962) proposed that lineages acquire mutations in a regular fashion, allowing sequence divergence to act as a proxy for time. Just five years later, Sarich and Wilson (1967) used this logic to conclude that humans and chimpanzees diverged ~five million years ago, overturning the widespread belief that humans and our sibling species are much more distantly related. This claim was soon backed up by new fossil evidence from East Africa, most notably the *Australopithecus afarensis* specimen "Lucy." Since then, the development of molecular-dating techniques has flourished, and the approaches used today are vastly more complex than those initial forays. However, many of the simplest questions in the field of molecular dating are still unanswered, and our ability to date molecular events with conviction remains surprisingly constrained.

Mathematics tells us that molecular dating will always be an uncertain science. Ideally, we would like to know how the ages of lineages within populations and species vary across the genome. A branch of mathematics known as coalescent theory has made this possible. One of its most important outcomes was the derivation of equations that describe the mean and variance of the time back to the most recent common ancestor (T_{MRCA}) of a population sample (Hudson 1990; Donnelly and Tavare 1995; Wakeley 2008a).

$$E(T_{MRCA}) = 2n\left(1 - \frac{1}{n}\right)$$

$$\text{var}(T_{MRCA}) = n\left(8\sum_{i=2}^{i=n}\frac{1}{i^2} - 4\left(1 - \frac{1}{n}\right)^2\right)$$

As the sample size increases, this expectation rapidly converges to 2 and the variance to 1.16 (in units of N generations). The take-home message is that the variance is large relative to the mean, which implies that even if we knew the ages of a set of genes perfectly, they would still vary considerably from each other. In practice, this raises concerns about our past reliance on individual loci, such as mitochondrial DNA or the Y chromosome. Assigning any one molecular date to a historical event has considerable potential to be misleading, which explains why molecular anthropology and, indeed, most subfields of genetics are now turning rapidly to genome-scale data. The entire human complement of DNA comprises thousands of small genetic regions, each representing an independent historical record due to recombination breaking the physical links between loci. With key exceptions, such as studying a particular gene or sex-biased process, looking at thousands of genes simultaneously across the autosomes allows us to "average across" histories at different loci, and thus generate more robust estimates of the dates of influential historical events.

Other evolutionary features also impose strong confounding effects when reconstructing dates from molecular evidence. For instance, both genes and lineages evolve at different rates (reviewed in Ho 2014). It is perhaps unsurprising that loci change at variable speeds, as some genetic regions are strongly constrained (such as the ubiquitin gene, which differs little between mammals and nematodes; Castillo-Davis et al. 2004), while other regions are free to vary. Indeed, the fast rate of the mtDNA control region, which offers improved resolution over the time scale of recent human history, is widely perceived as one of the key benefits of this locus (Endicott et al. 2009). That genes evolve at different rates has been known since at least the early 1960s, and was suggested by Dickerson (1971) to reflect variation in the proportion of sites that are free to vary. Such gene-specific effects demand a careful choice of locus when studying different aspects of human history, and a variety of loci are employed today precisely to study evolutionary processes at different time scales.

However, even for the same gene, rates may vary among different human lineages. Genes were originally considered to evolve according to a "strict" molecular clock, whereby all lineages change at some fixed regular rate, much like the ticking of a metronome. Today, there is greater awareness that rates can vary with generation time, metabolic rate, and the accuracy of DNA repair mechanisms (among other drivers). These influences can be pivotal at deeper time depths (such as comparisons between species; Bromham 2011), but are also important within species (Endicott and Ho 2008). Most contemporary methods now employ a "relaxed clock," in which mutation rates are free to vary among lineages (Heath and Moore 2014; Yang 2014). Within this framework, different branches in a gene tree can evolve at different rates. Although there is still considerable discussion around which relaxed-clock methods perform best (Lepage et al. 2007; Heath et al. 2012; Ho et al. 2014a), there is a general consensus that relaxed clocks are preferable to strict-clock alternatives (Drummond et al. 2006), even when dating recent divergences (Brown and Yang 2010).

Even within a lineage, different parts of a gene may evolve at substantially different rates. This has been studied in most detail for human mtDNA: the control region evolves more quickly than regions that encode genes, and even within genes, rates vary considerably between first, second, and third codon sites due to the redundancy of the genetic code (Endicott et al. 2009). Even more nuanced mutation patterns are still being identified in humans. Although the mechanism remains unknown, mutation events are clustered along the genome and therefore lack independence, perhaps due to localized failures of DNA repair (reviewed in Ségurel et al. 2014). Furthermore, mutation rates strongly covary with paternal (but not maternal) age, primarily due to the marked difference in the number of cell divisions required to produce eggs (an average of 31 cell divisions) versus sperm (~400 cell divisions at age 30 and ~650 at age 40) (Ségurel et al. 2014). Many existing molecular-dating tools do not readily accommodate these and other similarly complex determinants of mutation rates. As genome sequencing continues apace, it is becoming increasing clear that our understanding of mutation processes is still in its infancy, with all that implies for the accuracy required of molecular dating.

A crucial, but often underappreciated, feature of molecular dating is that all methods return only the order and relative ages of events. Real chronological dates only emerge from the process of calibration. At least within molecular anthropology, calibrated dates are most commonly obtained by rescaling a tree with a given mutation rate (Endicott et al. 2009). (This is not necessarily true for other disciplines; see Hipsley and Muller 2014.) Unfortunately, mutation rates are never known with perfect accuracy, which in turn increases the uncertainty of calibrated dates. Three primary methods have been developed to obtain mutation rates: (i) phylogenetic comparisons, (ii) archeologically dated events, and (iii) pedigrees. Each has its benefits and disadvantages, and importantly, the three approaches often do not return the same mutation rate (as discussed in greater detail in this chapter).

One of the most common calibration points used in molecular anthropology is the divergence time between humans and chimpanzees. The mutation rate (μ) is typically calculated as:

$$\mu = \frac{d}{2t}$$

where d is the sequence divergence (i.e. the number of differences per base pair [bp] between humans and chimpanzees for a given genetic region), and t is the divergence time (in years). While sequence divergence can be calculated with high accuracy (Prüfer et al. 2012), the date at which humans and chimpanzees diverged remains contentious. This separation event is often assumed to have occurred six million years ago (Endicott et al. 2009), but molecular estimates support values anywhere from four to eight million years ago (Bradley 2008), while the fossil record places the split at least six million years ago, in part due to a lack of clarity around which fossils are direct human ancestors versus sister taxa (Benton and Donoghue 2007). Phylogenetic rates also typically assume that humans, chimpanzees, and intermediate forms have broadly similar generation intervals (Langergraber et al. 2012) and that the human–chimpanzee split was a relatively discrete event (Patterson et al. 2006; Wakeley 2008b; Presgraves and Yi 2009). Both of these assumptions have been questioned. Nevertheless, the phylogenetic comparison is still among the most widely used methods for calibrating molecular rates for humans and their hominin cousins today.

An alternative approach favors calibration points closer to the present typically employs archeologically dated events, such as the arrival of modern humans in Australia, the settlement of the Americas, or other well-defined migration events (Endicott and Ho 2008; Henn et al. 2009). Because these events are younger, they are often known with more temporal precision than older events, such as human–chimpanzee divergence. Unfortunately, many potential calibration points are also highly contentious in their own right, and therefore often the preferred target rather than basis of molecular dating. For example, early and late dates have been proposed for the settlement of both Australia (O'Connell and

Allen 2004; Hudjashov et al. 2007; Rasmussen et al. 2011) and the Americas (Fagundes et al. 2008; Ho and Endicott 2008; O'Rourke and Raff 2010; Reich et al. 2012). Selecting the wrong age for a calibration point will obviously bias subsequent dating, and at times, archeologically defined mutation rates can introduce hidden circular arguments. However, even if a date is well accepted, determining whether a particular genetic lineage is associated with that historical event can still be problematic. An alternative, and increasingly feasible, approach employs ancient genomes as known calibration points. Sequence divergence between modern and ancient genomes is readily calculated, and archeological remains, particularly those within the radiocarbon limit, are often robustly dated. We even have increasingly extensive genome sequences from a number of our sister species, including Neanderthals (Briggs et al. 2009; Green et al. 2010) and Denisovans (Reich et al. 2010, 2011a; Meyer et al. 2012). This general approach is now used extensively with nonhuman systems (Drummond et al. 2003; Shapiro et al. 2004), and calibrations against ancient human genomes show considerable promise for molecular anthropology as well (Brotherton et al. 2013; Fu et al. 2013b; Rieux et al. 2014).

The third calibration method uses direct estimates of mutation rates from human pedigrees (1000 Genomes Project Consortium 2010; Roach et al. 2010).

$$\mu = \frac{d}{t_g}$$

Sequence divergence is determined between two or more generations (such as parent–child pairs), but it must be corrected for the mean generation interval (t_g), which can vary widely between different human communities (Fenner 2005; Matsumura and Forster 2008). As sequencing costs decrease, this approach is growing in popularity. Although it has the major limitation that sufficient new mutation events must be identified to estimate the mutation rate accurately, this concern has largely dissipated with the advent of genome-scale sequencing. Identifying false-positive and false-negative mutations remains an important consideration, as these can respectively inflate and decrease the observed mutation rate (Conrad et al. 2011). Nevertheless, pedigree methods are rapidly supplanting many earlier approaches of estimating mutation rates.

It is important to note that application of a single mutation rate, regardless of how accurately it is known, cannot account for rate variation among lineages. An alternative approach, which is implemented in some of the more popular software solutions (see Box 4.1), is to instead infer a gene tree and fit that tree to multiple calibration points (Duchêne et al. 2014). As described in this chapter, these calibration points might be archeologically dated events, ancient genomes, or some other well-dated feature. Because many modern dating methods have adopted Bayesian statistics, calibration points are now often provided as tight priors on particular nodes in the gene tree, thereby facilitating uncertainty in the inferred date rather than employing a single point estimate. This approach has recently

Box 4.1

A wide range of tools is now available for molecular dating, many of which target specific research goals. Recent years have seen a noticeable switch from relatively simple dating methods to far more computationally intensive likelihood and Bayesian approaches. Some early methods, such as the ρ statistic (http://www.fluxus-engineering.com/sharenet.htm; Forster et al. 1996), remain appealing due to their simplicity, but unfortunately they produce extremely biased dates under many real-world demographic scenarios (Cox 2008). In contrast, modern likelihood and Bayesian methods explicitly measure uncertainty using standard statistical frameworks and are frequently robust to minor violations of their assumptions. However, the range of demographic models offered by these methods is still often limited, especially models of population structure, and the software is often not trivial to run, typically requiring a high level of familiarity with advanced software and access to substantial computational resources.

Some older methods, such as Genetree (Griffiths and Tavare 1996), still exhibit excellent statistical properties but are limited in application (e.g. the genetic dataset can contain no recombination events or recurrent mutations; Woerner et al. 2007). Other software, notably DAMBE (http://dambe.bio.uottawa.ca/DAMBE/dambe.aspx; Xia 2013) and the "chronos" function of the "ape" package in R (https://cran.r-project.org/web/packages/ape/index.html; Paradis 2013), assign dates to a tree via calibration points at internal nodes, but do not consider uncertainty in the tree topology or branch lengths. Yet other programs, such as RevBayes, provide a general environment for implementing bespoke Bayesian phylogenetics (https://revbayes.github.io). However, the software perhaps most commonly used for molecular dating today is BEAST (http://beast.bio.ed.ac.uk; Drummond and Rambaut 2007), and increasingly its successor BEAST2 (http://beast2.org; Bouckaert et al. 2014). Using Markov chain Monte Carlo (MCMC) and Bayesian statistics, the primary aim of this program is to infer dates for rooted trees under a range of strict and relaxed molecular clock models. The variety of demographic models implemented, including a "skyline" model that allows population sizes to vary freely over time (Drummond et al. 2005), makes BEAST2 one of the most general solutions available for molecular dating today.

Finally, more specialized tools also exist. The IM suite of packages (https://bio.cst.temple.edu/~hey/software; Pinho and Hey 2010) can reconstruct divergence times between pairs of populations, as opposed to dating the divergence of genetic lineages. In most instances, determining the sizes, divergence times, and migration rates of populations is closer to real research goals than obtaining dates for particular genetic lineages on a gene tree.

been used to date (and place geographic paths on) the expansion of the Indo-European language family (Bouckaert et al. 2012), thus illustrating its utility beyond genetics. Multiple calibration points are required for this method to yield any accuracy, but as a subsequent benefit, inferred molecular dates are typically robust to one or a few calibration points that happen to be inaccurate or even simply wrong.

It might be thought that the three calibration methods will return the same mutation rate for a given gene, at least in the limit of large sample sizes, where substantial amounts of information can be brought to bear. Surprisingly, this is not true; systematic differences in estimated mutation rates have been observed using the three different approaches. This discrepancy has been explored in most detail for mtDNA. Considering just the noncoding control region, pedigree studies estimate a mutation rate around 7.95×10^{-7} mutations/bp/year ($4.30 \times 10^{-7} - 10.2 \times 10^{-7}$) (Sigurðardóttir et al. 2000; Heyer et al. 2001; Howell et al. 2003; Santos et al. 2005). Conversely, assuming a human–chimpanzee divergence time of six million years (and recalibrating published rates for this value, as in Henn et al. 2009), phylogenetic rates for this same genetic region fall around 2.14×10^{-7} mutations/bp/year ($1.15 \times 10^{-7} - 3.60 \times 10^{-7}$) (Hasegawa and Horai 1991; Vigilant et al. 1991; Ward et al. 1991; Stoneking et al. 1992; Tamura and Nei 1993; Horai et al. 1995; Forster et al. 1996). These two estimates are radically different; the pedigree rate is 3.7 times larger than the phylogenetic rate, and the two distributions do not overlap.

This difference has been attributed to the "time dependency" of molecular rates (Ho and Larson 2006). Rates do not simply differ between the pedigree and phylogenetic methods, but have been shown to change regularly through time, exhibiting an exponential decline from fast rates in the present to slow rates in the past (Henn et al. 2009; Ho et al. 2011). This pattern likely reflects the fact that not all changes we observe are neutral. Most genetic regions studied in molecular anthropology are assumed to evolve neutrally, and under neutral theory, the mutation rate is determined only by the rate of spontaneous mutation, independent of population size (Kimura 1968). In practice, however, most regions experience some level of selection: either purifying selection, which removes disadvantageous mutations that arise in the population (Endicott and Ho 2008), or positive selection, which sweeps advantageous mutations (and any mutations physically linked to them; Smith and Haigh 1974) to higher frequency. Most of the mutations that arise within a population are therefore transient and quickly lost through drift or selection. Pedigree studies observe and count these spontaneous mutations, but only a small proportion of them survive long enough to contribute to sequence comparisons at older time depths (such as between humans and chimpanzees). Put differently, pedigree studies measure the spontaneous mutation rate, while phylogenetic studies instead measure the substitution rate (i.e. fixed differences between species). This distinction likely explains the apparent discrepancy between pedigree and phylogenetic rates, at least for mtDNA. Ad hoc methods for correcting mitochondrial mutation rates for time dependence have been proposed (Soares et al. 2009; Gignoux et al. 2011), but because these methods do not account for other sources of uncertainty (such as rate variation among lineages; Endicott et al. 2009), alternative analytical solutions are actively being explored.

It is worth emphasizing again that mutations are independent of population size only when they are neutral. If selection pressures are acting, changes in population size can also lead to changes in the observed mutation rate. Even for neutral markers, various aspects of population demography can substantially alter the shape of the gene tree, and hence the regularity with which mutations arise and persist in the population (Cox 2008; Henn et al. 2009). Unfortunately, these demographic features – population structure, founder events, and bottlenecks – have all been extremely common throughout human prehistory, thus leveling additional uncertainty on dating estimates.

In the modern genomics era, molecular anthropologists are rapidly turning to genome-wide datasets, and in concert, questions about mutation rates have been moving away from mtDNA and focusing instead on the nuclear chromosomes (Scally and Durbin 2012). Genome-wide data have been available for some time – initially with datasets of short loci distributed across the genome (Wall et al. 2008), and increasingly with genome-wide sequencing of the entire DNA complement of individuals and populations (1000 Genomes Project Consortium 2012). Phylogenetic rates, calibrated by fossil evidence of the divergence between humans and orangutans, or humans and macaques, typically fall around 1×10^{-9} mutations/bp/year (Takahata and Satta 1997; Green et al. 2010). Conversely, pedigree studies of dominant disease loci, genome-wide coding regions ("exomes") and the whole genome suggest a rate around 13×10^{-9} mutations/bp/generation ($9.7 \times 10^{-9} - 18.5 \times 10^{-9}$) (Kondrashov 2003; 1000 Genomes Project Consortium 2010; Awadalla et al. 2010; Lynch 2010; Roach et al. 2010; Kong et al. 2012; O'Roak et al. 2012; Sanders et al. 2012). When corrected for a human generation interval of 30 years (Fenner 2005; Matsumura and Forster 2008), this equates to 0.43×10^{-9} mutations/bp/year ($0.32 \times 10^{-9} - 0.62 \times 10^{-9}$). Again, a 2.3-fold discrepancy is observed between the two rates, but in striking contrast to mtDNA results, the pedigree rate is lower than the phylogenetic rate. When calibrations are made against a range of primate divergences (in descending temporal order: macaques, orangutans, gorillas, and chimpanzees), a strong trend is observed of lower mutation rates toward the present (Scally and Durbin 2012), a process that has been termed the "hominoid slowdown" (Goodman 1961). Superficially, it appears that mutation rates have decreased through time during the evolution of humans and the great apes. There is still considerable discussion around the veracity of this explanation, and alternatives such as errors in fossil divergence estimates, higher effective population sizes, longer generation intervals, false negatives leading to missing mutations, and the sheer complexity of the mutation process have all been proposed (Ségurel et al. 2014; Veeramah and

Hammer 2014). While the driver of this pattern remains a matter of debate, its discovery highlights how much we have still to learn about mutation processes, even in extremely well-studied organisms like humans.

Molecular dating has also been performed with length-variable microsatellites instead of point mutations (Sun et al. 2009), and recently, new forms of molecular dating have been developed that do not require mutation rates at all. A dominant characteristic of the human past has been rampant admixture between long-separated groups (Cox et al. 2010; Pugach et al. 2013; Hellenthal et al. 2014; Lipson et al. 2014). As these previously separated populations interact and intermarry, they produce offspring that carry chromosomes from each of the parent groups. Recombination events create mosaic chromosomes with ancestry blocks from each parent population, and as recombination events accumulate over time, these ancestry blocks become progressively smaller (Winkler et al. 2010). If the recombination rate is known, the admixture time can be inferred by measuring the distribution of block sizes (Pugach et al. 2011; Xu et al. 2012; Sanderson et al. 2015). At present, this technique is restricted to dating admixture events, but it emphasizes that calculating dates with molecular data requires only that a given process changes in some regular way, and this is not the sole purview of mutation. Still, recombination rates provide no simple panacea to the issues raised here: like mutation rates, recombination rates are increasingly recognized as being variable between species (Jeffreys and Neumann 2009; Laayouni et al. 2011), populations (Laayouni et al. 2011), and individuals (Broman et al. 1998; Kong et al. 2002; Neumann and Jeffreys 2006; Calabrese 2007; Khil and Camerini-Otero 2009). There is also a growing body of evidence showing that recombination rates can change over time (Cox et al. 2013). While new dating methods are always welcome, attempts to date past events using patterns of recombination simply emphasize that most sources of uncertainty in molecular dating are generic and will presumably require broad universal solutions.

Molecular dating has come a long way from the strict molecular clocks and relatively simple tests of the 1960s. Sophisticated analyses are now routine, key sources of uncertainty in molecular dating have been identified, and in many cases, reasonable practical solutions have been developed to address them. Nonetheless, dating past events using molecular evidence is by no means a solved problem. Ongoing research shows that we know less about mutation rates than we once thought; new sources of variation are being discovered, and old sources are being revisited in the light of new genomic information. Uncertainty in the basic foundations of molecular dating, particularly with mutation rates or even whether a mutation rate for a given locus exists in any singular sense, is causing considerable upheaval for our interpretation of major anthropological questions. Many of these fundamentals of molecular dating are areas of active research and are likely to remain so for the foreseeable future.

ACKNOWLEDGMENTS

The author thanks Simon Ho (University of Sydney, Australia) and Phillip Endicott (Musée de l'Homme, France) for their constructive and detailed comments. The author was supported by a Rutherford Fellowship from the Royal Society of New Zealand (RDF-10-MAU-001), the New Zealand Marsden Fund (11-MAU-007), and the Allan Wilson Centre for Molecular Ecology and Evolution.

CHAPTER 5

Primate Genomics and Phylogenetics

Christopher A. Schmitt[1] *and Omer Gokcumen*[2]

[1]*Anthropology Department, Boston University, Boston, MA, USA*
[2]*Department of Biological Sciences, University at Buffalo, Buffalo, NY, USA*

Next-Generation Sequencing: The Paths to Strange, Unexplored Worlds

It is now cheap to produce tremendous amounts of genome sequence data. With the advent of massively parallel sequencing technologies in the last decade, the cost of sequencing has reduced at a truly unprecedented pace, dwarfing the already considerable reduction in cost of computational power (Koboldt et al. 2013; Wetterstrand 2015). The abundance of data made possible by next-generation sequencing (NGS) technologies has opened three avenues of investigation for biological phenomena that had previously been prohibitively expensive. First, the mere increase in the amount of data allowed for the investigation of genomic variation in unprecedented detail. Rather than focusing on just parts of the genome (e.g. mitochondrial DNA or hypothesized genomic regions of interest), investigations of detailed and deeply sequenced whole genomes have helped uncover previously hidden trends in several species' evolutionary history. For instance, the polar bear was thought to be an offshoot of the brown bear based on mitochondrial DNA sequence comparisons. In an exciting turn of events for the evolutionary biology community, this symbolic species rightfully gained its place in the phylogenetic tree as a distinct species, and its complicated connections with

A Companion to Anthropological Genetics, First Edition. Edited by Dennis H. O'Rourke.

brown bear lineages are now being elucidated thanks to the availability of autosomal, wide-ranging sequencing efforts (Hailer et al. 2012; Miller et al. 2012; Cahill et al. 2013; Liu et al. 2014).

The second major path to be opened by NGS is the ability to resequence whole genomes at the population level (i.e. sequencing multiple genomes from a single species or population). As exemplified best by the herculean effort of the 1000 Genomes Project in humans (1000 Genomes Project Consortium 2012), as well as smaller but equally valuable efforts in nonhuman primates (e.g. Gokcumen et al. 2013a; Prado-Martinez et al. 2013b), such sequencing studies are important on two levels: (i) they provide datasets that do not suffer from ascertainment biases inherent in *genotyping* chips and help identify a vast range of novel genomic variations (for an anthropologically oriented discussion in humans, see Lachance and Tishkoff 2013); and (ii) they allow investigating within-species genomic trends across the genome, such as identifying mutational and recombination hotspots, or regions that show unusually high population differentiation or linkage disequilibrium patterns, which help identify unique genome-wide trends within a species (for an integrative example of the methods, see Jones et al. 2012; Jasinska et al. 2013).

The third major path of research that NGS has enabled is related to the nature of the unbiased, random creation of short reads by the technology. These short reads have rendered the *a priori*, locus-specific amplification steps that were once essential for studies of ancient DNA (e.g. Prüfer et al. 2014), microbial DNA (e.g. David et al. 2013), and total RNA (e.g. Lappalainen et al. 2013). Consequently, ancient genomics, transcriptomics, and metagenomics have emerged as new fields, unearthing completely unexpected findings along with them. Studies of ancient genomes of now-extinct hominids have revised recent human evolutionary history by showing introgression from both Neanderthal and Denisovan lineages into modern Eurasians (Green et al. 2010; Reich et al. 2010), while the sequencing of more recent *Homo* remains has illuminated the complex migration and mating patterns of ancient humans in Eurasia to a level of detail that was unattainable using previous archeological and molecular genetic evidence (Brandt et al. 2013; Lazaridis et al. 2014; Allentoft et al. 2015). Moreover, some of these shared elements lead to unique, population-specific adaptations (Huerta-Sánchez et al. 2014; Cooper et al. 2015; Lin et al. 2015). Microbiome studies have discovered complex ecosystems in our organs that are highly regulated by evolutionary trends within and across species (Ley et al. 2008). RNAseq techniques have revealed thousands of long noncoding RNAs with mostly unknown functions (Guttman et al. 2009) with the promise of explaining cryptic regulatory variation across evolution (Necsulea et al. 2014).

Applying these exciting techniques to anthropological primate research has just opened a new and exciting chapter in our understanding of ourselves and our closest evolutionary relatives.

Comparing Primate Genomes

The last 10 years have witnessed the publication of more than a dozen nonhuman primate reference genomes, including all extant apes (for a thorough review, see Rogers and Gibbs 2014). A direct comparison of these datasets gives a comprehensive and direct look at the differences and similarities across primate genomes. Of course, one of the first questions often asked, albeit anthropocentrically, is: how much genetic difference is there between humans and other, nonhuman primates?

The standard analytical method used to compare sequences is to find orthologous sequences between each species' genomes, align them to each other, and calculate the differences between them. Central assumptions include that these sequences have evolved from a common ancestral sequence and that, after speciation, individual sequences accumulate mutations at a steady rate. Under these assumptions and looking at the structurally conserved, alignable regions of the genome, the genomes of primates look very similar to each other. For example, the estimated difference between humans and chimpanzees is roughly 1% (Chimpanzee Sequencing and Analysis Consortium 2005). Even Old World monkey genomes, such as that of the rhesus macaque, differ from human genomes by only about 6% (Rhesus Macaque Genome Sequencing and Analysis Consortium et al. 2007). Considering the fact that the aligned sequences between primates are mostly noncoding and that more conservation is expected in the coding sequences, the difference among primate genomes as measured by these analyses is proportionally very small. It is important to note, however, that a 1% variance between two primate genomes roughly corresponds to a staggering 30 000 000 base pairs, a small portion of which are linked to important phenotypic differences (reviewed in Varki and Altheide 2005).

While empirically establishing the paucity of differences between well-aligned sequences between primate genomes, reference genome assemblies have also helped establish previously underappreciated types of variation among these genomes. Specifically, in the last 10 years, several studies have documented large structural variations between species that involve deletions and duplications of entire genes (e.g. Perry et al. 2007) and regulatory regions (e.g. McLean et al. 2011), functional duplications of gene parts (e.g. Dennis et al. 2012), translocations (e.g. Stankiewicz et al. 2001) and inversions of large chromosomal segments (e.g. Ullastres et al. 2014), and expansions and contractions of gene families (e.g. Johnson et al. 2006), among other functionally relevant variations (for an anthropological review, see Eaaswarkhanth et al. 2014). Great apes, for example, had a major burst in segmental duplications (i.e. >1 kb pieces of DNA that are repeated across the genome two or more times) (Marques-Bonet et al. 2009). These segmental duplications include several important gene families, some of which are already related to human-specific traits, such as larger and more complicated brains (reviewed in Samonte and Eichler 2002). We now appreciate a largely unexplored realm of structural diversity with clear functional implications beyond single nucleotide variants in primate genomes.

Similar to structural variants, multiple primate genome assemblies allowed reassessment of distribution and overall impact of transposable elements in different primate species. These results revealed previously underappreciated complexity, lineage-specific activity, and functional relevance of these ancient viral remnants in primate genomes, including that of humans. Assembly of the gibbon genome assembly, for instance, unearthed a novel gibbon-specific retrotransposon, called a LAVA element. The study identified 1797 LAVA elements across the gibbon genome (Carbone et al. 2014). These elements are a composite of multiple repeat elements, including other retrotransposons and short tandem repeats. As such, with their multi-kilobase size range (compared to ~300 bp sizes of Alu elements), these elements stand out as unique gibbon-specific genomic features. More importantly, these elements may also induce further genome plasticity in the form of structural variants by interfering with genes that involve regulation of chromosomal segregation. In fact, similar lineage-specific mechanisms, such as increased rate of non-allelic homologous recombination in great apes and much higher activity of Alu retrotransposition in rhesus macaques, suggest that not only are there structural differences between primate lineages, but also the ways in which primate genomes evolve differ from each other (Gokcumen et al. 2013a). Overall, recent genome assemblies revealed a fascinating world of multidimensional genomic complexity that we are just beginning to understand.

Primate Genomes at the Population Level

It is now possible to sequence whole primate genomes at the population level. Such population-level resequencing allowed several important analyses that were previously not possible. The first of these analyses, of course, is excavating the population history of different primate groups with much higher depth and comprehensiveness than locus-specific studies. Population genomics studies have been especially powerful in determining genetic structure within species. For example, a clear structure in the chimpanzee population among the species' Eastern, Western, Central, and Nigeria–Cameroon groups has now been empirically demonstrated (Prado-Martinez et al. 2013b). In addition, whole-genome sequences enabled researchers to investigate the impact of recent inbreeding as well as older bottlenecks and other demographic trends with very high sensitivity. An especially intriguing example of a bottleneck is the recently revealed difference in genetic variation between Sumatran and Bornean orangutan populations, where the former group, despite its smaller contemporary population size, harbors seven times more genetic diversity than the latter (Locke et al. 2011). An unlikely example for inbreeding among wild primates comes from the whole-genome sequencing of an albino gorilla, Snowflake, where inbreeding in the wild was documented by an analysis of autozygosity across his genome (Prado-Martinez et al. 2013a).

One advantage of whole-genome sequences at the population level is the ability to identify genomic regions where the measures of genetic variation (e.g. linkage disequilibrium, allele frequency spectrum, and population differentiation) show deviation from other parts of the genome (for a short introduction, see Akey 2009). Such deviations may indicate the impact of adaptive forces and have been used to identify lineage-specific evidence of positive selection in multiple species. For example, a recent study showed, based on the whole-genome sequences of 17 gorillas, that genes related to taste perception and cardiac muscle function, among others, have been overrepresented in regions that show signatures of recent positive selection in gorillas (McManus et al. 2015). These selective sweeps may have implications for further research on dietary selection in the wild based on bitter taste receptors, as well as on health in captivity, where gorillas often die of cardiomyopathy.

Thanks to analysis of genome assemblies, we now know that the speciation events that define extant primate lineages are complex processes rather than instantaneous splitting of lineages. A remarkable study comparing chimpanzee, bonobo, and human genomes noted the existence of multiple genetic loci, corresponding to 1% of the whole genome, where the locus-specific phylogeny did not match the general species phylogeny (Caswell et al. 2008). In other words, there are loci in human genomes that share more alleles with the bonobo genome than they do with the chimpanzee genome. This is explained by incomplete lineage sorting, where the ancestral populations to humans, chimpanzees, and bonobos carry genetic variations that are carried into contemporary populations. An earlier paper by the same group also noted the complicated processes of speciation of humans and chimpanzees, putting forward evidence suggesting that chimpanzees and humans interacted *after* the lineages initially diverged (Patterson et al. 2006). Adding to the already complicated speciation history of our ancestors, a recent study involving whole-genome sequences of 10 chimpanzees showed that the genomic locations of chimpanzee recombination hotspots differ significantly from those of human recombination hotspots at the fine scale (Auton et al. 2012).

While these unexpected genome-wide trends of speciation and recombination in humans and chimpanzees were documented, population-level genomics also helped identify regions of the genome where trends in single nucleotide variation have been maintained since before human and chimpanzee ancestors and remained variable in extant human and chimpanzee populations (Leffler et al. 2013). The authors explained this observation by balancing selection acting on these loci, likely affecting the immunity-related function of membrane-bound glycoproteins as a response to constantly changing host–pathogen interactions. Others showed maintenance of hotspots of recurrent structural variation among humans, chimpanzees, and rhesus macaques, effectively spanning 25 million years of evolutionary time (Gokcumen et al. 2011). Similar to what was observed for maintained single nucleotide variation, the immune system genes are overrepresented in primate structural variation hotspots. Population-level genomic variation data provide strong empirical support for the complex and adaptive evolution of the immune system in primates. Indeed, follow-up studies are now dissecting individual immune system genes and gene families and their complicated evolutionary stories within a balancing selection–diversifying selection framework (Cagliani et al. 2010; Hiwatashi et al. 2010; Ferguson et al. 2012; Ségurel et al. 2012; Gokcumen et al. 2013b; Teixeira et al. 2014).

Beyond the Genome: Transcriptome, Methylome, and Microbiome

"Their macromolecules are so alike that regulatory mutations may account for their biological differences" is the simple, prophetic conclusion of King and Wilson's landmark paper in 1975 entitled "Evolution at Two Levels in Humans and Chimpanzees" (King and Wilson 2014). At one level, the authors observed a very small number of changes between a subset of chimpanzee and human proteins. At another level, they surmised that most of the observed phenotypic differences between the species, which in itself is a challenging and understudied field (Varki et al. 1998; Freimer and Sabatti 2003), stem from regulatory differences between their genomes. Now, with the advent of several comparative techniques scrutinizing transcriptome, epigenome, proteome, and microbiome among other *postgenome* variations, we can empirically study these claims. The initial glimpses of these variations are, as we have now learned to expect, "stranger than we imagine."

Gene regulation differences among primates, including humans, were shown to be regulated at multiple stages during the process leading from genome to transcriptome and finally to proteome. Direct measurement of expression levels across different primates using total RNA sequencing has revealed several pathways, most notably metabolic pathways, that showed tissue- and sex-specific expression levels and splicing patterns (Blekhman et al. 2008, 2010). Similar studies have unearthed patterns suggesting that expression differences may explain neoteny or delayed development in brain (Somel et al. 2009) and reproductive organs (Saglican et al. 2014). Neoteny, as a general concept, has been put forward for explaining major phenotypic differences among primates (for a thorough discussion of this concept, see Gould 1977). In addition, the genes involved in innate immune-signaling pathways have been shown to be differentially expressed among different primate species, even if the protein sequences remain conserved (Barreiro et al. 2010).

The comparative analysis of expression patterns in primates led to the identification of both inherited and environmental causes of observed differences. Multiple studies revealed the impact of variation in regulatory sequences among the primate genomes that significantly affect gene levels, consequently leading to species-specific phenotypes (Prabhakar et al. 2008; McLean et al. 2011; Iskow et al. 2012). Genome-wide expression data coupled with other *-omics*-level data led to preliminary glimpses into the functional impact of transcription factor binding (e.g. Scally et al. 2012), epigenetic patterns (Zhou et al. 2014), methylation (e.g. Pai et al. 2011), and chromatin accessibility (e.g. Shibata et al. 2012), among others (reviewed in Romero et al. 2012).

The posttranscriptional-level studies have enjoyed a renewed spotlight with the recent appreciation of the phenotypic impact of *noncoding* RNA molecules. For example, micro-RNAs, which are small RNA molecules that are involved in post-transcriptional regulation of messenger RNAs (mRNAs), were shown to evolve in a lineage-specific fashion in primates, likely contributing to the uniqueness of the human brain (Somel et al. 2011). Similarly, larger *long noncoding RNA* molecules have been discussed within the context of unique human brain development. For example, the human-specific incomplete duplication of *SRGAP2* produces a long noncoding RNA molecule (Dennis et al. 2012) that interferes with the expression of the ancestral *SRGAP2* gene, leading to denser synapse distribution in the human cerebral cortex (Charrier et al. 2012). As such, posttranscriptional and posttranslational modification is emerging as one of the major players in which phenotypes are fine-tuned. Concordantly, a recent genome-wide study found less cross-species variation in protein expression levels than what was observed for mRNA expression levels (Khan et al. 2013). This result suggests that protein levels, which are arguably more important for the phenotypic end results, are more strictly regulated than the mRNA levels.

It is now clear that only very few major phenotypic variants between primate species only involve nonsynonymous single nucleotide changes, which change the protein sequence directly. Instead, what we observe is a complex and mostly unexplored universe of often redundant and overlapping adaptive mechanisms, including structural and regulatory variants in the genome, as well as environment-dependent epigenetic mechanisms of tissue-specific, developmentally temporal RNA and protein regulation. We now know that it will take no less than disentangling the underlying principles of this unexplored regulatory universe to truly understand phenotypic variation within and among primates.

A Genomics Framework to Study Primate Behavior

The genetic bases of complex behaviors, ranging from aggression to novelty seeking, comprise a hotly debated area of inquiry. However, studying the behavioral genetics of primates has been notoriously difficult due to the challenges inherent to measuring and characterizing behavioral phenotypes (Bearden et al. 2004). Multiple genes may affect a single behavior, and environment has an observable influence on the behavioral phenotype (e.g. Chesler et al. 2005). The more affordable genomic technologies outlined in this chapter made it possible to design more powerful studies accounting for the multigene basis of behaviors and incorporating larger number of individuals.

Detailed research on molecular genetics in primates has demonstrated that individual behavioral decisions – mediated by both primate social and ecological variables in the wild – are able to influence allele frequencies and genetic structure at the group and population levels (for a review, see Difiore et al. 2011). Elements of social structure such as dominance rank (Altmann et al. 1996; Runcie et al. 2013), perceived kinship status (Pusey 2004), migration decisions (Di Fiore et al. 2009), the presence and strength of affiliative interactions (Huchard et al. 2009), and various individual mating strategies (Tung et al. 2012) and other aspects of mating preferences (e.g. Paul 2002) can all have an impact on how genetic variation is distributed within a population. Studying how genes in turn may influence social and individual behavior, however, has been far more difficult.

One promising approach to discovering genetic influences on behavior is to study known intermediate phenotypes that have a demonstrable behavioral effect. One field of study that uses this approach is research into the genetic basis of primate sensory perception phenotypes (Melin et al. 2013; Veilleux et al. 2013a; Matsushita et al. 2014). Primate sensory perception, such as how and what an individual's body may see or smell or hear, is integral as a mediator between the organism and how it interacts with its environment. Having the alleles of an opsin gene that permits red-green color vision, for example, has been hypothesized to influence foraging behaviors by allowing individuals to identify and favor the selection of riper, more nutrient-dense fruits (Melin et al. 2013). Such methods allow researchers to draw a clear line from genotype to behavioral phenotype while also elucidating relatively straightforward evolutionary mechanisms (Brent and Melin 2014).

With more complex behavioral phenotypes, perhaps the most prolific area of research within primate behavioral genetics has been the work done on the serotonergic system, particularly the promoter region of the 5-HTT serotonin transporter gene in the genus *Macaca* (for reviews, see Lesch et al. 1997; Suomi 2006). This work has shown that both sequence variation and gene expression variation caused by environment have effects on complex behavioral phenotypes. For example, sequence variations such as the "short" and "long" alleles in the promoter region of 5-HTT in *Macaca*, and many more variants in apes and humans, are associated with different levels of anxiety in social circumstances (Lesch et al. 1997), while further work has shown that the effects of these alleles on behavioral outcomes in monkeys and humans are highly dependent on environmental factors, suggesting a strong epigenetic effect on gene function (Suomi 2006). The studies on the serotonin transporter gene have proven invaluable to illuminate evolutionary aspects of behavioral genetics and provide translational research with implications for assessing human susceptibility risks across a number of behavioral issues, including alcoholism, depression, and anxiety disorders (Lesch et al. 1997). Similar comparative genetic and genomic work currently being done on the evolution of other well-characterized neurobehavioral systems in nonhuman primates, such as the prolactin and vasopressin pathways (Anestis et al. 2014; Baab et al. 2014), also holds translational and therapeutic potential (Meyer-Lindenberg et al. 2011).

Looking Deeper: Longitudinal Studies

With the increasing number of published primate reference genomes across the order, whole-genome sequencing efforts in both captive colonies and wild populations, and an increased recognition of the value of long-term behavioral databases as sources of deeply phenotyped and pedigreed individuals, the potential for NGS technologies in studying primate behavior is finally being realized (Tung et al. 2010; Bradley and Lawler 2011). Three ongoing research projects, in particular, represent two complementary frameworks by which next-generation technologies can be put to use in studies of wild primates: the Amboseli Baboon Research Project and the International Vervet Research Consortium.

The Amboseli Baboon Research Project is one of the longest running wild primate research projects in the world, and it has been the first such long-term field site to begin integrating genomics into their research program. Begun in 1971, the project has information on over 1500 yellow baboons (*Papio hamadryas cynocephalus*) throughout its history, including 300 animals that are currently being monitored under the collaborative eye of a number of key researchers and institutions (https://amboselibaboons.nd.edu). They made strides in understanding how two related species interact with each other in hybrid zones and how social behavioral differences between the two species translate into genetic variation in both populations (Tung et al. 2012). This rich behavioral within-species resource has also been used in a number of innovative ways to better understand how genes interact with both the social and ecological environment. In one striking study, for example, it was found that social trends affect female and male gene expression in different and unexpected ways: maternal dominance rank, social connectedness in adulthood, and adult group size influence allele-specific gene expression in male baboons, while social connectedness, group size in adulthood and early in life, and age affect allele-specific gene expression in females (Runcie et al. 2013). The researchers went on to suggest that genome-wide methods can lead to the design of powerful studies that can investigate whether socially mediated changes in gene expression could increase fitness under a heritable social regime (Tung et al. 2015b).

Like the Amboseli Baboon Research Project, the International Vervet Research Consortium is a multi-institutional collaborative group that focuses on better understanding the links between genes, environment, and behavior in a single taxon, the vervet monkey (*Chlorocebus* spp.). Unlike the Amboseli Baboon Research Project, there is no single long-term field site associated with the project. Rather, the International Vervet Research Consortium brings together a network of field sites, established and pedigreed captive colonies, and a series of broad-scale trapping efforts for phenotype collection across the worldwide range of vervet monkeys to build a massive research structure with considerable phenotypic and genomic resources (Jasinska et al. 2013). This framework has produced a number of studies on the genetics and genomics of phenotypes using quantitative genetics with the extensively pedigreed vervet cohorts. Results have found high additive genetic heritability for brain morphology (Fears et al. 2009, 2011), high heritability of impulsivity and serotonergic systems (Fairbanks et al. 2001), and links between heritable aspects of monoamine neurotransmission and behavior (Bailey et al. 2007; Fairbanks et al. 2012). More recent large-scale collections have led to broader studies of phenotypic variation in the wild, including genital coloration (Cramer et al. 2013) and allometry (Rodríguez et al. 2015), the evolution of simian immunodeficiency virus (SIV) resistance in West Africa (Ma et al. 2014) and South Africa (Ma et al. 2013), and factors associated with parasite loads (Gaetano et al. 2014). With the publication of the vervet reference genome (Warren et al. 2015), this work is now branching out to link these phenotypes and more – including heritable shifts in gene expression in relation to stress, and heritable aspects of obesogenic growth – to the genomic variation underlying them (Jasinska et al. 2013).

Finally, the rhesus macaque (*Macaca mulatta*) colony at the Caribbean National Primate Research Center on Cayo Santiago island in Puerto Rico is quickly becoming an invaluable resource for genomic studies. Founded in 1938 with 409 founders that range freely on the island, the colony pedigree now includes over 4424 individuals that have been both genotyped and consistently phenotyped since 1992 (Widdig et al. 2015). Since that time, numerous studies have been conducted on the genetics of the Cayo Santiago macaques, ranging from the early seminal work on blood groups (Duggleby et al. 1986) to short tandem repeat–based analyses of social influences on genetic structure, including kin bias and kin recognition (for a comprehensive review, see Widdig et al. 2015). More recently, the work on Cayo Santiago has expanded to include genome-wide data. For example, network analysis coupled with serotonergic gene variation revealed loci that may have an impact on the heritable basis of social traits in this population of macaques (Brent et al. 2013). Other studies highlighted the genetic basis of variation in important life history traits, such as maturation (Blomquist 2009). In addition, serotonin transporter promoter region research discussed in this chapter (Trefilov et al. 2000) and the role of the μ-opioid receptor gene on mother–infant behavioral interactions (Higham et al. 2011) have been investigated in this population. As with other projects, the expansion of research using NGS has exciting implications for translational research, including investigations into macular degeneration, glaucoma, and circadian rhythm disorders, as discussed at length by Widdig et al. (2015).

These longitudinal studies, when combined with the interrogation of genomic, transcriptomic, and metagenomics approaches, have the potential to finally elucidate the complex pathways from sequence variation through epigenetic endophenotypes to ecological conditions and the selective regimes that influence them. With the planned expansion of NGS techniques to other longitudinal wild research populations in more far-flung primate taxa, there is enormous potential for comparative work in the future.

Resolving the Interactions Between Primates and Their Habitat Using Genomics

Field-based studies of primates remain the most challenging and potentially the most rewarding when it comes to understanding primate populations. Two major fronts in field-based primate research are conservation and the gene–environment interactions. Most nonhuman primates are endangered. Conservation efforts increasingly depend on genetic information to identify pressing problems for a given population and prioritize the action. Several locus-specific studies have already made important headway on the current genetic diversity of different primate populations, defined and documented evolutionarily manageable population units, and made crucial assessments of species' long-term *viability* within the context of their contemporary habitat (Ferreira da Silva et al. 2012). Such studies have also empirically demonstrated the manmade decline of several primate populations (e.g. Goossens et al. 2006). Now, with NGS tools, these lines of conservation research have been conducted with unprecedented resolution (e.g. Prado-Martinez et al. 2013b).

In addition to increasing the resolution, NGS made possible several new fronts in conservation genetics. The most visible of these fronts are (i) the reconstruction of genetic variation of primates with no available reference genome using transcriptome sequencing, (ii) a genome-wide survey for adaptive and functional genetic variation within populations, and (iii) empirically inferring the pathogenic and dietary burdens on the primate populations using metagenomics tools. The impacts of NGS on conservation efforts have been reviewed extensively for a broad number of species by Allendorf et al. (2010). Briefly, short reads from NGS are still not adequate for affordable assemblies of whole genomes. However, given that they represent only a small fraction of the genome, coding sequences can be assembled de novo using RNAseq techniques. These techniques can be used to assess population properties of primate species without ascertainment bias and considering thousands of different loci. A recent example documented, using RNAseq methods, that several endangered lemur species, despite a reduction in their census population sizes, still harbor considerable genetic variation (Perry et al. 2012). As the whole-genome data for primates continue to accumulate at the population level, as discussed in the "Primate Genomes at the Population Level" section, it is plausible to identify genetic variation that gives fitness advantage amid the ecological pressures (reviewed in Bradley and Lawler 2011). Similarly, recent metagenomics studies have identified several interesting immunity, diet, and (surprisingly) social variations affecting the microbiome diversity within and among primate species (Yildirim et al. 2010; Tung et al. 2015a). All this implies that ecological and social adaptations are interconnected in primates, and multiple aspects of species' existence from genomes to transcriptomes to metagenomes have important impacts on shaping the complex, interconnected picture of primate behavior and adaptation. Understanding this picture may be the key for conservation efforts.

Toward an Anthropological Integration of Next-Generation Sequencing Technologies to Primate Research

The application of NGS technologies at the genomics, transcriptomic, and metagenomic levels to questions involving primate evolution has already yielded a number of remarkable discoveries. As these practices become cheaper and more accessible, the anthropological genetics community can capitalize on these powerful technologies to design unique and powerful studies. In particular, field-based and longitudinal studies where anthropologists have a historical presence can be frameworks for comprehensive and systems-level analyses of primate behavior and phenotypes. We believe that by doing so, the anthropologists, who have a robust integrative understanding of primate evolution, will also significantly contribute to conservation efforts, as well as to translational research by elucidating the evolutionary basis of biomedically important traits.

CHAPTER 6

The Neanderthal and Denisovan Genomes

Sloan R. Williams

Department of Anthropology, University of Illinois, Chicago, IL, USA

Two very different models have been proposed to explain the emergence and spread of anatomically modern humans (AMHs) across Eurasia. The Recent African Origins model (RAO model, also known as the Out-of-Africa model) proposes that AMHs originated in Africa relatively recently and dispersed into Eurasia, replacing resident archaic human groups (Stringer and Andrews 1988). The Multiregional model (Wolpoff et al. 1984; Frayer et al. 1993) posits that archaic hominins evolved into AMHs throughout the Old World in regional populations linked to each other through gene flow. The two models make different predictions about where the earliest AMH fossils would be found and how genetic patterns should correlate with geography in the Old World. RAO model proponents cite early AMH fossil finds in Africa and the seemingly abrupt transition from Neanderthal occupation to AMH sites in the European archeological record to support their model. Multiregionalists emphasize the shared morphological features in Neanderthals and modern Europeans and similarities between Asian archaic forms and modern Asians as evidence for regional continuity. Genetic studies generally supported the RAO model, but recent archaic DNA research has required re-evaluations of both models.

Genetic studies of living people generally support the RAO model (although see Templeton 2005, 2007). Cann, et al. (1987) and Vigilant, et al. (1989) studies of present-day human mitochondrial DNA (mtDNA) shortened the divergence times for modern humans and their nearest common ancestors,

A Companion to Anthropological Genetics, First Edition. Edited by Dennis H. O'Rourke.

relegating *Homo erectus* and archaic *Homo sapiens* to extinct side branches. They argued that modern humans evolved very recently and replaced earlier hominin forms throughout the world, which supports the RAO model. After some initial debate over analytical methods (Templeton 1993; Stoneking 1994), the root of the modern human mtDNA phylogenetic tree was placed in Africa. These original mtDNA studies were quickly supported by Y-chromosome data (Hammer et al. 1997). Current nuclear autosomal DNA research shows strong signatures of that recent AMH dispersal (Ramachandran et al. 2005; Jakobsson et al. 2008). The genetic pattern in modern humans is best interpreted as a series of founder effects, with human genetic diversity decreasing as a function of distance from East Africa. Thus, each new DNA dataset has continued to support a recent origin for AMHs in Africa.

Ancient DNA (aDNA) obtained from Neanderthal and other archaic hominin remains show a more complex pattern than either of the original models predicted. Consequently, one of the more intermediate models that incorporates some degree of admixture (Bräuer 1981), assimilation (Smith et al. 2005), or what Pääbo has called *leaky replacement* (Gibbons 2011) better explains the archaic DNA patterns and their similarities to haplotypes and allele patterns observed in contemporary human populations. Although the first mtDNA sequences obtained from the Neanderthal fossils showed no evidence of admixture (Krings et al. 1997), admixture calculations based on low-resolution Neanderthal nuclear genomes suggested small amounts of admixture (~1–4%; Green et al. 2010). This initial estimate has since narrowed to a still significant contribution of 1.8–2.6% in high-coverage sequence analyses (Prüfer et al. 2017). Researchers also found that Neanderthals made a 20% greater contribution to modern East Asian gene pools than to European ones, which would not have been predicted from the distribution of Neanderthal archeological sites in Eurasia (Meyer et al. 2012). Finally, aDNA researchers were able to identify a previously unknown archaic group based on genomic data recovered from a phalanx in Denisova Cave, Siberia (Reich et al. 2010). The origin and dispersion of AMHs are more complicated and involve interactions with more hominin forms coexisting in time and space than was previously thought.

Researchers did not accept that the sequences shared by Neanderthals and contemporary people necessarily proved admixture at first. Green et al. (2010) favored admixture as the most parsimonious explanation for their data, but they stated that ancient substructure in Africa could not be ruled out as an alternative hypothesis. In other words, the same genetic pattern might be observed if the ancestral African population was diverse and contained subpopulations with limited gene flow. If the ancestors of Neanderthals and Eurasians both came from the same subset of the ancestral population (which might be the case if they exited via the same route), but a different subset gave rise to most modern Africans, then Neanderthals and non-African populations might share DNA with each other that they did not share with African populations. Researchers (Durand et al. 2011;

Eriksson and Manica 2012) attempted to test the hypothesis that population structure in Africa could explain these alleles that Neanderthals and non-African contemporary people shared. They found that ancient African substructure could result in large numbers of shared derived alleles, but that it would not result in the same patterns of allele frequency distribution (Yang et al. 2012) or linkage disequilibrium (LD) observed in non-African people (Sankararaman et al. 2012). Recent admixture was a better fit for the entirety of the observed genetic patterns because it also explained the excess of rare alleles and the long stretches of LD observed in modern humans.

Archaic Hominin Genomes

Although aDNA studies have contributed to many fields of science, nowhere has their impact been more apparent than in the field of paleoanthropology. Much genetic information has been lost through drift, selection, and population and species extinctions in contemporary populations. As a result, the reconstruction of their demographic and evolutionary history can be difficult. These aDNA studies enable researchers to examine the genomes of individuals who lived in the past to examine how populations and species changed over time. We now know much more about extinct hominins such as the Neanderthals, and we have even identified groups previously unknown in the fossil record, such as the Denisovans.

DNA preserves best at low temperature and humidity. Caves where many Eurasian archaic hominin remains are found have yielded some well-preserved specimens (Heintzman et al. 2015). After an organism's death, enzymes such as lysosomal nucleases are released as the cellular compartments break down, and nucleases rapidly degrade the DNA (Pääbo et al. 2004, Sankararaman et al. 2014 and Serre et al. 2004). Oxidative and hydrolytic lesions, cross-linkage, and strand breakage all contribute to aDNA damage and fragmentation (~100–500 base pair strand lengths). Base deamination is probably the greatest concern, however, because it results in the mis-incorporation of bases by the polymerase enzyme used in the polymerase chain reaction (PCR). Polymerases recognize deaminated cytosine residues as uracil residues and transcribe them as thymines, so that aDNA sequences become sprinkled with C-to-T transition artifacts. Contamination with modern DNA through human contact in either the field or lab is almost unavoidable, and its impact is intensified by the PCR replication process. In the past, DNA damage and contamination made aDNA studies of hominin remains extremely difficult, slow, and sometimes unreliable.

In the mid-2000s, next-generation sequencing (NGS) allowed researchers to retrieve and study the nuclear genome from ancient remains, which transformed the field. Whereas traditional PCR requires multiple experiments, each one targeting a different region of DNA, in NGS all of the DNA is amplified and sequenced at the same time. The term *deep sequencing* refers to when the coverage, or the average number of times a genomic position is sequenced from separate templates, is greater than a couple of "reads" at each position. Such deep sequencing makes it possible to identify mis-incorporations and distinguish between modern and aDNA (Der Sarkissian et al. 2014). Although the analysis of NGS is computationally intensive, it provides an extremely efficient and cost-effective use of aDNA. Furthermore, NGS requires less starting material, and so less bone or teeth are destroyed than with other methods. NGS takes advantage of what had previously been obstacles when working with aDNA – strand breakage and nucleotide damage – and turns them into advantages. The already highly fragmented aDNA is well suited for NGS analysis, as it requires short fragments. In addition, the characteristic damage patterns caused by cytosine deamination make it relatively easy to separate aDNA sequences from modern contamination (Racimo et al. 2016).

From mitochondrial to high-coverage nuclear genomes

The first Neanderthal mtDNA sequences came from a 40000-year-old humerus from Feldhofer I in the Neanderthal Valley (Krings et al. 1997; see Table 6.1 and Figure 6.1). Krings and colleagues sequenced 379 bases from the Hypervariable region I (HVRI) control region of the d-loop in short overlapping fragments. They cloned 123 PCR products to remove degradation artifacts and generate a consensus sequence that contained 27 differences (24 transitions, 2 transversion, and 1 insertion) from the human reference sequence (Anderson et al. 1981). Human sequences included in the study differed among themselves by about 8 ± 3.1 substitutions, while the Neanderthal sequence differed from the human sequences by 27.2 ± 2.2 substitutions. The Neanderthal sequence branched off from the base of the phylogenetic tree, and it resembled European sequences no more than it did any other global population sample included in the study. The researchers estimated a Neanderthal–human divergence time of approximately 550000–690000 ya (years ago). Krings et al. (1999) next published Hypervariable region II (HVRII) from the same individual. Ovchinnikov et al. (2000) added a new individual, publishing HVRI mtDNA sequences from a Neanderthal infant buried in the Mezmaiskaya Cave in the northern Caucasus. The two geographically distant Neanderthal mtDNA sequences clustered tightly together, which showed that they probably were Neanderthal and not contamination. Their Neanderthal–human divergence estimate of 365000–853000 ya was similar to the Feldhofer estimate. Having two individuals also permitted them to calculate a measure of genetic diversity. Soon after, Krings et al. (2000) published mtDNA HVRI and HVRII sequences from a third individual, Vindija 75, in Croatia. Since then, Neanderthal genomes from over a dozen different sites have been published (see Table 6.1). The oldest sequenced individual is from Scladina Cave, Belgium, at 100000 ya (Orlando et al. 2006), but most of the samples date from 40000 to 50000 ya.

In 2006, Green et al. (2006) and Noonan et al. (2006) both published large stretches of Neanderthal nuclear DNA (nDNA) from the same specimen

Table 6.1 List of sites and specimens that have yielded archaic DNA.

Site and number of individuals	*Location*	*Hominin*	*Genomic information*	*References*
Feldhofer (2 = I, 2)	Germany	Neanderthal	mtDNA genome	Krings et al. (1997, 1999) Briggs et al. (2009); Schmitz et al. (2002)
Mezmaiskaya I (1)	Russia	Neanderthal	mtgenome, 0.5× nDNA	Briggs et al. (2009); Ovchinnikov et al. (2000) Prüfer et al. (2014)
Vindija (7 = 75, 77, 80, 33.25, 33.16, 33.26, 33.19)	Croatia	Neanderthal	mtgenome, low-coverage nDNA 30× coverage (33.19)	Krings et al. (2000); Serre et al. (2004) Green et al. (2006, 2008, 2010); Prüfer et al. (2017)
El Sidrón (4 = 441, 1252, 1253, 1351)	Spain	Neanderthal	mtgenome, low-coverage nDNA	Briggs et al. (2009); Burbano et al. (2012) Lalueza-Fox et al. (2005, 2006, 2008)
Engis 2 (1)	Belgium	Neanderthal	mtDNA I	Serre et al. (2004)
La Chapelle-aux-Saints (1)	France	Neanderthal	mtDNA I	Serre et al. (2004)
Monti Lessini (1)	Italy	Neanderthal	mtDNA I, nDNA	Caramelli et al. (2006); Lalueza-Fox et al. (2007); Lari et al. (2010)
Rocherrs de Villenueve (1)	France	Neanderthal	mtDNA I	Beauval et al. (2005)
Scladina (1)	Belgium	Neanderthal	mtDNA I	Orlando et al. (2006)
Teshik Tesh (1)	Uzbekistan	Neanderthal	Partial mtDNA I	Krause et al. (2007)
Valdegoba (1)	Spain	Neanderthal	mtDNA I	Dalén et al. (2012)
Okladnikov (1)	Russia	Neanderthal	mtDNA I	Krause et al. (2007)
Denisova (1 = 5)	Russia	Neanderthal	52× coverage nDNA	Prüfer et al. (2014)
Denisova (4 = 2, 3, 4, 8)	Russia	Denisovan	mtgenomes, nDNA 31× coverage (Den. 3)	Krause et al. (2010); Reich et al. (2010) Meyer et al. (2012); Sawyer et al. (2015); Slon et al. (2017b)
Sima de los Huesos (1)	Spain	Neanderthal	mtgenome, low-coverage nDNA	Meyer et al. (2014, 2016)

Figure 6.1 Map of sites that have yielded archaic DNA.

(Vindija 80; see Table 6.1 and Figure 6.1). The two studies used slightly different methods and analyzed different amounts of nDNA, but the researchers came to different conclusions concerning Neanderthal–human divergence times and admixture rates. Wall and Kim (2007) reanalyzed their data and found that the longer sequences in the first study often did not match the shorter reads and instead more closely resembled modern human sequences. This mismatch often occurs when ancient samples are contaminated with well-preserved modern DNA. The second dataset did not show this property, leading Wall and Kim to conclude that contamination, and perhaps higher sequencing error rates, likely explained the more recent divergence times and higher admixture rates reported by Green and colleagues. By 2010, a draft Neanderthal genome composite based on three females from Vindija Cave, Croatia, was published yielding 5.3 gigabytes of sequences with 1.3× coverage (Green et al. 2010). The researchers used three different methods to ensure that modern human contamination did not affect their analyses. First, they checked for mtDNA contamination and determined that it was less than 0.5%. Next, they looked for evidence of chromosome-Y contamination, since all three samples were female, and found levels of approximately 0.6%. Finally, they assessed autosomal DNA contamination by identifying sites where their five comparative modern human samples all differed from the ancestral allele by a transversion. They examined those sites in their Neanderthal sequences and counted how often pairs of Neanderthal sequences did not match each other at these sites. This count gave them a measure that combined both heterozygosity and contamination, providing an upper boundary for potential contamination. They estimated the contamination level to be about 0.7%, based on this method. Green and coworkers estimated the human–Neanderthal split to have occurred 270 000–440 000 ya. They found an excess of derived alleles shared with contemporary Europeans and Asians, but not with contemporary Africans.

Green and colleagues also found evidence of extended haplotypes (LD), which suggested recent admixture. The admixture rate between Neanderthals and AMHs was estimated at 1–4%. With adequate methods for identifying contamination in NGS data now developed, researchers began to increase the number of Neanderthal samples both to study their population structure and dynamics and to explore the nature of the relationships between Neanderthals and AMHs that resulted in the observed admixture patterns.

Nuclear genes, chromosomes, and low-coverage genome scan studies have been carried out on Denisovan, Feldhofer, Mezmaiskaya, and El Sidrón individuals in addition to the original Vindija individuals (see Table 6.1 and Figure 6.1). Most importantly, though, three high-coverage (30–50-fold) archaic genome sequences have been published: first, a Denisovan (Denisova 3; Meyer et al. 2012); next, an Asian Neanderthal from Denisova Cave (Denisova 5, also referred to the Altai Neanderthal; Prüfer et al. 2014); and, most recently, a European Neanderthal (Vindija 33.29; Prüfer et al. 2017). Based on nucleotide substitutions, the Vindija Neanderthal lived 52 kya, the Altai Neanderthal lived 122 kya, and the Denisovan lived 72 kya. The most recent divergence estimates (see Figure 6.2) indicate that the eastern Neanderthal Vindija and AMH populations separated 520 000–630 000 ya, the Vindija Neanderthal and Denisovan populations split 390 000–440 000 ya, the western Vindija and eastern Altai Neanderthal populations separated 130 000–145 000 ya, and the western Vindija

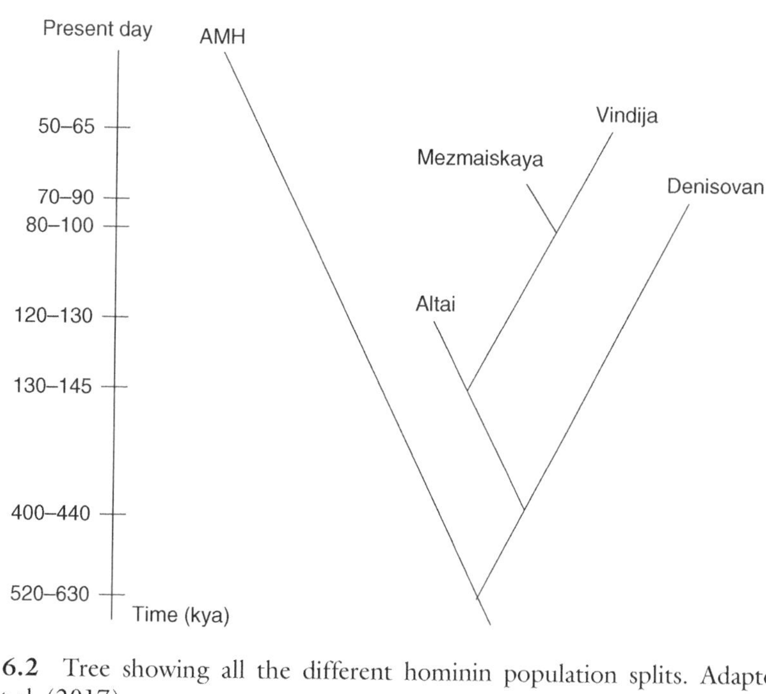

Figure 6.2 Tree showing all the different hominin population splits. Adapted from Prüfer et al. (2017).

and Mezmaiskaya populations split 80 000–100 000 ya (Prüfer et al. 2017). With these high-coverage genomes, researchers can undertake detailed studies of the entire genome and address more nuanced questions than were impossible with low-coverage genome analyses.

Neanderthals

Neanderthals lived in small groups spread across a large geographic area from the Iberian Peninsula to the Altai Mountains of western Siberia (Prüfer et al. 2014). The Neanderthal DNA sequences that have been obtained to date are remarkably similar, indicating that their effective population size was very small, about 3000 individuals (Prüfer et al. 2017). The admixture signals are stronger in Neanderthals from the Caucasus than those from Croatia and Siberia, which suggests some Neanderthal population structure and multiple admixture events (Vernot and Akey 2015). Effective population sizes declined over time throughout the Neanderthals' range (Kuhlwilm et al. 2016). The Vindija high-coverage sequence shows a long-term small population size, but the Altai Neanderthal sequence shows evidence of recent inbreeding as well. The Altai sequence contained long stretches of homozygosity that would be compatible with half-sibling or double first-cousin matings and may reflect a long period of isolation in the Altai Mountains (Kuhlwilm et al. 2016). Lalueza-Fox et al. (2011) have suggested that some Neanderthals were patrilocal because the three adult males found at the El Sidrón site shared the same mtDNA haplotype, while the three females all carried different mtDNA haplotypes.

Growing knowledge of Neanderthal phenotype provides insight into their dental morphology, body shape, and climate adaptations. A functional analysis of the miR-1304-mediated regulation of enamelin and amelotin can partially explain the thicker enamel and enlarged pulp cavities that characterize Neanderthal teeth (Lopez-Valenzuela et al. 2012). The allelic differences noted in the RUNX2 gene that is linked to cleidocranial dysplasia helps explain the bell-shaped thoracic cavity and the morphological changes in the clavicle that are seen in Neanderthal skeletons (Green et al. 2010). Lalueza-Fox et al. (2007) identified a mutation in the MC1R gene in two Neanderthals that would likely result in red hair and pale skin. Pale skin would maximize ultraviolet light capture in northern latitudes, where levels are lower than near the equator. Light skin is found in AMHs who live in these areas today, although the mutations found in Neanderthals and AMHs differ.

Neanderthal phenotypic information also enables us to identify the timing of phenotypic changes, which improves our general understanding of hominin evolutionary biology. Neanderthal and Denisovan sequences have provided insights into consumption, plus polymorphisms long of interest to biological anthropologists. Both Neanderthals and Denisovans share a MYH16 (sarcomeric myosin) frameshift deletion with modern humans that reduces temporalis musculature

(Perry et al. 2015). Perry and colleagues believe this reduction in chewing muscle size is linked to cooking, which implies that archaic humans controlled fire and cooked their food. Furthermore, the Altai Neanderthal and Denisovan high-coverage genomes contained only two copies of the salivary amylase (AMY1) gene, which is the ancestral number of copies found in chimpanzees (Perry et al. 2015). Multiple copies of AMY1 are linked to starch digestion. Copy number is variable in AMHs ranging from 4 to 14 copies, with an average of 6–7 copies. Present-day human populations who consume domesticated grains high in starch benefit from multiple copies, but some researchers had hypothesized that these gene duplications would have occurred early in human evolution if tubers were a critical component of early hominin diets. If correct, AMY1 variation would be observed in archaic hominins as well, so observing the ancestral condition in both the Neanderthal and the Denisovan genomes greatly weakens their argument. Lalueza-Fox et al. (2008) blood-typed two Neanderthals from El Sidrón and showed that the O allele of the ABO blood system arose before the Neanderthal–human split. Two groups of researchers have examined phenylthiocarbamide (PTC) bitter taste receptors. Lalueza-Fox et al. (2009) amplified the TAS2R38 allele from El Sidrón 1253 and found that the individual was a PTC taster and was heterozygous for one of the common human variants. Perry et al. (2015) analyzed the Neanderthal and Denisovan high-coverage genome sequences for TAS2R62, TAS2R64, and TAS2R38. The Neanderthal and Denisovan TAS2R62 and TAS2R64 sequences both matched the human reference sequence, which means that the functionality of these receptors was lost before their lineages diverged from ours 550 000–590 000 ya. The researchers found a unique polymorphism in the Altai Neanderthal, indicating that chimpanzees, Neanderthals, and AMHs all show variation in taste sensitivity in this receptor caused by different functional mutations. Research in phenotypic analyses from archaic genomic data is in its early stages, but these initial studies have already provided important information about when hominins began to cook their food, the components of their diets, and when some of our more well-characterized polymorphisms arose.

Denisovans

The Denisovans were first identified by DNA recovered in Denisova Cave, Siberia, from a juvenile distal manual phalanx (fingertip; Krause et al. 2010) (see Figure 6.1 and Table 6.1). The bone (Denisova 3) was later identified as female from the nDNA. When excavators first discovered the phalanx in 2008, they assumed the individual was a Neanderthal, but the mtDNA genome was completely unlike those of Neanderthals or modern humans (Krause et al. 2010). Phylogenetic analyses of the mtDNA genome indicated that the unknown hominin shared a common ancestor with both AMHs and Neanderthals one million years ago, identifying the latter two as sister groups who shared a more recent divergence time. When the nuclear genome (1.9-fold coverage) was sequenced, however, the researchers concluded that the Denisova genome shared a more recent common ancestor with the Neanderthals (Reich et al. 2010). They compared the sequence to nuclear genomes obtained from six Neanderthal fossils (three Croatian, and one each Spanish, German, and Russian) and argued that the Denisova sequence was not a Neanderthal, but rather a member of a sister group. The Neanderthal sequences, although found across a large geographic area and ranging in dates from 40 000 to 70 000 ya, were much more closely related to each other than any of them were to the Denisova sequence. Even the sequence from Mezmaiskaya Cave, Russia, which was the oldest fossil and the one geographically closest to Denisova Cave, clustered with the other Neanderthal specimens. The researchers argued that these differences were sufficient to merit placing the Denisova sequences into their own separate category.

Researchers have now recovered mtDNA genomes from four Denisovan individuals from the cave and nDNA from three individuals (Nos. 3, 4, and 8). The original specimen identified as Denisovan was recovered from layer 11.2 in the East Gallery, radiocarbon dates from the stratum cluster around 50 000 ya, but the sample itself could not be dated because there was not enough material. The stratum contained artifact assemblages with a mixture of Middle and Upper Paleolithic elements. Sawyer et al. (2015) added two more individuals (Denisova 4 and 8) to the first one by extracting DNA from molars. They published complete mtDNA and partial nDNA genomes. Denisova 4 (mtDNA first described in Reich et al. 2010) was recovered from layer 11.1 of the South Gallery. Denisova 8 was discovered in the interface between layers 11.4 and 12 in the East Gallery. The molars were quite fragmentary, especially those of Denisova 8, which were four fragments put back together; not unexpectedly, more contamination was identified, and, hence, the sequence inclusion criteria were strict. Most recently, Slon et al. (2017b) published a mtDNA genome from Denisova 2, a lower second deciduous molar (female based on DNA) recovered from layer 22.1 in the Main Gallery, which dates to 128 000–227 000 ya. Again, the sample itself cannot be dated, but the phylogenetic analyses place it as 50 000–100 000 years older than Denisova 3. According to their analyses, Denisova 3 and 4 are roughly contemporaneous at 50 000 years of age, while Denisova 8's age is intermediate in time at approximately 70 000–80 000 years of age. The Denisovan remains, although extremely fragmentary, span approximately 50 000–100 000 years of at least intermittent cave occupation.

Denisovans inhabited the Altai region for tens of thousands of years (Slon et al. 2017b). The sequence diversity among the three Denisovans is much lower than among contemporaneous humans today and is consistent with the small population size inferred from Denisova 3. The Denisovans are characterized by the same low population density and small effective population size as the Neanderthals, but they do not show evidence of the inbreeding found in the Altai Neanderthal individual from Denisova Cave. As all three Denisovan individuals are from the same cave, however, Denisovan diversity may increase when individuals from other parts of their geographic range are sampled. The Denisovan molars are extremely large and do not show the crown features that characterize Neanderthal and AMH teeth. Some Asian fossils have been identified as Denisovan candidates, but no DNA sequences have been published yet (Stringer and Galway-Witham 2017). In comparison to the Neanderthals, relatively little is known of the Denisovans.

The Denisovans' appearance is also poorly understood because the fossil remains are so fragmentary and incomplete. Meyer et al. (2012) determined that Denisova 3 carried alleles that in present-day humans are associated with dark skin, brown eyes, and brown hair. They also examined the EVC2 gene because it could explain the taurodontic features seen in archaic hominins. Denisovan genes that have been identified in present-day human populations indicate that the Denisovans were well adapted to high altitude. They had a highly developed sense of smell and could detect very subtle scents. They also carried genes involved in phospholipid transporter activity that improved fat metabolism (Sankararaman et al. 2016). More fossil remains with well-preserved DNA are needed to understand the range of both genotypic and phenotypic variation among Denisovans.

One of the more surprising findings reported is that approximately 3–6% of the genomes of modern Melanesians, New Guineans, and Aboriginal Australians appear to be of Denisovan origin (Reich et al. 2011a; Meyer et al. 2012; Sankararaman et al. 2016). Continental Southeast Asians and Mainland Asians share only 0.1–0.3% of their genome with Denisovans (Skoglund and Jakobsson 2011; Prüfer et al. 2014). Sankararaman et al. (2016) detected Denisovan ancestry in South Asia, both in the Himalayan region and in South and Central India. The highest estimate was found in Sherpas, notable because the EPAS1 allele that confers high-altitude adaptation in contemporary human populations was contributed by Denisovans (Huerta-Sánchez et al. 2014). Little is known of the Denisovan range from the fossil record, but their contributions to contemporary humans provide some clues and suggest that their geographic range was confined to Asia.

Hominin Admixture

True introgression (alleles being passed between two distantly related groups or species through gene flow) must be distinguished from shared ancestral variation (i.e. incomplete lineage sorting [ILS]) in order to identify and measure admixture. Ancestral alleles, which are not candidates for introgression, are identified through comparisons with outgroups (i.e. other species who are closely related to hominins phylogenetically; researchers usually use chimpanzees or a group of closely related primates). The alleles of interest in admixture research are the derived alleles, those restricted to a single group: in this case, Denisovans, Neanderthals, or AMHs. Researchers assume that introgression among AMH, Neanderthals, and Denisovans occurred outside Africa, so they also consider geographic distribution when identifying archaic introgression. The alleles found in archaic genomes that are also common in African population samples are more likely to be examples of ILS. Large, unbiased samples are necessary to accurately represent populations and avoid misidentifying alleles, as Gokcumen and colleagues (2013b) demonstrated in their NE1 locus study. Researchers also consider time to the most recent common ancestor (TMRCA) because an introgressed haplotype in contemporary human populations will have a more recent TMRCA with an archaic population than with other human haplotypes. Introgressed haplotypes will also show LD, and the haplotype blocks will be much longer than ILS haplotypes because there will not have been as much time for recombination to break up the genes. Allele distribution patterns across the genome and dramatic differences in allele frequencies among populations can provide evidence of selection, while LD and haplotype length provide time estimates. Allele occurrence in archeological samples also provides information about allele distribution in space and time.

As more Neanderthal and Denisovan genomes become available, the evidence for admixture increases in both the archaic and AMH genomes. To date, most research has focused on Neanderthal and Denisovan introgression into AMHs and evidence of that exchange in living populations, but admixture occurred among the archaic groups as well. Denisovans show evidence of introgression from both eastern Neanderthals and an unknown, deeply diverged archaic group that diverged from present-day humans long before the Neanderthals and Denisovans (Kuhlwilm et al. 2016). The Altai Neanderthal genome shows evidence of introgression from AMHs (Kuhlwilm et al. 2016). Currently, Neanderthal and Denisovan ancestry has not yet been seen in AMH Y chromosomes and mitochondria DNA (Vernot and Akey 2014; Mendez et al. 2016), but there is clear evidence of introgression in the autosomal DNA. However, the pattern of introgressed alleles across the nuclear genome is non-uniform (Sankararaman et al. 2014, 2016; Vernot and Akey 2014, 2015; Vernot et al. 2016). Positive selection is posited where archaic haplotypes show higher than expected frequencies, while negative selection is assumed in so-called *archaic deserts*, where archaic alleles are absent. Vernot et al. (2016) note that other mechanisms such as structural variation could contribute to the appearance of archaic deserts, and these possibilities would require further study. Genomic regions of both higher and lower than expected amounts of introgression provide important information.

Adaptive introgression

Positively selected allele frequencies increase over time. A telltale signature of adaptive introgression is the presence of mutations in strong LD that exist at high frequencies in a single population. Adaptive introgression is likely when those mutations are only present in the archaic source population, while being absent or at very low frequencies in other present-day human populations. Researchers look for evidence of adaptive introgression by two means. Researchers can either perform genome-wide screens to identify regions of DNA that fit the above pattern and then attempt to identify genes that might be targets of selection nearby, or they can use a single-locus approach to search for archaic haplotypes in genes likely to have been adaptive in the physical environments in which archaic hominins lived. Vernot and Akey (2014, 2016) and Sankararaman et al. (2014, 2016) used genome-wide association studies to compare the Altai Neanderthal and Denisovan genomes with present-day European, Asian, and Oceanian genomes using slightly different contemporary population samples and methods. Kuhlwilm et al. (2016) found 163 segments likely to have introgressed from AMHs into the Altai Neanderthal. One segment overlapped with a region suspected to be under positive selection in modern humans that carried a transcription factor (NR5A2) involved in liver development. Another segment was located within the FOXP2 gene that encodes a transcription factor associated with language acquisition. To date, most of the introgressed haplotypes and alleles where evidence for positive selection is strong can be grouped into four categories based on selective pressure: immune response/pathogen defense, pigmentation, altitude adaptation, and metabolism.

Archaic introgression into AMHs linked to immune response and pathogen defense has been identified by both genome-wide screen and single-locus approaches. Five identified regions either span or are adjacent to immune-related genes, including a haplotype encompassing the GBP4 and GBP7 genes that are induced by interferon as part of the innate immune response (Vernot et al. 2016). Targeted studies of the HLA region and STAT2 also show evidence of archaic introgression. Abi-Rached et al. (2011) studied the highly polymorphic *HLA class I genes* (HLA-A, -B, and -C) of the major histocompatibility complex (MHC) on chromosome 6. These genes are subject to strong multi-allelic balancing selection. Maintaining their diversity is vital in immune defense and reproduction, as they code for ligands to T-cell and natural killer (NK)-cell receptors. Abi-Rached and colleagues showed that a deeply divergent haplotype (HLA-B*73:01) in strong LD with HLA-A and -C alleles, which was found in the Denisovan genome, is present at high frequencies in populations in West Asia but absent or infrequent in the rest of the world. Another study found that the DPB1*0401 allele in the HLA-DPB gene closely matches the Neanderthal genome, but it seems more likely to be evidence of ancient population structure because the divergence time is 2.2 mya, and the allele is found in Africans (11%). The innate immune gene, STAT2, encodes a protein that plays an important role in interferon response after viral infection. Mendez and colleagues (2012b) found a long (130–260 kb) haplotype called "N" that overlaps this gene and is broadly distributed across Eurasia. The haplotype is absent in Sub-Saharan African populations and has a very recent TMRCA with the Neanderthal genome (78 kya), suggesting that it introgressed from this archaic hominin group. Although found throughout Eurasia at a frequency of ~5%, this haplotype, is present at substantially higher frequencies in Papuans (~54%).

Two other genes found in the introgressed N haplotype, ERBB3 (encoding a receptor tyrosine kinase involved in cell growth and apoptosis) and ESYT1 (encoding extended synaptotagmin-like protein 1, a transmembrane protein with a role in fibroblast differentiation), could also be targets of selection. Finally, Mendez and colleagues (2012b) this haplotype identified another potential example of adaptive introgression, the 2′–5′-oligoadenylate synthetase (OAS) gene cluster that encodes for proteins that help to inhibit viral replication as part of the innate immune response. Archaic contributions to AMH defense against pathogens through admixture would have enabled AMHs to adapt quickly to new environments filled with unfamiliar disease-causing organisms.

Genome-wide studies have identified several introgressed stretches of DNA where pigmentation genes are located. Vernot and Akey (2014) identify Basonuclin 2 (BNC2 on chromosome 9), which is associated with skin pigmentation and freckling in Europeans. The archaic haplotype is present at 70% frequency in Europeans but is absent in Asians. They also found a strong adaptive introgression signal in a cluster of keratin genes on chromosome 12 in both Europeans and Asians. Sankararaman et al. (2014) note that the introgressed keratin genes are members of hair keratin groups in East Asians, while they are skin keratins or without annotated function in Europeans. In East Asians, both HYAL2 on chromosome 3 and POU2F3 on chromosome 11 are examples of putative adaptive introgression. Ding et al. (2014) identified a 200-kb haplotype of Neanderthal origin common (>49%) in East Asians that contains HYAL2, a gene involved in the cellular response to ultraviolet radiation. Sankararaman et al. (2014) also identified HYAL2 as potential evidence of adaptive introgression in East Asians using the CRF framework. POU2F3 encodes a transcription factor that mediates keratinocyte differentiation and proliferation. Vernot and Akey (2014) indicate that modern humans acquired adaptive regulatory sequences at BNC2 and POU2F3 loci from Neanderthals. AMHs adapted more quickly to cold climates with reduced ultraviolet exposure through admixture with archaic hominins.

Altitude adaptation has provided clear evidence of adaptive introgression from Denisovans into AMHs (Huerta-Sánchez et al. 2014). EPAS1 encodes a transcription factor induced under hypoxic conditions that is associated with differences in hemoglobin concentration at high altitudes. Previous studies had shown this gene to be under strong positive selection in Tibetans, but it did not fit with a simple model of selection from de novo mutation or from standing variation. Huerta-Sánchez and colleagues examined sequence variation in the EPAS1 gene in a sample of Tibetans and Han Chinese and identified a 5-SNP (single nucleotide polymorphism) motif (AGGAA) in a 32-kb window of the EPAS1 gene that is both present in the Denisovan genome and at high frequency (~80%) in Tibetans. When they searched for the motif in the 1000 Genomes Project sample, it was absent in all but two Han Chinese individuals. A haplotype network of present-day human and Denisovan haplotypes revealed that the Tibetan haplotype most closely matches the Denisovan sequence. The unusually long haplotype length (32.7 kb) also makes ILS unlikely. The putatively selected Tibetan variants in EPAS1 are significantly associated with hemoglobin concentrations, a phenotype that distinguishes Tibetans from lowland populations and confers an adaptive advantage under hypoxic conditions at high altitude.

Several studies have identified potential evidence of archaic introgression in genes involved in metabolic pathways. Khrameeva et al. (2014) proposed that Neanderthal alleles in lipid catabolism genes have been targets of recent positive natural selection in Europeans. The authors located regions of Neanderthal-like sequences and searched within them for functional groups of genes they had identified using the gene set enrichment analysis (GSEA) algorithm. They then computed composite of multiple signals (CMS) scores (which test for positive selection) along the genome based on three signatures of selection: long-range haplotypes, differentiated alleles, and high-frequency-derived alleles. They found introgressed functional groups in immune and hematopoietic pathways in Asians, while the strongest signal of Neanderthal-like sequence was found within the lipid catabolism gene set in Europeans. Even though they could not rule out ILS, phenotypic information also supported their argument for adaptive introgression. They observed significantly different concentrations of lipid (by mass spectrometry) and altered lipid metabolic enzyme gene expression (by RNA sequencing) in European brain tissue samples. They compared the putatively introgressed regions with Denisovan sequences and did not find the same pattern, which also makes ILS less probable. Williams et al. (2013) identified putatively introgressed haplotypes from Neanderthals in two genes encoding solute carriers (SLC16A11 and SLC16A13). In a genome-wide association study for type 2 diabetes in 8214 individuals from Mexico and other parts of Latin America, they found the strongest association in a five-SNP haplotype within SLC16A11. The haplotype is at ~50% frequency in Mexican populations and at 10% frequency in Asians, but rare or absent in Europeans and Africans. They argued that the haplotype was of archaic origin because the haplotype is long (73 kb), the TMRCA with Neanderthals is very recent (250 000 ya vs. 677 000 ya for other human haplotypes), and all five SNPs are present and homozygous in the high-coverage Altai Neanderthal sequence. They demonstrated that expressed SLC16A11 in heterologous cells alters lipid metabolism, most notably causing an increase in intracellular triacylglycerol levels. Most recently, Vernot et al. (2016) have identified high-frequency archaic haplotypes that overlap several metabolism-related genes, including GCG (a hormone that increases blood glucose levels) and PLPP1 (a membrane protein involved in lipid metabolism). Sankararaman et al. (2016) report a phospholipid transporter related to fat metabolism that introgressed from Denisovans. Prüfer et al. (2017) report a Vindija Neanderthal introgressed haplotype that influences low-density lipoprotein (LDL) cholesterol levels. Racimo et al. (2015) speculate that the changes in metabolism caused by introgression may have conferred an evolutionary advantage on the recipient population, possibly in relation to an altered diet.

Archaic deserts

As just discussed in this chapter, adaptive introgression provides important insight into archaic–AMH interactions. Archaic deserts also are quite informative. Neanderthal ancestry is lowest in regions of high gene density or low recombination rate, where selection can act efficiently (Sankararaman et al. 2014, 2016; Vernot and Akey 2014). Vernot et al. (2016) found significantly more overlap in areas depleted of both Neanderthal and Denisovan DNA in a sample of 35 Island Melanesians than would be expected by chance. They noted that the regions depleted of archaic DNA are those significantly enriched for genes expressed in specific brain regions, particularly in the developing cortex and adult striatum. Sankararaman et al. (2016) identified 18 Neanderthal deserts in non-African individuals and 63 Denisovan deserts in a population sample that included 20 Oceanians. Both research groups noted the lack of archaic alleles, both Neanderthal and Denisovan, on the FOXP2 gene on chromosome 7, which has been associated with speech and language. Sankararaman et al. (2016) emphasized that both Neanderthal and Denisovan ancestry decreased in regions of the genome under strong selective pressure, particularly those in proximity to functional elements. Numerous studies report the absence of archaic alleles on chromosome X, which supports the hybrid sterility argument. But Vattathil and Akey (2015) caution that AMH and Neanderthal X chromosomes are less diverged, which may reduce our power to detect Neanderthal sequences. Fu et al. (2016a) note that the genome-wide fraction of Neanderthal introgression has decreased over the past 40 000 years, especially near genes. Kuhlwilm et al. (2016) report that the AMH deserts in the Altai Neanderthal genome show the same pattern as those observed for archaic deserts in AMHs. The AMH deserts in the Neanderthal genome are also concentrated in areas of high gene density, especially areas related to reproduction.

Sankararaman et al. (2014) were the first to focus on the areas of the human genome where Neanderthal ancestry was rare or absent. They suggested that these deserts were examples of Haldane's rule of incompatibility between two groups in the process of speciating. Haldane's rule (1922) states that when members of two different species or subspecies mate and one sex is rare, absent, or sterile in their offspring, it will usually be the heterogametic sex (i.e. the male in mammals). This pattern could mean that the underlying alleles were not deleterious in Neanderthals, but rather the deserts represent genetic incompatibilities that became deleterious once they were introduced to AMH populations. Juric et al. (2016) and Harris and Nielsen (2016) instead emphasize the effects of dramatic differences in effective population size. They suggest that there was insufficient time to form the reproductive barriers required for Haldane's rule to apply, arguing that demographic factors could explain the observed pattern without requiring genetic incompatibility.

A fivefold reduction of Neanderthal ancestry on the X chromosome has been observed in contemporary human populations. Chromosome X is known from studies of diverse species to be especially dense in male hybrid sterility genes (Sankararaman et al. 2014). The genes that are more highly expressed in testes than in other tissues are especially reduced in Neanderthal ancestry. Sankararaman and colleagues concluded that Neanderthal alleles caused decreased fertility in males when moved to a modern human genetic background. The same research group later found similar patterns of Denisovan DNA deserts on the X chromosome and around genes more highly expressed in testes than in other tissues. Modern humans exhibited 16–34% less Neanderthal DNA on chromosome X when compared to the autosomes, and Oceanians showed 21% less Denisovan DNA on chromosome X than on autosomes. The researchers suggest that reduced male fertility may be a general feature of mixtures of archaic populations diverged by more than 500 000 years. Kuhlwilm et al. (2016) found that introgressive AMH DNA decreased in regions of the Altai Neanderthal genome under strong purifying selection, including the X chromosome, which suggests that AMH alleles were not tolerated well in these regions on Neanderthal backgrounds either.

Juric and coworkers (Juric et al. 2016) argue that the observed patterns of Neanderthal ancestry in modern humans can be explained by the difference in effective population size between Neanderthals and humans. They find that on average, selection against individual Neanderthal alleles is very weak. This is consistent with the assumption that Neanderthals over time accumulated many weakly deleterious alleles that were effectively neutral in their small populations. Drift is dominant in small populations, making it harder for purifying selection to act. However, after introgressing into larger human populations, those Neanderthal alleles became exposed to purifying selection. Thus, rather than being the result of hybrid incompatibilities, differences between human and Neanderthal effective population sizes would have played a key role in shaping present-day shared ancestry. Harris and Nielsen (2016) made a similar argument using simulation data.

Time and place of interactions

When evidence of archaic introgression into AMHs was first reported, most researchers assumed that the admixture occurred in the Middle East, a likely migration route out of Africa where the archeological record documents overlapping Neanderthal and AMH ranges. The Altai Neanderthal sequence shows introgression from AMHs that would have occurred about 100 000 ya (Kuhlwilm et al. 2016). Prüfer et al. (2017) find similar patterns in the high-coverage Vindija genome, so early admixture with AMHs must have occurred before the western and eastern Neanderthal lineages separated from each other. The admixture would probably have occurred along the two most likely migration routes, either in the Levant where the Tabun and Shkul sites are located (dating to about ~120 000 ya), or in southern Arabia around the Persian Gulf (Armitage et al. 2011).

This Middle Eastern admixture scenario does not explain why admixture rates are 20% higher in East Asians than in Europeans, though (Meyer et al. 2012;

Wall et al. 2013). Various authors have outlined three possible explanations for higher admixture rates in Asians than in Europeans: (i) purifying selection removed more archaic alleles in Europeans because their populations were larger than those in East Asia, (ii) Neanderthal ancestry was diluted in Europeans by admixture with other populations that contained less Neanderthal ancestry, or (iii) further admixture events occurred between Neanderthals and Asians after the European and Asian populations separated (Nielsen et al. 2017). Fu et al. (2016b) found that the proportion of Neanderthal DNA ancestry decreased from 3–6 to 2% over a 40 000-year-period in their study of 51 ancient modern Eurasians. They attribute this decrease over time to selection acting against Neanderthal DNA in Eurasian AMH populations. Kim and Lohmueller's (2015) simulations showed that a model that combined purifying selection on larger European populations with greater drift in smaller East Asian populations could not explain the observed asymmetry in Neanderthal contributions to each group, however. The two remaining hypotheses each have some support, so the history of interactions among these Eurasian hominins may be a combination of more recent gene flow into Europeans from AMH populations who had less interaction with Neanderthals, along with multiple admixture events between archaic and AMH populations across Eurasia. Europeans today share a substantial genetic component with Near Eastern populations that began to appear in ancient modern Europeans about 14 000 ya. This component may be one source of new European DNA source material (Fu et al. 2016b). Both Kuhlwilm et al. (2016) and Prüfer et al. (2017) argue that the Altai and Vindija Neanderthal genomes support the view that European Neanderthals contributed more alleles to present-day people than did the eastern Neanderthals. The Altai Neanderthal genome shows evidence of admixture with AMHs, but the Vindija and Mezmaiskaya sequences share more alleles with non-Africans than do the Altai Neanderthal, which implies an admixture event after Neanderthals left the Middle East and split into eastern and western populations. Both the western Vindija and Mezmaiskaya Neanderthal lineages share alleles with AMHs in approximately equal numbers, which suggests that the admixture occurred before those lineages diverged 80 000–100 000 ya. Admixture between AMHs and Denisovans occurred more recently than it occurred between AMHs and Neanderthals because the introgressed Denisovan fragments are longer than the Neanderthal fragments (Sankararaman et al. 2016). The researchers estimate that the Denisovan and AMH admixture dates to 44 000–54 000 ya and assume it occurred in Asia. A single-event admixture model now looks overly simplistic, and simulation studies suggest two or more "pulses" fit the currently available information.

The timing and location of AMH and archaic group interactions provide insight into the movement of AMHs as they migrated out of Africa and into Eurasia and Oceania. Current debate centers on how many times AMHs migrated out of Africa, where they went, and how the AMHs who settled in Europe, Asia, and Australia are related to each other. Three recent studies address these questions with NGS data from large population samples focused on Asia and Oceania. Malaspinas et al. (2016) and Mallick et al. (2016) both favor single-wave models, but differ on when and where that wave splits into two groups. Malaspinas' group argues that modern humans split into two groups immediately after leaving Africa, with one group moving south to Australasia while the other group expanded north to settle mainland Eurasia. Mallick and colleagues propose a split between western and eastern Eurasians, with present-day Australians and Papua New Guineans being descendants of the eastern wave, along with the mainland East Asians. Pagani et al. (2016) support a multiple-dispersal model, with 2% of modern Papua New Guinean DNA coming from what they describe as an early wave of AMHs that left Africa ~120 000 ya. Rasmussen et al.' (2011) report of the sequencing of an Australian Aborigine from a hair collected in the 1920s indicated that the ancestors of Australian Aborigines split from mainland Eurasian populations about 62 000–75 000 ya, which correlates well with archeological data. A multiple dispersal model makes it somewhat easier to explain how AMHs could admix with Denisovans, but then leave little evidence of that admixture in modern East Asians while showing significant Denisovan contributions to modern Oceanian populations.

Conclusion

The relationship between archaic and anatomically modern humans in Eurasia is complex and falls somewhere between the extremes of the early Recent African Origins and Multiregional models, but closer to the RAO model. Neanderthals were once found across Eurasia living in small groups. The Denisovans' range is less well understood, but it was probably confined to Asia and Oceania. The Neanderthal and Denisovan genomes were much less diverse than those of modern people, and their effective populations are likely to have been relatively small. Sikora et al. (2017) have postulated that the early AMH groups were also small, but they had an evolutionary advantage because their social organization resembled that of modern foragers who practice exogamy and participate in extended social networks that facilitate gene flow across large areas. As the AMHs migrated out of Africa and settled throughout Eurasia, they admixed with both Neanderthals and Denisovans. Evidence of introgression from both archaic groups is observed in present-day human populations, but admixture from AMHs into Neanderthals and between archaic groups has been reported as well. European Neanderthals contributed more genetic material to present-day populations than the eastern ones. Nevertheless, living Asian populations have about 20% more Neanderthal DNA in their genomes than Europeans.

Archaic hominin genetic research is advancing rapidly, but several recent studies highlight gaps in our knowledge and suggest fruitful research directions. The search continues to locate more complete Denisovan remains to better understand their anatomy, geographic range, and behavior. Known Asian archaic fossils show great morphological diversity, and none has yet yielded publishable aDNA sequences for comparison. New specimens are promising (Li et al. 2017), but researchers have not yet been able to integrate DNA and skeletal information from Denisovans as they have done with Neanderthals (Stringer and Barnes 2015). Slon et al. (2017a) have used hybridization capture to retrieve DNA from Late and Middle Pleistocene cave sediments, which suggests that it may be possible to detect the presence of archaic hominins in areas where no skeletal remains are found. Finally, new methods are enabling researchers to recover DNA from specimens far older than would previously have been thought possible. Single-stranded sequencing is permitting researchers to study extremely small fragments of DNA from fossil material as old as 780000 years under permafrost conditions (Orlando et al. 2013), and perhaps as old as one million years old in deep-frozen environments (Allentoft et al. 2012). The single-stranded process permits sequencing of fragments as tiny as 25 base pairs. PCR, enzymatic manipulation, and repeated purification steps are not required, which eliminates PCR bias and minimizes enzyme incompatibility with chemically modified templates. Meyer et al. (2014, 2016) have used this method to sequence both mtDNA and nDNA from archaic individuals found in Sima de los Huesos de Atapuerca in Spain. Their results indicate that these Middle Pleistocene hominins, whose remains have been dated to about 430000 ya, were most closely related to Neanderthals. Paleoanthropologists are working to locate the Denisovans in the Asian fossil record, even as geneticists develop methods to identify archaic remains where no fossils remain and to expand their ability to recover and analyze DNA from even older material.

Researchers have just begun to explore the benefits of archaic introgression to human health and adaptation. Analyses of archaic genomes and comparisons with contemporary human samples have already provided evidence of archaic introgression of alleles that protect against pathogens, cold, ultraviolet radiation, hypoxia, and starvation in new environments. Researchers also have identified archaic alleles that are risk factors for certain diseases (Simonti et al. 2016). The recent addition of a high-coverage western Neanderthal genome (Prüfer et al. 2017) has permitted researchers to identify even larger archaic DNA blocks in modern human genomes because the western European Neanderthals contributed more genetic material to modern humans. DNA sequences can also provide information about epigenetic variation and predict patterns of gene expression and variation (Gokhman et al. 2014; Orlando and Willerslev 2014; Pedersen et al. 2014). Epigenetics is another field that is just starting to have an impact on aDNA studies.

Archaic genomes are contributing important new information about human evolution and hold the potential for even greater advances in the future. Technical improvements continue to promote the field by expanding the range of DNA sources that can be used. aDNA is a direct source of information about our nearest relatives and, when combined with morphological and archeological information, is an extremely powerful tool for understanding both our past and present.

CHAPTER 7

Evaluating Correspondence in Phenotypic Variation, Genetic Diversity, Geography, and Environmental Factors

Noreen von Cramon-Taubadel

Department of Anthropology, University at Buffalo, Buffalo, NY, USA

Introduction

It has long been noted that phenotypic data may be employed as useful proxies for genetic data when analyzed under specific and explicit conditions (e.g. Lande 1977; Cheverud 1982, 1988; Lynch 1989; Relethford and Blangero 1990; Roseman and Weaver 2007). Hence, parameters such as population structure, and the past effects of microevolutionary forces such as gene flow, genetic drift, and natural selection, can be estimated from patterns of within- and between-group patterns of phenotypic variation. This is an important endeavor within anthropology more generally for two major reasons. First, it is necessary to better understand how phenotypic covariation patterns evolve both within and among

A Companion to Anthropological Genetics, First Edition. Edited by Dennis H. O'Rourke.

species. This in turn allows us to infer the past actions of stochastic evolutionary processes such as genetic drift and identify the potentially deterministic effects of past directional selection on phenotypic traits. Second, and perhaps more crucially, there are many questions in anthropology concerning deeper hominin evolutionary history as well as modern human antiquity for which phenotypic data are the only direct biological evidence we will ever possess. Therefore, understanding how phenotypic variation evolves is a fundamentally important issue within anthropology.

Quantitative genetics extends the basic principles of evolutionary population genetics to model the inheritance and evolution of quantitative (continuous) phenotypic traits (see also Falconer and Mackay 1996; Lynch and Walsh 1998). Here, I will focus on methods used in anthropological studies to directly assess the congruence between phenotypic, genetic, geographic, and environmental sources of variation. At a very basic level, quantitative genetics is concerned with understanding how phenotypic variation may be partitioned into heritable ("genetic") and nonheritable ("environmental") sources of variation. At the within-population level, this can be achieved by assessing narrow-sense heritability (i.e. the proportion of overall phenotypic variance attributable to additive genetic variance) (Falconer and Mackay 1996). This information can then be employed to determine the potential evolvability of a particular continuous trait or integrated suite of traits, given the direct relationship between heritability and the potential response to evolutionary forces such as selection (Falconer and Mackay 1996; Roseman and Weaver 2007; von Cramon-Taubadel 2014a).

An example of anthropological interest in this regard is the study by Roseman et al. (2010), who investigated the relative contributions of genetic and environmental variance to patterns of cranial trait variance in a large sample of pedigreed baboons. Almost all of the cranial traits assessed were at least moderately heritable ($h^2 \sim 0.44$). Genetic and phenotypic covariance matrices were strongly correlated ($r = 0.83$), but there was also a strong correlation between overall phenotypic covariance and estimates of environmental covariance ($r = 0.74$). Interestingly, levels of genetic integration (due to pleiotropic genetic effects) between traits varied from one cranial region to another, suggesting that some cranial regions are more constrained (via stabilizing selection) than others in terms of their potential evolvability (termed *conditional evolvability*) (Roseman et al. 2010). This study, therefore, illustrates how it is possible to estimate the apportionment of phenotypic variation into constituent additive genetic and environmental variation in an anthropologically relevant animal model (see also Cheverud 1988, 1995). In turn, this information can be usefully employed to make predictions about the potential evolvability of particular phenotypic traits and/or their potential reliability for phylogenetic or evolutionary reconstruction in the absence of genetic data (Roseman et al. 2010).

However, as can be seen from the example given here and from other studies directly assessing trait heritability in primate models (e.g. Cheverud 1982; Hlusko et al. 2004, 2007; Sherwood et al. 2008), to estimate these within-group parameters accurately, detailed genetic and phenotypic data from large pedigrees of related individuals are required. In the case of modern humans, the relatively few heritability analyses of three-dimensional cranial traits that have been conducted (Sjøvold 1984; Carson 2006; Martínez-Abadías et al. 2009) all used the same pedigreed skeletal population (from Halstatt, Austria). However, 2D measures of cranial traits obtained from lateral cephalographs collated by the Fels Longitudinal Study have also been used to obtain heritability estimates and identify potential quantitative trait loci (QTLs) contributing to the genetic architecture of the human cranium (e.g. Sherwood et al. 2008, 2011). Nevertheless, the lack of suitable skeletal collections with well-defined pedigree information limits the ability to obtain direct heritability estimates for anthropologically relevant phenotypic traits. Moreover, even in cases where heritability estimates can be obtained, it is inappropriate to extend these beyond the specific population used to generate them (Vitzthum 2003). Therefore, in the absence of direct genomic estimates of additive genetic variation, many quantitative genetic studies proceed by using the phenotypic covariance matrix (**P**), which can more easily be estimated, as a proxy for the underlying additive genetic covariance matrix (**G**). This approach is justified because **P** and **G** have repeatedly been shown to be proportional for a wide range of phenotypic traits across different animal models (e.g. Cheverud 1988, 1995; Roff 1995; Marroig et al. 2009; Porto et al. 2009; Roseman et al. 2010).

Comparing genetic, phenotypic, and environmental sources of variation among groups

The examples discussed above illustrate how phenotypic, genetic, and environmental data can be integrated and compared at the within-group level. However, most anthropological research questions concern patterns of among-group variation: how are these groups related? Is there a relationship between the form of a particular phenotypic trait and particular environmental conditions? Did these groups diverge under neutral (stochastic) or adaptive influences? Hence, addressing such questions requires the use of methods designed to quantify and compare between-group patterns of variation.

Tree-building methods In the case of analyses at supraspecific taxonomic levels, evolutionary questions have typically been addressed using various phylogenetic tree-building methods such as cladistics (Hennig 1966), phenetics (Sneath and Sokal 1973; Saitou and Nei 1987), and/or phylogenetic comparative methods (Felsenstein 1985; Harvey and Pagel 1991; Martins and Garland 1991). Each of these methods differs in terms of how the phenotypic data are coded and the model algorithms used to construct the branching patterns (tree topology) of the taxa in question (for a review, see Nunn 2011). In very basic terms, phylogenetic methods can be employed to compare statistically the topologies of tree(s) estimated using phenotypic data and those based on molecular genetic data for the same taxa. A stronger similarity (or congruence) between these tree topologies indicates that the phenotypic traits in question provide relatively reliable estimates of phylogenetic history and display relatively low levels of homoplasy among taxa (for select anthropologically relevant examples, see Collard and Wood 2001 2007; Strait and Grine 2004; Lycett and Collard 2005; Smith and Grine 2008; Gilbert et al. 2009; Bjarnason et al. 2011).

While tree-building phylogenetic methods are routinely used in anthropology to address questions about phylogenetic signal in phenotypic data at higher taxonomic levels, they are generally deemed inappropriate for testing analogous phylogenetic questions at the intraspecific level (von Cramon-Taubadel 2014b). The primary reason cited is that among-population gene flow would render the implementation of a bifurcating tree model inaccurate due to potentially high levels of reticulation between population lineages as a result of lateral gene flow (e.g. Sherry and Batzer 1997; Rohlf 2000; Smouse 2000; Athreya and Glantz 2003).

While the probability of gene flow between geographically contiguous demes is undeniable, the use of tree comparison methods might be a useful addition to the toolkit of analytical techniques used to study intraspecific population history (e.g. Pickrell and Pritchard 2012; Kopelman et al. 2013; Hunley et al. 2016). Tree-building methods have been employed to reconstruct modern human phylogenies for both craniometric (e.g. Howells 1973; Guglielmino-Matessi et al. 1979; Lahr 1995; Athreya and Glantz 2003; Schillaci 2008) and genetic data (Cavalli-Sforza et al. 1994a; Li et al. 2008; Pemberton et al. 2013), under the assumption that a hierarchical model of shared ancestry is a valid means of explaining the deep population history of the human species. The assumption that supraspecific phylogenetic trees are free from potential reticulations and hybridizations is also not upheld (e.g. Ackermann 2010; Guevara and Steiper 2013). In addition, while phylogenetic techniques are routinely employed to assess the potential patterns of relatedness between fossil hominin and primate taxa, we cannot know for certain whether we are dealing with biological species (*sensu* Mayr 1942) or indeed subspecific relationships in many of these cases. Moreover, despite criticisms of the application of a bifurcation model to the divergence of intraspecific demes (e.g. Smouse 2000), population gene flow and admixture can be reconciled within a tree-like hierarchical model as simply another potential source of phylogenetic "noise." For example, Cavalli-Sforza et al. (1994a, p. 54) view modern human population diversity as resulting from a large number of population divergences, where complete population separation did not transpire in every instance, but rather genetic isolation occurred periodically throughout human history due to the imposition of physical or cultural (e.g. linguistic) barriers. Subsequently, geographic contact between populations may have resulted in increased gene flow, especially among populations occupying contiguous geographic areas. Therefore, deeper historical migration processes, such as the out-of-Africa dispersal event(s) (Eriksson et al. 2012; Henn et al. 2012; Reyes-Centeno et al. 2014), together with more recent gene flow among populations combine to establish contemporary genetic (and phenotypic) patterns of modern human variation (Hunley et al. 2009; Pickrell and Pritchard 2012).

Matrix comparison methods Given the reluctance to apply tree-based techniques at the subspecific level, the most common methods used to assess the correspondence between among-population phenotypic, genetic, and environmental distance patterns are matrix comparison methods. In the case of biological data such as genetic allele frequencies or quantitative phenotypic traits, these matrices generally comprise pairwise between-population distances based on measures such as Wright's F_{st} (Wright 1951) and the phenotypic analog (Q_{st}; Spitze 1993) or derivatives of such measures (e.g. Harpending and Jenkins 1973; Relethford and Blangero 1990; Relethford 1991; Roseman 2004; Holsinger and Weir 2009). Smaller values indicate that populations are relatively similar in their genetic or phenotypic structure, while larger values indicate that populations are biologically more distinct. One of the major advantages of comparing population matrices is the ability to compare biological patterns of population affinity directly against nonbiological data such as geographic distance, temporal distance, climatic variation, dietary differences, as well as any number of model ("design") matrices (Sokal et al. 1997) intended to parameterize particular hypothetical phenomena (e.g. Waddle 1994; Pinhasi and von Cramon-Taubadel 2009; Hubbe et al. 2010; von Cramon-Taubadel and Pinhasi 2011; Reyes-Centeno et al. 2014).

The statistical congruence between matrices can be assessed using the Mantel (1967) test. Matrices violate the basic statistical assumptions of traditional association tests (e.g. Pearson's correlation) due to the non-independence of the off-diagonal elements, such that their associated probability statistics cannot be accurately estimated. Hence, the Mantel test assigns *p*-values through a randomization procedure where the observed correlation between two or more matrices is assessed against a distribution of values obtained by random permutation of the rows and columns of the matrices (Smouse et al. 1986; Smouse and Long 1992). The commonly used version of the Mantel test is equivalent to the Pearson correlation between the standardized elements of the two matrices being compared (Diniz-Filho et al. 2013), although nonparametric extensions are also possible. Extensions of the Mantel test include the partial Mantel test, whereby the effects of a third matrix (or multiple potentially confounding matrices) can be controlled while assessing the correspondence between two matrices, and the multiple Mantel test, where multiple independent matrices can be regressed against a single dependent matrix (Smouse et al. 1986).

Mantel tests have been used extensively by anthropologists interested in assessing the congruence between craniometric, genetic, and environmental estimates of among-population affinity matrices for globally distributed human groups (e.g. González-José et al. 2004; Roseman 2004; Harvati and Weaver 2006; Smith 2009; von Cramon-Taubadel 2009a, 2009b, 2011). Much of the impetus for these studies was to test the neutral evolutionary expectation that craniometric distance matrices should be proportional to genetic distance matrices if diversifying natural selection had not played a substantial role in shaping global craniometric diversity patterns (e.g. Roseman and Weaver 2007). The logic is that if among-population craniometric patterns are largely the consequence of stochastic evolutionary forces such as mutation, gene flow, and genetic drift, then estimates of among-group neutral genetic and craniometric distances should be significantly correlated. The ability to test a neutral evolutionary expectation using this empirical approach became feasible with the availability of neutral genetic datasets (especially microsatellites and single nucleotide polymorphisms [SNPs]) for a large number of globally distributed human populations through the human genome diversity cell line panel (the Human Genome Diversity Project and Centre d'Etude du Polymorphisme Humain [HGDP-CEPH]) (e.g. Cann et al. 2002; Rosenberg et al. 2002; Li et al. 2008; Pemberton et al. 2013). This meant that large craniometric datasets such as that collected by William Howells (1996) could be matched with large neutral autosomal genetic datasets for the same (or very similar) human populations (e.g. Roseman 2004). Although the degree of congruence between craniometric and genetic population matrices is limited by the quality and accuracy of the population matches made, it did provide, for the first time, the opportunity to assess the congruence of among-population neutral genetic and phenotypic patterns directly. Moreover, it was also relatively easy to collate matching environmental distance matrices for the same populations based on between-population geographic distance and/or estimates of climatic or dietary differences between populations (e.g. Roseman 2004; Harvati and Weaver 2006; Smith 2009; von Cramon-Taubadel 2011).

Roseman (2004) presented the first comparison of matched neutral genetic (microsatellite loci; Rosenberg et al. 2002), craniometric (Howells 1996), and temperature distances for 10 globally distributed human populations. He constructed a neutral genetic affinity matrix based on R_{st} (Slatkin 1995), a craniometric matrix based on the Relethford–Blangero (1990) estimator of multivariate quantitative F_{st} (termed Q_{st}), and a temperature distance matrix based on the squared pairwise differences in mean annual temperature between populations. Full and partial Mantel tests were employed to assess the congruence of these three matrices. He found that the craniometric matrix was significantly and relatively strongly correlated ($r \sim 0.48$) with both neutral genetic and temperature distance. Moreover, the correlation between craniometric distance and temperature remained significant and strong ($r = 0.58$) even when genetic distance was corrected for in a partial Mantel test. On the face of it, the results are confusing as they align with the predictions of neutral expectation, yet they suggest the action of a potentially selective force of global temperature in driving patterns of craniometric diversity. However, a careful analysis of the effects of individual populations in the dataset clarified what was going on. Removal of the single arctic population ("Siberian") both substantially increased the correlation with the neutral genetic matrix ($r > 0.8$) and substantially reduced the correlation between craniometric distance and temperature, controlling for neutral genetic distance ($r \sim 0.1$), while removal of the other nine populations did not appreciably alter the findings. Therefore, this study suggested that global patterns of craniometric distance were largely consistent with neutral expectation, but that in the case of populations living in extreme cold climates, the directional selective effects of climate on the cranium could not be ruled out.

It is worth pointing out that the results of Roseman (2004) and others using the genetic-craniometric matrix comparison method (e.g. Harvati and Weaver 2006; von Cramon-Taubadel 2009a, 2009b, 2011; Smith 2009) are also in agreement with other tests of neutral expectation for craniometric evolution based on different methodologies (Roseman and Weaver 2007; von Cramon-Taubadel and Weaver 2009). For example, multiple studies have shown that estimates of among-population apportionment based on neutral genetic loci and craniometric data are similar (González-José et al. 2004; Relethford 1994, 2002, 2004a, 2004b; Roseman and Weaver 2004; Hubbe et al. 2009) and show the same pattern of low among-population and high within-population variance apportionment (Wright's F_{st}) expected for neutrally evolving genotypes and phenotypes. This empirical finding is best appreciated when contrasted with the apportionment of skin color reflectance data at a global level (Relethford 2002), which shows the very different pattern of low within-population and high among-population apportionment of variance expected for phenotypes that have been subject to strong diversifying selection. Another approach to testing the neutral theory of phenotypic evolution (Roseman and Weaver 2007) for human craniometric data has been to fit a model of iterative founder effects from a single (African) origin to such datasets. The logic underlying this approach is that if craniometric variation has evolved largely under neutral parameters, then within-population cranial variation should exhibit the same pattern of ever-decreasing variation with increased distance from Sub-Saharan Africa, as has been repeatedly shown to be the case for global neutral genetic datasets (e.g. Prugnolle et al. 2005; Ramachandran et al. 2005; Liu et al. 2006; Li et al. 2008; Hunley et al. 2009; Reyes-Centeno et al. 2014). This same empirical pattern has been shown using several different global human craniometric datasets (Manica et al. 2007; von Cramon-Taubadel and Lycett 2008; Betti et al. 2009).

Given the wide range of methods and datasets to which this question has now been applied, it would appear justifiable to assert that global patterns of modern human craniometric diversity are *largely* the result of neutral forces of evolution such as mutation, gene flow, and genetic drift. However, this still leaves open the question of whether some cranial regions have indeed been subject to non-neutral diversifying selection, and, if so, whether this affects all or only some populations (Roseman 2004; Roseman and Weaver 2004). These questions have also been addressed using Mantel matrix comparison approaches (e.g. Harvati and Weaver 2006; Smith 2009; von Cramon-Taubadel 2009a, 2009b, 2011). In these studies, invariably the cranium has been divided into constituent regions (such as the face, basicranium, or vault) based on a whole host of different criteria, depending on the hypotheses being tested (for a recent review, see von Cramon-Taubadel 2014a). The empirical findings themselves are not relevant here, but the study design generally follows the following procedure: a comparison of phenotypic data matrices against a neutral genetic matrix (a general test of neutrality); a comparison of phenotypic data against environmental data, such as climate or dietary behavior (a test for potential non-neutral diversification); followed by partial Mantel tests of the congruence of phenotypic data and environmental data, controlling for the potentially confounding effects of population history (using either genetic data or geographic distance as a proxy). This latter approach using partial Mantel tests is usually designed to check whether two matrices are actually "causally" related or whether the apparent congruence of these two matrices is simply reflecting the fact that both matrices are spatially or genetically structured in similar ways (Figure 7.1; see also Diniz-Filho et al. 2013).

In cases where matched genetic data are not available for the populations under study (as is also the case for many prehistoric population samples), researchers have used geographic distance as a general proxy for the neutral genetic relatedness of the populations in question (Relethford 2004b; Pinhasi and von Cramon-Taubadel 2009; Betti et al. 2010, 2014). This approach has been justified on the basis that a significant correlation between genetic and geographic distance at a global level has been demonstrated (Relethford 2004a; Manica et al. 2005; Ramachandran et al. 2005; Liu et al. 2006; Romero et al. 2009). Indeed, such a relationship is to be expected (Figure 7.1) given that contiguous populations are more likely to both share more recent common ancestry and experience more intensive bilateral gene flow than distantly separated populations (e.g. Kimura and Weiss 1964; Relethford 2004a). However, what cannot be discerned from this overall empirical congruence between genetic and geographic distance is the relative contribution of more ancient hierarchical population history related to a model of nested serial founder effects as humans dispersed out of Africa (Hunley et al. 2009; Reyes-Centeno et al. 2014), and the more recent among-population gene flow patterns that some ascribe to a classic model of "isolation-by-distance" (*sensu* Wright 1943; Malécot 1973). Both of these geographically mediated processes undoubtedly have contributed to the contemporary pattern of human genetic and phenotypic variation observed today. What is less clear, however, is the relative potency of more ancient structured population dispersal versus more recent population gene flow in generating the nested hierarchical genetic structure observed at a global level (Li et al. 2008; Pemberton et al. 2013). This is an issue that will be returned to later in this chapter, when a specific case study is examined in more detail.

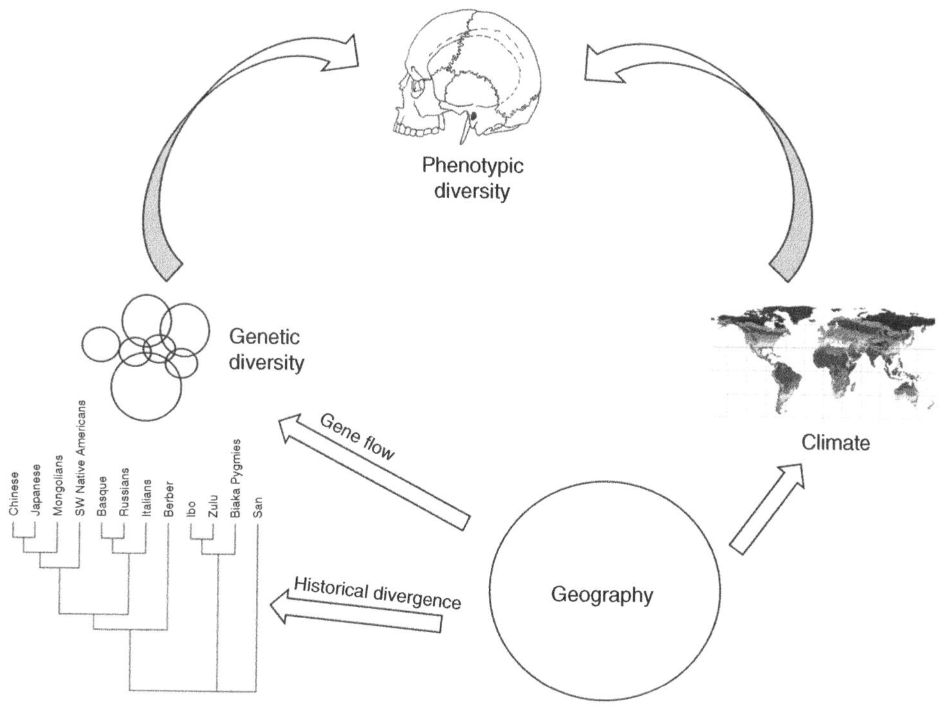

Figure 7.1 Evaluating the correspondence between phenotypic, genetic, geographic, and environmental sources of variation is complicated by the fact that all of these factors are interrelated. Geography, in the sense of the spatial pattern of landmass arrangement, mediates the demographic processes that generate patterns of genetic, and ultimately phenotypic, diversity. These comprise deeper historical processes, such as long-range dispersals and migrations, which may include population isolation and divergence, as well as gene flow among contiguous populations. Geography can also mediate, or correlate with, other environmental patterns such as climatic and ecological variation, which in turn may correspond with differences in mobility, diet, or behavior. Any of these environmental factors may causally influence phenotypic (and by extension genetic) diversity patterns.

Dow–Cheverud test In studies where the primary objective is to compare the relative genetic congruence of different phenotypic regions or trait sets, it is often imperative to be able to assess statistically whether one phenotypic matrix is indeed *more* genetically congruent than another. This raises the importance of not taking matrix correlation coefficients (r-values) too literally or attempting to compare r-values reported in one study with those obtained in another. The reason being that each matrix comparison conducted will be limited by the reliability with which that particular genetic, phenotypic, or environmental affinity matrix can be estimated. For example, even in the highly controlled case of the pedigreed baboon colony employed by Roseman et al. (2010), the authors found that although phenotypic covariance matrices could be estimated with high (>90%) repeatability, genetic and environmental matrices contained some error in their estimation (covariance repeatability ~0.7). This limited the highest potential correlation between these matrices to less than r = 1.0 and highlights the need to consider correlation statistics obtained between matched datasets to be minimum estimates of congruence in light of both imperfect population matches and imperfect matrix estimations. For all of these reasons, one should caution against assuming that a correlation coefficient of r = 0.5 is automatically indicative of a lower strength correlation than a value of, say, r = 0.6.

One methodological approach to assessing whether a correlation of a coefficient of 0.6 is indeed significantly higher than one of 0.5 is to employ the Dow–Cheverud test (Dow and Cheverud 1985). The Dow–Cheverud test was designed to assess whether the correlation between two matrices, A and B – ($r(AB)$) – is significantly different from the correlation between matrix A and a third matrix, C; ($r(AC)$). During this procedure, matrices B and C are *z*-transformed to make their elements comparable, and a difference matrix, B–C, is created that can be compared against matrix A using a standard Mantel permutation procedure. If $r(A,B–C)$ is significant and positive, then B is more congruent with A than C is, and vice versa if $r(A,B–C)$ is negative and statistically significant.

Criticisms of matrix comparison methods While the Dow–Cheverud test provides a convenient means of assessing whether two or more matrix correlations are actually statistically different from one another, they have been criticized on the grounds of being vulnerable to spuriously high rejection rates if two or more of the matrices in question are spatially autocorrelated (Oden 1992; Konigsberg 1997). Similar criticisms concerning elevated type 1 error rates and low statistical power have been levied at the use of full and partial Mantel tests (e.g. Oden and Sokal 1992; Harmon and Glor 2010; Legendre and Fortin 2010; Guillot and Rousset 2013) in cases where one or more of the matrices in question are spatially or phylogenetically autocorrelated. In a simulation study investigating the integrity of the full and partial Mantel tests, Guillot and Rousset (2013) suggest that while the tests perform well in the absence of autocorrelation in both matrices, considerably deflated *p*-values result when both matrices exhibit autocorrelation. Guillot and Rousset (2013) also discuss some alternative strategies that may be developed in the future to allow the comparison of two or more variables, while mitigating against elevated type 1 errors. However, as they note, there is currently no computer software available to perform such hypothetical analyses in an automated way. Alternative methods, in the form of Bayesian generalized linear mixed models (e.g. Roseman and Auerbach 2015) and mixed models derived from Bayesian sparse factor analysis (e.g. Katz et al. 2016), are currently being tested with human phenotypic data, and may provide more robust alternatives to Mantel tests in the future.

Defense of matrix comparison methods Diniz-Filho et al. (2013) recently defended the use of full and partial Mantel tests for the assessment of evolutionary hypotheses, especially in cases where the relationships among groups can best be expressed as pairwise distances (see also Legendre and Fortin 2010). They make the important point that type 1 errors can easily be dealt with by assigning lower alpha levels than the conventional $\alpha = 0.05$. They also advocate for the use of multiple lines of evidence in favor of particular hypotheses, and make a point that most empiricists know to be true: that robust results from empirical studies employing different techniques and approaches "usually converge" (Diniz-Filho et al. 2013, p. 483). This resonates with a point made earlier in this chapter: in the case of modern human craniometric data, multiple studies conducted by different teams of researchers, utilizing different datasets and methodological approaches, all converge on the overall notion that *global* patterns of craniometric shape diversity are largely the results of neutral evolution due to mutation, migration, and genetic drift, with directional selection playing a relatively minor role (e.g. González-José et al. 2004; Roseman 2004; Roseman and Weaver 2004; Harvati and Weaver 2006; Smith et al. 2007; Betti et al. 2009, 2010; Smith 2009; von Cramon-Taubadel 2009a, 2009b, 2011). Hence, results obtained from the use of Mantel's test and its derivatives have not proven to be at odds with those obtained using other analytical approaches.

Matrix comparison methods as a complement to tree comparison methods Despite the popularity of tree-based methods when addressing phylogenetic questions at the supraspecific level, in recent years, some researchers have applied Mantel test approaches to such questions (e.g. Polly 2001; Monteiro and Dos Reis 2005; Macholán 2006). Cardini and Elton (2008) presented such a study using a primate dataset of guenon taxa. They used geometric morphometrics to characterize the shape of various skull regions representing functionally and developmentally integrated modules (FDMs) in primates (Cheverud 1995). Matrices representing the Procrustes distances for different FDMs were compared against a genetic distance matrix based on a published molecular tree for the same species (Tosi et al. 2005). A conservative alpha level of 0.01 for the Mantel tests was employed, and the FDMs with the absolute strongest correlation coefficient were then subjected to further phylogenetic tree-building methods such as UPGMA (Michener and Sokal 1957), neighbor joining (Saitou and Nei 1987), and maximum likelihood (Felsenstein 2004). Cardini and Elton (2008) concluded that the basicranium of guenons was the most phylogenetically informative of all cranial regions tested. This study serves as an example of how Mantel matrix methods, which have typically been employed in the analysis of interpopulation diversity, can be used at higher taxonomic levels and integrated with more traditional tree-building phylogenetic methods. Following on from Cardini and Elton (2008), a similar study design was implemented by von Cramon-Taubadel and Smith (2012) and von Cramon-Taubadel and Lycett (2014) to investigate the relative genetic congruence of cranial regions in hominoid taxa as well as compare the relative genetic congruence of cranial shape data with postcranial data. Hence, it is possible to use Mantel tests to bridge an empirical and comparative gap between population-based intraspecific studies and those comparing higher taxa. Moreover, this brief review reinforces the point made earlier that researchers should investigate their hypotheses using several methodologies (e.g. matrix comparisons and tree-building methods) and different datasets, where possible, to arrive at robust empirical conclusions.

Using matrix comparisons to assess explanatory power: A worked example The following simple case study is designed to illustrate how Mantel tests can be used to assess the strength of correlation between phenotypic, genetic, and environmental datasets, and it also attempts to assess the relative explanatory power of different sources of variation on phenotypic (craniometric) among-population distances. The craniometric data employed represent the 12 globally distributed populations previously analyzed by von Cramon-Taubadel (2009a, 2009b) that could be matched with microsatellite allele frequency data from the HGDP-CEPH database (Rosenberg et al. 2002). The details of how the craniometric shape data (based on 141 landmarks) were obtained and transformed to a between-population distance matrix can be found in the aforementioned publications. Alongside the craniometric and microsatellite distance matrices, three additional distance matrices were constructed: (i) a geographic distance matrix based on great circle distances between populations, using waypoints to avoid unrealistic connections across oceans (for a similar approach, see Ramachandran et al. 2005). This matrix was calculated in kilometers using the Geographic Distance Matrix Generator (v. 1.2.3) freely available through the American Museum of Natural History (Ersts 2015). (ii) A Euclidean distance matrix describing average differences in mean annual temperature based on data collated from the Intergovernmental Panel on Climate Change database (New et al. 2002) and available online at http://www.ipcc-data.org. (iii) A "design" matrix reflecting the hierarchical bifurcating relationship between these 12 populations. This latter matrix was based on the tree topology produced by Li et al. (2008, see also Pemberton et al. 2013) using neutral genetic data for the HGDP-CEPH dataset. This tree topology, which reflects the hierarchical relatedness between populations (Figure 7.1), was coded such that populations sharing a single node (common ancestor) were coded as 1, those separated by two nodes coded as differing by 2, and so forth. To confirm that the resultant "history" matrix accurately reflected the branching relationship described by Li et al. (2008; Pemberton et al. 2013), it was subjected to neighbor joining (Saitou and Nei 1987) in the software Phylip 3.69 (J. Felsenstein, http://evolution.genetics.washington.edu/programs.html). The topology that resulted is that depicted in Figure 7.1.

Mantel tests were performed in *R* (R Development Core Team 2008), using the *mantel* function from the *vegan* package, to test for overall correlation between

Table 7.1 Results of Mantel test correlations between matched craniometric and genetic matrices and three potential explanatory factors (geographic distance, historical divergence, and temperature).

	Craniometrics	*% Exp.*	*Genetics*	*% Exp.*
Genetics	0.544 (0.005)	29.6%	—	—
Geography	0.788 (<0.001)	62.0%	0.480 (0.023)	23.0%
History	0.520 (<0.001)	27.1%	0.653 (0.002)	42.6%
Temperature	0.222 (0.081)	4.94%	0.149 (0.222)	2.22%

Results are correlation coefficients with significance values in parentheses. % Exp. = coefficient of determination expressed as a percentage.

the craniometric matrix and the genetic, geographic, history, and temperature matrices, based on 9999 permutations. The results (Table 7.1) show that all correlations were significant at $\alpha = 0.005$, except for craniometrics/temperature, which resulted in a *p*-value of 0.081. In each case, the percentage of among-group craniometric variance explained by each of the four factors was calculated as the square of the correlation coefficient. If we were to assume that each of these factors (neutral genetics, geography, historical divergence, and temperature) were ***independent*** causal sources of variation, the total variance explained would be 123.64%, which is obviously nonsensical. Even if we dismiss the possible causal effects of temperature due to the much higher *p*-value, the total variance explained would still exceed 100%. Therefore, these are obviously not mutually exclusive explanations for overall patterns of craniometric diversity.

As illustrated in Figure 7.1, all of these factors are or could be interrelated in either a correlative or causative fashion. Geography (*sensu* spatial arrangement of landmasses) mediates genetic diversity, both in terms of historical divergences among populations (e.g. due to the out-of-Africa migration[s]) and via potential ongoing gene flow among geographically contiguous populations. Geography also mediates climatic variation to a certain extent, especially given latitudinal correlates with global temperature (Roseman and Auerbach 2015). The same could be true for other environmental factors, such as dietary practices, given the ecological link between climatic conditions and net primary productivity (e.g. Eriksson et al. 2012). In turn, genetic diversity patterns can be strongly correlated with phenotypic diversity patterns, under a largely neutral model of phenotypic diversification. Indeed, this relationship could, in theory, be causative, if the genomic system included the polygenes underlying the particular phenotypic in question. In addition, environmental factors, such as climate or dietary behavior, could be causally influencing patterns of phenotypic diversity due to diversifying selection acting on some or all populations. What Figure 7.1 emphasizes, therefore, is that these biological, environmental, and spatial factors must be considered in a single integrated framework, with any perceived deterministic relationship between selective forces and phenotypic variation being carefully examined in light of possible underlying geographically mediated correlates.

Table 7.1 also shows that the matched microsatellite genetic distance matrix is positively correlated with geographic distance and the historical divergence matrix, emphasizing the geographically mediated processes shown in Figure 7.1. In this particular example, the environmental factor under consideration – temperature – is not significantly correlated with either genetic or craniometric distance, so there is no need to investigate this relationship further. However, if this were a significant relationship, the next analytical step might be to use partial Mantel tests to examine whether the relationship between temperature and craniometric distance remains significant once genetic distance (or geographic distance) has been controlled for. Thereafter, following Roseman (2004), one might wish to use a jackknifing procedure and remove each population in turn to investigate whether this relationship is being determined by just one or a few "extreme" climatically adapted populations.

de Campos Telles and Diniz-Filho (2005) employed a method derived from Legendre and Legendre (1998) to further deconstruct the relationship found between genetic distance and geographic distance, by investigating what proportion of this relationship could be attributed to long-term historical divergence as opposed to more recent gene flow. As shown in Figure 7.1, they argued that if divergence patterns among lineages (using a bifurcating tree model) were also geographically structured, then this could result in the spurious conclusion that the patterns found were simply the result of isolation-by-distance (IBD) due to spatially mediated gene flow.

The same might be true in the cranial case study described here, given that both the genetic and craniometric distance matrices are correlated with the historical divergence model and geographic distance. Following de Campos Telles and Diniz-Filho (2005), the percentage of craniometric (or genetic) among-group distance due to different factors can be apportioned as:

a = "pure" historical divergence
b = "pure" IBD
c = overlapping variance due to both a + b
d = unexplained variance

where

$$c = (a + c) + (b + c) - (a + b + c); \text{ and}$$
$$d = 1 - (a + b + c)$$

(a + b) is equal to the R^2 of the Mantel test between craniometrics (or genetics) and "History"; (b + c) is equal to the R^2 of the test between the biological matrices and "Geography." The term (a + b + c) is found by a multiple regression Mantel test combining both "History" and "Geography" as predictor matrices. Multiple Mantel tests were conducted in *R* using the *multi.mantel* function from the *phytools* package, implementing 9999 permutations.

The results for the craniometric data were as follows:

$c = R^2_H (0.270) + R^2_G (0.621) - R^2_{HG} (0.656 [p = 0.0001])$
$c = 0.235$
Pure history (a) = 0.035 = 3.5%
Pure IBD (b) = 0.386 = 38.6%
Overlap Hist*IBD (c) = 0.235 = 23.5%
Unexplained (d) = 34.4%

The results for the matched microsatellite data were as follows:

$c = R^2_H (0.426) + R^2_G (0.230) - R^2_{HG} (0.470 [p = 0.0062])$
$c = 0.186$
Pure history (a) = 0.240 = 24.0%
Pure IBD (b) = 0.044 = 4.4%
Overlap Hist*IBD (c) = 0.186 = 18.6%
Unexplained (d) = 53.0%

These results suggest that the matched craniometric and microsatellite datasets differ in terms of the proportion of their variance that can be attributed to the bifurcating model shown in Figure 7.1, versus that attributable to gene flow mediated by geographic distance. The microsatellite data are better explained using the model of historical divergence, with relatively little additional variation explained by gene flow patterns, while the opposite is true of the craniometric data matrix. It is worth pointing out that the overlap between these two processes also accounted for ~20% of the overall variation in the case of both biological datasets. The majority of the overall pattern of craniometric variation was explained by these two geographically mediated processes, leaving only ~35% variation unaccounted for. In contrast, over 50% of the variation in microsatellite diversity remained unexplained by the two geographically mediated processes.

The results are likely to be affected by several methodological issues, such as the construction of the historical divergence model and uncertainties in the quality of the matches made between the populations from which microsatellite frequency data and craniometric data were collated. Nevertheless, the results suggest that geographically mediated processes such as population divergences, long-range migrations, and gene flow may affect different biological systems differently. This is despite the fact that in the case of modern human craniometric data, the overall fit to a neutral microevolutionary pattern of expectation is relatively strong. This is again emphasized here by the relatively strong correlation ($r \sim 0.5$) between craniometric distance and both genetic distance and historical divergence distance.

In any case, the results emphasize the importance of probing the association between biological, geographic, and environmental variation using different methodological approaches. The case study presented here showcases how Mantel tests may be used to investigate such associations, and highlights the need to pay attention not only to the significance values obtained, but also to the relative explanatory power of these different potentially causative sources of variation in an internally consistent manner.

Conclusions

Mantel's (1967) matrix tests and their derivatives have proven a useful tool for evolutionary morphologists interested in the relative congruence of phenotypic, genetic, and environmental patterns of among-population variation. The empirical examples given here have revolved around modern human craniometric variation, but these methods are now also being increasingly applied to questions concerning the relative neutrality and signals of adaptation in postcranial data (e.g. Betti et al. 2013, 2014; von Cramon-Taubadel et al. 2013; Roseman and Auerbach 2015). Also, researchers are increasingly combining matrix comparison methods with more traditional tree comparison methods to assess phylogenetic hypotheses at higher taxonomic levels (e.g. Cardini and Elton 2008; von Cramon-Taubadel and Smith 2012). Little attention was paid here to the use of design or model matrices as a means of testing how well phenotypic and genotypic diversity patterns fit competing hypotheses of migration and dispersal (see e.g. Waddle 1994; Pinhasi and von Cramon-Taubadel 2009; Hubbe et al. 2010; von Cramon-Taubadel and Pinhasi 2011; Reyes-Centeno et al. 2014). These studies illustrate the flexibility of the Mantel testing framework to include hypothetical distance matrices that quantify prehistoric population movements and dispersals. Moreover, it is relatively easy to integrate the dimension of time (temporal distance) (Konigsberg 1990; Pinhasi and von Cramon-Taubadel 2009) by using such an approach, such that changes in among-population biodistance patterns can be investigated in both space and time.

Despite all of the obvious advantages of employing a matrix comparison framework, as noted in this chapter, it has not been without its critics. Some researchers employ Bayesian inference methods as alternatives to Mantel tests as a means of assessing the effect of environmental factors on phenotypic variation, while accounting for spatial autocorrelation and/or population history (e.g. Betti et al. 2014; Roseman and Auerbach 2015; Katz et al. 2016). Another issue is the difficulty in separating out the geographically mediated processes of gene flow from a hierarchical model of shared ancestry. The simple case study presented here suggests a matrix comparison approach to address this problem. While an empirical fit between biodistance patterns and geography confirms that one or more of the processes of long-range migration, serial founder effects, and gene flow have been at play, it does not, by itself, shed light on the degree to which these different processes have been important in structuring biological variance patterns. Modeling methods, such as generalized hierarchical modeling and coalescent simulations, have also been used to assess the relative fit of biological data to these different hypothetical scenarios (e.g. Hunley et al. 2009, 2016; Roseman 2016). Ultimately, it is important to recognize that the use of Mantel tests and their derivatives is but one form of analytical tool in the empiricist's toolbox. All anthropological hypotheses require testing using a range of study designs, employing different statistical and methodological procedures, and being applied to as wide a range of datasets as possible. In the case of the microevolutionary history of modern human craniometric variation, Mantel matrix methods have proven a versatile approach, yielding internally and externally consistent results across a range of studies.

CHAPTER 8

Simulating Human Demographic Processes and Patterns of Genetic Variation

Aida T. Miró-Herrans

Department of Anthropology, University of Texas at Austin, Austin, TX, USA

Human evolutionary history has been characterized by a series of population growths, contractions, splits, and migrations. Each of these events has had an effect on patterns of human genetic variation. Mutations in the DNA of individuals increase the genetic variation of a population as, over generations, mutations accumulate in the population. However, different demographic processes have varying effects on the fixation or extinction of mutations. For example, population growth increases genetic variation in a population, with each new individual increasing the opportunity for mutation and fixation. In contrast, population contractions and splits tend to decrease genetic variation, as mutations can be lost from a population when individuals die or leave the population. Population splits can further reduce variation through founder effects, where the individuals who leave a population carry only a subset of the genetic diversity found in the original population. Over time, the original population and the new population accumulate mutations independently, increasing genetic difference between the two populations. This difference can be further increased by subsequent population growth or contraction. In contrast, migration can reduce the genetic difference between populations through the exchange of genetic variants (gene flow), while

A Companion to Anthropological Genetics, First Edition. Edited by Dennis H. O'Rourke.

increasing the genetic diversity within each population. The individual effects of these simple demographic processes on genetic variation are well understood and can be accurately inferred. However, the joint effects of these processes over time, or when more than one occur at the same time, are less clear.

Humans have dispersed into almost every region of the world, producing a complex evolutionary history for our species. Anatomically modern humans' appearance and persistence in Africa, our global expansion, subsequent local migrations, and gene flow between modern and archaic hominins have led to multiple and diverse demographic processes in different geographic regions. Each of these complex scenarios has generated diverse patterns of genetic variation in current populations.

Genetic anthropologists examine and interpret patterns of genetic variation within and between populations to reconstruct the demographic processes that have characterized human evolutionary history. Differences in the number and location of mutations between individuals and between populations are used to understand their relationship and to estimate parameters related to their history, such as the timing of a population split or levels of gene flow between populations. Inferences of past processes from current genetic variation can be challenging because different demographic scenarios can produce similar patterns of genetic variation and because more recent processes can obscure genetic patterns from older events. Genetic simulations provide a tool that can overcome some of these challenges and improve demographic reconstruction.

Generally, DNA simulations generate synthetic genetic data with patterns of genetic variation expected under a population genetics framework, such as coalescent theory (Kingman 1982). Programs that simulate genetic variation simplify complex scenarios into a set of parameters that describe "simple" and better understood demographic processes. For example, a migration can be defined by the time of population split and the proportion of the population that is moving. Thus, the evolutionary history of a population (or populations) can be expressed as a model for which genetic data can be simulated. A single demographic model can include many demographic processes over a specific time frame, allowing the explicit modeling of genetic variation for both older and more recent population events. By jointly modeling the different demographic processes over time, we can gain an understanding of the effects of these processes on current genetic variation and identify the extent to which recent events mask older genetic variation and limit our ability to reconstruct older processes.

Furthermore, a single model can be represented by different demographic scenarios, where the same set of model parameters has different values for each scenario. Comparisons of simulated DNA generated under varying demographic scenarios offer the possibility of more broadly understanding how each demographic parameter contributes to the genetic variation of a particular model. Under a simplified model for the migration of modern humans out of Africa, for example, I simulated and compared 42 demographic scenarios, each including different values for combinations of three different parameters: time of colonization, colonization size, and rate of subsequent gene flow between the African and non-African populations (Miró-Herrans and Mulligan 2013). These comparisons showed that for the migration of modern humans out of Africa, colonization size and rate of gene flow explained most of the mitochondrial genetic variation. When these parameters had large values, many demographic scenarios generated similar patterns of genetic diversity, and demographic scenarios with a colonization time of 50 000 years ago (kya) were not distinguishable from those of 100 kya. These findings support that subsequent gene flow has reduced our ability to detect the genetic signal of older events from mitochondrial DNA (mtDNA). These results suggest that parameters that explain a large portion of the genetic variation can reduce the genetic signal for parameters that explain less of the genetic variation, particularly if the larger impact parameters represent more recent events.

DNA simulations can be combined with statistical methods of inference to reconstruct demographic histories. Simulated DNA from different models can be tested against observed genetic data to identify which model best explains the observed patterns of current genetic variation. These model-based methods of statistical inference improve demographic reconstruction by allowing explicit comparisons of competing hypotheses. Furthermore, parameter values can be estimated from simulated data and further refine an evolutionary history by defining, for example, when an event occurred or the extent of a population bottleneck.

In this chapter, I explain how genetic simulations can be used to make statistical inferences about demographic histories and help us understand human evolutionary processes. While there are many approaches and programs to implement these model-based methods of statistical inference, I specifically discuss the advantages and challenges of approximate Bayesian computation (ABC). ABC has gained popularity in the last decade because it offers flexibility to test complex models with the convenience of summary statistics. I briefly describe the rationale of ABC, but focus on practical considerations for its application by using examples from a simplified model for modern humans' migration out of Africa. I also describe different approaches in which DNA simulations and ABC can be applied to reconstruct demographic histories, and discuss new directions in which simulations are contributing to our understanding of human evolutionary history. The intention is that this chapter may serve as a guide for researchers interested in implementing DNA simulations in their own research.

Simulations and Model-Based Statistical Inference for Human Demographic Reconstruction

DNA simulations have long been used to validate methods developed for genetic analyses. More recently, however, simulations have begun to be incorporated as part of the actual genetic analyses. Model-based methods for demographic reconstruction compare simulated genetic data generated under a specific demographic model to observed genetic data to identify how well a model explains the observed data (Marjoram and Tavaré 2006).

To achieve this, each demographic model is expressed as a set of parameters. For example, a population split can be described by a parameter for the time of the split and a parameter for the proportional size of the "new" population. Each parameter is defined by a single value or a range of values specifying the magnitude of the parameter. In the out-of-Africa model, the "new" population migrating out of Africa could be defined with values from 1 to 33% of the original population size (Tenesa et al. 2007; Atkinson et al. 2008; Deshpande et al. 2009). Different models can be developed by varying parameter combinations. Simulations from competing models can then be tested against the observed genetic data to identify which model generates data that most closely resemble the observed data and, therefore, which demographic history best explains the genetic data. This approach also offers the flexibility to test multiple values for the same combination of parameters and estimate parameter values from the model that best explains the genetic data.

Model-based methods present multiple advantages over traditional approaches in which individual population parameters are estimated and interpreted to infer a demographic model that is consistent with other sources of evidence (e.g. archeological or geological data). For example, one advantage of these model-based methods is that they allow the inclusion of data from multiple disciplines. Data from genetic or nongenetic sources, such as archeological evidence, can be directly incorporated to define the parameter values for each model. In this way, prior knowledge is explicitly incorporated, eliminating the need to find post-hoc explanations for the observed genetic data that are consistent with the other sources of evidence.

Incorporating prior information also offers the possibility of developing and comparing plausible alternative models. This ability to test competing hypotheses represents an additional advantage when reconstructing demographic processes because different parameter combinations can cause similar effects on the genetic variation of a population. For example, population growth and gene flow can both increase genetic diversity within a population, and both events could be consistent with high measures of genetic diversity and archeological evidence of technological change. Testing competing models that differ only in this event is therefore an important step to identify which process(es) most likely occurred. Jointly estimating demographic parameters given a specific model can also further refine our knowledge of the process.

How well a model explains the genetic data depends on the fit of the model (i.e. how closely the simulated data resemble the observed genetic data). Different models can be compared to determine which has the overall best fit to the genetic data using statistical model selection approaches, such as a likelihood ratio test (LRT) or Aikaike information criterion (AIC; Aikaike 1974). Parameter values can then be estimated for the best-fit model from the values of the simulations that are most consistent with the observed data. Multiple approaches have been developed to compare simulated and empirical data, differing primarily in how the simulated data are represented. For example, the isolation with migration (IM) model simulates different genealogies under a coalescent model and compares them to the observed genealogy (Hey 2005, 2010; Hey and Nielsen 2007). The program ∂a∂I uses a diffusion approximation model to simulate allele frequency spectra (AFS; i.e. the distribution of alleles in a population) and compares them to the observed AFS (Gutenkunst et al. 2009). ABC generally relies on a coalescent model to simulate genealogies and requires the genetic diversity of the genealogies to be represented through summary statistics (Beaumont et al. 2002). This allows comparisons of the simulated and observed data to be performed with much less computational requirements.

ABC builds on previous algorithms that find the smallest difference between the simulated and empirical data (Tavaré et al. 1997; Weiss and von Haeseler 1998; Pritchard et al. 1999). The foundations of ABC consist of subtracting the summary statistic calculated for each simulated dataset (s') from the summary statistic calculated for the empirical dataset (s) (Pritchard et al. 1999). The simulated datasets that have summary statistics with the smallest distance (i.e. that have the smallest absolute difference after being subtracted) represent the model that best explains the empirical data. Specifically, a cutoff value $\boldsymbol{\delta}$ is selected to determine the simulated datasets that will be used to select the best model (i.e. datasets are accepted when $||s' - s|| \leq \delta$). The model with the most support from the simulated datasets within the cutoff value represents the best-fit model. These accepted values can be used to estimate demographic parameters because this posterior distribution is assumed to contain the values that best represent each parameter in the empirical data. ABC expands on this algorithm with two main modifications (Beaumont et al. 2002). The first modification improves on simulation acceptance by weighing the simulated datasets based on the value of $||s' - s||$ and giving more weight to datasets with smaller differences. Second, instead of calculating the absolute distance to select the simulated datasets, ABC performs local-linear regression to calculate the distance, which improves parameter estimation.

Performing Simulations for ABC

In practice, ABC requires that each model is represented by thousands of simulated datasets. Each simulated dataset reflects both the pattern of genetic variation generated by the specified demographic model and the stochasticity of the mutation process. Therefore, independent datasets simulated under a specific model each have a slightly different pattern of genetic variation that is still representative of the model.

A single model can include multiple parameters where each parameter has specific values that reflect knowledge from other sources. For example, in our simplified out-of-Africa model, I included specific values for three parameters (the colonizing population size, the time of colonization, and the rate of gene flow between the African and non-African populations after colonization) to describe 42 distinct demographic scenarios. The 42 scenarios included parameter combinations with three values for colonization size (1, 10, and 30% of the African population), two values for time of colonization (50 000 and 100 000 years ago), and seven values for rate of gene flow (10^{-6}, 10^{-5}, 10^{-4}, 10^{-3}, 10^{-2}, 0.1, and 0.5 proportion of migrants).

Alternatively, a parameter can be defined by a range of values, where the specific values can be randomly selected from a predetermined distribution of values and a few simulations can be performed for each selected value. This allows a model to be developed that includes some uncertainty in the values of specific parameters. In these cases, thousands of simulations must be performed, sampling across the distribution of values to account for the uncertainty. As the number of parameters defined by prior distributions increases, so does the required number of simulations, leading to millions of simulated datasets required per model. In performing ABC, the goal is to strike a balance between the complexity of the model (i.e. the explored parameter space) and maintaining a feasible number of required simulations.

The DNA simulations can be generated based on different population genetic frameworks. Two of the more commonly used simulation frameworks are coalescent and forward-time simulations. Coalescent simulations are a backward-time simulation approach that involves tracing the genealogy of alleles in the current population back to the most recent common ancestor of all alleles (Kingman 1982). Forward-time simulations trace the genealogy of all alleles (even those that don't persist to the present) from a given point in the past (Carvajal-Rodríguez 2010). Forward-time simulations are useful for developing models that incorporate alleles that are no longer present in the population under study, but these simulations come at a cost of increased computational power and time.

Programs that perform DNA simulations generally generate DNA sequence data. The genetic variation among the sequence data must then be summarized into numbers that can be included in the ABC analyses. Summary statistics provide a quick and easy way to calculate values that summarize these patterns of genetic variation, and multiple summary statistics can be calculated for each simulated dataset. The number and types of summary statistics considered can have important effects on ABC analyses (discussed further in this chapter).

Incorporating Data into ABC

ABC analyses allow the inclusion of different types of data to help researchers make inferences about demographic processes. Specifically, in addition to the genetic data used directly in ABC analyses, nongenetic data can be included to help develop models and improve inferences. Incorporating knowledge from other sources of data (e.g. archeology and demography) to define some of the parameter values, and even ranges of values, becomes important to reduce uncertainty, delimit the explored parameter space, and limit the number of needed simulations.

Incorporating data from other sources (e.g. demography) can make it possible to set fixed values for some specific parameters across different models. This can be useful to define parameters that are necessary for the model, but may not be of interest in reconstructing a demographic process. For example, a population's average generation length is important in defining time between population events, but may not be a parameter of interest in a model. Fixing the same value (e.g. 25 years per generation) across all models frees the parameter space to explore values for other parameters of interest.

Sometimes little is known about an event, such that there is no prior information about the specific value of a parameter. However, data may exist that can help limit the range of plausible values. For example, timing for the dispersal of modern humans out of Africa can be delimited by combining the dates of archeological remains out of Africa and the divergence dates of African and non-African haplotypes estimated from previous genetic studies. Archeological evidence suggests that the successful migration of modern humans out of Africa happened at least 120 kya (Grün and Stringer 1991; Mercier et al. 1993; Liu et al. 2015). Analyses of mtDNA suggest that the L3 haplogroup, which is ancestral to all non-African haplogroups, diverged between 60 and 90 kya (Forster 2004; Gonder et al. 2007; Soares et al. 2012), and the most ancestral non-African haplogroups diverged as late as 50–65 kya (Macaulay et al. 2005; Atkinson et al. 2008; Behar et al. 2008). These data narrow the time during which modern humans could have migrated from Africa to a range of about 60–120 kya. Defining ranges for parameters like this allows the analysis to explore values for a particular parameter of interest while narrowing the parameter space to some degree.

In some cases, it can be useful to define the prior distribution of values around point estimates obtained from related fields of study. For example, in Miró-Herrans et al. (2014), we posited that migration parameters estimated from the dispersal patterns of a contemporary population with reduced technology and reduced mobility might provide values that could inform parameters for ancient migrations. Our demographic data showed that individuals four generations ago (~60–80 years ago) moved significantly less than more recent generations, so we estimated migration parameters from this older generation. We found that 10% of this older generation migrated to a new location, suggesting that this value may represent a realistic estimate of the proportion of migrants in a colonization event. Our results also supported 0.10% as the percentage of individuals moving between the same two locations, which could represent the proportion of migrants exchanged during gene flow between populations. Narrowing priors around a point estimate like this increase the likeliness that realistic values are included in the simulations.

Although ABC is robust to including parameters with wide uninformed ranges of values, the inclusion of any informed prior data to reduce the parameter space will help improve inference. Including narrow informative priors allows testing many "likely" values for a specific parameter. In this way, informed priors can increase the probability of developing models that will generate simulated data similar to the observed data.

Selection of Summary Statistics

Genetic summary statistics provide numeric values that summarize genetic diversity and are used to infer the historic relationship within and between populations (Holsinger and Weir 2009). For example, F_{st} measures the proportion of genetic diversity due to allele differences between populations (Wright 1951), with large F_{st} values indicating that populations are highly differentiated. When reconstructing demographic histories, high F_{st} values can be interpreted as indicating little gene flow between populations. Such summary statistics can be used in ABC analyses, as they provide an easy way of comparing genetic variation in empirical and simulated datasets and can facilitate inferences about demographic processes.

The quantity and types of summary statistics included in ABC analyses can have an important effect on model selection and parameter estimation (Csillėry et al. 2010). Including a large number of summary statistics can lead to so few simulated datasets being accepted within the cutoff threshold ($\boldsymbol{\delta}$) that it may become impossible to make any conclusions about the empirical data, or it may be necessary to increase the cutoff threshold, which reduces the accuracy of the inferences (Beaumont et al. 2002). Additionally, it is important to note that different summary statistics summarize different proportions and aspects of the genetic variation (Hamilton 2005; Hickerson et al. 2006). The challenge in ABC of simultaneously estimating multiple parameters (or selecting a model) is that a summary statistic may be effective to estimate a specific parameter, but work poorly for another parameter. For example, F_{st} is informative for estimates of gene flow, but less so for estimates of colonization size (Miró-Herrans and Mulligan 2013). Thus, the optimal set of summary statistics depends on the specific evolutionary model and parameters of interest, and may be different for each model or study.

It is therefore critical to identify the most informative summary statistics for statistical inferences in model testing and parameter estimation. Within a single model, summary statistics may be selected based on their utility for estimating the parameters of interest. For example, in our out-of-Africa model, we tested 12 summary statistics to identify the most informative summary statistics for making inferences about the three main demographic parameters (colonization size, gene flow between the African and non-African populations, and time of colonization). The optimal summary statistics were selected by identifying the statistics that explained the highest percentage of variation for a given demographic parameter and that showed statistically significant differences among the 42 parameter combinations tested (in pairwise *t*-tests). F_{st} and Tajima's D were the most informative summary statistics for this out-of-Africa model.

The main challenge in selecting summary statistics for model selection lies in identifying the smallest number of summary statistics that capture the largest proportion of genetic variation across all tested models in an ABC analysis. With ABC becoming more widely used, more researchers are focusing on developing methods to identify optimal summary statistics. Various methods have been proposed for selecting optimal summary statistics that include statistical approaches, such as reducing the information of multiple summary statistics into a few components (Wegmann et al. 2009; Blum et al. 2013; Prangle et al. 2014). Other methods use mathematical algorithms (Joyce and Marjoram 2008; Nunes and Balding 2010) and machine learning techniques (Aeschbacher et al. 2012; Barnes et al. 2012).

DNA Simulations for Hypothesis Testing

Genetic simulations and statistical inference can be used in a variety of ways to reconstruct demographic processes. In one approach, genetic data are simulated under a null demographic model. The probability of observing the empirical data, given the distribution of simulated data, is calculated by testing whether the empirical value of the summary statistic falls within the range of values for the summary statistic simulated under the null model. If the empirical value is statistically different from the simulated values, then the null model does not explain the genetic data and is rejected. This approach is useful for testing a model's ability to explain genetic data when alternative models are not included. For example, Mata-Miguez et al. (2012) used simulations to test whether the genetic difference (F_{st}) between two temporally separated populations in Xaltocan, Mexico, was greater than expected, given a null model of mutation and genetic drift alone. Through a comparison of the empirical F_{st} value with the distribution of simulated F_{st} values, they rejected the null model of genetic continuity between the populations.

Simulations can also be used to test a null model against an alternative model, where the alternative model represents a new hypothesis to explain the observed genetic data. For example, Vernot and Akey (2015) used this approach to identify a demographic model that explains the higher percentage of Neanderthal ancestry in East Asians than in Europeans. Their null model consisted of a single pulse of Neanderthal introgression into the common ancestor of East Asians and Europeans, while their alternative model added a second pulse of additional Neanderthal introgression into either East Asians or Europeans after the populations split. After performing ABC, only simulations under the two-pulse model were accepted, thus rejecting the one-pulse (null) model. Additionally, they compared the null model with another plausible scenario, a single pulse of Neanderthal introgression followed by admixture between Europeans and a third modern human population that had never interbred with Neanderthals, representing a dilution of Neanderthal ancestry in Europeans. In the second test, the alternative model was also accepted. By testing the null model against different plausible scenarios, they effectively rejected a single-pulse model as an explanation for the observed patterns of genetic variation.

Another common use for ABC is to test multiple alternative models to identify which model best explains the observed genetic variation. For example, Ghirotto et al. (2013) tested three models to determine whether modern Tuscans are descendants of ancient Etruscans. They tested all three models on each of four modern populations from Tuscany and found that two populations had high support for a model of population continuity between Etruscans and Tuscans, while the other two populations had high support for a model of population differentiation from the Etruscans. In this case, there were two different best-fit models for the four different populations, indicating that Etruscan ancestry is present in isolated localities.

Genetic simulations and ABC are also used to estimate values for parameters of interest. This can be achieved by estimating the parameters from the posterior distribution of the best-fit model. Specifically, the posterior distribution of a parameter is used to generate a new set of simulations, and ABC is performed on this dataset to identify the values of the parameter that best explain the genetic data. A point estimator, such as the mean or the mode, and the associated confidence interval are calculated from this new posterior distribution. For example, Nater et al. (2015) used this approach to infer the evolutionary history of orangutans. They tested increasingly complex models describing the demographic history of orangutans in Borneo and Sumatra through ABC analyses. Upon identifying a 10-population model as best describing the orangutan genetic variation, they performed 10 million more simulations under this model to estimate parameters. These parameter estimates further refined the demographic model to show that Sumatran orangutans exhibited a deep and temporally stable population structure before experiencing a strong recent (~24 kya) decline. In contrast, Borneo orangutan population structure seems to have been established more recently (~80 kya), probably due to a population decline and restructuring that occurred at that time, associated with the emergence of a glacial refugium.

ABC can also be used to validate results obtained through other analyses. Scliar et al. (2014), for example, performed ABC analyses to validate parameter estimates they obtained using the IM program. IM uses the full genetic data to make inferences, instead of summary statistics, at the cost of less flexible models (in this case, a model that didn't include intralocus recombination). Scliar et al. performed ABC analyses for models with and without recombination to estimate the same parameters obtained with IM. They generated similar results from the ABC analyses, indicating that recombination had little effect on their IM estimates.

Although ABC constitutes a powerful tool for studying human evolutionary history, researchers should be aware of various challenges that need to be addressed. Importantly, as noted in this chapter, different demographic processes can lead to similar patterns of genetic variation, such that different models may produce similar simulated data. Second, summary statistics can produce a biased representation of the genetic variation of a model depending on the summary statistics that are used and how they capture genetic variation. Third, loss of information can occur when complex patterns of genetic variation are reduced to a set of numbers (summary statistics). This loss of information can also bias the selection of accepted simulations. Altogether, these factors can lead to overlap in the posterior distributions of different models, making it difficult to identify the best model (Robert et al. 2011).

However, there are various ways to address these challenges to improve inference with ABC. As mentioned in this chapter, selecting optimal summary statistics that capture the greatest percentage of genetic variation can reduce some bias. Different approaches can also be included to validate the model selection and parameter estimation processes. For example, one method of model selection validation involves identifying if the tested models generate distinguishable differences in genetic variation (i.e. do they generate summary statistics that are significantly different in statistical comparisons?). This can be achieved by simulating a subset of data from each model and treating it as empirical data to perform ABC (Wegmann et al. 2010). Because the model and corresponding parameter values are known, these pseudo-observed datasets can be used to identify how often the ABC procedure misclassifies the model that generated the data. Another method of model selection validation is to test if the best-fitting model can reproduce the observed genetic data (Wegmann et al. 2010). In this approach, parameter values from the posterior distribution are used to generate a subset of simulations under the best-fit model. The summary statistics for the empirical data are checked to see if they fall within the distribution of these simulated summary statistics.

Other considerations to keep in mind when using ABC to reconstruct demographic processes include striking a balance between the amount of genetic data and the number of parameters to be estimated in a model. More genetic markers are necessary to estimate parameters that explain a smaller proportion of the genetic variation, such as the time of migration in our out-of-Africa model (Miró-Herrans and Mulligan 2013). Also, while there are many programs that generate simulated genetic data and many others that calculate summary statistics (see Hoban et al. 2011 for a comprehensive list of programs and capabilities), the best choice will depend on the type of empirical data that is being tested. Multiple programs and R packages have been developed to implement ABC approaches (Csilléry et al. 2010; Wegmann et al. 2010; Jabot et al. 2013; Nunes and Prangle 2015), many of which can calculate some summary statistics as part of the analysis pipeline.

Moving Forward: DNA Simulations beyond Demography

Both environmental adaptations and sociocultural practices have played important roles in human evolutionary history. As computational capabilities have increased and performing simulations has become quicker and easier, simulations can now be used to develop more complex models of human evolutionary history that include factors such as natural selection, the role of geography, and sex-specific demographic history. Notably, natural selection and demographic events can generate similar patterns of genetic variation and can occur simultaneously, making it difficult to disentangle the effects of the two on the genome (Li et al. 2012). One increasingly common approach to test for selection and account for demographic events involves identifying the "best" demographic model for explaining the overall patterns of genetic variation, and then using that model as a null model to test for signals of selection (Andres et al. 2009). Alternatively, an increasing number of simulation programs are allowing demographic and selection events to be jointly modeled (Hernandez 2008; Ewing and Hermisson 2010; Messer 2013). This development allows models of selection and selection-related parameters, such as the selection coefficient, to be tested under a given demographic scenario using ABC (e.g. Peter et al. 2012). Additionally, forward-in-time simulations can be particularly useful when ancient DNA can be used to represent the ancestral genetic variation, and selective processes can then be modeled into the present (e.g. Wilde et al. 2014).

Furthermore, the environment also causes ecological effects on demographic processes. For example, a geographic barrier may limit the amount of gene flow that can occur between two populations. Similarly, an environment's carrying capacity limits how large a population can become and may affect the rate at which new colonizations occur in search of environmental resources. Spatially explicit modeling incorporates environmental heterogeneity and allows us to study ecological effects on genetic variation, such as geographic limitations to gene flow or colonization rate. For example, SPLATCHE 2 (Ray et al. 2010) allows a cost to be specified across geographic boundaries to define the probability of a colonization event or gene flow occurring through a specific area. Alternative models with different geographic costs and distances can allow different hypothesized migration routes to be tested. Spatially explicit models that incorporate the effects of geography on genetic variation therefore allow us to improve our understanding of human interactions with the environment.

Genetic simulations are also being used to improve our understanding of human behaviors that may leave a mark on genetic variation. Simulations of Y chromosome (YC) and mtDNA data have been performed to represent sex-biased demographic processes, such as postmarital residence practices (e.g. Rasteiro et al. 2012). Notably, although sex-biased dispersals have generally been studied through YC and mtDNA, these haploid systems' patterns often do not reflect the processes experienced by autosomal DNA. In contrast, the X chromosome (XC) undergoes recombination like autosomal DNA, and comparisons of XC and autosomal diversity have been shown to reflect sex-biased demography (Ramachandran et al. 2004; Hammer et al. 2008; Keinan et al. 2009; Heyer et al. 2012). Recent developments of a framework that models varying female and male genetic contributions over multiple generations (Goldberg et al. 2014; Goldberg and Rosenberg 2015) may soon allow simulations of XC and autosomes to provide insight into local sex-specific processes.

Thus, with the increasing availability of genome-wide data and computational power, genetic simulations and model-based methods of statistical inference will allow a finer scale view of human evolutionary processes. Importantly, with these methods, integrating data from nongenetic sources will be critical to make accurate inferences from genetic data and to improve our interpretations of ancient events. Interdisciplinary approaches will be essential as we continue to move forward in disentangling human evolutionary history. The interdisciplinary nature of genetic anthropologists places us in an ideal position to take advantage of these approaches.

ACKNOWLEDGMENTS

Thanks to Deborah Bolnick and members of the Bolnick lab for insightful comments throughout the writing of this chapter.

PART III Microevolution: Colonization, Dispersal, and Differentiation

CHAPTER 9

Human Population Structure and History

John H. Relethford

Department of Anthropology, State University of New York at Oneonta, Oneonta, NY, USA

Introduction

Two major areas of research in anthropological genetics are population structure and population history. Both of these fields deal with analysis of the patterns of genetic variation within and among populations, with a scale of analysis that ranges from small, local populations (or subpopulations) to the entire human species. Defining the study of population structure and population history is sometimes tricky because many studies combine aspects of both approaches to understanding genetic variation.

Typically, the term *population structure* has been used in connection with a variety of approaches to human population dynamics, ranging from demographic to genetic. In population genetics, the study of population structure (often specifically referred to as the *genetic structure of populations*) has often been defined in terms of deviations from panmixia (random mating) in a population of infinite size, and the effect of these deviations on allele and genotype frequencies (e.g. Yasuda and Morton 1967; Morton 1969, 1973a; Jorde 1980). One obvious difference from a theoretical state of panmixia is inbreeding. Another deviation concerns population size; in the Hardy–Weinberg equilibrium model, population size is assumed to be infinite. Finite population size leads to both genetic drift (causing a change in both allele and genotype frequencies)

A Companion to Anthropological Genetics, First Edition. Edited by Dennis H. O'Rourke.

and increased mating between individuals who are related, even if mate choice is random with respect to consanguinity. Random inbreeding will not change allele frequencies, but it will change genotype frequencies. The effects of random inbreeding can be increased or decreased by cultural preferences (or avoidances) of close consanguineous mating (what is often referred to as ***nonrandom inbreeding***).

In addition to drift and inbreeding, another focus of models and studies of population structure concerns the division of a population into a set of subpopulations rather than a panmictic total population. This subdivision is often a function of geography (people more often choose mates from nearby than from far away), but it can also occur due to differences in language, religion, ethnicity, or any number of cultural barriers to mating and gene flow. In addition to studying genetic variation within subdivided populations, the study of population structure examines genetic relationships *among* populations, often in terms of the impact of geographic, ecological, demographic, and cultural influences on population subdivision and gene flow. A major focus of many studies of subdivided human populations has been on the impact of geographic distance on genetic variation (Jorde 1980), where individuals are more likely to choose mates from close by than from farther away, and hence demonstrate genetic similarity.

From the late 1960s through the early 1980s, the study of human population structure was perhaps the central focus of anthropological genetics, with numerous studies examining the genetic structure of local populations in a variety of ecological, demographic, and cultural settings (e.g. Crawford and Mielke 1982). By the late 1980s, much of the focus in anthropological genetics had shifted to the study of the history of human populations, with a focus on origins and migrations. As with population structure, the study of population history is concerned with patterns of genetic variation among individuals and populations. Whereas population structure is concerned more with the genetic impact of mate choice and population subdivision, population history deals more with the impact of historical events, such as population origins and migration between populations, on patterns of genetic distance among populations (Relethford 2012). Again, distinguishing between studies of population structure and history can be difficult at times, as many studies incorporate aspects of both. For convenience, studies described in this chapter will be categorized depending on whether the primary objective is an examination of population composition or population origins.

Today, most studies of population structure and history use DNA markers. There are many analyses that have been made possible that were not available in the past. However, it is important to realize that important insights have also come from (and continue to come from) other sources of data, including classical genetic markers, historical demographic data, surnames, and metric traits, among others.

Studies of Human Population Structure

There are so many models and studies of human population structure that it is not possible to sample more than a small number in a single chapter. The focus here is on illustrating some examples of the types of models and studies.

Analyzing mate choice

An obvious interest in anthropological genetics is the biocultural aspects of mate choice, looking to determine the effect of cultural and demographic influences on mate choice and the subsequent genetic impact. Levels of inbreeding in human populations are often determined from genealogical analysis on both contemporary and historical populations. Surname analysis has also played an important role in studies of human inbreeding, developing from the observations of Crow and Mange (1965) that the frequency of marriages between spouses having the same surname (isonymy) is often directly proportional to the level of inbreeding expected for their children.

An interesting aspect of studies of mate choice is the partitioning of random and nonrandom components of inbreeding. Random inbreeding is the level of inbreeding expected in a population of finite size even if mate choice were random with respect to the level of consanguinity. In a small population, most available mates might be related to each other, generating a certain level of inbreeding. If we were to take the level of inbreeding as an indication of a preference for choosing a close relative as a mate, we might be in error, as some level of inbreeding would be expected by chance. In general, the smaller the population, the higher the level of random inbreeding would be. Nonrandom inbreeding, however, would be a reflection of mate choice based on consanguinity, where a positive value suggests a preference for consanguineous mating and a negative value suggests the avoidance of such matings. Crow and Mange (1965) developed a method of separating the total amount of inbreeding into random and nonrandom components based on the surname distributions of spouses. A more detailed analysis of mate choice is possible through potential mates analysis, where computer simulation is used to compare observed patterns of mate choice with those expected under different demographic models (Leslie 1985).

Although a number of human populations have relatively low levels of inbreeding (Jorde 1980), others have elevated levels due to a preference for consanguineous marriage (typically cousin marriage, as most societies have prohibitions against closer inbreeding). Bittles and Black (2010) estimate that roughly 10% of humanity marries their second cousins or closer. Such marriages tend to be more common in parts of Africa, the Middle East, and western, central, and southern Asia. Inbreeding has been found to affect morbidity, mortality, and reproductive success in human populations (Bittles 2004; Bittles and Black 2010; Weller et al. 2012).

Cultural subdivision

Mate choice is often influenced by a variety of cultural and demographic factors such as ethnicity, social class, religion, and language, among others. Some studies of population structure have examined one or more of these types of population subdivision. These studies may examine the *level* of genetic differentiation among subpopulations using Wright's F_{ST} or a similar statistic, and some type of genetic distance measure to examine the *pattern* of variation among subpopulations. As noted in this chapter, different types of data (e.g. demographic, genetic, and metric) may be used, and a fair amount of attention has been given in studies of population structure to determine if these different windows into human variation all track the same patterns of differentiation.

Isolation by distance

By far, the most important influence on patterns of genetic variation among local human populations is geographic distance. Throughout the world, populations are subdivided by geographic distance – we do not live in a single panmictic population, but instead live in localized populations (village, town, city) separated by geographic distance. Populations tend to be most genetically similar to their neighbors. The isolation-by-distance model of population genetics predicts that geographic distance limits gene flow, which acts to counter genetic drift. According to formal models of isolation by distance (e.g. Morton 1973b; Relethford 2004), genetic similarity between pairs of populations decreases exponentially as geographic distance increases, reaching a plateau at large distances, in part due to long-range gene flow.

The isolation-by-distance model predicts a positive relationship between the genetic and geographic distances between all pairs of populations under study. Here, the most common approach is to compute the correlation between the two distance measures, testing for significance with a Mantel test (designed to deal with the problem that pairs of distances in any matrix of between-group distances are not statistically independent). A more formal approach to the relationship of genetics and geography is to fit the parameters to a specific model of the isolation-by-distance model (see Morton 1973b and Relethford 2004 for examples and formulae). In the 1970s, there was considerable effort to compare parameters in the attempt to derive some demographic and genetic inference. Although there are some overall generalizations that can be made, these tend to be more limited than once thought (Jorde 1980). In many cases, little additional insight is available beyond the basic test of statistical correlation.

Innumerable studies of human populations over the last 50 years have examined the relationship between genetic and geographic distance, some using the correlational approach and others examining the fit to an explicit model of isolation by distance. Although results obviously vary from case to case, it is a fair statement that many, if not most, studies of human population structure show a moderate correlation between genetics and geography in agreement with the expectation of isolation by distance. Although other factors, such as religion and language, also affect population structure, geography is frequently the primary determinant of genetic variation among populations. It is not surprising to find a close relationship between genetics and geography given the influence of geographic distance on gene flow. What is often more useful is to examine where the model does *not* fit, as such residuals can provide us with information on population-specific history or other reasons for deviation from the expected model (e.g. Workman et al. 1973; Relethford 2010).

Global Human Population History

In the early days of anthropological genetics, the focus was on genetic variation in small, often isolated populations in a specific region. The population-specific nature of much of this research led Harpending (1974, p. 229) to comment that such work has "not advanced our understanding of human evolution in a global sense." This was to change dramatically in later decades, when increasing attention turned to the analysis of human population history, a focus made possible with new sources of data and methods that allowed precise reconstruction of population history. As a result, increasing attention was brought to the study of not only the genetic history of specific populations but also the *global* history of our species.

Today, Harpending's concern has been more than adequately met. Population history has always been a focus in anthropological genetics, but there has been increased attention since the 1980s, in large part because of developments in molecular genetics that have provided new ways of examining questions regarding the origin, dispersion, and admixture of human populations. There is a wealth of new genetic markers, including those with haploid inheritance such as mitochondrial DNA and Y-chromosome DNA, eliminating problems due to recombination. In addition, new analytic methods have been developed, such as coalescent and phylogeographic analysis, that inform us about the origin and dispersal of populations.

The origin of modern humans

Debate began among paleoanthropologists in the 1980s about the nature of the transformation of earlier, "archaic" humans, such as *Homo heidelbergensis* and *Homo neanderthalensis*, into anatomically modern *Homo sapiens*. Some advocated that *H. sapiens* arose as a new species in Africa some 200 000 years ago, who then spread out across the world replacing preexisting archaic human populations (the African replacement model). Others supported the view that there was no single point of origin in time and space for modern humans, but instead a gradual evolution of our species across the Old World (the multiregional model). Other models, such as the assimilation model, included aspects of both views, such as an initial African origin, combined with admixture of modern humans with other human populations outside of Africa.

Over time, the central importance of Africa in the origin of modern humans led to consensus among many (though not all) anthropologists. This emerging consensus has been shaped by both fossil and genetic evidence. Modern human fossils appear first in Africa (200 000 years ago), much earlier than in other geographic regions. Furthermore, phylogeographic analyses based on the genetics of present-day populations point most often (though not always) to an African origin, combined with a series of dispersions out of Africa, primarily over the past 70 000 years or so (see Jobling et al. 2014 for a recent review). However, the African origin of modern humans seen in both fossil and genetic analyses does not support a model of complete replacement. Ancient DNA, from both Neanderthal fossils as well as an enigmatic population known as the Denisovans, has provided evidence that there was gene flow from "archaic" populations into an expanding modern human gene pool. To date, there is ancient DNA evidence that roughly 1–4% of the ancestry of present-day Eurasians is Neanderthal, likely deriving from interbreeding in the Middle East following the initial dispersion of modern humans out of Africa (Green et al. 2010). Another archaic human group, the Denisovans are known primarily from the genome extracted from fragmentary fossil remains at Denisova Cave in Siberia. The Denisovans appear to be a sister group of the Neanderthals, but different genetically from both modern humans and Neanderthals. Denisovan ancestry has been detected in present-day Melanesians, Australian aborigines, and Polynesians, and in some parts of Indonesia (Reich et al. 2010, 2011b). Recent analysis suggests the possibility for additional ancestry from other archaic populations (Prüfer et al. 2012). As further ancient genomes are sequenced, it is likely this will continue. We are seeing a view of the origin of modern humans as being "mostly out of Africa," but with varying amounts of additional ancestry from a number of archaic populations. To date, some present-day human populations show ancestry from Neanderthals and Denisovans, some from Neanderthals only, and some with neither.

The genetic legacy of the expansion out of Africa

The African origin and subsequent dispersion of modern humans provide us with a useful framework for interpreting global genetic variation in the human species. A number of recurrent observations of genetic variation both within and between human populations make sense in terms of our African origin and dispersion. Elsewhere, I identify several major features of global genetic variation that tie into our species' African origin (Relethford 2012, 2013). The first of these is the fact that the human species shows low levels of genetic diversity relative to the great apes, perhaps related to an out-of-Africa bottleneck.

The genetic effects of an African origin and subsequent dispersion are seen when examining within-group genetic variation in different geographic regions. A number of studies of DNA markers have shown that levels of genetic diversity are typically greater in Sub-Saharan Africa when compared to other parts of the world. In addition, genetic diversity outside of Africa is often a subset of the diversity within Africa. Furthermore, the level of genetic diversity within populations decreases as the geographic distance from Africa increases. The three observations listed here are perhaps best explained by an out-of-Africa dispersion. Modern humans existed in Africa since 200 000 years ago, followed by dispersals after 100 000 years ago. Assuming standard models of mutation, genetic diversity would have increased over that time. When populations began to disperse out of Africa, they were likely initially small, and a genetic bottleneck would cause non-African populations to have lower levels of genetic diversity. This genetic drift would also explain why some African alleles have been lost in non-African populations (Tishkoff and Gonder 2007).

Founder effect can also be used to explain the observed decline in genetic diversity with geographic distance out of Africa. Ramachandran et al. (2005) developed the serial founder model, where a small founding population grows over time until a new founding population splits off, with a subsequent decrease in genetic diversity. This model is essentially a series of founding events spread out over space as our species dispersed out of Africa, step by step, with subsequent declines in genetic diversity. This model has since been modified to allow for the effect of likely gene flow between neighboring populations as well as admixture with archaic humans (DeGiorgio et al. 2009).

One of the most notable features of global human genetic variation is that it is geographically structured. Populations that are geographically close to each other tend to be more genetically similar than those that are farther away (e.g. Cavalli-Sforza et al. 1994a; Relethford 2004, 2010; Ramachandran et al. 2005). This relationship is very clear when likely routes of migration are used to compute geographic distances, such as connecting African and Eurasian populations to the New World via travel from northeast Asia rather than computing straight-line distances across oceans. This close correspondence of genetics and geography is in part expected for a dispersing species under the serial founding model. Each population is related genetically to its parent population, which in most cases would be geographically proximate. Furthermore, continued gene flow would more often be with neighboring populations because of isolation by distance, which would further make neighboring populations more similar genetically.

All of these patterns of global genetic variation are most clear when looking at DNA marker variation. However, many are also apparent when looking at quantitative traits, such as craniometrics, which also show higher levels of variation in Africa, a decline in variation out of Africa, and moderate correlations of geography and biology on a global scale (e.g. Relethford 2004, 2010, 2013). Although most of anthropological genetics today focuses on DNA markers, it is noteworthy that the field is not confined or defined by one source of data.

Studies of Regional Human Population History

There have been numerous studies of human population history around the world. No attempt is made here to summarize the scope of this research – see Cavalli-Sforza et al. (1994a) and Jobling et al. (2014) for thorough reviews. Instead, the purpose here is to provide several examples of two broad themes in the study of human population history: (i) the origin and dispersion of regional populations, and (ii) admixture and culture contact.

The origin and dispersion of human populations

One of the classic examples of genetics and population origins concerns the peopling of the New World. For many years, the conventional wisdom was that the first Americans crossed over from Siberia into the Americas via a "land bridge" that connected Asia and North America during glacial times when the sea levels dropped. These migrants would then have had to move down into the Americas via an ice-free corridor. Because the traditional model required movement during a window of time when both the land bridge and ice-free corridor were in existence at roughly the same time, this meant that the first Americans must have arrived roughly 13 500 years ago. However, archeological evidence accumulated for an earlier entry date (perhaps 15 000 years ago), as well as the increasing likelihood that some groups moved down into the Americas along the west coast, has led to the questioning of the traditional model (Goebel et al. 2008).

Biological variation of Native American populations supports the northeastern Asian origin hypothesis. Early analyses of skeletal and dental measures as well as red and white blood cell polymorphisms confirmed the hypothesis that Native Americans have an origin in northeast Asia (Cavalli-Sforza et al. 1994a; Crawford 1998). This conclusion has been supported by DNA analyses. All Native Americans (excluding those with known admixture) have one of five mitochondrial DNA haplogroups – A, B, C, D, and X. The first four of these haplogroups have been found only in the New World and in East Asia, further supporting an Asian origin of Native Americans. The analysis of haplogroup X was initially a bit confusing, as it is rare in East Asia but has been found in low frequencies in Europe. However, closer analysis shows that the European variant of X is different from that found in Native Americans. Although haplogroup X is rare in Siberia and absent in eastern Siberia (where the first Americans came from), such differences can be explained easily due to genetic drift, given its low frequency in both Native American and Asian populations (Fagundes et al. 2008). Analysis of Y-chromosome DNA markers show the same connection between East Asia and the Americas – both regions are characterized by the presence of Y-chromosome haplogroups C and Q (Goebel et al. 2008).

The northeast Asian origin of Native Americans has also been supported (and elaborated on) by ancient DNA analysis. Prehistoric Native Americans also have one of the five mitochondrial DNA (mtDNA) haplogroups found in present-day populations (e.g. Chatters et al. 2014). Genome analysis has confirmed the Asian connection, as demonstrated by Rasmussen et al.'s (2014) analysis of the genome of a male infant from the Clovis site dating back to 12 500 years ago.

Although the northeast Asian origin of Native Americans is clear, there are a number of remaining issues that will require additional genetic and archeological investigation, including questions about whether there were multiple dispersals into the Americas, the possibility of alternative routes of entry, and the timing of various events.

Another example of the genetic analysis of the history of a colonizing population comes from Polynesia. Archeological evidence shows that the first Polynesians, associated with the Lapita culture, spread out from either South Asia or Southeast Asia, reaching the Bismarck Archipelago roughly 3200–3500 years ago. Seafaring skills and technological innovations such as outrigger canoes led to relatively rapid expansion into the Pacific over the next two millennia, reaching as far as New Zealand, Hawaii, and Easter Island. The west-to-east expansion of the first Polynesians is interesting because their path of expansion took them near New Guinea and other parts of Melanesia, which had been inhabited as much as 35 000 years ago. The question here is whether there was any gene flow with Melanesian populations during this expansion, and to what extent there is any Melanesian ancestry in Polynesians. The two major hypotheses of Polynesian origin are the *express train model* and the *slow boat model*. According to the express train model, the first Polynesians moved rapidly past Melanesia, and there was little, if any, genetic mixture with Melanesians (Diamond 1988). On the other hand, the slow boat model proposes that there was significant genetic mixture with Melanesians during expansion into the Pacific (Kayser et al. 2000).

Early genetic analyses were based predominately on mitochondrial and Y-chromosome DNA and gave mixed results. The geographic distribution of mtDNA frequencies fits the express train model, with no evidence of Melanesian gene flow. On the other hand, the analysis of Y-chromosome DNA suggested a large Melanesian component, supporting the slow boat model. Such mixed results are clear from a study by Kayser et al. (2006), which found that 94% of Polynesian mtDNA markers were of Asian origin, whereas 66% of the Y-chromosome markers were of Melanesian origin. Although such a large difference suggests the possibility of sex-specific patterns of admixture, it must also be remembered that both mtDNA and Y-chromosome DNA essentially represent single loci. Here, the picture has been clarified by large studies of autosomal DNA markers. Friedlaender et al. (2008) examined hundreds of STR (short tandem repeat) and insertion/deletion markers and found a larger contribution from Asia than from Melanesia. Kayser et al. (2008) analyzed hundreds of STR loci and estimated that 79% of Polynesian autosomal ancestry came from East Asia and 21% from Melanesia. Although the East Asian ancestral component is larger, the contribution from Melanesia is significant and supports the slow boat model. Against this background, the differences between mtDNA and Y-chromosome DNA results indicate that most of the Melanesian admixture was from men, consistent with evidence that early Polynesian societies were matrilocal.

Admixture and culture contact

The examples of the peopling of the Americas and Polynesia represent cases where modern human populations expanded into areas that had been previously unoccupied. Throughout much of human history, however, there have been cases where humans expanded into places where other humans were already living, or where multiple groups have moved into new environments and mixed with each other. An example of the former is the expansion of Middle Eastern farmers into Europe, mixing with European hunter-gatherer populations. An example of the latter is the early colonial history of the United States, where migrants from Western Europe and enslaved peoples from Western Africa moved to the Americas and contributed to the ancestry of African Americans.

The human species began developing agriculture 12 000 years ago. Although this change occurred independently in many parts of the world, a notable exception was the origin of agriculture in Europe. Here, we find evidence for the spread of agriculture out of the Middle East, starting about 9000 years ago, spreading northwest to reach northwestern Europe about 3000 years later. Although the spread of agriculture is clear from the archeological record, there was debate over whether this spread also involved gene flow. One hypothesis, known as the cultural diffusion model, argued that the practice of agriculture spread, but not the early farmers. As such, the idea moved across Europe, but with no corresponding movement of people, and the early European farmers were descended almost completely from earlier European hunter-gatherers. The alternative hypothesis, known as the demic diffusion model, proposed that early farmers spread out from the Middle East, with a movement of *both* their agricultural methods and their genes. Here, early European farmers genetically represent the admixture between earlier European hunter-gatherers and dispersing farmers from the Middle East. The key difference between these models is the expectation of present-day genetic variation, where southeast-to-northwest genetic clines (paralleling the dispersion of agriculture) are expected under demic diffusion but not under cultural diffusion.

An early test of these hypotheses was conducted by Cavalli-Sforza and colleagues, who applied principal components analysis to a large spatial database of allele frequencies based on classical genetic markers of red and white blood cell polymorphisms. They found that the first principal component (representing the greatest amount of spatial covariation in allele frequencies) showed the same spatial distribution as the carbon-14 dates for the spread of agriculture. The correspondence of archeology and genetics demonstrated that the demic diffusion model, with the spread of both farming *and* farmers, was correct (e.g. Menozzi et al. 1978; Cavalli-Sforza et al. 1993).

These results have been criticized. Preliminary investigation using haploid DNA markers showed mixed results; there was no evidence of clinal variation in studies of mitochondrial DNA, but there were clines for Y-chromosome markers. However, there was still debate over the cause of the latter clines (Jobling et al. 2014). More recently, insights have been gained through the analysis of ancient DNA, which allows researchers to examine the relationship of present-day populations to prehistoric foragers and farmers. For example, the mitochondrial DNA sequences of skeletal material from an early farming population in Central Europe (the Linear Pottery Culture, also known as LBK) show a genetic relationship with present-day populations from the Middle East, consistent with the demic diffusion model (Haak et al. 2010). However, although such studies point to genetic input from a dispersion of farmers, there is growing evidence that the pattern of mixture was likely more complex, with gene flow between foragers and farmers varying over time, including more than one dispersal of farmers out of the Middle East (Bollongino et al. 2013; Brandt et al. 2013).

There are now studies of ancient DNA in prehistoric Europe based on complete genomes. One study examined the genomes of a farmer from Germany dating back roughly 7000 years ago and eight European hunter-gatherers dating back roughly 8000 years ago (Lazaridis et al. 2014) and compared them with a number of present-day genomes in order to reconstruct the dynamics of population mixing in Europe. They found evidence of three ancestral groups: Western European hunter-gatherers, early European farmers, and an ancient north Eurasian population related to ancient Siberians. The relative contribution of these populations varies across Europe; for example, the amount of ancestry from early European farmers ranges from roughly 30% in Baltic populations to about 90% in the Mediterranean, consistent with dispersal from that region. Such studies will continue to inform us about the relative contributions of the earliest hunter-gatherers in Europe, the first farmers there, as well as the Neanderthals.

The second example of admixture and culture contact is the genetic history of African Americans. Hundreds of thousands of Africans, primarily from Western Africa, were enslaved and brought to the Americas starting in the early seventeenth century. Over time, gene flow took place from those of European ancestry into the African American population. Early in the history of the United States, much of this gene flow likely occurred from men of European ancestry who had offspring with enslaved African women. After slavery ended, gene flow continued as social barriers to interracial marriage began to diminish. During the 1950s and 1960s, researchers used classical genetic markers of the blood to estimate the overall amount of European ancestry in the African American gene pool (e.g. Reed 1969). It is important to remember that these estimates refer to the total level of accumulated ancestry over the course of many generations, and are not the per-generation admixture rates (although these can be estimated from total ancestry given an estimate of the number of generations of admixture). It soon became apparent that the actual level of European ancestry varied considerably from one population to the next, ranging from very little admixture to a moderate level (Chakraborty 1986).

Later work using DNA markers provided greater resolution of the genetic history of African American populations. Parra et al. (1998, 2001) estimated European admixture in 16 African American communities in the United States using autosomal markers, finding rates that ranged from 3.5% in the Gullah of South Carolina to 23% in New Orleans. They also looked at admixture estimates based on mitochondrial and Y-chromosome DNA markers, finding consistently that the level of European admixture was greater in the paternal line (assessed via the Y-chromosome) than in the maternal line. The reason for these sex-specific differences is the fact that for much of American history, any interracial matings were between a man of European descent and an enslaved African or African American woman. Although in the past two generations, the frequency of marriages between a husband of African ancestry and a wife of European ancestry is greater than the reverse, this social shift has not offset earlier generations of male European ancestry.

The use of DNA markers also allows ancestry estimation of *individuals*, which covers a wide range of individual histories that are obscured by estimates based on entire samples. For example, Parra et al. (2001) estimated a value of 18% European ancestry for an African American sample from Columbia, South Carolina. The distribution of individual ancestry estimates is skewed; roughly half of the estimates are 10% or less European ancestry, and a moderate number of individuals have more than 50% European ancestry. Tishkoff et al. (2009) also found a large range in the level of European ancestry among African American individuals, as well as variation in the contributions of different regional groups in Africa to their African ancestry. Bryc et al. (2010) performed a genomic analysis of 365 African Americans, and found an average amount of 77% West African ancestry and 23% European ancestry. However, there was a great amount of variation across individuals, with some African Americans having less than 1% European ancestry, and others having less than 1% West African ancestry. All of these findings have broader implications in how we view ancestry in terms of racial classification, including construction of cultural identity, as there are individuals whose genetic ancestry is primarily European, but whose cultural identity is African American. As noted by Bryc et al. (2010), there are also implications for medicine, where treatments might be based on racial identity without consideration of actual genetic ancestry.

Conclusion

Over the past half century, there have been enormous strides in describing and analyzing the genetic variation of the human species, including local population structure, local and regional history, and the history of the entire species. Early work built upon a large body of data on classical genetic markers and physical traits, many of which continue to be useful in unraveling population structure and history (e.g. Relethford and Crawford 2013). However, the most recent advances in molecular genetics and genomics have pushed the envelope much further, and have allowed even greater precision and depth of view into historical processes. The advances in the study of ancient DNA have provided us with the ability to tie together past and present genetics in new ways. Newer methods, such as haplotype tree analysis, phylogenetics, and coalescent modeling, have given us different ways of testing evolutionary hypotheses. There are also new challenges, including the continued development of computational methods that protect us from drowning in data and the need for further developments in population genomics and genetics. Overall, the future is bright for studies of human population structure and history.

CHAPTER 10

Regional Patterns of Variation: Concepts and Applications

Graciela S. Cabana

Department of Anthropology, University of Tennessee, Knoxville, TN, USA

Introduction

Anthropological genetics concerns itself with human evolutionary studies at multiple scales, from global to local. It is at local and regional scales, however, where the discipline has made exceptional contributions. This is largely due to the fact that the field emerged in the post-1950s landscape, which consisted of a "new" physical anthropology (Washburn 1951) that was inspired by a newly synthetic evolutionary biology (Huxley 1943). Starting at that time, anthropological geneticists leveraged anthropology's traditional focus on isolated human groups along with expertise in molecular-based methods to broadly enhance human-based studies of microevolutionary change (Crawford 2007a).

Because human evolution has taken place over millennia and across a broad swath of the globe, both temporal and global scales are vital to understanding the human evolutionary context. Whereas global-scale analyses tend to shed light on long-term evolutionary processes, regional-scale analyses inform us about human adaptations to local environments in the shorter term. At the regional and local scales, the combination of biological and cultural influences on local demography takes on greater importance: how people have opted to organize themselves regarding group size, composition, and structure is of interest, as are the ways in

A Companion to Anthropological Genetics, First Edition. Edited by Dennis H. O'Rourke.

which individuals move and mate within and between groups. At the local scale, these are the factors most likely to account for most or all observed variation in biological traits. Natural selection, conversely, is expected to exert an even-handed pressure throughout any given region and is therefore rarely explicitly analyzed at this scale (e.g. Stojanowski 2004, 2005). Likewise, because regional studies also typically involve short time scales, mutation is assumed to be a rare enough event so as to not contribute significantly to microevolutionary change.

This chapter complements Chapter 9 ("Human Population Structure and History") in that it provides comprehensive information on how to conduct regional studies on both modern (i.e. currently or recently living) and past populations using archeological mortuary samples. A *region* is defined here as a geographical area of interest in which selection pressures are assumed to be relatively uniform, and therefore do not significantly contribute to intra- and intergroup variation. The chapter begins with a consideration of the ways in which the key concepts of "population," population size, gene flow, and genetic drift are operationalized. It then covers theoretical approaches to studying regional patterns of variation based on classic models whose foundations derive from the modern synthesis era of the 1920s to 1950s. It moves on to a discussion of conceptual gaps between modern and past contexts. In an effort to close those gaps, it explores analogical bridges from ethnographic and archeological sources and discusses two recently published regional studies of genetic variation to exemplify the points.

Concepts and Definitions

Classic models, described further below, evaluate regional microevolutionary change by measuring and tracking gene frequency changes within and between populations over time. *Gene frequency* is a catchall term that, depending on the study, can refer to frequencies of gene variants (i.e. alleles) or of molecular markers derived from blood groups, proteins, or genetic sequences (e.g. single nucleotide polymorphisms [SNPs], short tandem repeats [STRs], and copy number variants [CNVs]).

All models rely on "population" as a core concept. The broadest definition of the term *population* can rely on any combination of cultural, ethnic, or biological criteria. For biologists, a population most generally refers to "a group of interbreeding individuals that exist together in time and space" (Hedrick 2011, p. 66); but, in the context of classic models, a population refers more specifically to a "breeding population," or a group of individuals who are more likely than not to choose mates from within their same group. To measure and track gene frequency changes within and between populations, classic models also consider the following key concepts: random genetic drift, effective size, and gene flow. These are reviewed in this section.

Random genetic drift

The concept of random genetic drift refers to the process by which stochastic changes in genetic frequencies occur within populations. As elaborated by Wright (1931), genetic drift is thought of, and modeled, as the outcome of random and independent samples of gametes drawn from populations that are finite in size. At the population level, this sampling process can be shaped by a variety of factors such as variance in reproductive success, population sizes, population subdivision (explained further in this chapter), and gene flow. The concept of drift is highly amenable to statistical modeling and, indeed, often has been included in theoretical human population genetic models (Rogers and Harpending 1986; Rogers 1988; Relethford 1996).

Random genetic drift becomes an extremely important evolutionary force in the context of small populations. Genetic drift alone would lead genetic variants to eventually either become extinct or rise to 100% frequency, or *fixation*, in small populations. In a large population, stochastic changes in allele frequencies from one generation to the next do not lead as easily to the complete loss or fixation of alleles, simply because there are more chances for recovery and long-term stability of those alleles.

Effective size (N_e)

One way to evaluate the effects of genetic drift on populations of finite size (as opposed to infinite size, as is commonly assumed in theoretical models) is via the concept of effective population size, or N_e. In finite populations (as all real populations are), not all members of a population contribute equally to future generations because of unequal sex ratios, variance in reproductive success, and/or population size fluctuations (Templeton and Read 1994). Therefore, the rate of drift in finite populations is not affected by all members of a population, but by the population of individuals who reproduce.

The concept of *inbreeding effective size* captures this population of interest (the reproducing population) by referring to the size of an idealized population that would have the same genetic properties as those observed for a real population. This idealized population is assumed to be composed of members who are randomly mating, have discrete generations, and have a Poisson-distributed number of offspring (a Poisson distribution is used to model systems in which the probability of an event occurring is very low, but the number of opportunities for such an occurrence is high). Using this concept of inbreeding effective size, a measure of genetic drift can be estimated (Wright 1931; Templeton and Read 1994).

Gene flow (m)

The degree to which random genetic drift affects a population's genetic frequencies – especially small populations – depends on gene flow, because gene flow prevents either the loss or the fixation of gene variants. *Gene flow* refers specifically to the movement of alleles. These in turn travel within organismal contexts, so that *migration*, in population genetic terms, refers to the movement of alleles via the movement of organisms. Gene flow is expressed in population genetics as a migration rate *m*, defined as the proportion of alleles in a population per generation that is of migrant origin (Avise 1994).

Especially in the absence of natural selection, mutation, and drift, gene flow can connect otherwise isolated populations to effectively form larger populations (thus also affecting effective size). Depending on the rate of gene flow as well as the amount of time passed, gene flow will tend to homogenize gene frequencies across connected populations. If gene flow is asymmetric, it can also create patterning and subdivision of the genetic frequencies of populations. Asymmetries in gene flow can depend on population sizes (Relethford 1990), geographical distance (as captured by Wright's isolation-by-distance model [1943], as discussed further in this chapter), variation in the natural environment, as well as other culturally relevant factors such as language, ethnicity, and residence rules.

Population geneticists estimate gene flow in several ways. Gene flow is often evaluated through the product $N_e m$, or simply *Nm*. *Nm* is most commonly estimated through the use of *F*-statistics, a measure of genetic differentiation in an "island model" context (see below), in which migration occurs at random among a group of small populations. This relationship is described more fully later in this chapter.

Classic Models

The classical population genetic models discussed in this section address, among other things, the hypothesized relationships among gene flow, genetic drift, population size, geographic distance, and level of population subdivision.

The island model

The island model, as developed by Sewall Wright (1951, 1969), assumes a basic population structure that underlies much subsequent modeling of the effects of gene flow on that hypothesized structure. The model envisions two or more populations as isolated (but internally panmictic) islands that receive migrants from a large panmictic source population. The source population is assumed to be large so that genetic drift would not affect gene frequencies of the migrants, and those frequencies can remain constant. In addition, the migrants from the source population maintain a gene frequency equal to that of the subpopulation(s).

The island model is used to estimate the degree of heterogeneity or homogeneity within a subpopulation based on its population size and some amount of gene exchange. The formula

$$f = 1/4N_e m + 1$$

is used to evaluate the balance between random genetic drift and gene flow, where f is a measure of the relative population heterogeneity, or "inbreeding coefficient"; m is the rate of "migration," or gene flow; and N_e is the effective size of the population.

The formula predicts that when the gene flow rate (m) is low relative to the effective population size, we would expect a relatively high amount of differentiation of the island population from the main population. In other words, the island population's gene frequencies would be more a product of random genetic drift than gene flow. On the other hand, the island model also predicts that when gene flow rate is high relative to the island's effective size, the island's gene frequencies will quickly approach those of the main population. In this idealization, gene flow quite powerfully moderates the effects of random genetic drift.

Isolation by distance

In general, isolation-by-distance models predict that genetic similarity between any two populations will decrease with increasing geographic distance. Wright (1943) proposed the original and classic isolation-by-distance model, which has encouraged variations or elaborations, like Malécot's (1973) version or Kimura and Weiss's (1964) stepping stone model.

Wright's (1943) isolation-by-distance model views populations not as relatively isolated islands, as with his island model, but as forming part of a continuously distributed, uniform population. For Wright, isolation by distance is an ecological process: "[r]emote populations may become differentiated merely from *isolation by distance*" (1943, p. 117; emphasis in original). It describes a process of mating among spatially close individuals due to short-distance movements (Ishida 2009). Thus, distance is the key to the analysis of isolation by distance (Fix 1999). Populations are distributed continuously as if in a single line across a landscape. Gene flow is modeled as normal distribution of the differences between parent and offspring birthplaces. Wright later developed variations on this basic model, considering the case in which populations are distributed in two-dimensional space, or if the distribution of parent–offspring distances is not normal (Wright 1969). However, the basic assumptions remain the same. Malécot (1967, 1973) later used "isolation by distance" to refer to the outcome of the process envisioned by Wright, namely, a particular pattern of local differentiation in which genetic relatedness between individuals increases as the geographical distance between them increases (Ishida 2009).

A widely used variation of Wright's basic isolation-by-distance model is Kimura and Weiss's (1964) stepping stone model. Unlike classic isolation by distance, the stepping stone model sees human populations as discretely bounded groupings. Like isolation by distance, the stepping stone model does assume that a significant variable affecting genetic variation among populations is geographic distance. This model also does not tend to limit itself to a one-dimensional scheme of human populations; rather, populations can exchange migrants with neighbors on four sides.

Migration matrices

Instead of assuming a gene flow model, Bodmer and Cavalli-Sforza (1968, 1974) constructed migration matrices from actual data of peoples' short-range movements. The rate of movement, and therefore of gene flow, is commonly derived from the pattern of birthplaces of parent and offspring.

Migration matrices differ from idealized population genetic models because they do not rely on estimates of patterns of gene flow, population sizes, or the structure of the populations (islands, stepping stones, or panmictic populations). Migration matrices represent the pattern of gene flow over a limited time, and cannot be used to predict changes in the pattern (Fix 1999).

Classic Models: General Issues for Real (Versus Theoretical) Modern and Past Contexts

In population genetic models, each core concept, or parameter, used to construct the models contributes problematic assumptions. One problematic assumption has to do with population structure: human populations are rarely "islands." Literal exceptions exist, such as the Tasmanian (Pardoe 1991) or the Isles of Scilly populations (Raspe 1988), but they are likely not internally panmictic as typically assumed by classic models. Additionally, human populations do not tend to neatly receive migrants from linearly aligned neighbors. A second assumption implicates population sizes – these models presuppose infinite population sizes to bypass the effects of random genetic drift that are triggered by small population sizes. A third assumption is non-overlapping generations. That is, in models of evolving populations, the entire population dies at once and is replaced by the descendant population, whereas in reality generations overlap (Rogers and Prügel-Bennet 2000). A fourth assumption involves the nature of the migrants from whom these populations receive genes. In the island model, migrants come from a large, panmictic population whose gene frequencies are essentially fixed. Over time, genetic variability is maintained. In isolation-by-distance models, the pattern of gene flow is assumed to be a normal distribution composed of the differences between parent and offspring birthplaces; it is unclear whether this pattern approximates reality (Fix 1999).

A fifth problematic assumption concerns the analytical unit, the breeding population. Ideally, human populations would fall into social units that conform to breeding population units. In reality, mating structure is influenced by so many factors that the overall breeding population structure does not map onto an easily identifiable socially defined unit, such as a "village" or a "tribe" (Fried 1968, 1975). The problem is further magnified when trying to isolate the long-term characteristics of a breeding population. Unfortunately, this is a problem that is rarely, if ever, addressed within population genetic models because of its apparent intractability.

Finally, a sixth issue involves gene flow rates. Simulation studies and ethnographic surveys of the past decades have issued the following caveats: (i) numbers of migrants might depend on population sizes of both recipient and donor populations. For example, Relethford (1986) argues that short-range migration patterns depend on population size. (ii) Smaller populations go farther to find their mates (Relethford 1990). (iii) Migration distance increases with smaller population densities, specifically among foragers (MacDonald and Hewlett 1999). (iv) Gene flow is often sex-biased due to residence rules upon marriage. (v) Gene flow may be kin-structured. Fix (1999, 2004) argues that, based on ethnographic analogies and simulations, relatives tend to emigrate together. This means that migration for gene flow purposes does not consist of a series of independent events, as is commonly assumed.

These issues are compounded when applied to mortuary samples from archeological contexts, which have their own set of problems. One problem with using skeletal remains from archeological contexts concerns sampling. Archeological deposits usually represent a palimpsest of cultural and skeletal remains that are additionally subject to variation in cultural burial practices, such as spatial segregation of individual members of society, and site formation processes, such as erosion or disturbance by animals, looters, and so on. This means that skeletal remains recovered from archeological deposits will probably be a poor sample of multiple generations of populations.

The realities of archeological samples contrast with how biological anthropologists often treat the skeletal remains from archeological samples – as "populations" whose morphologies are representable as variation centered around a mean (Mayr 1959; Cadien et al. 1976). To make this leap in logic, biological anthropologists must assume that (i) skeletal remains represent adequate samples that capture that variability, and (ii) the skeletal population remained unchanged through time so that sampling across generations is equivalent to sampling at any specific point in time. However, at best, skeletal samples from archeological sites represent evolutionary lineages, not synchronic populations (Cadien et al. 1976). Unfortunately, up to this point, no satisfactory body of theory exists to deal with this question of the concordance between living populations and skeletal samples.

Clearly there are challenges posed by, first, the construction of generalized models to adequately predict the regional microevolutionary consequences of population sizes, gene flow, and random genetic drift; and, second, the application and suitability of such models in both real life and past contexts. Given these challenges, how do anthropological geneticists go about evaluating microevolutionary change in regional contexts? The next section details two examples, the first from a study of living populations in Turkey, and the second from an archeological context from the Great Basin area of the United States.

Regional Patterned Variation: Two Examples

DNA-based studies became increasingly common in anthropological genetic studies beginning in the 1990s. With the use of modern DNA over skeletal morphology or other molecular (e.g. blood groups, proteins) markers to assess microevolutionary histories, it became possible to quickly type relevant genetic information from larger numbers of samples at a lower cost, and therefore either more densely sample individuals from a localized area, or target individuals who are distributed over a wide geographical area.

Of course, the choice of sampling scheme depends on the research question, although the nature of the research question has been historically somewhat biased depending on whether modern or ancient genetic variation is being assessed. On one hand, much anthropological research effort has gone toward tracing biological origins via the genetic relationships of broadly defined human groups and therefore does not comprise "regional" studies as defined in this chapter. More localized, regional studies of modern groups are relatively rare, but they hold greater promise to provide insights into the specific factors that may structure human variation. In this section, I provide an example of one such study that explicitly sought a "fine-scale" analysis through a regional study. On the other hand, anthropological genetic research that focuses on ancient DNA tends to be regionally focused, if not much more locally focused on a single archeological mortuary site or continually occupied locale due to the time and expense of retrieving damaged DNA. The second example discussed here provides an example of a regional analysis that uses a combination of ancient and modern DNA, along with other data sources, to test a hypothesis of population continuity in a region (versus replacement) via microevolutionary change.

Four geographically proximate modern central Anatolian settlements and the importance of paternal kinship

Gokcumen et al. (2011) conducted a focused study of localized settlements in Anatolia (the Asian part of modern-day Turkey) to study fine-scale historical and cultural processes shaping genetic variation. Their goal was to see whether there are structuring factors operating at the local scale that tend to be obscured in larger scale studies. To capture the "fine scale," the researchers initiated their project via three fieldwork excursions in the mid-2000s. They visited 30 settlements in the Central, Eastern, and Northeast regions of Anatolia, where they interviewed community elders, collected local oral histories, and documented the social, religious, political, and cultural affinities of the communities. They used this fieldwork period to reflect on local population dynamics and subsequently develop hypotheses with expectations for the structure of genetic variation in four Central Anatolian settlements. The settlements fall within a relatively limited, 2500 square kilometer region the authors called "Yuksekyer." This region name is a pseudonym due to the political sensitivity of the subject of origins and belonging.

Specifically, Gokcumen et al. (2011) predicted that kinship connections, particularly paternal kinship, will primarily shape genetic variation. Through their extensive fieldwork, the researchers noted that Anatolian villages are made up of patrilineal families that are connected through marriages. Marriage is an important social glue in these villages. For example, families relocating from other villages or regions into a new village cease to be considered foreigners once a family member marries a village member. This social rule seems to operate despite geographic location, religious or ethnic affiliation, ideological leaning, or claimed ancestral origins. Conversely, the researchers hypothesize that national elements of identity, such as religion or language, will have little effect on the structure of genetic variation. This hypothesis also derives from their initial fieldwork, which indicated to them that these elements seemed to have been selectively, and therefore inconsistently, incorporated into local social dynamics.

In their study, the unit of analysis that more closely corresponds to a "breeding population" is the village, a socially and geographically circumscribed entity. The question is: in what ways, if any, is this unit also genetically circumscribed, and what factors may explain the observed patterning? Sample sizes per village were around 30 individuals, for which they typed mitochondrial haplogroups and first hypervariable region (HVI) haplotypes, Y-chromosomal haplogroups and STR-based haplotypes (using 17 loci), as well as autosomal STR genotypes (15 loci). The four settlements are at most 15–20 km from each other. Due to their relative proximities, the researchers did not consider geographic distance as a parameter that would explain patterns of genetic diversity in their analyses. Additionally, because they found no correlation between genetic diversity and the population size of the settlements, they did not consider population size as an important factor in their analyses. Finally, neither gene flow nor genetic drift were directly estimated, but rather were inferred from the overall pattern of distribution of genetic lineages (maternal, paternal, and biparental) within and between settlements, versus the region as a whole.

To test their hypotheses, the researchers traced the relationships between similar maternal, paternal, and biparental genetic lineages within and between settlements as well as for the region as a whole. They tested the correlations between the observed within- and between-settlement genetic variation (as well as for the whole region) and religious traditions, language, and self-claimed ancestry.

They found that none of those factors correlated with the structure of genetic variation very well, if at all. They did find, however, that paternal genetic diversity was high in the Yuksekyer region, both within and between settlements. Within villages, they found significant shared paternal genetic variation, but this was much less so between villages. Within-village kinship coefficients (a measure of endogamy, or within-group mating, normalized by the identity-by-descent) observed among individuals were high. Additionally, they observed high frequencies of similar Y-STR haplotypes with low pairwise differences between them, indicating that men remain in the villages to perpetuate their paternal lineages (with occasional changes from generation to generation). Interestingly, whereas paternal genetic variation showed significant structure (i.e. settlements were genetically subdivided as ascertained via the Y-chromosome), maternal (mtDNA) and autosomal (biparental) did not. Interview data suggested that about a third of men obtained wives from different villages, whereas almost all men remained in their natal settlements. Taken together, these findings suggest that maternal gene flow has been the main force shaping autosomal genetic diversity in the Yuksekyer settlements. Despite its strong influence on paternal diversity, patrilocality had no significant observable effect on autosomal variation.

A test of regional continuity using an ancient and modern population sample

Cabana et al. (2008) test a hypothesis of regional continuity through the construction of a simulation model that they call *Nsitu*. *Nsitu* draws on ethnoarchaeological, paleodemographic, and population genetic studies to structure simulations of ancient populations experiencing genetic drift and gene flow over time. They apply the model to a long-standing question of whether or not the so-called Numic Expansion actually occurred. In other words, is the modern distribution of Numic speakers throughout the Great Basin (as well as the Colorado and Columbia plateaus) due to an ancient migration of these peoples into the region, and subsequent replacement of the previous inhabitants? Great Basin archeologists differ in opinions as to if, when, and how the Numic expansion occurred, and they have proposed various migration (as well as continuity) scenarios (see Madsen and Rhode 1994). Kaestle and Smith (2001) test one migration scenario: the modern-day distribution of Numic speakers is a result of migrations into the Great Basin approximately 1000 years ago (Steward 1940; Lamb 1958; Bettinger and Baumhoff 1982).

To test this hypothesis, Kaestle and Smith (2001) compare the genetic frequencies of ancient populations to those of modern Great Basin populations. Their ancient sample comes from two archeological sites in western Nevada: Pyramid Lake and Stillwater Marsh. They compare the combined mtDNA haplogroup frequencies of individuals from these prehistoric sites to those of modern Numic speakers (the Northern Paiute), and evaluate the statistical difference between sets of frequencies using a χ^2 test for homogeneity (with Yates's correction for small sample sizes). Kaestle and Smith (2001) reject a hypothesis of population continuity in the Great Basin under the assumptions of a χ^2 model.

Cabana et al. (2008) reevaluate the significance of their observed mtDNA haplogroup frequencies using only the Stillwater Marsh sample, under the assumptions of *Nsitu*. *Nsitu* tests a null hypothesis of genetic continuity: the genetic change observed between two population samples separated by time can be explained solely by the processes of random genetic drift and gene flow in a single breeding population. The original study also considered population growth, but it will not be considered here. The null hypothesis is based on the contention that the balance of random genetic drift and gene flow are the sole factors affecting genetic variation over time (Kimura 1983).

Cabana et al. (2008) based *Nsitu* on assumptions about prehistoric population structure based on the socioeconomic organization of small-scale hunter-gatherers. The researchers' first task was to decide on an analytical unit (a breeding population) and a model of breeding population structure. Both concepts were addressed by turning to previous studies that modeled the spatial organization of small-scale hunter-gatherers – and therefore the breeding population structure – as hexagonally shaped spaces on the landscape (Birdsell 1958, 1988; Wobst 1974, 1976). A single hexagon represents a breeding population, and neighboring hexagons/breeding populations can exchange mates. In the western Great Basin, a breeding population corresponds to a Northern Paiute "district." Historically, the Northern Paiute territory was divided into districts that were commonly referred to as "eaters" (Fowler 1992). For example, the Stillwater Marsh region was occupied by peoples within the *Toedökadö* ("Cattail eaters") district of the Northern Paiute Districts, which were loosely bounded geographically and consisted of camping places close to water and food resources (Stewart 1939). The simulations equate a district with a breeding population and therefore a hexagon. All simulation runs created 19 total hexagons or breeding populations to model one large regional mating network. This regional network represents the immediate area in and around the *Toedökadö* district (where Stillwater Marsh is located), along with its prehistoric inhabitants.

The researcher's second task was to build three important parameters into *Nsitu*: population sizes, genetic drift, and gene flow. To incorporate population sizes, the model estimates female effective size (N_{ef}), given that only maternally inherited mtDNA is used in this prehistoric Great Basin example. N_{ef} estimates were derived from early census data (Stewart 1939; Shimkin and Reid 1970; Pendleton et al. 1982). Most simulations allowed N_{ef} values to fluctuate around an estimated mean of 90 (range of 15–190 females). Random genetic drift was taken care of via *Nsitu*'s model of reproduction, which in turn is modeled with reference to a Poisson distribution. Conceptually, the model sees reproduction and survivorship as Poisson-distributed with a mean of 1 (Hudson 1990). *Nsitu* consults a Poisson distribution to determine the number of daughters each female produces in the next generation.

As for gene flow, to obtain a reasonable estimate of gene flow rates for the prehistoric Great Basin, Cabana et al. (2008) turned to ethnohistoric records from the area and from other small-scale societies. At the time of European contact, interdistrict mating patterns in the western Great Basin were fluid. Fowler (1992) reports that for the *Toedökadö*, kinship was reckoned bilaterally and marriages were prohibited between both paternal and maternal relatives. Thus, mates were found outside the local group. Unfortunately, interdistrict marriage rates at or before the time of European contact are unknown. To estimate interdistrict gene flow, Cabana et al. (2008) drew from ethnographic sources (Tindale 1953; Birdsell 1958; Hill and Hurtado 1996) to estimate gene flow rates of 1, 8, and 15% per generation. In the simulations, all females within a breeding population have an equal probability of moving to a neighboring breeding population. Gene flow takes place as a swap of an equal number of migrants between neighboring breeding populations (hexagons) each generation.

The significance of the model is estimated statistically using Monte Carlo simulations, a method that allows you to generate pseudo-samples from a small sample size that theoretically behave similarly to the underlying distribution of data. The simulations use the following logic and are graphically depicted in Figure 10.1.

1. The observed initial sample is expanded to a given breeding population size through random sampling of the initial haplogroup frequencies.
2. One to 18 neighboring populations are generated in the same manner.
3. All populations undergo a round of reproduction, leading to stochastic changes in genetic (in this case, mtDNA haplogroup) frequencies. (This is the "drift" part of the simulations.) Each female has the same probability of mothering any female child.
4. Neighboring populations exchange migrants, leading to gene flow.
5. Steps 1 through 4 are repeated, based on the number of generations estimated to separate the initial and final observed samples.
6. A sample equal to the size of the observed sample is taken from the simulated final population.
7. The test statistic – a measure of the differences between the initial and final simulated mtDNA haplogroup frequencies – is calculated as the sum of absolute differences of the frequency of each mtDNA haplogroup.
8. Steps 1 through 7 are repeated at least 1000 times to ensure close enough estimation of a predicted probability distribution for the test statistic.
9. The significance of difference between the initial and final *observed* populations is determined using the probability distribution estimated in Step 8 using *simulated* populations.

Nsitu uses genetic data derived from two samples that temporally bracket the proposed period of migration. In the case of the Great Basin, data from the

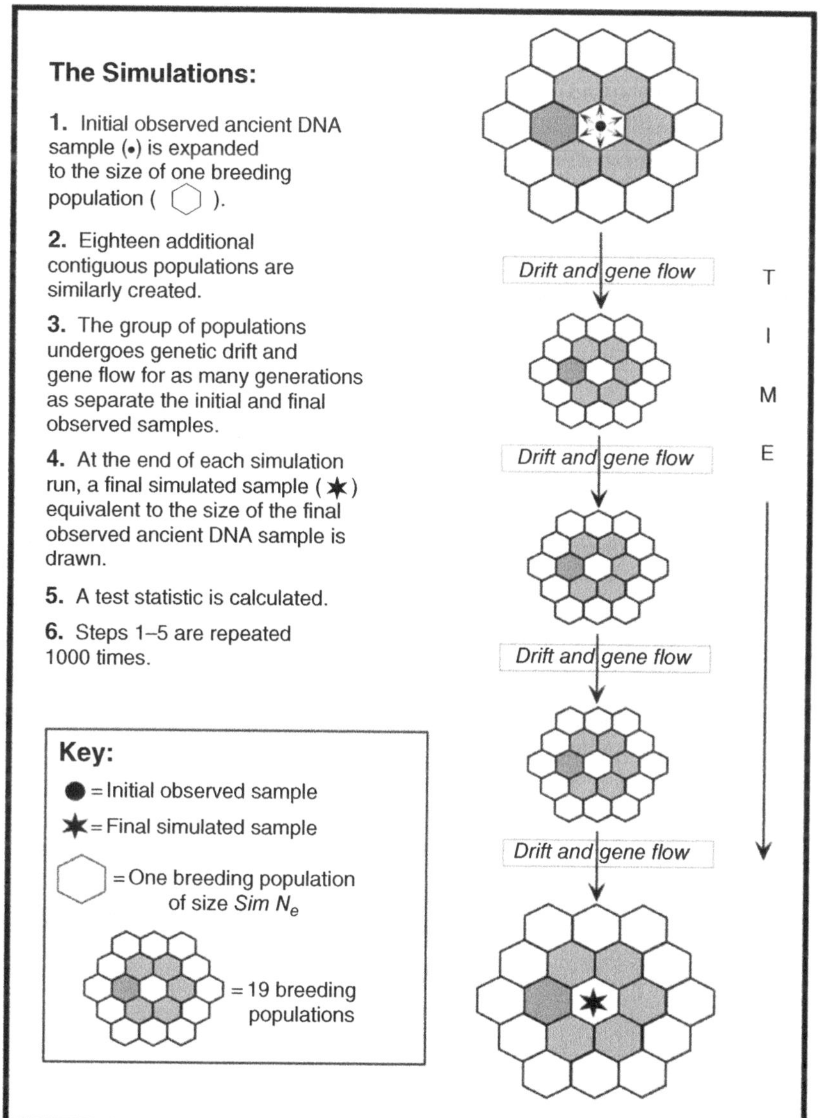

Figure 10.1 Simulations. *Source:* © John Wiley & Sons, Inc. Reproduced with permission from Cabana et al. (2008).

Stillwater Marsh archeological sample and from a modern sample of Northern Paiute individuals were input into the program. The breeding population structure underwent as many rounds of genetic drift and gene flow as there are generations separating the two samples. The number of generations used by the model was estimated by dividing the number of years separating the ancient Stillwater Marsh sample from the modern sample by a 27-year generation length

(Weiss 1973; Fix 1977; Fenner 2005). This estimate assumes that sampling of modern Northern Paiute individuals occurred at 1970 CE. Generations are modeled as non-overlapping.

Cabana et al. (2008) ran multiple simulations under multiple conditions – that is, varying female effective sizes, gene flow rates, and the number of generations separating ancient and modern populations in the area. They conclude that, depending on the specific conditions, a null hypothesis of population continuity through time cannot be readily rejected. In general, however, they demonstrate that hypothesis rejection can only occur with a relatively high (8%) rate of mate exchange between groups.

Conclusion

The studies by Gokcumen et al. (2011) and Cabana et al. (2008) represent much more than passive applications of microevolutionary theory developed in a prior era by modern synthesists. Rather, they are excellent examples of the ways in which anthropologists trained in biology and genetics conduct regional studies to understand the human-specific ways that microevolutionary change may take place, at least at a regional level. While understanding patterned variation and microevolutionary process in modern contexts has its challenges, these challenges increase by 100-fold when attempting to do the same in ancient contexts. However, the combined works of anthropological geneticists since the 1950s show that the past decades have seen considerable progress in this area. We can only expect these studies to continue advancing as new generations of DNA-sequencing technologies become increasingly accessible in terms of efficiency and cost, along with relevant bioinformatics tools to process the resulting data.

CHAPTER 11

Genetic Variation and the Colonization of Eurasia

Toomas Kivisild

Department of Human Genetics, Katholieke Universiteit Leuven, Leuven, Belgium

Hominins with diverse skeletal morphologies and material culture have inhabited Europe and Asia over the last two million years. Yet, considering this deep history of the territorial occupation of the continents, the genetic diversity in modern-day Eurasian populations is unexpectedly low. The overall low genetic diversity shows a further decline in regions that are geographically distant to Africa and supports a model that suggests that most present-day genetic diversity derives from a recent out-of-Africa dispersal and a serial effect of drift and founder events (Ramachandran et al. 2005). According to this model, the relative contribution of the archaic humans to the modern day has been minor, although consistently all non-African populations have been found to share more derived alleles with archaic humans than populations from Africa (Green et al. 2010). Although the levels of allele sharing between modern populations and the archaic humans have been found to be as low as 2–6%, at least two independent archaic admixture events have been inferred to have taken place shortly after the out-of-Africa migration – one early before the split of regional populations, and the other specifically at the route to Oceania (Reich et al. 2010). Analyses of whole-genome sequences of modern samples and ancient DNA (aDNA) evidence from late Pleistocene human remains suggest that East and West Eurasian populations started to diverge genetically from each other 40–50 kya (kya = thousand years ago) and that major changes in the genetic structure of local regional populations have continued to take place until very recent times (Fu et al. 2013a, 2014; Seguin-Orlando et al. 2014). Two

A Companion to Anthropological Genetics, First Edition. Edited by Dennis H. O'Rourke.

of the key events that have further shaped the genetic diversity of human populations in Eurasia in the past few tens of thousands of years have been the climate change, in particular the Last Glacial Maximum (LGM), and changes associated with the shifts to sedentary life in the Neolithic. As a result, genetic differences within and among populations at local geographic scales in different subterritories of Eurasia tend to be small.

Early and Middle Pleistocene Background to the Peopling of Eurasia

The earliest evidence of the presence of hominins outside Africa comes from a cave at Dmanisi, Georgia. The five Dmanisi skulls, which are morphologically highly diverse, have been dated to ca. 1.8 mya (millions of years ago) (Lordkipanidze et al. 2013). They are morphologically assigned either all to a single *Homo* species (Lordkipanidze et al. 2013) or, in fact, to as many as four different species (Schwartz et al. 2014). For how long these early colonizers of Eurasia continued to survive locally and how far their descendants spread are not precisely known because of the large gaps in the fossil record. However, similarities between the Dmanisi 5 and the large and robust remains of early *Homo erectus* from Java (Trinil/Sangiran) suggest morphological and possibly genetic continuity across large geographic distances in Eurasia (Lordkipanidze et al. 2013). Some of the Middle Pleistocene fossil findings from Eurasia suggest the preservation of significant proportions of the original diversity of the African *H. erectus* in the Eurasian descendant populations over more than a million years. It is possible that the latest descendants of *H. erectus* with characteristically small body size survived in the Indonesian island Flores up to 18 kya, also showing the extent of morphological diversity and the flexibility of adaptive response of our genus (Brown et al. 2004). Nevertheless, despite the widespread and notable morphological diversity of these early Eurasian genus *Homo* populations, no detectable genetic traces of them have been found in living populations.

Claims for unusually divergent mitochondrial DNA sequences of LM3 from Lake Mungo, Australia, that could have stemmed from a late *H. erectus* population (Adcock et al. 2001) are more easily explained by postmortem damage (Gilbert et al. 2003).

An increasing number of Middle Pleistocene (0.3–0.7 mya) fossils from Europe, Africa, and western Asia (see also Figure 11.1) are commonly classified as belonging to the same loosely defined species called *Homo heidelbergensis* (for reviews, see Stringer 2002; Bae 2010). The 300–400 kya estimate, which has been inferred from autosomal genome-wide data as the split time of the populations that gave rise to anatomically modern human (AMH) and Neanderthal populations (Meyer et al. 2012; Prüfer et al. 2014), suggests that *H. heidelbergensis* is indeed the most likely candidate species to be called the

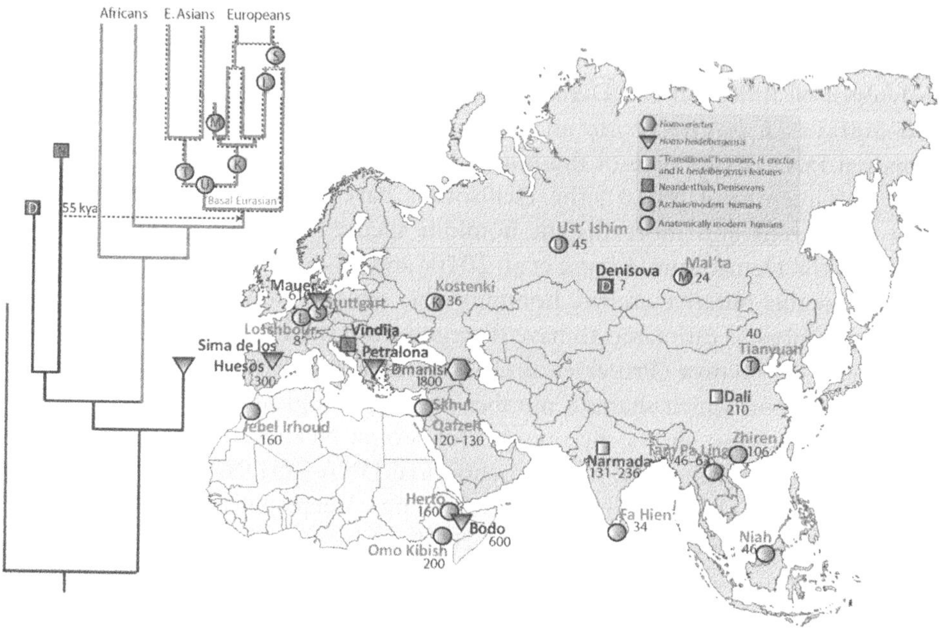

Figure 11.1 Map of Eurasia with key sites and a phylogeny relating modern and ancient DNA diversity. Approximate age estimates of the skeletal remains, in thousands of years, are indicated near the label of each site in black font.

direct ancestor to both modern humans and Neanderthals, as suggested earlier by a synthesis of paleoanthropological, archeological, and mitochondrial DNA data (Stringer 2002; Endicott et al. 2010). The phylogenetic relationship of the South and East Asian hominins from the Narmada River Valley in India and Dali Cave in China to the hominins of West Eurasia has not been clear – both Narmada and Dali specimens show combinations of ancestral and derived features that relate them with *H. erectus* as well as *H. heidelbergensis* fossils (Bae 2010; Wu and Athreya 2013).

The currently emerging model, which is now supported by genetic evidence from a large number of independent loci from across the whole genome, suggests that while the majority of the genetic variation of present-day Eurasian populations derives from a recent African dispersal, a minor fraction of it is due to genetic introgression from two distinct sources of archaic humans: Neanderthals and Denisovans (Green et al. 2010; Reich et al. 2010). Studies comparing archaic and modern human genomes have estimated that 1–2% of genetic variation in all non-African populations is derived from an admixture with Neanderthals (Green et al. 2010), whereas another 6% fraction, specifically of Oceanian genomes, was found to have genetic ancestry in another archaic species, the Denisovans (Reich et al. 2010; Meyer et al. 2012). In the light of this general model implying two archaic introgression events and a common descent from a *H. heidelbergensis* ancestor, the mitochondrial DNA evidence from the Denisovans has, however, been puzzling (Krause et al. 2010). In contrast to the model derived from the analyses of Neanderthal and Denisovan autosomal genomes, suggesting their ca. 300–500 kya *H. heidelbergensis* ancestor, which is supported also by the mitochondrial DNA–based date of the human and Neanderthal lineage split at 250–650 kya (Green et al. 2010), the Denisovans have additional genetic ancestry from a different and more distant hominin species. The mitochondrial DNA lineage of the Denisovans (Krause et al. 2010) and ~3–6% of their nuclear genome come from this putative archaic hominin who, on the basis of autosomal data–based inferences, is estimated to have diverged from the human and Neanderthal ancestors 0.9–1.4 mya (Prüfer et al. 2014). Interestingly, the 300 kya Sima del Huesos hominin, which shares a number of morphological features with *H. heidelbergensis* and also shows the presence of several Neanderthal-specific derived traits of cranial morphology, had mitochondrial DNA (mtDNA) that was closely related to that of the Denisovans while showing a deep split time (~1 mya) with humans and Neanderthals (Meyer et al. 2014).

The autosomal chromosomes of non-African populations show significantly more archaic ancestry than their X chromosomes (Meyer et al. 2012; Prüfer et al. 2014). This difference could be caused either by sex-specific admixture, implying predominantly male contribution from Neanderthals and Denisovans, or by selection of archaic against genes residing on the X chromosome. Further systematic scans of 1004 present-day human genomes have shown that regions of the genome that have decreased amounts of Neanderthal ancestry are enriched for genes, consistent with selection, and, furthermore, that testis-expressed genes in particular show significantly lower Neanderthal ancestry (Sankararaman et al. 2014). All this suggests that selection against genetic variants that decrease male fertility may have been a reason that explains these patterns. It is possible, therefore, that purifying selection explains the lower counts of Neanderthal-derived segments in the X chromosome of non-African populations, because X is known to host a number of male hybrid sterility genes. Some other gene variants, such as the innate immunity gene *STAT2* haplotype (Mendez et al. 2012a) derived from Neanderthals and the Denisovan-derived *EPAS1* haplotype in Tibetans (Yi et al. 2010), have been argued to have reached high frequencies in modern-day populations due to positive selection.

Out of Africa and the Early Split of East and West Eurasian Populations

Genomic features that characterize all Eurasians in comparison to Africans, such as lower overall genetic diversity and specifically the fact that Eurasian diversity can be seen as a subset of African diversity, higher linkage disequilibrium, and an almost identical proportion of Neanderthal ancestry, are most easily explained by a model according to which all Eurasians derive their genetic ancestry from a single African deme that recently, within the last 50–100 kya, spread out of Africa and almost entirely, bar the few percent of archaic contribution, replaced preceding hominin populations (for a review, see Sousa et al. 2014). The earliest skeletal evidence of AMHs in Africa comes from the ~200 kya Omo Kibish and ~160 kya Herto sites in Ethiopia and a ~160 kya site, Jebel Irhoud in Morocco. The first AMH sites outside Africa in the Levant, Skhul, and Qafzeh have produced human remains dated to 120–130 kya (for a review, see Shea 2013) (Figure 11.1).

Outside the Levant, the earliest direct genetic evidence from AMHs in continental Eurasia has been obtained from the skeletal remains of the Ust'-Ishim cave in Western Siberia (Fu et al. 2014). Radiocarbon dated at 45 kya, the Ust'-Ishim boy shared more derived alleles with genomes derived from present-day East Asian samples than European ones, while being equidistant to modern-day East Asian and pre-Neolithic Europeans. This evidence suggests that he was derived from a population that had not yet split into the East and West Eurasian ancestral pools and that the West Eurasian genetic pool had received gene flow in the Holocene from the basal Eurasian lineage. In contrast to the Ust'-Ishim genome, the other two early modern human DNA sequences uncovered from the 36 kya Kostenki cave in Russia and 40 kya Tianyuan cave near Beijing, China (Fu et al. 2013a), sites show higher affinities to European and East Asian populations, respectively, suggesting altogether that the split between the two gene pools, ancestral to present-day European and East Asian populations, had taken place 40–45 kya. The homozygosity tracks of Neanderthal-shared alleles have been found to be substantially longer in the Ust'-Ishim and Kostenki genomes, consistent with the argument that they were derived from populations who had only recently admixed with the Neanderthal. On the basis of the lengths of the homozygosity tracks of Neanderthal-shared alleles, the admixture event was dated at 55 ± 7 kya (Fu et al. 2014). Altogether, the evidence from the ancient genomes of the first inhabitants of Eurasia suggests that they derive from a single out-of-Africa dispersal event that occurred at some point after 55 kya, and the split into two distinct gene pools occurred within the first 5000–15 000 years after that. Analyses of coalescent events within the genomes of European and East Asian individuals, using the pairwise sequentially Markovian coalescent model, have suggested that at that critical time period, the effective population size of the ancestors of Eurasian populations was close to its lowest (*Ne*~1200) point of a bottleneck (Li and Durbin 2011), which by the analyses of multiple genome sequences (Schiffels and Durbin 2014) had started long before 50 kya.

Analyses of whole-genome low-coverage sequence data of HapMap Eurasian and African samples together with high-coverage exonic data of 800 genes using a site frequency spectra approach, ∂a∂i, have estimated the split time of European and East Asian populations at 23 kya (Gravel et al. 2011). This estimate appears to be more than 15 000 years younger than the date of the split of the Eurasian ancestral pool into its eastern and western compartments, derived from the genetic analyses of the late Pleistocene remains referred to in this chapter. One of the possible reasons for the discrepancy in the results of modern and ancient DNA analyses and the unusually young split dates obtained is likely in our understanding of the mutation rate (Scally and Durbin 2012). Several earlier studies had converged on a phylogenetic mutation rate of ca. 1×10^{-9}/bp/year, which was derived from comparisons of human and chimpanzee sequences; however, direct, pedigree-based estimates (Lynch 2010; Roach et al. 2010; Kong et al. 2012) and aDNA (Fu et al. 2014) support a rate estimate that is twice as slow and would thus also provide more consistency between the genetic data–based split date of AMH and Neanderthal lineages and the evidence from the fossil record (Scally and Durbin 2012). The other reason that could contribute to the discrepancies between the fossils and genomic divergence between modern-day populations is their more complicated demographic history.

Human population histories typically involve splits of ancestral populations, followed by migration events or long-term gene flow among neighboring groups. Considering the combined effect of these processes, a bifurcating tree can often be a misleading model to represent the whole complexity of evolutionary history of a population. Several quantitative approaches have been developed, including those that can be scaled up to the analyses of whole-genome genotype data of worldwide populations, to infer the history of population splits and mixtures. One such approach, TreeMix, when applied on a global dataset of human genotype data, revealed a tree with an early split between East and West Eurasian populations, followed by a number of further admixture events (Pickrell and Pritchard 2012). The admixture events that were inferred by this approach showed that Hazara and Uyghur populations from Central Asia, consistent with the results of some other analytical approaches (Rosenberg et al. 2002; Xu et al. 2008; Lawson et al. 2012), derive approximately one-half of their genetic ancestry from West and the other half from East Eurasia. This admixture event has been estimated to have occurred 800 years ago (Patterson et al. 2012). More surprisingly, however, the TreeMix analyses also suggested that the Cambodian population draws approximately 16% of its ancestry from an unknown source that is predicted to derive from the basal branch of the tree, that is, before the split of the Eurasian ancestral population into its East and West Eurasian branches. Analyses based on another approach, Globetrotter, have associated, however, a similar admixture proportion with a source population related to modern-day Central, South, and East Asians, and estimated that this admixture dates back to recent historical times, 1362 CE, to the end of the Indianized Khmer Empire, whose fall has been hypothesized to relate to an influx of Tai-speaking ethnic groups (Hellenthal et al. 2014).

While all Eurasian populations show equal Neanderthal ancestry by the standard Patterson's D statistic, the application of a modified or "enhanced" D statistic (Meyer et al. 2012) reveals a difference between East and West Eurasian populations. Because the sharing of derived alleles with the Neanderthal can be due to either incomplete lineage sorting (a majority of the sites, including Africans and non-Africans) or introgression (a minority of the sites, only to be expected in non-Africans), Meyer et al. devised the enhanced D statistic that controls for the presence of only ancestral alleles in the African populations. They estimated the Neanderthal ancestry to be 24% lower in Europe than in eastern Asia and South America, suggesting either that there were two independent admixture events or that the Neanderthal ancestry has been diluted in the ancestors of Europeans by a more recent gene flow from Africa. Analyses of the genomes of Southeast Asian populations also have revealed an excess of Denisovan ancestry compared to West Eurasians, suggesting that the human and archaic interactions may have been more complex than currently understood (Skoglund and Jakobsson 2011). Nevertheless, the genetic contributions of the archaic humans to the genomes of present-day populations of Eurasia are estimated at a total scale of 1–2%, and the majority of the genetic diversity that is observed derives from a recent common source with African populations.

When placing the present-day patterns of low genetic diversity of Eurasian populations in the deep-time perspective, it is notable that mtDNA genetic diversity in early modern humans who were the first colonizers of Europe has been estimated to have been about twice lower than in present-day Europeans, while still ca. 1.5 times higher than that measured in Neanderthals (Fu et al. 2013b). Similarly, low genetic diversity over genome-wide loci has been observed in the remains of the oldest AMH from Eurasia, the Ust'-Ishim genome, pointing to the genetic bottleneck associated with the out-of-Africa dispersal (Fu et al. 2014). Archeological evidence from the time period following the out-of-Africa dispersal, between 45 and 20 kya, has been relatively sparse, yet it is supportive of higher population size and numerical supremacy of AMHs as compared to Neanderthals in Western Europe (Mellars and French 2011). Inferences of effective population sizes of major geographic regions in Eurasia, based on mtDNA (Atkinson et al. 2008) and Y-chromosome data (Karmin et al. 2015), have highlighted South and Southeast Asia as regions with consistently the highest genetic diversity over tens of thousands of years, preceding the phase of global population growth in the Holocene.

Genetic Diversity in South and Southeast Asia

South and Southeast Asia have been important gateways for human colonization of East Asia and Oceania. Archeological sites in the Jurreru and Middle Son River Valleys in India have provided evidence of the presence of hominins over a time period spanning from 140 kya to the Holocene (Petraglia et al. 2012). It is, however, unknown whether and to what extent there was a continuity of occupation of these sites by the same groups or species of hominins. It has been argued that archaic human groups may have continued to exist in India before and after the eruption of Mount Toba 74 kya (Petraglia et al. 2007). The earliest skeletal evidence of AMHs from the La Fien Cave in Sri Lanka at 34 kya, Tam Pa Ling in Laos at 46–63 kya (Demeter et al. 2012), and Niah Cave in Borneo (Barker et al. 2007), and arguments based on analyses of archeological and genetic data, have suggested, however, that all earlier hominin species inhabiting South Asia would have been replaced by the ca. 60 kya dispersal of AMHs with their distinctively African microlithic and "backed-segment" technologies (Mellars et al. 2013). Unlike the findings of 1–2% Neanderthal introgression in all non-African populations and around 6% contribution of Denisovans into the Oceanian populations, no genetic evidence has been found so far in the genomes of present-day Indian populations that would indicate additional introgression from the local archaic humans that existed in South Asia (Figure 11.1). However, the presence of some morphological features of modern humans, such as the chin in the Zhirendong skull dated to 106 kya (Liu et al. 2010), in combination with other more archaic features (Demeter et al. 2012), suggests that the colonization history of the region may have been more complex than currently recognized.

At present, four different language families – Indo-European, Dravidian, Austro-Asiatic, and Tibeto-Burman – are spoken in South Asia. The latter two, plus three additional language families (Tai-Kadai, Hmong-Mien, and Austronesian), are, in turn, widely spoken in Southeast Asia. Despite the overlaps in the distribution of two of the language families, similarities in demographic profiles, and high ethnic and genetic diversity (Atkinson et al. 2008), the genome-wide profiles of these two regions are quite different from each other (Li et al. 2008; Consortium et al. 2009). South Asian populations are genetically more closely related to West than to East Eurasian populations, and their genetic diversity can be broadly assigned to two major ancestry components, one of which (ASI) is more specific to South India and Bangladesh, while the other (ANI) to North India and Pakistan (Reich et al. 2009; Metspalu et al. 2011). The ANI component forms a gradient of declining frequencies from ~70% in the northwest to ~30% in southeast India, being the most common genomic feature in populations who today speak Indo-European languages. This ancestry component has been inferred to have shared origins with populations who have contributed to the genetic diversity of Europe and the Caucasus (Reich et al. 2009).

Mundari speakers of the Austroasiatic languages, in contrast to Indo-European and Dravidian-speaking populations of South Asia, carry a notable proportion (~23% in autosomal genes) of East Asian ancestry in their genomes, consistent with the model that suggests Southeast Asian origins for this language group (Chaubey et al. 2011). Notably, the mtDNA lineages of Indian Mundari speakers do not contain evidence for East Asian ancestry, whereas ~60% of their Y chromosomes belong to haplogroup O2a, which is common in East Asia and almost absent in Indian populations who speak either Dravidian or Indo-European languages. Haplogroup O2a2 variation, which is characteristic to Indian and

Southeast Asian Austroasiatic speakers, has been dated, on the basis of the whole Y-chromosome sequence analyses, to derive from the most recent common ancestor, who lived 5.3–7 kya (Karmin et al. 2015). These findings suggest that the Austroasiatic languages were likely brought to India from the east during the Neolithic and that their introduction and spread in India involved sex.

While in Southeast Asia, the spread of Austroasiatic languages today is restricted to the mainland, analyses of genome-wide genotype data have inferred a cryptic ancestry component among Island Southeast Asian populations, particularly those from Sumatra and Borneo, suggesting a wider spread of the Austroasiatic speakers in the past (Lipson et al. 2014; Pugach and Stoneking 2015). It is likely that the Austroasiatic-speaking populations were not completely replaced but rather were involved in the admixture during the more recent spread of the Austronesian speakers in Island Southeast Asia.

Peopling of Northeast Asia

Consistent with the autosomal data, both mtDNA and Y-chromosome variation show notable differences between the West and East Eurasian populations. Populations from Europe, the Caucasus, and West Asia draw their mtDNA variation mainly from out-of-Africa founder haplogroups N and R, while South and East Asians, along with Papuans, Australians, and Americans, have in addition to N and R also a significant proportion of their mtDNA diversity deriving from haplogroup M. Similarly, Y-chromosome Europeans and West Asians are characterized predominantly by two basal haplogroups (F and K), whereas East Asians also carry haplogroups C and D (for a review, see Underhill and Kivisild 2007). Consistent with the central position in the autosomal principal component (PC) plot of the Ust'-Ishim sample, his mtDNA sequence falls to the root of basal haplogroup R, and his Y chromosome to the base of haplogroup NO. Both of these haplogroups are common in East and West Eurasian populations, being represented by younger and region-specific lineages such as mtDNA haplogroups HV, JT, and U in West Eurasia, and haplogroups A–G in East Eurasia and the Americas. The fact that Native Americans draw their genetic ancestry from a mixed pool of East and West Eurasian autosomal chunks (Raghavan et al. 2014b) is not reflected in their mtDNA and Y-chromosome haplogroup profiles, which are derived only from the East Eurasian source. This shows that assessment of genetic variation in a small number of individual genetic loci, such as mtDNA and the Y chromosome, may not be sufficient to detect old admixture events among relatively closely related human groups, even if the genetic contribution of the minor group was at the scale of ~40%. This is in part due to the effect of random genetic drift and partly because of the issues with phylogeographic inferences that, for example, may be ambiguous in the assignment of Native American haplogroup Q variation as being more affiliated with either the East or West Eurasian genetic variation.

Distinction between archaic versus modern and African versus non-African variation is relatively easier to make than the distinction between East and West Eurasian variation. In Eurasia, the frequency of any other mtDNA lineages (i.e. those not belonging to haplogroups M and N) is, in general, extremely low and restricted to areas that have received historic gene flow from Africa, such as Mediterranean Europe and West Asia. The African lineages from haplogroups L0–L6 are found with a combined frequency of <1% in Europe, and it has been estimated that two-thirds of these rare African lineages may have been brought to Europe recently, since the Roman times (Cerezo et al. 2012). Analyses of genome-wide data have shown that all individuals from Southern Europe harbor 1–3% ancestry in their genomes that has been derived from Africans within the last 55 generations (Moorjani et al. 2011).

Archeological evidence suggests that Western and Southern Siberia was inhabited by anatomically modern humans soon after the earliest signs of their presence in the Levant and Europe, possibly as early as 45 kya (for a review, see Goebel et al. 2008). Further evidence from sites as far north as the Yana River Valley suggests human presence in the Arctic Circle by 30 kya (Pitulko et al. 2004). The oldest human remains in Siberia, from the Mal'ta site near Lake Baykal and dating to 24 kya, have successfully yielded DNA for genome-wide analyses. The analyses of the autosomal DNA sequence of the Mal'ta boy (Raghavan et al. 2014b), as noted in this chapter, have shown that his genome entails features that are ancestral to the genomes of the majority of modern-day West Eurasian populations, while at the same time containing chunks that are closely related to those found in modern Native Americans, with no close affinities to East Asian populations. These findings suggest that populations ancestral to the modern-day Europeans were already widely spread in Eurasia by 24 kya.

Present-day populations of Siberia, however, derive their genetic ancestry, in a deep-time perspective, from two distinct sources, with most populations of Eastern Siberia showing higher affinities to East Asian rather than European populations both in their autosomal genes (Rasmussen et al. 2010) and in mitochondrial and Y-chromosome markers (Karafet et al. 2002; Derenko et al. 2010). Four distinct population groups have been identified in Siberia (Cardona et al. 2014), with the Northeast Siberian group being closest to the Native American populations who, on the basis of aDNA analyses, have been inferred to have dual (38% West and 62% East Eurasian) ancestry (Raghavan et al. 2014a, 2014b; Rasmussen et al. 2014).

Recent Changes of Eurasian Genetic Diversity in the Holocene

The falling global temperatures and the wide spread of ice sheets during the LGM, which occurred 14–27 kya, had a major impact on the demography of many different forms of life on Earth, including human populations in Eurasia

(Hewitt 2000). Increase in global temperatures since the LGM and the lifestyle shifts related to the spread of farming are associated widely throughout Eurasia with increasing population sizes. These changes in the Holocene period have, since the use of classical genetics data, been further associated with the spread of the main language families and also substantial changes in the genetic architecture of regional populations (Cavalli-Sforza et al. 1994a).

Genetic similarities that can be observed between modern-day populations of Europe and the Near East have historically been interpreted in light of several opposing models, including those predicting either extremely large or small genetic contributions of Pleistocene and Mesolithic Europeans with respect to the incoming farmers during the Neolithic period (e.g. Richards et al. 1996, 2000; Cavalli-Sforza 1997; Chikhi et al. 2002). PC analyses based on the allele frequency data of classical genetic markers had revealed a genetic map of Europe that resembled the map drawn from the distribution of the earliest Neolithic sites (Cavalli-Sforza et al. 1994a). In contrast, phylogeographic reconstructions based on the distribution of mtDNA haplogroups suggested that a vast majority of mDNA founding lineages from haplogroups U, HV, I, H, and T were brought to Europe in the Upper Paleolithic or Mesolithic period, while only a minority of variation embedded in haplogroups J and T1 can be assigned to Neolithic gene flow from the Near East (Richards et al. 2000). Thanks to the advancements of new aDNA-sequencing methods over the past decade, the analysis of prehistoric remains has offered the opportunity to test these models and to assess precisely the dynamics of genetic change in a geographic area, such as Europe, in time.

aDNA studies have shown that the genetic landscape of Europe before the onset of the Neolithic was highly homogeneous, with just two subclades of mtDNA haplogroup U (U4 and U5) characterizing ~80% of variation in geographically distant hunter-gatherer groups of Iberia, Scandinavia, and Central and Eastern Europe (Haak et al. 2005; Bramanti et al. 2009). The haplogroup U4 and U5 lineages are still widely spread even today across Europe, albeit they occur at much lower frequencies of 5–20%, which along with autosomal evidence bears witness to some but not major levels of continuity between the Mesolithic hunter-gatherers and the present-day populations of Europe (Skoglund et al. 2012, 2014; Lazaridis et al. 2014). A major part of the genetic diversity in Europe, however, appears to derive from more recent gene flow during and after the Neolithic. Analyses of genomic DNA extracted from ~5000-year-old remains of three hunter-gatherers and one farmer from Sweden showed significant differences among the groups, with the hunter-gatherers clustering closely with the present-day North Europeans, while the farmer showed the highest affinity to South Europeans and clustered closely with present-day Sardinians (Skoglund et al. 2012). Further analyses of Scandinavian hunter-gatherer and farmer genomes have revealed that, similarly to the high mtDNA homogeneity, the hunter-gatherer groups of Northern Europe were characterized by low genetic diversity; and, while the Scandinavian farmers showed evidence of assimilation of hunter-gatherer genetic components on the background of general affinities to South European populations, no significant evidence of gene flow from farmers was observed in hunter-gatherers (Skoglund et al. 2014).

The spread of the Neolithic in Europe and the interactions between the first farmers and the hunter-gatherer groups have turned out to be more complex than suggested initially by the proponents of the wave-of-advance models, which emphasized steady population growth and gradual assimilation of the local foragers. Instead of a view whereby the wave of advance of the early farmers would have established their numerical supremacy and their genetic profile in Europe since the first appearance of the farmers, analyses of mtDNA variation from a large number of samples associated with the Neolithic in Central Germany suggest that the Neolithic was a period of at least four major genetic changes (Brandt et al. 2013). mtDNA lineages from the LBK horizon, which was one of the earliest widespread Neolithic cultures in Central Germany, are significantly different in their composition from the DNA extracted from the remains from the Middle and Late Neolithic periods. These results suggest that the genetic profile of the earliest farmers became substantially altered in the Middle Neolithic and that only since the very Late Neolithic period one can observe similarities in haplogroup frequencies between individuals sampled in the same geographic area, such as Central Germany or Hungary (Brandt et al. 2013; Szecsenyi-Nagy et al. 2015).

Evidence for the complex pattern of genetic admixture wrought into the genetic history of European populations has also been obtained from the genomes of present-day populations. Analyses of genotype data of the HGDP panel with f-statistics (Patterson et al. 2012) have, consistent with the aDNA results (Skoglund et al. 2012), revealed a clear signal of admixture in Northern Europeans involving at least two distinct sources, one of which was shown to be closely related to extant Basque and Sardinian populations, with the other showing highest affinities to present-day populations of Northeast Asia and the Americas. The signal of Northeast Asian and American affinities in modern North Europeans is likely to reflect the continuity of the genetic ancestry inherited from the local hunter-gatherer groups, who, in turn, had derived their ancestry from the groups who were widely spread in Northern and Eastern Europe while being related to the ancestral groups in Siberia from which the Mal'ta boy was derived. Further analyses of the whole genomes derived from the aDNA of Mesolithic and Neolithic remains from Germany, Luxembourg, and Sweden have revealed that most present-day Europeans draw their genetic ancestry from at least three distinct source groups (Lazaridis et al. 2014). Besides the component ascribed to the Neolithic farmers, it appears that the hunter-gatherer contribution also derives from two highly differentiated sources, one having more western and the other more eastern distribution in Europe. The early European Farmers were inferred to have had ancestry from both West European hunter-gatherer and Near Eastern sources, and it was estimated that 44% of their ancestry was drawn from a "basal Eurasian" source before its split into East and West Eurasian populations.

Analyses of aDNA of 69 skeletal remains from the period of 8 to 3 kya in Europe have revealed that a further major genetic change took place in the Late Neolithic (Haak et al. 2015). According to the aDNA evidence, two genetically distinct pools of populations could be discerned at the end of the Middle Neolithic: one in Central and Western Europe drawing ancestry from West European hunter-gatherer groups and the early European farmers, and a second group of East European farmers whose ancestors were East European hunter-gatherers on one hand and Near Eastern farmers on the other. At the onset of the Late Neolithic, apparently a major change took place in the genetic composition of the inhabitants of Central and Western Europe. The results are based on the analyses of autosomal DNA of the skeletal remains associated with the Corded Ware culture from Germany, and they show that, in contrast to individuals buried in the same region before, roughly three-quarters of the genome of individuals buried after 4.5 kya shared their genetic ancestry with the remains from the spread area of the Yamnaya in the plains of Eastern Europe. Comparisons of the autosomal genes of the Late Neolithic farmers from Central Europe with those of present-day populations suggest continuation of a substantial proportion of their genetic legacy. It has been argued that the influx of cultural and genetic elements from the steppe belt in the Late Neolithic may have been associated with the spread of Indo-European languages to Europe according to the Gimbutas model. It is possible, though, that other dispersal events in the Early and Middle Neolithic may have, consistent with the Renfrew model, brought Indo-European languages to Europe from their Anatolian homeland.

The strongest impact of the Middle to Late Neolithic change has been left on the patterns of European Y chromosomes. The majority of European men today have Y chromosomes belonging to haplogroup R1. Two subclades of R1 differ in their geographic spread patterns: R1a occurs at frequencies of up to 50% in Eastern Europe (Underhill et al. 2015), while R1b is found at frequencies of up to 90% in Western Europe (Myres et al. 2011; Busby et al. 2012; Batini et al. 2015). Variation in both subclades has been estimated to derive from ancestral sequences that are 4.5–5 kya (Karmin et al. 2015). According to the aDNA evidence, haplogroup R1 lineages were extremely rare in Europe before the Late Neolithic, with only a single Spanish Early Neolithic individual detected in a combined sample of 70 (Haak et al. 2015). In contrast, all nine samples from Russia dating to 4.7–9.5 kya were affiliated with haplogroups R1a and R1b, and also 60% of Late Neolithic men from Central and Western Europe belonged to these two clades. At the same time, the change detectable in the mtDNA lineages seemed to have been smaller in its magnitude. Interestingly, this period of change in Central Europe also corresponds to the time when the male versus female effective population sizes have been estimated to have reached their minimum of 1/17, arguably due to the effect of culturally inherited male fitness (Karmin et al. 2015).

Analyses based on ancestry-painting methods that are designed to have maximum power for detecting and dating recent admixture events, such as

Globetrotter (Hellenthal et al. 2014), have also detected three major ancestry components in Europe – one related with Northern European ancestry, one with Southern European, and one with East Asian. However, along with the demographic processes that may have unfolded in the Neolithic, the ancestry-painting methods when applied on modern-day European data also reveal admixture dates that are consistent with more recent episodes of gene flow, such as the Slavic expansion; dispersals of Huns, Magyars, and Bulgars in Europe; and admixture processes related to the spread of the Mongols in Russia and Central Asia, and potentially recent Greek ancestry in the Kalash of Pakistan (Hellenthal et al. 2014). Application of these methods on genome-scale data also allows for the detection of significant elements of substructure within lowly differentiated population groups, such as the people of the British Isles (Leslie et al. 2015).

The development of new bioinformatics methods in combination with whole-genome sequence data from projects such as the 1000 Genomes Project, along with the dawning genomic era of aDNA studies, are likely to continue in the near future to provide crucial new insights about the complexities of the demographic history in Eurasia.

CHAPTER 12

Homo sapiens Dispersal and the Peopling of the Americas

Rafael Bisso-Machado[1] *and Nelson Jurandi Rosa Fagundes*[2]

[1] *Universidad de la República, Tacuarembó, Uruguay*
[2] *Universidade Federal do Rio Grande do Sul, Porto Alegre, Brasil*

Introduction

A recurrent question concerning human evolution is something like "What exactly makes us human?" This is an astonishingly complex question, for which no simple answer can be given (see Calcagno and Fuentes 2012). Nonetheless, it seems safe to propose that, among extant animals, only humans are curious about their origins. Only humans wonder where they come from, and, indeed, creation beliefs are nearly universal among human cultures. In modern societies, the quest for such knowledge have a strong ally in science and its several subdisciplines, such as archeology, paleoanthropology, linguistics, and, more recently, genetics and molecular biology.

Biological anthropology is a dynamic research field with the far-reaching goal of understanding *Homo sapiens* as a biological entity (for a more formal definition, please see Jurmain et al. 2013). In other words, human ecology, its evolutionary history, and the interaction among human populations can all be studied under

A Companion to Anthropological Genetics, First Edition. Edited by Dennis H. O'Rourke.

the lens of biological anthropology, based on different datasets and theoretical backgrounds that underlie the different subdisciplines that compose this field. The biological anthropological approach is particularly useful when we try to understand aspects of ancient or extinct populations, or for contemporary groups for which no written record is available.

The aim of this chapter is to present a brief review on the arrival of *H. sapiens* in the Americas, the last landmass reached by our species, and its subsequent colonization (for other recent reviews on this subject, please see O'Rourke 2011; Raff et al. 2011; Salzano 2011; Marangoni et al. 2014; Bisso-Machado et al. 2016; Bolnick et al. 2016; Hoffecker et al. 2016).

Homo sapiens' Origin: Setting the Stage for the Peopling of the Americas

Fossil evidence indicates that modern *H. sapiens* first appeared in Africa, with the oldest fossil of a specimen identified as modern *H. sapiens* found in Omo Kibish, Ethiopia, dated at ~195 kya (thousands of years ago; Day 1969; McDougall et al. 2005; Pearson et al. 2008a, 2008b). The recognition of older *H. sapiens* fossils (reviewed by Stringer 2016) – including the remains from Jebel Irhoud, Morocco, recently dated from 315 kya (Hublin et al. 2017; Richter et al. 2017) – could indicate a more gradual origin of the "modern suite" of the derived craniofacial morphological traits associated with *H. sapiens* (Stringer 2016; Stringer and Galway-Witham 2017).

The traditional view of an African origin of *H. sapiens* is also supported by genetic studies using DNA samples from contemporary individuals (for a critique of the African origin model, see Árnason 2016). Early studies based on the mitochondrial DNA (mtDNA) and relying on the molecular clock theory (for a review, see Bromham and Penny 2003) showed that the root of human mtDNA tree lies in Africa, at about 160 kya (Cann et al. 1987; Vigilant et al. 1991). This value did not change much when the first estimates based on complete mtDNA genomes were published (i.e. 177 kya; Ingman et al. 2000). Because the root of the mtDNA tree represents the maternal ancestor of all humankind, this ancestor was called "Eve," somewhat infamously. However, the existence of a single common ancestor to all modern lineages is merely the consequence of finite population size, which guarantees that modern lineages will find (coalesce) common ancestors at a rate inversely proportional to the effective population size (Hartl and Clark 2006). Thus, the mitochondrial Eve was neither the first living woman nor the only woman in a given human population, but the woman carrying the ancestral mitochondrial lineage from us all (Stoneking 1994). Curiously, the most recent common ancestor for the human Y chromosome ("Adam," by analogy, given that the Y chromosome has exclusive patrilineal inheritance) is much more recent than Eve, dated about 60 kya (for a review, see Jobling and Tyler-Smith 2003), reflecting a smaller population size for men, rather than a temporal reproductive isolation of the biblical couple. Further studies with different sets of markers, including those of the nuclear genome, have been systematically supporting the Out-of-Africa hypothesis (see as examples Gravel et al. 2011; Li and Durbin 2011; McEvoy et al. 2011; Lukic and Hey 2012). Linguistic data have also been taken in support of an African expansion (Atkinson 2011), although this interpretation has been criticized by other researchers (Cysouw et al. 2012; Jaeger et al. 2012; Van Tuyl and Pereltsvaig 2012; Wang et al. 2012).

The first migration of anatomically modern *H. sapiens* out of Africa probably occurred between 140 and 52 kya during an interglacial period (Ingman et al. 2000; Gutenkunst et al. 2009; Armitage et al. 2011; Lukic and Hey 2012). In Europe, the first evidence for *H. sapiens* presence is dated between 40 and 30 kya (Dodge 2012; Haber et al. 2013). However, other hominins had been living in Europe, Asia, and Oceania by the time *H. sapiens* left Africa and first arrived in these regions. Therefore, understanding how modern humans interacted with other hominid species became a fundamental study subject in paleoanthropology. While the fossil and archeological record is not clear about the consequences of such probable contacts (Schwartz and Tattersall 2010; Dodge 2012), genetic studies based on complete genomes of extinct hominids revealed that *H. sapiens* admixed with both Neanderthals and Denisovans (Green et al. 2010; Reich et al. 2010). According to available genomic data, non-Africans (non-Sub-Saharan Africans, to be precise) have 1–7.3% nuclear DNA of Neanderthal origin (Green et al. 2010; Sánchez-Quinto et al. 2012; Lohse and Frantz 2014), while contemporary Melanesians and populations from Papua New Guinea inherited 4–6% of their DNA from Denisovans (Reich et al. 2010).

This evidence of hybridization between modern humans and other representatives of the genus *Homo* raises questions about whether these are distinct species (Gibbons 2002). While there are some compelling examples of archaic alleles favoring adaptation in modern humans (Racimo et al. 2015), genomic data suggest that natural selection acted against genetic introgression, at least in the case of hybrids between modern humans and Neanderthals (Harris and Nielsen 2016; Juric et al. 2016). Thus, the complete population replacement of archaic hominids postulated by the classic Out-of-Africa model is not compatible with genomic data, which offers much better support to "assimilation" models (Alves et al. 2012). It is therefore curious that while it is true that only modern humans colonized the Americas, genes from archaic lineages survived long enough to be part of this story.

The Peopling of the Americas

The geographic origin(s) of the migration

Already in 1589, the Spanish Jesuit José de Acosta suggested that some hunter-gatherer groups had come to America from Asia, through what now is the Bering Strait, following big-game animals (Mann 2006; Santos et al. 2007), and many similar theories were widespread in the eighteenth century (Kodama 2009). Centuries later, studies with classical genetic markers (blood groups and other protein systems having two or more known alleles in human populations) corroborated the Asian affinities of Native Americans (Spuhler 1972; Salzano et al. 1986; Cavalli-Sforza et al. 1994; Crawford 1998; O'Rourke 2006).

Compared to classical markers, the study of mtDNA sequences provide much higher resolution, because (i) given the non-recombining nature of mtDNA, different mtDNA "alleles" (haplotypes) can be arranged into genealogical clades (haplogroups); and (ii) the mutational distance between two haplotypes could be inferred given their genetic similarity. Based on similarities of haplotype and haplogroup composition among populations, investigators began to search for the putative geographic "origin" of Native American populations, as well as which Asian population would more closely "resemble" Native Americans. The first studies based on mitochondrial haplogroups postulated that they would have come from Siberia and/or Mongolia (e.g. Ballinger et al. 1992; Torroni et al. 1993a, 1993b; Forster et al. 1996; Merriwether et al. 1996). For the Y chromosome, which also has haplotypes analogous to those in mtDNA, some authors have argued that the original migrants originated in the area surrounding Lake Baikal (Karafet et al. 1999), in southwestern Siberia. It has also been suggested that the people living in the Yenisei river basin and in the Altai mountains would be expected to share common ancestry with Native Americans (Santos et al. 1999; Wells et al. 2001; Seielstad et al. 2003; Zegura et al. 2004; Starikovskaya et al. 2005; Dulik et al. 2012a). Bortolini et al. (2003) suggested that the origin of Native Americans could be attributed to migrants from the north of Mongolia or even southwestern Siberia, and confirmed the putative Eurasian connection previously identified by Santos et al. (1999).

From a more modern perspective, however, it might be argued that looking for *the* specific ancestral Native American population among current Asian populations is naïve due to several reasons. First, human populations are not static over time. They evolve. Even if we leave mutation (due to the limited time scale) and natural selection (as most of the human genome seems to fit neutral expectations) aside, genetic drift and migration (gene flow) may be sufficiently strong to change the genetic composition of populations across space and time (e.g. Schurr et al. 1999; Raff et al. 2011). Second, mtDNA and the Y chromosome represent only two unique genealogies among thousands of alternative genealogies that, collectively, "tell" geneticists about the true population history. Coalescent theory has taught us that the story of a single gene may not be representative of the history of the populations where it was drawn (for a review, see Degnan and Rosenberg 2009). In other words, why should we focus so much on mtDNA and Y chromosome? Rather than interpreting *the* mtDNA genealogy (or *the* Y chromosome genealogy), we should interpret a set of genealogies derived from our samples (Fagundes 2016). What about a genomic perspective? At this point, it is important to highlight that while haploid uniparental markers had their "golden age" due to the reasons highlighted in this chapter, the genomic era had to await technological developments to show its power, because a single (or a few) autosomal marker is less informative than uniparental markers. Until it became possible to screen hundreds, or thousands, of autosomal genetic markers at a reasonable cost, uniparental markers provided a better option. The development of next-generation sequencing methods allowed the study of full genomes and the screening of hundreds of thousands of genome-wide genetic markers. This definitely changed the game in favor of genomic autosomal markers, giving investigators with sufficient statistical power to detect very subtle genetic diversity patterns (see Destro-Bisol et al. [2010] for a general review, and Bolnick et al. [2016] for a focus on the peopling of the Americas).

The analysis of a recently published genome from a 24 kya specimen known as the Mal'ta boy (MA-1), from south-central Siberia, provided interesting results (Raghavan et al. 2014a). The authors found that MA-1 is more closely related to western Eurasian populations, but also contributed to the makeup of the Native American genome (about 14–38% of Native American ancestry would come from MA-1) by admixture with a Native American ancestral population after its divergence from modern eastern Eurasian populations. These results should be taken with some caution since the MA-1 genome has been sequenced with low coverage. However, if confirmed by further analyses, these results illustrate that no modern East Asian population appropriately represents the most likely ancestor of Native Americans. Another cautionary note concerns statements that Native Americans have "strong affinities with Europeans," which seems to imply that the data support a direct migration from Europe or ancient affinities with modern European populations, which is clearly not the case (e.g. Yong 2013).

There is, however, an alternative hypothesis implying that Native Americans descended partially from Europe. The *Solutrean hypothesis* proposed that migration would have occurred across the glaciers that connected Europe and North America during the Last Glacial Maximum (LGM, from 26.5 to 20–19 kya; Clark et al. 2009), and was based on similarities between the lithic material found in sites in Europe and in eastern North America (Bradley and Stanford 2004; Meltzer 2009). From the genetic point of view, the discovery of the rare mtDNA haplogroup X in Native Americans was originally taken as possibly compatible with a "Caucasian" founder population (Brown et al. 1998). Almost all Native American mtDNA belongs to haplogroups A, B, C, and D, lineages that are also common in East Asian populations (even though Native Americans and Asians have different subhaplogroups; for a review, see Bisso-Machado et al. 2012). Haplogroup X, on the other hand, is rare in both Native Americans and Asians, and is more frequent in Europe, thus making the Solutrean hypothesis a possible alternative to explain its appearance among Native Americans. Recently, the "West Eurasian ancestry" identified in genomic studies (Raghavan et al. 2014a) has been interpreted by adherents of this hypothesis as evidence in favor of an ancient European migration to the Americas (Oppenheimer et al. 2014). According to them, the presence of

West Eurasian ancestry in the genome of the Anzik-1 individual (Rasmussen et al. 2014), whose burial is associated with the (presumably Solutrean-influenced) Clovis lithic culture, corroborates the Solutrean hypothesis.

Given the current evidence, however, the Solutrean hypothesis faces severe problems (Raff and Bolnick 2015). First, the similarities between the technologies have been interpreted as cultural convergence (Straus et al. 2005). Second, complete mtDNA genome analyses show that Native Americans and Eurasians have different X haplogroup lineages, and that Native American lineages display the expected diversity patterns for a typical Native American founder haplogroup (Achilli et al. 2008; Fagundes et al. 2008). Third, genomic studies that have identified West Eurasian ancestry in Native Americans have never interpreted their findings as representing a direct migration from Europe, but rather that an ancient population contributed to the ancestral populations of both modern Native Americans and modern Europeans (as discussed in this chapter). Furthermore, if the Mal'ta population contributed ancestry to all Native American populations, then it is quite trivial that the Clovis-culture Anzik-1 individual would also show this pattern. Thus, it seems clear that there is currently no clear genetic evidence in favor of the Solutrean hypothesis.

A few common mutation patterns in mtDNA haplogroup B between Polynesians and Native Americans led some researchers to consider the Pacific Ocean as another possible alternative route of colonization (Cann 1994), which was refuted by Bonatto et al. (1996) on the grounds that Native American and Polynesian haplotypes are only distantly related. However, the hypothesis of a direct migration from Polynesia to the Americas gained new strength with the finding that Botocudos, an extinct Native American group from central Brazil, had mitochondrial lineages characteristic of Polynesians (Gonçalves et al. 2013). To explain this unexpected result, the authors suggest that, rather than a trans-Pacific contact, the most likely scenario would involve the introduction of these lineages during the African slave trade, since they also occur in Madagascar, which is neighbor to Mozambique, a former Portuguese colony and known warehouse of African slaves to Brazil. Eventually, this story turned out to be even more complicated, as a genomic study of the two Botocudo individuals carrying Polynesian mtDNA lineages revealed that they have 100% Polynesian ancestry (Malaspinas et al. 2014). The authors suggest two alternative scenarios to explain this data. The first is that this finding represents genuine Polynesian expansion into South America. The second is that these individuals have been brought to Brazil in European trade ships. Both scenarios, however, fail to explain how (and why) these individuals were living among the Botocudos. A third alternative is that these crania represent genuine Polynesian individuals that were somehow mislabeled by the Museu Nacional (National Museum) staff. Even though this last possibility is considered unlikely by the authors, based on the available museum records (Malaspinas et al. 2014), it should be considered a plausible possibility as long as no trace of substantial Polynesian admixture is found in any other Native American population (for a discussion on the morphological variation of these individuals, see Strauss et al. 2015).

It is noteworthy that two recent genomic studies found traces of Australo-Melanesian ancestry in a few Native American populations from both North and South America (Raghavan et al. 2015; Skoglund et al. 2015). However, both studies offer slightly different interpretations for this genomic signal. Skoglund et al. (2015) identified Australo-Melanesian ancestry only among South American populations, and proposed that two major founder populations account for Native American ancestry. The first resulted from an ancient "admixture" between Mal'ta and an East Asian component (as discussed in this chapter). The second, called *Population Y* (from the Tupi word *Ypykuéra*, or ancestor) by the authors, resulted from a further "ancient admixture" event between the first founder population and another population closely related to Australo-Melanesians, which contributed only 1–2% ancestry to Population Y. On the other hand, Raghavan et al. (2015) were able to identify the Australo-Melanesian ancestry component in a few North American populations, including Aleuts and Athabaskans. These authors hypothesize that the Australo-Melanesian component could have reached the Aleuts through an East Asian continental route, and that further migration into North and South America could have resulted in the weak Australo-Melanesian ancestry signal scattered across the continent. Whatever is the correct interpretation for this surprising finding, understanding the peopling of the Americas will require a deep understanding of the evolutionary dynamics of human populations living in Northeast Asia in general, and Beringia in particular.

The role of Beringia

What is now known as the Bering Strait was a land mass of about 1 000 000 km^2 called Beringia, which remained exposed during most of the last glacial cycle in the Late Pleistocene, when sea level was 120 m lower than today (Peltier 2002). Despite its high latitude, Beringia was not covered by glaciers, even during the LGM ~25–15 kya (Brigham-Grette et al. 2003), allowing effective occupation by humans, especially in ecological areas covered by shrub tundra (Hoffecker and Elias 2007). Humans might have arrived in this area more than 30 kya (Pitulko et al. 2004), and, given the lack of evidence favoring earlier occupation, it is usually hypothesized that this region had not been inhabited by archaic hominids (Santos et al. 2007). Its configuration as a natural bridge between Asia and the Americas puts Beringia in a strategic position for the peopling of the Americas. However, it is getting increasingly clear that Beringia was much more important than as a mere corridor of passage (for a recent review on this subject, see Mulligan and Szathmáry 2017).

From the genetic point of view, an important role for Beringia was first devised by Szathmáry (1993) as a way to explain genetic differences between North American Amerind populations (sensu Greenberg et al. 1986), on one hand, and

Eskimo and Na-Dene on the other. According to her, these later groups would have remained longer in Beringia, turning them genetically more distinct than the lower latitude Northern Amerinds. This seminal idea was later invoked to accommodate the reduced mtDNA variation in Eskimo and Na-Dene (Bonatto and Salzano 1997a) and, more recently, gave rise to the *Beringian standstill model* (Tamm et al. 2007). In this model, genetic isolation in Beringia for ~10 kya would be necessary and sufficient to explain the mutational origin of autochthonous mtDNA haplogroups (Bonatto and Salzano 1997a; Tamm et al. 2007; Fagundes et al. 2008; Mulligan et al. 2008; Tackney et al. 2015; Llamas et al. 2016). The Beringian standstill model may also explain the accumulation of exclusive Native American substitutions in the Y chromosome (e.g. Pena et al. 1995; Bortolini et al. 2003). The most common Y chromosome lineage in Native American men, Q1a2a1a1 (Pena et al. 1995; Santos et al. 1999; International Society of Genetic Genealogy 2014), for example, is defined by a C→T transition at position 181 of the locus DYS199/M3 that must have originated in Beringia (Underhill et al. 1996; Lell et al. 1997; Santos et al. 2007). Similarly, isolation in Beringia may also explain the genetic differentiation in the nuclear genome between Native American and other Eurasian populations (Schroeder et al. 2007; Raghavan et al. 2015). It seems clear that the settlement of Beringia plays a crucial role in all current models for the peopling of the Americas, even though the precise duration of the "standstill" is under debate. Recent estimates are slightly shorter than 10 kya. An estimate based on complete mtDNA genomes suggest values between 2.4 and 9 kya (Llamas et al. 2016), while another estimate based on genomic data indicates an upper bound of 8 kya for the Beringian stage (Raghavan et al. 2015).

Finally, the recently sequenced Mal'ta boy genome (as discussed here; see also Raghavan et al. 2014a) and the inference of a minor genomic component of Australo-Melanesian ancestry among Native Americans (Raghavan et al. 2015; Skoglund et al. 2015) may indicate Beringia as a crossover of different populations. It was in Beringia that a western Eurasian group (represented by MA-1) met an eastern Eurasian group coming from Southeast Asia, to originate the ancestral population from whom most Native American genomic ancestry comes (Raghavan et al. 2014a). Following this event, this ancestral population might have partially admixed with other human populations migrating from the south and bringing this ancient Australo-Melanesian ancestry component, which is now scattered throughout the New World (Raghavan et al. 2015; Skoglund et al. 2015). Although these scenarios seem highly speculative at present, new genomic data from ancient Siberian individuals may corroborate this new role for Beringia.

Out of beringia: colonization routes

Even if the role of Beringia as the only (or, at least, major) route to the Americas is almost consensual, there is debate on which specific route the Beringian settlers followed on their way into the New World. Two alternative migration models include a coastal route following the Pacific Coast, and a "continental" route following the ice-free corridor that ultimately existed between the Cordilleran and Laurentide ice sheets. These alternative models, of course, are not mutually exclusive.

The strongest candidate for the most likely path followed by the earliest Native American ancestral migrants is the Pacific Coastal route (Braje et al. 2017). This represents an interesting change in perspective. When the Clovis archeological sites dated ~13.2 kya (Waters and Stafford 2007), in New Mexico, were thought to represent the oldest archeological sites in the Americas, the interior route through the ice-free corridor was highly popular (e.g. Meltzer 1993; Cann 1994; Starikovskaya et al. 1998; Santos et al. 1999). However, the increasing recognition of a myriad of sites older than Clovis (reviewed by Goebel et al. 2008; see also Waters et al. 2011; Halligan et al. 2016) led to a rethink regarding the likelihood of available routes. In particular, the Monte Verde site, the oldest uncontested archeological site in the Americas (between 14.5 and 18.5 kya; Dillehay et al. 2008, 2015), in southern Chile, posed a problem for interior routes. Apparently, Native Americans traveled (very) fast southward, and following the Pacific Coast was deemed a reasonable alternative (e.g. Mandryk et al. 2001; Fix 2002; Hoffecker and Elias 2003; Fix 2005; Waguespack 2007; Misarti et al. 2012). Indeed, genetic data from mtDNA (Tamm et al. 2007; Fagundes et al. 2008; Llamas et al. 2016), Y chromosomes (Roewer et al. 2013), and autosomes (Wang et al. 2007) support the Pacific Coast route. For example, Wang et al. (2007), based on microsatellite data, estimated that traveling through coastal lines was 10× less costly than traveling though the continent, and much more compatible with the available genetic data.

Given the current priority of the Pacific Coastal route, does this mean that the idea of human migrations occurring through the ice-free corridor should be completely abandoned? The ice-free corridor (also known as the Alberta corridor) is the result of the retraction of two major glaciers, the Laurentide and the Cordilleran, which coalesced near the eastern foothills of the Rocky Mountains during the LGM. After the LGM, a warming climate led to the retreat of the glaciers, opening an "ice-free corridor" between them (Meltzer 2009). However, the "opening" of the corridor is a minor issue concerning its ability as a viable biological corridor between Beringia and the unglaciated North America. Recent studies based on ancient DNA suggest that the bison (*Bison priscus*) would have used that route about 13 kya (Heintzman et al. 2016). This would be evidence, albeit indirect, that people also could have used this route at the same time. The glaciers also had been a real obstacle for the movement of other game animals, such as the elk (*Cervus elaphus canadensis*), until 15 kya (Meiri et al. 2013). In agreement with these estimates, a study based on macrofossils, pollen remains, and metagenomics found evidence of a sustainable steppe environment, including bison, by 12.6 kya (Pedersen et al. 2016). This later study tested four models of human occupation in North America and found that all models, including an ancient (older than 12.6 kya) north-south crossing of the corridor, were not compatible with data. However, the ice-free corridor could have served as a secondary route (later than 12.6 kya) in both north-south and south-north directions.

The consequences for the peopling of the Americas are immediate. It is highly unlikely that the first Native American populations used the ice-free corridor when moving south of the glaciers, as the corridor was inhospitable for humans by the time associated with the earliest archeological findings, including Monte Verde, other pre-Clovis sites, and the earliest Clovis localities (Pedersen et al. 2016). This is compatible with the idea that the Clovis culture expanded northward not earlier than 12.4 kya (Goebel et al. 2013). The spatial variation in the distribution of mtDNA haplogroups also has been interpreted by some researchers as evidence that the ice-free corridor may have served as a secondary route into the Americas (Perego et al. 2009; Hooshiar Kashani et al. 2012; Achilli et al. 2013). It may be tempting to associate this putative "secondary route" into the Americas to specific patterns of genetic heterogeneity in Native American populations, such as the signal of a scattered Australo-Melanesian ancestry discussed in this chapter (Raghavan et al. 2015; Skoglund et al. 2015), even though Raghavan et al. (2015) clearly favor a coastal route following the Aleutian Islands. Determining the precise importance of the ice-free corridor during the earliest movements of founder populations during the settlement of the New World must await better archeological and genetic evidence.

Time of entrance in the Americas and the number of migration "waves"

After discussing the geographic context for the peopling of the Americas, the next question is defining both the timescale for the peopling of the Americas and how human populations left Beringia going south of the North American glaciers. Traditionally, answering these questions, especially the latter, requires postulating migration "waves" (also called migration "streams" or "flows") from Asia into the New World. While we keep the wave/stream analogy, it is important to keep in mind that this is a simplification of the real story: the one about the range expansion of a human metapopulation whose subpopulations interact in complex ways.

Direct estimates The timing of the peopling of the Americas can be inferred directly, from the archeological record, or indirectly, based on linguistic or genetic methods for example. As discussed in this chapter, the oldest well-accepted archeological site in the Americas is Monte Verde, in southern Chile (between 14.5 and 18.5 kya; Dillehay et al. 2008, 2015). This is puzzling, given that the Americas must have been settled following a north-south route (e.g. Fix 2005). There is, however, accumulating evidence of other early dated sites between 14 and 15 kya (Halligan et al. 2016, and references therein), suggesting human populations were widespread in the Americas before the iconic Clovis culture emerged by

13.2 kya (Waters and Stafford 2007). On the other hand, there are no sites unequivocally older than the LGM (Dillehay 2009), which is problematic for hypotheses postulating a very early settlement of the Americas. These dates are in good agreement with scenarios based on paleoclimatological data, which indicate two time windows when migration to America would have been ecologically possible. The first is from 20 to 40 kya, and the second from 12.5 to 16 kya (Pielou 1992; Hoffecker and Elias 2007), with the vast majority of findings indicating that human populations effectively colonized the Americas during the later time window. An earlier estimate was recently published by Fariña et al. (2013), who obtained a dating of ~30 kya for megafaunal bones showing evidence of human modification (cut marks) from an archeological site in Uruguay. As with all early (pre-LGM) sites in the Americas, the interpretation is controversial (e.g. Fariña et al. 2014; Suárez et al. 2014). Other earlier sites in South America include Toca da Tira Peia (Lahaye et al. 2013), Vale da Pedra Furada (Boëda et al. 2014), and Santa Elina (Vialou et al. 2017), all in Brazil, dating between 22 and 24 kya. As will be further discussed here, it is challenging to accommodate these dates with current genetic models. One alternative is that molecular models have been underestimating the antiquity of the peopling of the Americas. Another possibility is that if such early migration took place, it was carried out by individuals who eventually did not contribute to the genomic ancestry of Native American individuals studied to date.

Indirect estimates: the Greenberg et al. Model and uniparental molecular studies The linguist J.H. Greenberg, along with physical anthropologist C.G. Turner II and the geneticist S.L. Zegura, proposed a model of colonization for the Americas that sought to integrate data from different disciplines (Greenberg et al. 1986). They suggested that the Americas had been settled after three separate migration waves, each corresponding to three distinct linguistic phyla: Amerind, Na-Dene, and Eskimo-Aleut. The first migration would have occurred around 12 kya, giving rise to Amerinds, while the second would have led to Na-Dene, around 9 kya, and the third to Eskimos and Aleuts, around 4 kya. Even though Greenberg's linguistic classification has been heavily criticized and discredited by linguists (e.g. Bolnick et al. 2004), it has been continuously cited by other authors (mostly geneticists) to the point that it became some sort of a null hypothesis for the peopling of the Americas. However, differently from statistical null hypotheses, it is uncertain if Greenberg et al. have any ability to account for the observed genetic data.

Since the first mtDNA studies on Native American populations, which were based on restriction site polymorphisms, it has been clear that the mitochondrial genomes present in contemporary Native Americans could be arranged in four (A, B, C, and D) pan-American haplogroups shared with Eurasian populations (for a review, see Schurr and Sherry 2004). The use of a wide set of restriction enzymes, together with control-region sequencing (the fastest mutating portion of the mtDNA), allowed a more thorough characterization of the genetic diversity pattern for each haplogroup, leading some authors to suggest independent migrations for different haplogroup subsets. For example, Torroni et al. (1992, 1993a) reported higher genetic diversity for haplogroups A (in Amerinds), C, and D; intermediate values for haplogroup B; and lower values for haplogroup A among Na-Dene. This was interpreted as different migrations occurring around 20–40, 9–18, and 6–10 kya, respectively. A more extreme interpretation of the mtDNA genealogical structure was made by Horai et al. (1993), who suggested that each haplogroup represented an independent migration from Asia between 14 and 21 kya.

It might be argued that an important change in perspective came with the rescue of Beringia as the homeland of the Native American ancestral population, and the recognition that this ancestral population was polymorphic for its mtDNA. This led to interpretations of the mtDNA pattern as suggesting a single migration wave from Asia, a relatively early population expansion in Beringia around 20–30 kya (Merriwether et al. 1995; Forster et al. 1996; Bonatto and Salzano 1997a, 1997b; Stone and Stoneking 1998), including haplogroup X (Brown et al. 1998). Under this view, the differences between Na-Dene and Amerinds reflect the long-term isolation *in situ* rather than by different migrations (Forster et al. 1996; Bonatto and Salzano 1997a). However, the lower diversity shown by haplogroup B was still interpreted by some authors as a later arrival in the Americas (~13–16 kya) compared to the first Native American migration to Beringia, estimated at ~34 kya (Starikovskaya et al. 1998).

In contrast to the high genetic diversity exhibited by mtDNA, the first studies on Y chromosome variation in Native Americans suggested a major founder haplotype for all Native Americans (Pena et al. 1995; Santos et al. 1996), tentatively dated between ~30 and ~23 kya (respectively, Underhill et al. 1996; Bianchi et al. 1998). Adding more Y-chromosome loci revealed additional variation within Native American populations (e.g. Ruiz-Linares et al. 1999) and allowed Karafet et al. (1999) to propose two major founder haplotypes representing two separate migrations into the Americas, in line with hypotheses based on mtDNA data (Torroni et al. 1993a). Lell et al. (2002) also interpret the Y-chromosome data as more compatible with two separate migrations, one occurring ~20–30 kya and associated with Amerind-exclusive mtDNA haplogroups, and another much more recent one (between ~7 and ~9.5 kya) associated with Na-Dene mtDNA lineages. Other authors interpreted the Y-chromosome data in light of an alternative view of the mtDNA variation, suggesting that differences in Y-chromosome diversity among Native American populations could reflect an ancient isolation in Beringia from a single founder population (Santos et al. 1999; Tarazona-Santos and Santos 2002). Further studies using Y-chromosome markers suggested much later dates (between 10.1 and 17.2 kya) irrespective of the number of migrations to the New World (two or one; respectively, Bortolini et al. 2003; Zegura et al. 2004). If these revised dates were more congruent with the archeological record, they were quite distinct from the estimates based on mtDNA.

The development of faster and cheaper DNA-sequencing platforms in the early 2000s allowed researchers interested in mtDNA to move from a *control-region perspective* to a *full mtDNA genome perspective* (for a global study of human mtDNA variation, see Ingman et al. 2000). The full mtDNA genome has ~16 000 base pairs (bp), compared to the ~1000 bp of the mtDNA-CR, providing better resolution to understand the true genetic diversity within haplogroups, and allowing a more accurate calibration of the evolutionary rate for mtDNA, which is essential in molecular clock analyses (for a critical view of the impact of calibrations on molecular clock estimates, see Endicott et al. 2009).

The studies based on complete mtDNA sequences initially confirmed some of the findings of the earlier studies, such as the importance of Beringia in the development of specific Native American mutation motifs. They also suggested that coalescence times were actually more recent than previously thought (between 15 and 20 kya) and, therefore, that the population expansion previously associated with the standstill in Beringia was actually better reconciled with an expansion into the New World following the LGM around 16 kya (Tamm et al. 2007; Achilli et al. 2008; Fagundes et al. 2008; Mulligan et al. 2008; Achilli et al. 2013; Llamas et al. 2016). Finally, it was also clear that there were more founding mtDNA lineages, as many as 15 (Perego et al. 2010; De Saint Pierre et al. 2012; Achilli et al. 2013). Several of these minor lineages have distributions restricted to North America and the Arctic, possibly representing secondary migration streams from Northeast Asia into the Americas after the major expansion occurring after the LGM (see Achilli et al. 2013; Bortolini et al. 2014). This is in line with previous interpretations of the pattern of mtDNA genetic variation in Arctic populations (Starikovskaya et al. 1998; Saillard et al. 2000; Tamm et al. 2007; Gilbert et al. 2008a; Smith et al. 2009; Zlojutro et al. 2009; Malyarchuk et al. 2011; Dryomov et al. 2015; Raff et al. 2015). New findings for Y-chromosome diversity lend support for the inferences of further migrations after the initial movement of people into the Americas (Dulik et al. 2012b). While these revised dates and genetic variation patterns made mtDNA and Y-chromosome estimates much more similar, the idea of a single migration begged for some additions.

Craniofacial morphology variation in the Americas and a new synthesis with molecular data Another early obstacle for the idea of a single migration, and one affecting the whole continent, was interpreting the wide morphological craniofacial variation in Native Americans. It is common to summarize such variation by describing the extremes of variation across the cranial shape morphospace. One extreme shows a generalized, ancestral suite of traits, including narrow and long neurocrania, prognatic low faces, and relatively low and broad orbits and noses. The other extreme exhibits a derived, specialized combination of characters, including short and wide neurocrania; high, orthognatic, and broad faces; and relatively high and narrow orbits and noses. These two extremes have been associated with labels such as *Paleoamerican* and *Mongoloid*, respectively (Neves and Pucciarelli 1991; Steele and Powell 1993; Neves et al. 1999; Brace et al. 2001; González-José et al. 2001; Jantz and Owsley 2001; González-José et al. 2003; Neves et al. 2003; González-José et al. 2005; Neves and Hubbe 2005; Pucciarelli et al. 2006; Neves et al. 2007a, 2007b; Perez et al. 2009; Hubbe et al. 2010, 2015). While some modern Native Americans and Asians share the derived morphospace, and some early Native American populations occupy the generalized and ancestral morphospaces, most Native American groups are not on the extremes of cranial shape space (González-José et al. 2008).

One possibility is that this pattern indicates two migration waves from Asia, sometimes referred to as the *two (biological) components* hypothesis (see e.g. Marangoni et al. 2014). An initial migration during the Pleistocene was associated with the so-called Paleoamerican (ancestral) pattern, and the second, in the Holocene, was associated with the Mongoloid (derived) pattern (Neves and Pucciarelli 1991; Steele and Powell 1993; Neves et al. 1999, 2003, 2007a, 2007b; Brace et al. 2001; González-José et al. 2001; Jantz and Owsley 2001; González-José et al. 2003, 2005; Neves and Hubbe 2005; Pucciarelli et al. 2006; Perez et al. 2009; Hubbe et al. 2010, 2015). An alternative hypothesis is that microevolutionary processes such as natural selection and genetic drift, operating within Native American populations, changed craniofacial morphology over time (see Steele and Powell 1993; Pucciarelli et al. 2003; Bernal et al. 2006; Little et al. 2006; Sardi et al. 2006; Perez et al. 2007, 2009).

González-José et al. (2008) suggested that the Paleoamerican and Mongoloid categories were too simplistic to account for the full Native American craniofacial variation, which should be viewed as a morphological continuum. In their model, a morphologically heterogeneous population founded the Americas (probably after the LGM, according to the molecular data) and evolved a high craniofacial morphological diversity as the result of adaptation, stochastic effects, or a combination of both. However, the extreme derived phenotype seen today only in Siberians and Aleut-Eskimos must represent the result of more recent (Holocene) gene flow among circumarctic populations. Therefore, this *recurrent gene flow* (RGF) model, which is an attempt to bring together genetic and craniofacial evidence, suggested that neither the single-migration nor two-component hypotheses were accurate depictions of the process of peopling the Americas.

The RGF model has been tested directly using both molecular and morphological data. Using a set of ~400 microsatellite (short tandem repeat [STR]) markers typed in 10 Asian and 20 Native American populations, Ray et al. (2010) compared three alternative hypotheses: the single-wave, the two-wave, and the RGF hypotheses. Based on a Bayesian model choice procedure, these authors concluded that the RGF model had much higher support than alternative models, even when only South American populations were used in the analysis, which is curious because the recent gene flow was originally proposed to account for the extreme derived craniofacial morphology in the Arctic. De Azevedo et al. (2011) also found a better performance of the RGF model compared to both single-wave and two-wave models in explaining craniofacial variation in the Americas. Indeed, several recent findings may be compatible with the RGF model, from either a morphological (De Azevedo et al. 2015) or genetic perspective (Dulik et al. 2012b; Reich et al. 2012; Achilli et al. 2013; Raghavan et al. 2014b). A complication, however, is that these studies did not test the RGF model explicitly against alternatives based on discrete migration episodes.

A final remark on the discussion concerning the morphological variation found in the Americas is: does the recurrent gene flow in the RGF model explain the apparent morphological change in the Americas? Although this is a difficult question, the answer apparently is no – or, at least, that no Australo-Melanesian/Polynesian gene flow seems to be required to explain Native American craniofacial morphology. First, early ancient DNA studies from Pleistocene Native American individuals showed the typical mtDNA haplogroups found in modern individuals (e.g. Gilbert et al. 2008b; Jenkins et al. 2012; Rasmussen et al. 2014, 2015). Second, Chatters et al. (2014) found that a Paleoamerican skull from Hoyo Negro, Mexico, dated to 12–13 kya had mtDNA haplogroup D1, which is typical of modern Native Americans. Third, genomic data from the Kennewick man, an iconic Paleoamerican skeleton from Washington State dated to 8.4–8.7 kya, suggest that its genetic profile is typical of other northern Native Americans (Rasmussen et al. 2015). Finally, Raghavan et al. (2015), who found a small fraction of Australo-Melanesian ancestry scattered throughout a few Native American populations, did not find an Australo-Melanesian contribution in Pericúes or Fuego-Patagonians, two extinct populations from Mexico and Chile/Argentina, respectively, with allegedly Paleoamerican morphology. We should note, however, that the presence of generalized cranial shapes in the Americas (and in Australo-Melanesian populations) is unsurprising, since generalized cranial shapes were ubiquitous in Late Pleistocene remains all across the world (Lahr 1996).

Settlement models in the genomic era In the last decade, genomic studies have been reshaping the discussions about the peopling of the Americas based on an unprecedented amount of data. The set of STR markers characterized by Wang et al. (2007) and also analyzed by Ray et al. (2010) supported the RGF model, in which a first major migration into the New World would have occurred by 13.2–15.3 kya and followed a Pacific Coastal route. Gene flow between Native Americans and Asians (represented by two native Siberian populations) would have occurred early (>7.5 kya), even though the data had little power to estimate this parameter with precision (Ray et al. 2010). The next generation of genomic studies used single nucleotide polymorphism (SNP) data obtained from predefined polymorphisms or from full genomes.

Reich et al. (2012), for example, studied ~350 000 SNPs from 52 Native American and 17 native Siberian individuals. While these numbers are not particularly important in themselves, they illustrate the gain of power that molecular anthropologists had in the last years. Based on a method to identify secondary ancestry connections among populations, they proposed that Native Americans descend from three streams of Asian gene flow, similarly to the 25-year-old Greenberg et al. (1986) hypothesis. They suggest that a first gene flow stream ("First Americans") was completely associated with most Native American populations (from Amerind languages). The two other streams would have admixed with First Americans to contribute 10 and 43% of Na-Dene and Eskimo-Aleut populations' ancestry, respectively (Reich et al. 2012). A fourth migration into the New World was then proposed to account for the genomic distinctiveness of a Paleo-Eskimo individual dated to ~4 kya (Rasmussen et al. 2010; Raghavan et al. 2014b). These analyses indicated that this individual is closely related to modern Siberians, should have arrived in the Americas after a migration occurring ~5.5 kya, but apparently did not contribute to the makeup of modern Native Americans (Rasmussen et al. 2010; Raghavan et al. 2014b; Flegontov et al. 2016).

The genome of a male infant, buried approximately 12.6 kya in association with Clovis artifacts at Anzick, Montana, indicated a divergence between the two major founder Native American Y-chromosome haplogroups by 16.9 kya. This genome also showed a closer genetic relationship with central and southern Native American populations compared to northern Native Americans, indicating an early genetic structure among major "First Americans" branches (Rasmussen et al. 2014). The much more distant relationship between this individual with Na-Dene and Eskimo-Aleut individuals was taken in support of the additional pulses of gene flow in northern North America (sensu Reich et al. 2012). These results were mirrored by the analysis of the genome of the Kennewick man (Rasmussen et al. 2015), even though this individual was not as closely related to central and southern populations as the Anzick child. This study also proposed a further stream of gene flow in populations from the Pacific Northwest that could be related with the Paleo-Eskimo or Inuit migrations into North America, even though this seems unlikely in the case of Paleo-Eskimo (Raghavan et al. 2014b; Flegontov et al. 2016).

Finally, an additional migration stream (or, alternatively, a long, drawn-out period of gene flow from a structured Beringian or Northeast Asian source) has been suggested to account for the Australo-Melanesian ancestry detected in some Native American populations scattered throughout the Americas. This inference had been proposed independently by Skoglund et al. (2015), who characterized 63 individuals from 21 Native American populations for ~600 000 SNPs; and by Raghavan et al. (2015), who sequenced an impressive 31 genomes of modern individuals and 23 genomes of ancient individuals, and genotyped an additional 79 new individuals from 28 populations using SNP chip arrays. Raghavan et al. (2015) estimate the divergence between Native American and Siberian populations to have occurred ~22 kya, followed by a maximum 8 kya isolation in Beringia (as discussed in this chapter), and leading to a first major population migration into the New World no later than 15 kya. These authors also suggest that this migration brought ancestors of both First Americans and Athabaskans (a major Na-Dene branch), who diverged from each other around 13 kya, most likely in lower latitude North America rather than in Beringia. The Inuit and, possibly, an Australo-Melanesian related population may represent two independent migrations into the New World. The Inuit, occupying the extreme northern edge of North America, would have admixed with some northern First American populations, as implied by previous analysis (Rasmussen et al. 2015). On the other hand, the signal of Australo-Melanesian ancestry may have reached the Americas via the Aleut Islands, ultimately reaching parts of South America (Raghavan et al. 2015). It is tempting to establish a parallel between this signal and the support for the RGF model when only South American populations were compared with Siberian populations (Ray et al. 2010), but some caution is needed as these studies used different methods to infer gene flow among populations.

A tentative synthesis Considering the accumulated evidence, what can we say about the peopling of the Americas concerning the time and number of migratory streams? The number of migratory streams seems to vary from three to five. Figure 12.1 summarizes most of the findings discussed in this chapter.

Any scenario should start with a Beringian population that expanded into the New World ~16 kya, being associated with most (or all) of the genetic ancestry of the First Americans. On the other extreme, any scenario should end with the Neo-Eskimo (Thule/Inuit) replacement of Paleo-Eskimo cultures ~700 years ago. However, the evidence of gene flow between Neo- and Paleo-Eskimos as old as 4 kya suggest that the ancestors of the Thule/Inuit were already living in the Arctic, either in the Old World or in Beringia, during this time (Raghavan et al. 2014b). It seems safe to invoke another migration to explain the emergence of the Paleo-Eskimo genetic pattern, even though its precise date is uncertain. Possible values for this migration are 5.5 kya, based on the origin of the Arctic Small Tool tradition (Harritt 1998); 4.4–6.4 kya, based on the divergence with the Siberian Chukchi (Rasmussen et al. 2010); or up to 9 kya, assuming a common genetic background with Aleutians (Davis and Knecht 2010), with whom the migrants share some genetic features (Volodko et al. 2008; Rasmussen et al. 2010; Reich et al. 2012). The emergence of the full set of derived "East Asian" craniofacial traits by 7800 year BP can also be tentatively bound to this event (Brown 1999; González-José et al. 2008). A fourth migration stream would bring, as previously discussed, the Australo-Melanesian component into the New World following an Aleutian route. This stream would be more recent than 9 kya, constrained by the earliest dates for the settlement of the Aleutians (Davis and Knecht 2010). Finally, some authors propose another migration stream, whose date is uncertain, contributing to the genomic ancestry of Na-Dene populations (Reich et al. 2012). However, this has been challenged by more recent data that suggest a common origin with First Americans, followed by admixture with Inuit

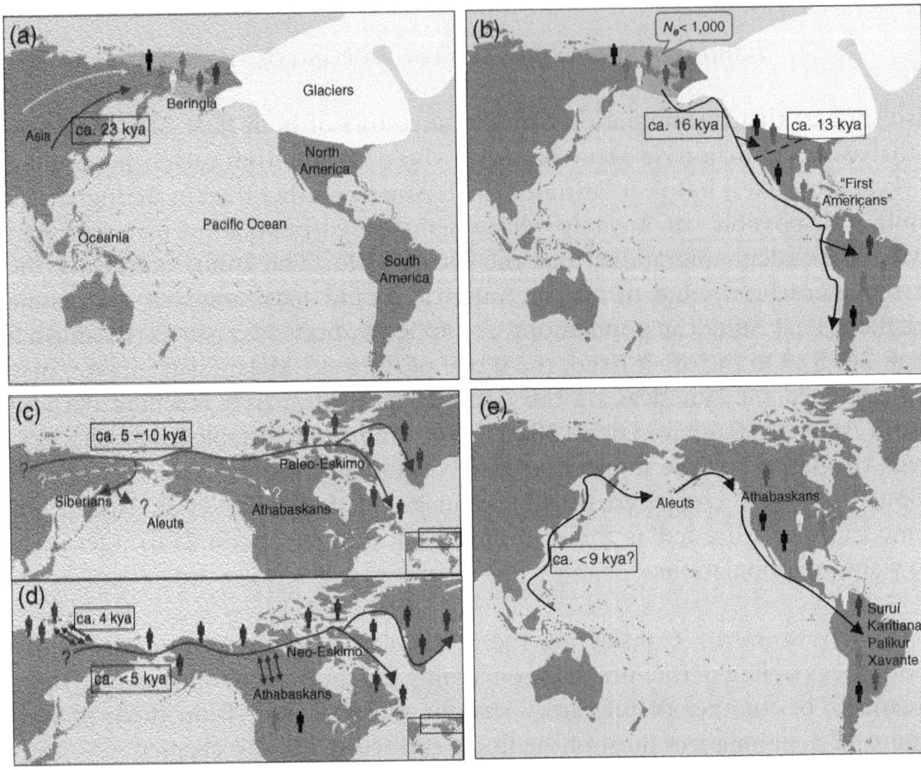

Figure 12.1 Summary of the number and timing of independent "migration streams" into the Americas. The drawings simplify several features, such as the precise area occupied by Beringia and the extent and dynamics of the North American glaciers, and are intended to capture the most relevant aspects of the peopling of the Americas, as discussed in the chapter. (a) Divergence between East Asian and Native American populations (red arrow) is estimated to have occurred ca. 23 kya, and can be linked to the occupation of Beringia. In Beringia, there was an admixture event with a western Eurasian population represented by the Mal'ta individual (yellow arrow), originating a morphologically diverse ancestral population. (b) A major migration stream into the Americas occurred ca. 16 kya (dark blue arrow), from a small but morphologically diverse founder population, originating most of the ancestry found in modern Native American (Amerind and Athabaskan, sensu Greenberg) populations. This migration probably used the Pacific Coast as the route south of the North American glaciers. The dashed line represents the divergence between northern (Amerind and Athabaskan) and southern populations (Amerinds), which occurred ca. 13 kya. (c) Paleo-Eskimo cultures represent a secondary migration stream (purple arrow), possibly associated with some of the ancestry found in modern Siberian and Aleut populations. Another, questionable migration stream has been proposed to account for part of the Athabaskan genomic ancestry (dashed light gray arrow), as an alternative to the scenario proposed in (d). The precise origin and dating of these migration streams are unknown. (d) The arrival of Neo-Eskimos represents a third robust migration stream into the Americas (dark green arrow), leading to a replacement of Paleo-Eskimo in the Arctic. The precise origin and dating of this migration is unknown. Two admixture events have been associated with this migration stream (thin bidirectional arrows): one with Paleo-Eskimos, and one with Athabaskans. (e) A last migration stream (black arrow), associated with Australo-Melanesian ancestry, has been detected in a few populations from North and South America, but its precise dating is unknown. This migration is unlikely to be connected with the diverse craniofacial types found in Native Americans. Different authors may interpret processes represented in panels (c–e) either as supporting a model of "additional migration pulses" or as supporting the recurrent gene flow (RGF) model. Events (c–e) likely postdate the existence of Beringia and the glaciers as an important barrier to migration.

ancestors (Raghavan et al. 2015). Curiously, the admixture graph presented by Reich et al. (2012) seems compatible with this interpretation.

An almost philosophical question is: how many "secondary streams" are needed for this process to be called "recurrent gene flow"? In part, methodological issues about how to deal with migration and admixture may determine one or another interpretation. Population genetics models based on continuous migration will be easily translated as recurrent gene flow, while admixture graphs will be easier to interpret as indicating discrete migration streams. The take-home message, nonetheless, is clear. The single-migration model can no longer explain the observed biological patterns for Native Americans.

The size of the founding population and natural selection patterns in native Americans

An important parameter for understanding genetic diversity is the effective population size, which determines if nearly neutral alleles behave as selective or neutral (e.g. Ohta 1992), and the size of linkage blocks, which is an important quantity in mapping studies (e.g. Hartl and Clark 2006). The different levels of genetic variation between African and non-African populations, for example, reflects the strong population bottleneck during the Out-of-Africa movement (e.g. Yu et al. 2002; Long et al. 2009). Did something similar happen during the peopling of the New World? Estimating the effective size of the founder population of Native Americans is difficult for several reasons, but above all is the fact that the Americas have been settled in a complex fashion, such that there is no single "founder population of Native Americans." However, there is evidence for a major migration wave into the Americas, which can be associated with most genetic ancestry of most Native American populations (as discussed in this chapter). Therefore, understanding the size of this ancestral population, which is often equated as the Native American ancestral population, may offer interesting insights on Native American evolution.

The first quantitative approach to infer the effective population size of the founder Native American population was developed by Hey (2005), who did a meta-analysis of nine sequence loci, and assumed the isolation-with-migration (IM) population model to suggest an extreme population bottleneck with an effective population size of ~70 individuals. Since this pioneer work, other groups tried to replicate this result using multiloci autosomal data, with partial success. In general, estimates including complete mtDNA genomes usually find larger values, ranging from 1000 to 5400 individuals (Fagundes et al. 2008; Kitchen et al. 2008), which may be due to a larger effective population size for women. However, most estimates based only on autosomes find values of around a few hundred individuals, with point estimates of 87 (ranging from 42 to 140 individuals; Ray et al. 2010) or 450 (between 71 and 1280 individuals; Fagundes et al. 2007). Recent autosomal data generated from admixed Latino populations and filtered for non–Native American ancestry also resulted in heterogeneous estimates. Gutenkunst et al. (2009), based on a dataset of ~13 000 SNPs, suggested a value of 800 effective individuals (between 140 and 1600 individuals), while Wall et al. (2011) estimated a bottleneck effective population size not larger than 150 individuals using resequencing data. An intermediate estimate of 514 individuals was proposed by Gravel et al. (2013), ranging between 316 and 2264 individuals.

This relatively small founder population of (most) Native Americans, associated with a pattern of dispersal and colonization along the continent that also favored the role of genetic drift, might explain why some possibly adaptive genetic variants do not show any signature of selection in the Americas (e.g. Paixão-Côrtes et al. 2011; Augusto et al. 2013). These findings may also predict an increased genetic load due to a higher frequency of slightly deleterious variants compared to a larger Eurasian population, similar to the relationship between African and non-African populations, in which a higher load of deleterious polymorphisms is found in the latter (Lohmueller et al. 2008; Subramanian 2012).

However, signals of genetic adaptation can also be found in the genome of Native Americans. An interesting example came from a group of genes called *fatty acid desaturases* (FADS), which were suggested to have evolved by positive selection in the Inuit as an adaptation to a diet rich in omega-3 polyunsaturated fatty acids (Fumagalli et al. 2015). A further study conducted in several Native American populations showed that this signal was not restricted to the Inuit, but could be observed throughout the Americas, reflecting a single and strong instance of local adaptation in the Beringian ancestral population (Amorim et al. 2017). It is quite curious that the Inuit are usually assumed to represent an independent migration into the Americas. This pattern may suggest either some adaptive introgression in the Inuit from a First American source (for evidence of gene flow between these groups, see Raghavan et al. 2015) or a common origin of the Inuit in the Beringian population (González-José et al. 2008).

The study of similar environments occupied by Native American and other human populations allows the identification of shared or unique genetic adaptations. One example comes from one of the most challenging environments in the Americas: the Andes Altiplano, where high-altitude hypoxia results in a significant physiological stress to the human body due to the decreased barometric pressure. An early study, comparing genetic adaptation to high-altitude hypoxia in Tibetans and Andeans, found a single gene (EGLN1) involved in convergent evolution in both populations (Bigham et al. 2010). Based on a different and more powerful method, Foll et al. (2014) suggested that of all SNPs found to be potentially under selection for high-altitude hypoxia, 31% of them showed evidence of convergent adaptation in Tibetans and Andeans. This is interesting because it suggests that the population bottleneck experienced by Native American populations did not affect their ability to adapt to new environmental conditions. More recently, Jacovas et al. (2015) suggested that some genes in the TP53 pathway, a major cell cycle regulator, had been co-opted for high-altitude adaptation in the Andes. On the other hand, a study designed to capture patterns of convergent selection in forest dwelling hunter-gatherer populations from Africa and the Americas found a much weaker signal of convergent selection (Amorim et al. 2015). The authors suggest that for this forest-adapted "Pygmy" phenotype, nonconvergent positive selection signals were more pervasive than signals of convergent evolution, indicating that the similar selective pressures in Africa and the Americas have targeted different genes and pathways, possibly reflecting the different genetic background in these two continents.

One last example that indicates adaptive potential despite reduced genetic diversity in Native Americans involves an exclusive Native American allele. Hünemeier et al. (2012) showed that the allele frequency of 230Cys*ABCA1 in Central America was significantly correlated with the archeological data on maize (*Zea mays*) pollen, and that a neutral model was unable to fit the observed patterns of genetic variation in Mesoamerica. The ABCA1 gene (ATP-binding cassette transporter A1) is associated with cholesterol metabolism, and the 230Cys*ABCA1 allele was shown to have a 27% cholesterol efflux reduction (Acuña-Alonzo et al. 2010). It was suggested that this allele could be advantageous during periods of food scarcity in the early stages of the implementation of the maize culture in Mesoamerican populations (Acuña-Alonzo et al. 2010; Hünemeier et al. 2012).

Concluding Remarks

This chapter summarizes the dispersion of *H. sapiens* into the American continents. The search for enlightenment about this fascinating theme comes from several areas of knowledge: archeology, paleoanthropology, geo-climatology, linguistics, and genetics, which was the focus of this review. Not only have demographic processes shaped the genetic landscape of Native American populations, but also genetic adaptation has played an important role, as shown by recent studies. It is inspiring to realize how much we have learned about Native American evolution, and equally inspiring to realize how much we still have to learn. The task is far from complete.

ACKNOWLEDGMENTS

We would like to thank Dennis O'Rourke for inviting us to write this review. We are also grateful to Daniela Quinsani, Guillermo Reales, Maria Cátira Bortolini, and Rolando González-José for comments on a preliminary version of the text. Finally, we would like to apologize to all the colleagues whose work has not been cited.

CHAPTER 13

Ancient DNA and Bioarcheology

Jennifer A. Raff

Department of Anthropology, University of Kansas, Lawrence, KS, USA

Introduction

Bioarcheology takes an integrative approach to studying human prehistory, drawing upon sociocultural and biological data to understand human lives throughout time. The scope of bioarcheology is broad, seeking to understand the prevalence and impact of diseases, the nature and frequency of violence and trauma, the effects of nutrition upon growth and development, social structure and mortuary practices, and demography and biological relationships between individuals and groups in past populations.

The majority of work in bioarcheology has historically been based on the study of human skeletal remains, which reflect the processes of growth and development and the interaction between an individual's environmental and cultural milieus. However, ancient DNA (aDNA) is increasingly contributing to bioarcheology research, complementing osteological analyses and testing hypotheses that skeletal data cannot (Hagelberg et al. 2015).

In this chapter, we will go through practical strategies for integrating aDNA work into bioarcheology. We will also examine the types of bioarcheological research questions that aDNA can contribute to, and look at illustrative case studies in the published literature.

A Companion to Anthropological Genetics, First Edition. Edited by Dennis H. O'Rourke.

What Is Ancient DNA, and How Is It Studied?

The term *ancient DNA* is typically used in reference to the properties of the DNA, rather than its exact age. aDNA is retrieved from the remains of deceased organisms, and as it is no longer subjected to the protection of the cellular environment and repair mechanisms, it is invariably degraded and scarce. The extent of aDNA degradation depends on a number of complex factors, including age, temperature, environmental pH, and the presence of exonucleases/endonucleases, water, and microorganisms (Lindahl 1993; Misner et al. 2009). The scarcity of aDNA is influenced by all of those factors, by the tissue source of DNA (Pinhasi et al. 2015), and by the macropreservation status of that source. With regard to human aDNA (the topic of this chapter), two genomes are available for analysis. The mitochondrial genome (mitochondrial DNA, or mtDNA) is maternally inherited, fast-evolving, non-recombining, and present in hundreds to thousands of copies per cell, making it more likely to be successfully recovered from ancient sources, and relatively simple computationally to work with. For these reasons, aDNA studies have historically focused on the mitochondrial genome.

The nuclear genome, present in only two copies per cell, has a lower chance of successful recovery from ancient contexts, and it is thus much less often a target of aDNA research. However, the advent of next-generation sequencing, coupled with new extraction methods (Rohland and Hofreiter 2007; Rohland et al. 2010; Dabney et al. 2013b), has improved access to this genome in ancient samples, as libraries can now be made of single fragments and completely sequenced. Endogenous aDNA fragments in genomic libraries can be distinguished from contaminating modern DNA by characteristic damage patterns: the overrepresentation of purines (particularly guanines) near the ends of ancient fragments, and cytosine deamination leading to a high C-to-T substitution rate near the ends of fragments (Briggs et al. 2007; Sawyer et al. 2012; Dabney et al. 2013a).

Coupled with the greater information about ancestry and phenotype obtainable from nuclear genomes (assuming preservation is adequate), whole-genome approaches provide significant advantages over single-locus studies. However, samples with low amounts of endogenous DNA must be sequenced to high depth of coverage, and the expense of retrieving a single genome, therefore, usually limits the sample size. As much of bioarcheology is focused on population-level questions, the cost and computational requirements of shotgun sequencing limit its routine application. At present, therefore, the majority of aDNA laboratories continue to do targeted sequencing analyses as their main focus, often collaborating with larger groups in instances where genomic analysis is desired (Pickrell and Reich 2014; Bolnick et al. 2016).

Approaches for aDNA in Bioarcheology

We begin with some general suggestions to be considered at different stages of a bioarcheological research project.

Collaboration and consultation

Collaboration between anthropological geneticists and other experts is essential in developing a well-designed, anthropologically informed project. While the aDNA researcher brings unique methodological and analytical skills to the table, the identification of useful questions and nuanced interpretation of results are enhanced by the expertise of members from the other subdisciplines of anthropology. Therefore, projects typically require collaboration between aDNA investigators and osteologists, historians, bioinfomaticians, linguists, paleontologists, or archeologists.

If there is a suspected cultural affiliation between the ancient individual or group to be studied and a contemporary community, consultation with that community (even if not required by Institutional Review Board regulations) is highly recommended prior to the beginning of research. The advice and assistance of culturally affiliated communities are critical for ensuring that research is done in an appropriate and ethical manner, and the cultural knowledge that stakeholders may bring can also be extremely beneficial in designing and interpreting bioarcheological research (Claw and Garrison 2015; Bolnick et al. 2016).

aDNA research cannot be done in regular molecular biology laboratories. While next-generation methods have made it much easier to distinguish contamination from endogenous DNA on the basis of DNA damage patterns (Jónsson et al. 2013), preventing contamination during the extraction process is still extremely difficult (Cooper and Poinar 2000; Yang and Watt 2005). Contamination is such a pervasive problem that the work requires specialized facilities and workers trained in aDNA protocols. Thus, archeologists who wish aDNA work to be done in their investigations should seek out collaborators at universities with dedicated aDNA laboratories. Such laboratories will be positively pressurized, with HEPA-filtered air, polymerase chain reaction (PCR) hoods, separation from post-PCR or modern DNA laboratories, and strict access protocols, and they will be staffed with researchers trained in the specialized laboratory and analytical methods necessary for ancient biomolecules.

Sample choice and extraction method

The process of aDNA extraction generally involves the destruction of a small (<0.5 g) amount of bone or tissue. This adds a further layer of ethical concern to those considerations that exist already in the study of human remains, particularly those of indigenous peoples. While nondestructive extraction methods have been developed (Bolnick et al. 2012; Hofreiter 2012), they are not always successful in retrieving DNA from some sources, and they do not always leave the sample unscathed (unpublished observation). Thus, the aDNA researcher should ideally select material for extraction that is not required for osteological research. For example, nonpathological ribs have been frequently used as sources of aDNA, as they are not typically used in estimates of stature, biological sex, or other osteological approaches. Dental calculus can also be a minimally destructive source of aDNA (Black 2011). The petrous bone from the inner ear, and hair, have both been shown to be exceptional sources of well-preserved, high-quality aDNA; if such elements are available, they are optimal extraction targets (Rasmussen et al. 2010; Gamba et al. 2014; Pinhasi et al. 2015).

Sample decontamination is essential before beginning extraction. Specific protocols will vary depending on the material type (e.g. hair vs. bone vs. soft tissue), but they will include one or more of the following steps: drilling or removal of surface material, soaking or rinsing in bleach, and UV irradiation to cross-link surface DNA (Kaestle and Horsburgh 2002).

Several different types of aDNA extraction methods are available (see Gamba et al. 2015). The most commonly used protocols make use of an overnight digestion of the sample in a solution of EDTA (to decalcify the material) and proteinase K (to digest proteins). DNA in the remaining solution is bound to silica in an extraction buffer containing guanidinium thiocyanate, then washed with an ethanol-based buffer; after that, the purified DNA is eluted with a low-salt buffer (typically either water or Tris–EDTA [TE]). For genomics applications, a modified protocol that makes use of silica spin columns has been shown to improve the recovery of short (<40 base pair) DNA fragments (Dabney et al. 2013a). Older phenol–chloroform-based methods are less commonly employed, but they may still be optimal for some difficult samples (Kaestle and Horsburgh 2002; unpublished results). As it is impossible to know in advance which method will yield results, test extractions and protocol optimizations are recommended before beginning large-scale projects.

Analysis and case studies

aDNA research encompasses different scales of analysis. At the broadest level, it can be used to reconstruct human evolutionary history, revealing large-scale migrations, genetic bottlenecks, and population replacements. It can also be used on a more regional level to reconstruct local population genetic prehistories and evolution, and test hypotheses regarding social structure, kinship, and postmarital residence patterns. On an individual level, aDNA can be used to identify the biological sex of a skeleton, make phenotypic inferences from an individual's genotype, and identify specific pathogens within an infected individual.

There are a variety of analytical approaches developed for aDNA analyses. The choice of approach necessarily depends on questions to be asked. Here, we discuss case studies illustrating each of these approaches in the context of bioarcheological research.

Sex and gender Sex determination of human remains is crucial for studying demography, differential patterns of disease, injury, resource consumption, as well as the conceptualization of gender and gender roles in past populations (Kaestle and Horsburgh 2002). While morphological-based methods for sexing well-preserved adult remains are relatively straightforward, accurate sex determination of infants, prepubescent children, adults lacking strong dimorphism, and fragmentary adult remains is extremely difficult (Mays and Cox 2000). aDNA can, in cases of sufficient preservation, reliably identify the biological sex of any individual.

One method of genetic sex determination relies upon differences in the amelogenin gene located on the X chromosome and the amelogenin pseudogene on the Y chromosome (Mannucci et al. 1994; Faerman et al. 1995; Stone et al. 1996). This approach has been used recently to investigate the demographics of infanticide victims in ancient Roman societies. Roman males typically had a higher status than females, and it has been hypothesized that the widespread practice of infanticide would have been directed more often at females than at males, effectively allowing families to decide the sex ratios of the children they subsequently raised (Hassan et al. 2014). However, a molecular investigation of the sex of infanticide victims at the Roman Ashkenlon brothel site revealed a greater proportion of males than females among the perinatal infants (Faerman et al. 1998). The authors interpreted this result to mean that females were being raised with the eventual aim of working in the brothel as soon as they were old enough, whereas males would have been killed as they would have had no use in that context (Faerman et al. 1998). Similar aDNA study of perinatal infants buried in several Romano-Britain nonbrothel sites revealed balanced sex ratios, suggesting that at least in Roman Britain, gender-directed infanticide was not a common practice (Mays and Faerman 2001; Hassan et al. 2014).

However, the amelogenin-based method can be problematic due to allelic dropout. That is, if only the copy present on the X chromosome is preserved, and not the copy on the Y chromosome, the individual will be falsely typed as female, not male. Another concern is modern contamination from a male source, which could lead to the erroneous typing of an ancient individual as male. High-throughput shotgun sequencing has been proposed as a more accurate means of sex assignment, as it simultaneously avoids the problem of allelic dropout by testing many more discriminating loci, and allows the detection of contaminating fragments via assessment of DNA damage patterns (Skoglund et al. 2014). Skoglund et al. (2013) recently evaluated the feasibility of this approach for applications in population-scale aDNA investigations. They found that approximately 100 000 reads are required for accurate sex determination, so that indexed genomes of up to 13 individuals could be pooled and simultaneously sequenced at low coverage at an estimated cost at <$300 per sample. While considerably more expensive than a simple amelogenin size-based analysis, it does have the advantage of being more accurate. It is likely that as shotgun sequencing costs continue to decrease, this approach will become much more routinely applied to archeological populations.

Inferring individual identity In rare cases, anthropologists seek to ascertain the identity of a specific deceased individual. This occurs most often where a historically significant individual is suspected to be present in a particular burial due to written records, characteristic artifacts, or oral traditions. In this case, simple comparisons between genetic lineages can be sufficient to ascertain identity, enhancing conclusions from osteological and archeological approaches.

One recent prominent example of aDNA helping to identify a historic individual occurred in 2014, when geneticists at the University of Leicester published findings from an investigation of the remains of a man purported to be the English Plantagenet king Richard III (King et al. 2014). Historical records and oral tradition led investigators to believe that Richard III's remains were buried at the Grey Friars medieval church in Leicester following his death in battle in 1485. Excavations at a likely location for the Grey Friars church recovered the skeleton of a man aged approximately 30–34 years old with extensive pathologies, including scoliosis (which would have made one of his shoulders higher than the other), and perimortem trauma, consistent with warfare-related injuries sustained at the time of death. The remains (Skeleton 1) were radiocarbon-dated to 1456–1530 CE. These lines of evidence were all consistent with the remains belonging to Richard, but aDNA analysis of the individual was required in order to confirm his identity. Mitogenomes were sequenced from known descendants of Richard III's maternal line, and compared to the mitogenome recovered from the remains. The ancient mitogenome was a perfect match with one maternal line descendant and differed by only a single substitution in another, strongly supporting the identification of Skeleton 1 as Richard. Y-chromosome comparisons between Skeleton 1 and known paternal-line relatives revealed a likely false-paternity event subsequent to Richard III's death (King et al. 2014).

Another case of the use of molecular data to identify known historical figures is Gill et al.'s (1994) study of the remains of nine individuals suspected of being the Russian Romanov family and retainers, executed in 1918 by Bolshevik revolutionaries. These remains, recovered from a suspected (although unmarked) burial site of the Romanovs, had wounds consistent with perimortem violent trauma matching descriptions of the executions of the family. After forensic study, genetics methods, including the analysis of mitochondrial, Y-chromosome, and short tandem repeats (STR) markers, were employed to test the relationship of the deceased individuals to each other, as well as to living relatives of the Romanovs. Originally developed for forensic use, STR loci are highly variable in the number of repeats present from individual to individual. Typed in combination (up to 13), they can offer the precision necessary to reliably identify paternity and sibling relationships, assuming preservation is high. STR analysis revealed that two of the adults were the parents of the three juvenile individuals present, and comparisons of mitochondrial DNA recovered from the mother and children with that of a living maternal relative of the Tsarina (Prince Philip, Duke of Edinburgh) identified a shared lineage. A similar comparison of mtDNA recovered from the father of the family group and living maternal relatives of the Tsar revealed similar sequences except for a heteroplasmic mutation at np 16169 in the putative Tsar. Together with paleopathological and archeological evidence (such as the precious metals used in the dental work of some of the individuals), the genetic results strongly supported the hypothesis that the remains were those of Tsar Nicholas II, Tsarina Alexandra, three of their five children, three servants, and the family physician.

The cases of Richard III and the Romanov family are excellent examples of the use of multiple, complementary lines of evidence – genetic, historical, archeological, and osteological – to solve a problem in bioarcheology that was of great interest and importance to the peoples of Britain and Russia. However, it is important to note that the identification of the remains of historical figures is generally biased to include elite individuals. It becomes vastly more difficult to ascertain the individual identity of skeletal remains for whom fewer historical records exist.

Inferring kin relationships Kinship studies are an important component of bioarcheological research. Understanding the biological relationships between individuals in a mortuary context can give insights into marriage patterns (e.g. exogamy vs. endogamy), postmarital residence patterns (e.g. matrilocal or patrilocal), and the relationship between lineage and social structure and mortuary treatment. Kin relationships have been frequently inferred from nonmetric skeletal and dental traits, but the degree to which these are a reflection of biological relatedness (vs. environmental influence) is unclear (reviewed in Larsen 2015).

However, kin relationships between specific individuals within a burial population can be determined effectively through aDNA approaches if preservation is sufficient (Kaestle and Horsburgh 2002) and can provide a powerful tool for studying kin-based mortuary practices and social structure. For example, Bolnick and Smith (2007) used this approach to study the relationships between 39 burials within the Pete Klunk mound group, a Hopewell site in west-central Illinois dating to 1775 ± 75 years BP. Several types of mortuary practices existed simultaneously within this site, interpreted as reflective of the differential status of individuals included. Bolnick and Smith investigated the distribution of mitochondrial haplogroups in burials across the site. They found no relationship between mortuary treatment (including location in the mounds) and mitochondrial haplogroups, suggesting that social status did not depend on maternal ancestry.

This pattern had also been observed by Lisa Mills in burials at the Hopewell Mound group (2003) and in Mississippian and Late Woodland burials at the nearby Schild site (Raff 2008), suggesting that conceptions of kinship and social organization may have been similar throughout the Eastern Woodlands over time. Whether social status was patrilineally based, or arranged according to some other criteria (such as achieved status), has yet to be clarified.

Interestingly, Bolnick and Smith found greater mtDNA lineage diversity among males than among females buried in the Pete Klunk mounds. This result fits the expected pattern of communities that practice matrilocal postmarital residence, where husbands travel to live with their wives' families. Because females stay with their kin group, their mtDNA diversity is predicted to be lower than that of males joining their communities. In patrilocal societies, where wives join their husbands' families, both males and females are predicted to have approximately equally high mtDNA diversity (because the male children will inherit their "outsider" mothers' mitochondrial lineages).

Paleopathology aDNA approaches can be used to identify the presence and characterize the genomes of specific pathogens within ancient populations, and they can put a time stamp on a pathogen's evolution by capturing a snapshot of genetic diversity at a certain point in the past, as has been done with *Yersinia pestis* as the causative agent of the Black Death (Raoult et al. 2000), the 1918 Spanish Flu (Taubenberger et al. 1997, 2005), and tuberculosis (Bouwman et al. 2012; Bos et al. 2014). Here, too, the application of genomics approaches has both improved the reliability of pathogen detection (i.e. discrimination from other soil microbes) and allowed more complex questions to be addressed, such as host–pathogen co-evolution, and the effects of changing subsistence and habitation methods on infection patterns.

One recent paper serves as a good example of the kinds of questions that aDNA approaches are making possible to address. Warinner et al. (2014) shotgun-sequenced DNA recovered from dental calculus from four adult individuals from a medieval monastery in Germany dated to 950–1200 CE. They characterized the ancient oral microbiomes present in these individuals, who exhibited various degrees of periodontal disease, and successfully identified many bacterial taxa down to the species level, including so-called *red complex pathogens* associated with periodontal disease: *Tannerella forsythia*, *Porphyromona gingivalis*, and *Treponema denticola*. Complete sequencing of the genome of the medieval isolate of *T. forsythia* showed that it did not harbor tetracycline-resistant genes present in contemporary strains. Analysis of the genes present in the paleo oral microbiomes further revealed homologs to modern antibiotic-resistant genes, suggesting that there is great potential for this approach in reconstructing the evolutionary history of human pathogens. Finally, the researchers were able to reconstruct elements of the ancient diets of these individuals from DNA preserved in the dental calculus. This new approach promises to complement microfossil and stable isotope analyses, and provide more precise taxonomic information about the plants and animals consumed by individuals in the past.

Identifying differences in lived experiences Phenotypes are the result of a complex interplay between genetic, developmental, and environmental factors. Differences in lived experiences, such as extreme stress or nutritional deprivation, can impact a person's development in profound ways. Bioarcheologists often study the relationship between skeletal manifestations of physiological disruptions and cultural factors such as warfare, violence, famines, and social structure (Larsen 2015). A new development in aDNA within the last few years – paleoepigenetics – is likely to significantly expand this field of study. Gene expression, often in response to an organism's environment, is tightly regulated. One means of gene regulation is chemical modifications to DNA, such as the addition of methyl tags to cytosines, which can silence or induce gene expression. Collectively, these patterns of cytosine methylation are known as the *epigenome*, and there have been successful instances of the recovery of these methylation patterns from some ancient genomes (Briggs et al. 2010; Llamas et al. 2012a; Gokhman et al. 2014; Pedersen et al. 2014). In a very recent development, Smith et al. (2015) were able to use bisulfite sequencing to reconstruct methylation patterns in multiple individuals from five ancient human populations. This step was important because it demonstrated that paleoepigenetic approaches could be applied on a population scale, critical for understanding the effects of environmental stresses in the past. Paleoepigenetics may also prove to be a valuable new tool in biological age determination, as methylation status is highly correlated with age (Gokhman et al. 2016). Although still in its nascence, and therefore prone to exaggerated claims about its utility, this field of study nevertheless has the potential to offer new directions in bioarcheology research in the coming years.

Population history One major question that bioarcheologists address is the extent to which demographic changes accompany (or are responsible for) cultural change. Historically, bioarcheology has utilized biodistance studies of metric and nonmetric skeletal traits to elucidate how cultural and biological forces affect populations over time. However, because phenotypic traits are an indirect measure of genetic relatedness, aDNA has emerged as a more precise tool in studying past demographic events such as population expansions or decreases, migrations, and gene flow from other populations.

Temporal changes in the genetic makeup of ancient populations can be assessed through single loci (e.g. mtDNA or Y-chromosome DNA), which track matrilineal or patrilineal histories. For example, Mata-Míguez et al. (2012) investigated how Aztec conquests in the Basin of Mexico might have changed the genetic composition of residents at the site of Xaltocan. Xaltocan was a city-state that emerged in CE 1100 and grew to importance over the next several centuries. However, Xaltocan was defeated by the Tepanec Cuauhtitlan kingdom and its allies in 1395, and was depopulated. The expansion of Aztec imperialism into the Basin of Mexico resulted in the annexation of Xaltocan in 1428. In 1435, according to written documents of the time, Xaltocan was repopulated by settlers from multiple ethnic groups across central Mexico. But at least some archeological evidence suggests that Xaltocan was still occupied between 1395 and 1435 (Overholtzer 2012). Mortuary practices and house locations were similar before and after the repopulation event, suggesting at least some degree of cultural – if not biological – continuity (Overholtzer 2011).

Mata-Míguez et al. (2012) tested alternative hypotheses of demography at the Xaltocan site by sampling burials within three households that predated and postdated the conquest of Xaltocan in 1395. They recovered mitochondrial DNA from 25 burials, and tested a null hypothesis that pre- and postconquest populations would be genetically similar. A repopulation event, on the other hand, would likely have resulted in significant differences of genetic variation between the two temporal groups, particularly if – as written records suggest – the postconquest population was derived from multiple ethnic groups.

Mata-Míguez et al. (2012) found differences in the haplogroup frequencies (A2, B2, C1, and D1) and standard measures of genetic variation (haplotype diversity, nucleotide diversity) between the preconquest and postconquest occupants of Xaltocan. Because the individuals at Xaltocan were buried within three houses used over the course of the preconquest and postconquest periods, the authors were able to test for a possible correlation between individual lineages and household membership over time. Several lineages were shared within contemporaneous burials in the same preconquest houses, which was consistent with a practice of multiple generations of the same maternal families occupying the houses. However, there were no shared matrilines between preconquest and postconquest groups in the households. This finding might be explained by (i) a complete population replacement, which would be consistent with the written accounts of Xaltocan; (ii) demographic changes that included an extensive influx of new matrilines; or (iii) a change in kinship and inheritance practices of the households following the Tepanec and Aztec conquests, which might reflect patrilineal inheritance instead of matrilineal inheritance. These three interpretations are equally plausible given the available data; Y-chromosome and autosomal markers are necessary to distinguish between them.

Demographic modeling using genome-wide markers or complete genomes can yield much more information about population histories. The recent developments in laboratory and computational methods for ancient genomes are transforming the discipline of anthropological genetics. Although the use of complete genomes or genome-wide markers is not yet routine in aDNA studies, owing to a lack of financial and technical resources, it is extremely likely that the next few years will see these approaches become much more common (Bolnick et al. 2016; Slatkin and Racimo 2016). Already, these approaches have proven to be powerful tools in the study of long-standing hypotheses about population history. For example, Haak et al. (2015) and Allentoft et al. (2015) tested alternate models to explain the spread of Indo-European languages into Europe. Using ADMIXTURE, principal component analysis (PCA) plots, and outgroup f_3 statistics, they analyzed genome-wide markers (Haak et al.) and whole genomes (Allentoft et al.) from individuals belonging to a number of populations throughout Eurasia dating from the Neolithic through the Iron Age. The two research groups independently found that Corded Ware peoples of Central Europe likely had ancestry from the pastoralist Yamnaya peoples of the Pontic-Caspian steppe, suggesting that the latter likely migrated westward into Europe during the late Copper and early Bronze Age. This migration broadly fits what would be predicted by the *Steppe hypothesis* for the spread of Proto-Indo-European into Europe (Novembre 2015).

Conclusion

aDNA approaches have already proven to be useful in bioarcheological studies of kinship, identity, sex and gender, paleopathology, and demography. The continual improvements in laboratory and computational methods are improving the ability of researchers to reliably retrieve more ancient genomes, allowing more precise modeling of population history and biological relationships. Research emerging from aDNA studies in recent years has challenged many assumptions about past populations, and has provided robust tests of previously unresolved hypotheses. When done in an ethical manner, aDNA work can benefit both researchers and stakeholder communities. However, the bioarcheologist should be cautious about expecting too much of aDNA work, as its expense, inconsistent preservation, and intensive labor requirements place certain limitations on its use. It is critical that bioarcheologists work in close collaboration with experienced aDNA laboratories that have both the facilities and technical expertise to maximize the likelihood of the successful retrieval of endogenous DNA.

PART IV Human Adaptation: Natural Selection and Physiological Variation

CHAPTER 14

Tempo of Recent Human Evolution

Michael D. Weight[1] *and Henry C. Harpending*[2]

[1] *Anthropology and Mathematics Departments, University of Utah, Salt Lake City, UT, USA*
[2] *Emeritus, Yale University, University of New Mexico, The Pennsylvania State University, and University of Utah*

Introduction and Historical Perspective

Since the nineteenth century, there have been two prominent streams in evolution and genetics. The biometricians, principally associated with Victorian naturalists Francis Galton and Karl Pearson, collected, measured, and classified everything from butterflies and snails to humans. The natural haunt of human biometricians was the field where they measured quantitative traits like height, weight, metabolic rate, and school grades. Biometricians then understood inheritance through statistical analysis of their vast datasets. The other stream proceeded from the rediscovery of Mendel's rules of inheritance. Under the Mendelian view, discrete particles of genetic information persisted through time via diploid organisms that were opportunistic alliances between two sets of such particles. Mendelians searched for the laws of inheritance from the laboratory bench by identifying the chemical and physiological effects of more and more Mendelian traits.

Although Fisher (1919) reconciled these two intellectual streams in "The Grand Synthesis" (Provine 2001), a shadow of this divide persists. On the one

A Companion to Anthropological Genetics, First Edition. Edited by Dennis H. O'Rourke.

hand, quantitative genetics, a descendant of the biometrical tradition, has strongholds in agriculture, plant and animal breeding, much of ecology and evolutionary biology, and other biological subfields like taxonomy. Research emphasizes measurable traits that are modeled as outcomes of a large number of genes each with small effect. These models rely on (some version of) the Central Limit Theorem to guarantee that the sum of a large number of small random effects is normally distributed. On the other hand, the Mendelians discovered the structure of DNA, established large projects to sequence the genomes of humans and other species, and made great advances in understanding the effects of many major genes.

Descendants of the Mendelian tradition attempt to identify the "genes for" one thing and another. Descendants of the biometricians doubted that many such major genes would ever be found. Indeed, the last few years have revealed that searching for such genes (e.g. the common disease–common variant hypothesis) has had limited payoff and that alleles of large effect should be rare (Hill et al. 2008; Visscher et al. 2008; Yang et al. 2010; Visscher et al. 2012). The theory is that genes of large effect should either rapidly become fixed or disappear from the population (Otto and Whitlock 1997). Genes of minor effect can persist much longer so we expect that the direction of evolution in general is toward the model envisioned by the quantitative geneticists. Plant and animal breeders, despite the Mendelian fixation throughout the twentieth century, have thus been able to generate valuable plants and animals like Jersey cattle, Labrador retrievers, Berkshire hogs, maize, wheat, domestic chickens, and many more without any knowledge of any of the specific genes involved.

Culture has been imagined to be a pervasive force that makes natural selection on either Mendelian or quantitative traits irrelevant to recent human history and change (Mayr 1963; Cavalli-Sforza and Cavalli-Sforza 1995; Gould 2000). Under this view, humans simply respond to environmental pressures with culture as if it is a new force substituting natural selection.

However, recent discoveries about human genetic variation suggest this view of an autonomous cultural force and passive genetic history is incorrect. Natural selection has not only been acting throughout our history, but the rate of positive selection has likely *increased*: genetic data suggest that sudden demographic and environmental changes induced by population expansions gave more mutation targets and novel social and ecological environments for natural selection to act on (Hawks et al. 2007; Hughes et al. 2008; Cochran and Harpending 2009; Laland et al. 2010). Agriculture is likely responsible for much of this change in our evolutionary tempo (Larsen 1995, 2006; Armelagos and Harper 2005a, 2005b; Hawks et al. 2007; Cochran and Harpending 2009; Foley and Gamble 2009; McCullough et al. 2015).

Agriculture changed humans through several mechanisms. First, we underwent a rather sudden change in diet with the advent of cereal grain dependence (Larsen 1995, 2002; Cordain et al. 2000, 2002; Naugler 2008). Although cereal grains supplied abundant calories, they lack many nutrients important to humans

(Hurrell 2003). Deficiencies followed in turn, with widespread health consequences observed in the skeletal record (Larsen 1981; Stuart-Macadam 1998; Steckel et al. 2002; Wittwer-Backofen and Tomo 2008; Meiklejohn and Babb 2011). Second, these cereals can be boiled to a sticky mush that is an ideal food for toddlers. Foragers typically do not have such a source of soft food. However, this same mush adheres to teeth, apparently leading to dental disease observed in many remains of early agricultural people (Lukacs 1992; Tayles et al. 2000; Larsen 2003; Eshed et al. 2006; Temple and Larsen 2007; Watson 2008).

New sources of abundant food also vastly increased human population densities (Bogaard 2004). If a population in a region increases 100-fold, then there are 100 times as many targets for possible new advantageous mutations to occur. This means more opportunities for the origin and spread of advantageous forms and an increase in the overall rate of evolution. The overall nature and severity of infectious disease threats also change with more humans. Microparasites like measles and many others cannot persist in low-density human populations, like those seen before agriculture (Scott and Duncan 1998; Schlossberg 2008; Cochran and Harpending 2009). A large number of new infectious threats appeared after human populations grew following agriculture (Dobson and Carper 1996; McMichael 2004; Weiss and McMichael 2004; Wolfe et al. 2007). While foragers likely suffered from parasites, zoonoses, and persistent infections (Morse 1995; Hewlett and Hewlett 2007; Schlossberg 2008), the large number of viral and other microparasites that are known in human written history appeared and persisted only when our numbers exceeded a threshold density. Thus, infectious diseases have become a significant selective pressure in our recent evolutionary past.

The discovery of these changes and their tempo rewrote several beliefs about recent human history, health, and the relationship between culture and genetics. Traditional objects of sociocultural interest like social organization, mating patterns, economy, and subsistence have recently engineered some of the most substantial genetic changes in our species. In return, many of these genetic changes have social and historical consequences. The Mendelian and biometrical approaches to evolutionary biology are also coming together through studying these recent human evolutionary changes.

Why Speed Matters

It seemed clear in the latter twentieth century that evolution happens slowly, since available skeletal data tracking morphological changes in the fossil record spanned thousands to millions of years. However, it became apparent that the rate of change increased as the observational timescales became shorter (Gingerich, 1983): evolutionary changes developed more rapidly as the interval of study shortened. Most of these changes on shorter (i.e. faster) scales are approximately random and directionless and disappear over longer perspectives. When we speak of weather, we usually refer to a fast timescale of hours to months. The vagaries of weather are blended out over longer timescales so that, on a slow timescale, we speak of climate rather than weather. If we wonder what tomorrow will be like, we look at a weather forecast. If we wonder what the next decade will be like, we inquire about climate.

Evolution therefore occurs on a continuum of scales. We look at differences between various flavors of *Homo erectus* and anatomically modern humans over hundreds of thousands to millions of years. On the other side, inquiries about skin color or body mass span centuries to millennia. Lumsden (1980) suggests visible evolutionary change should occur in human populations over a single millennium, giving a "thousand years rule." Although controversial, the heritability of a number of human characteristics, like cognitive and personality traits, suggests Lumsden is correct: even mild selection can cause substantial changes in humans.

Human skin color, although viewed as a pervasive and ancient variable among human populations, rapidly changed in our recent past and is perhaps continuing to change. For example, ancient DNA indicates that anatomically modern invaders of Europe had dark skin and blue eyes, although they are popularly depicted with white skin. White skin apparently came thousands of years later with the first farmers of Middle Eastern origins into Europe (Mathieson et al. 2015). White European skin color and its north–south cline may be only a few millennia old (Berg and Coop 2014).

Recent Selection of LCT and EDAR

In the last several decades, volumes of DNA sequence data have accumulated, showing many episodes of recent and rapid evolution in human populations. The technology for inference is fundamentally simple. If a new gene appears with a strong selective advantage, then many copies of that gene spread rapidly. The rapid spread means that copies in the population today have a recent common ancestral gene. Little recombination of the chromosomes bearing that gene has occurred, so chromosomal regions with high linkage disequilibrium are taken as signatures of recent rapid evolution (Huff et al. 2010). Scans have detected many such regions (Pritchard and Przeworski 2001; Nielsen et al. 2007; Pickrell et al. 2009; Oleksyk et al. 2010). Following these developments, *recent* thus means that there has not been enough time for recombination to erase the signature of rapid evolution. The timescale for these methods ranges from the past few centuries up to 10 000–20 000 years ago.

A well-studied case of a recent advantageous mutation that has reached high frequency in Northern Europe and elsewhere is LCT, which creates lactose tolerance or, equivalently, lactase persistence. Ordinarily, the enzyme lactase responsible for the ability to digest the milk sugar lactose is produced after birth in mammalian infants and then turned off about the time of weaning. Two possible hypotheses may explain why lactase production is turned off. First, lactase production may be metabolically costly. Weaned mammals do not gain from its continued production, so natural selection has favored ceased lactase production after weaning. Alternatively, mothers and their infant conflict over weaning. The metabolic and fitness costs of lactase persistence may be too great to mammalian mothers, so natural selection favored ceased lactase production in her offspring.

However, several mutations in the human population break the regulatory region that shuts down lactase production around the time of weaning. The most common lactase persistence mutation, known as 13910, is found in Europe and extends to Pakistan and India. Since the chromosomal region around the Eurasian version of LCT is identical throughout Europe and northern India, the mutant versions all descend from a single mutation. This Eurasian mutant is one of the fastest evolving genes known in humans (Bersaglieri et al. 2004).

It is useful to take estimates of the selective advantage of this new mutant, inferred from its rate of spread, and use computer simulations to "rerun" history. Figure 14.1 was generated by simulating a population of 20000 individuals in which, at time 0, a single mutation with a selective advantage of 5% per generation to the new LCT allele occurred. The mutation is dominant, so both heterozygotes and mutant homozygotes have the same reproductive advantage.

Figure 14.1 illustrates the course of evolution of the new mutation in 100 reruns of history. In 94 of the 100 simulated histories, the new mutant was soon lost.

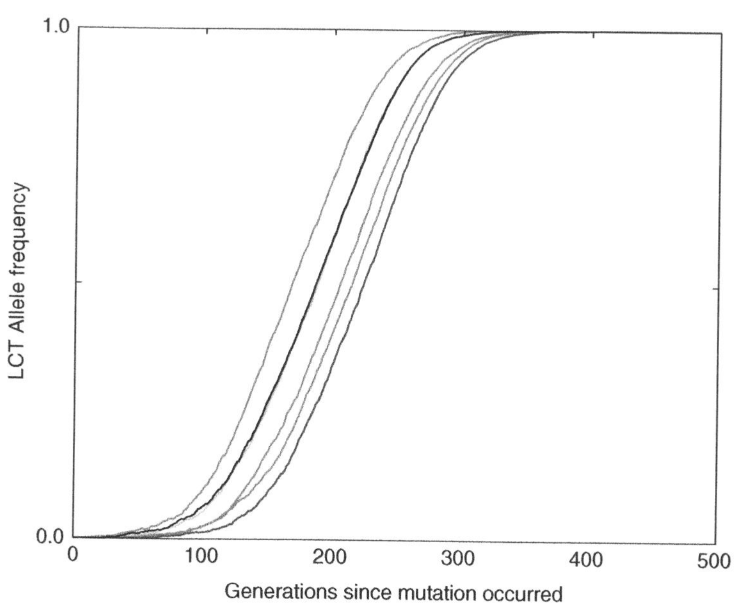

Figure 14.1 Results of single LCT mutations introduced at time 0 in 100 reruns of history.

In 6 of the 100 histories, the mutation became common enough that the selective advantage overcame the vagaries of drift when it was initially rare.

In each of these six histories, the mutation proceeded to fixation, but they became common at slightly different times. In the middle of progress toward fixation, the slowest and fastest histories reached frequencies of 0.5 at about 40 generations apart: the fast one was about 1000 years ahead of the slow one, even though they all started at the same time. In the other 94 simulated histories, the new allele disappeared even with its selective advantage. In some of these histories, the new allele was lost after a few generations. In others, it persisted at low frequency for a handful of generations before eventually disappearing.

The converse problem of estimating the age of a mutation suffers from a corresponding uncertainty. Lactose tolerance must have had a major effect on population growth and dynamics, more so once it became relatively common at, say, a gene frequency of 0.5. This occurs in the neighborhood of 200 generations, or 5000 years, after its appearance. For less advantageous mutations, we expect a longer "time to significant impact." For example, the same simulation conferring mutants a selective advantage of 0.01 rather than the 0.05 we have chosen for simulated LCT requires about 1000 generations, or 25 000 years, to reach an allele frequency of 0.5. Advantageous dominant mutations with small effects like 1% must have appeared before the Last Glacial Maximum and are unlikely to be responses to environmental change associated with agriculture.

There are several lessons to be taken from this locus. First, the course of history can depend on unpredictable random events. LCT arose perhaps 6000 years ago, was essentially insignificant for another thousand or so years, and then spread rapidly. In other runs of the simulation, it never happened, and Eurasia would have had a completely different history. Second, there is a constant stream of new mutations into the human population, but most are lost almost immediately. In order for a new allele to persist and become common, it requires both a substantial selective advantage *and* good luck.

While we have solid evidence of recent rapid evolution in this regulatory gene, we have no indication from genetic data of what LCT's advantage might have been or, perhaps, still is. Harpending (2013) suggests lactase persistence into late childhood and adulthood allowed better nutrition in a dairying culture. Over half the energy in mare's milk, for example, is from lactose. People who could digest that sugar had twice the energy available from mare's milk as did those lacking lactase. This advantage could have been critically important for survival in times of famine or other disturbances.

EDAR V370A, or EDAR for brevity, also appears to be a gain-of-function mutation (Bryk et al. 2008; Chang et al. 2009; Cluzeau et al. 2012; Voutilainen et al. 2012; Kamberov et al. 2013) like LCT (Enattah et al. 2002; Tishkoff et al. 2007). It has been hypothesized that it originated and underwent strong and rapid selection (Sabeti et al. 2007; Bryk et al. 2008; Grossman et al. 2010), in Central Asia, between 11 000 and 39 500 years BP (Bryk et al. 2008; Kamberov et al. 2013; but see Hlusko et al. 2018). EDAR is nearly fixed in East Asians, Han Chinese, and Pima Indians; present at intermediate frequencies in the Japanese; and at only modest to low frequencies in populations like Uygers, Hazara, and Melanesians (Sabeti et al. 2007; Bryk et al. 2008). Despite this relatively limited geographic distribution, Kamberov et al. (2013) propose EDAR conferred a 3–18% selective advantage, or potentially two or three times that of LCT (as discussed in this section).

During embryonic development, proteins coded by EDAR and its variants are part of a signaling pathway instructing ectoderm and mesoderm development (Laurikkala et al. 2002; Pummila et al. 2007). EDAR's known phenotypic effects – coarse (Mou et al. 2008), straight (Tan et al. 2013), and thick (Fujimoto et al. 2008a, 2008b) hair; small and dense breasts (Chang et al. 2009; Voutilainen et al. 2012); in addition to a higher density of sweat glands (Kamberov et al. 2013), shovel-shaped incisors (Kimura et al. 2009; Park et al. 2012), and changes in ear morphology (Adhikari et al. 2015) – reflect EDAR's regulation of ectoderm structural tissues like skin (Aswegan et al. 1997; Monreal et al. 1999), hair, nails, teeth, sweat glands, the pituitary gland, noses, and eyes (Headon and Overbeek 1999; García-Martín et al. 2013). EDAR also regulates inflammation when stimulated by bacterial endotoxins and viral infections with double-stranded RNA (Courtois et al. 2001). Comparisons between EDAR and the ancestral allele, *EDAR370V*, show the former requires chemical stimulation (e.g. bacterial endotoxins and viral infections) before it produces substantially more gene product than the latter (Bryk et al. 2008; Fujimoto et al. 2008a; Kimura et al. 2009; Cluzeau et al. 2012; Kamberov et al. 2013; Kowalczyk-Quintas et al. 2014).

The functional advantage of EDAR is, and all of its phenotypic effects are, currently unknown. However, several hypotheses emphasizing each phenotypic effect have been suggested. For example, Bryk et al. (2008) and Fujimoto et al. (2008b), respectively, predict thick (or, following Tan et al. [2013], straight) hair was sexually selected or was a form of cold tolerance. Others (e.g. Fujimoto et al. 2008b; Kimura et al. 2009) alternatively suggest shovel-shaped incisors were favored for their adaptability to new diets or as tools given their overall increased structural strength. Less visible traits, like increased sweat gland density, could more efficiently thermoregulate those bearing EDAR in East Asia's hot and humid climate (Tan et al. 2013; Lu and Fuchs, 2014). The pleiotropic effects themselves of EDAR's gene products could also make it advantageous in fluctuating environments (Kamberov et al. 2013).

Both LCT and EDAR illustrate that evolution can happen much more rapidly than it does in our familiar stereotypes. It is appropriate, when thinking about evolution in humans, to think in terms of chunks of time of 500–5000 years. Human populations can easily change on this short timescale, so the biology of people 500 years ago can easily be very different from the biology of their descendants today.

Search for selection stalls

Our LCT simulation illustrates a mutation that reaches fixation through natural selection and historical good luck. Yet, many single nuclear polymorphisms (SNPs) are at intermediate frequencies (Voight et al. 2006a; Andrés et al. 2009; Hancock et al. 2010; Pritchard et al. 2010). Despite strong selection, these SNPs never reached (near) fixation.

We understand a major reason that SNPs at intermediate frequencies should be common following rapid environmental change. Rare alleles favored under new environmental conditions enter the population through heterozygotes. These rare alleles likely increase in frequency only if they confer heterozygotes selective advantages greater than those of wild-type homozygotes (Haldane 1927). In other words, they passed through "Haldane's Sieve" (Charlesworth 1998). Selection favors alleles that alter gene expression (i.e. *broken genes*) in rapidly or significantly changing environments (Kleinjan and van Heyningen 2005; Carroll 2008). Modifications in gene expression by these alleles have phenotypic effects ranging from relatively trivial changes to changes of several orders of magnitude (Daborn et al. 2002; Denver et al. 2005; Rifkin et al. 2005; Sellis et al. 2011).

Variation in the effects of broken alleles has implications for fitness over time. Those with large phenotypic effects are more likely to invade (Orr 1999, 2007; Otto and Gerstein 2008). We therefore expect selection on single locus changes after agriculture to be greatest on broken genes with a large effect in heterozygotes. Those producing less of a gene product in a drastically changed environment would enjoy a selective advantage over others producing an amount of the gene product adaptive in the older environment (Wilson et al. 2006; Cagliani et al. 2008; Gokcumen et al. 2013b). Heterozygote carriers of these broken genes accrue these advantages only if sufficiently rare. Furthermore, the wild-type allele at the locus shields heterozygotes against any fitness-reducing effects of the rare allele if the latter's phenotypic effect is large enough (Sellis et al. 2011).

Over time, mutant homozygotes become more common as heterozygote advantage increases the frequency of the broken allele. These mutant homozygotes unfortunately do not enjoy the mitigating effects of the broken genes' dramatically altered expression that heterozygotes do: they suffer all the costs without any of the benefit. Selection thus simultaneously acts *against* homozygotes and *favors* heterozygotes. In other words, the allele has become a balanced polymorphism (see Cochran et al. [2011], Sellis et al. [2011], and Connallon and Clark [2014] for theoretical arguments of how general this process is). Incomplete sweeps are, in return, characteristic of balancing selection, especially for alleles recently subject to strong selection (Hermisson and Pennings 2005; Charlesworth 2006; Fijarczyk and Babik 2015).

Balanced polymorphisms are therefore an expected outcome of recent and rapid evolution. The hemoglobinopathies noted in this chapter, like sickle cell (HbS), hemoglobin C (HbC), HbE, alpha- and beta-thalassemia, Melanesian ovalocytosis, or anemias like G6PD, are well-known examples (Kwiatkowski 2005; Hedrick 2011). These malaria defenses give heterozygote advantage while causing problems of varying severity in homozygotes. These adaptations are thus "quick fixes" offered by sloppy evolution that is characteristic of initial responses to rapid environmental changes (Harpending and Cochran 2006; Cochran et al. 2011). Other candidate balanced polymorphisms are also responses to infectious agents besides malaria (although see further in this chapter). Cystic fibrosis is probably the most common serious genetic disease among Europeans because heterozygote carriers of the main mutation, Δf508, enjoy a selective advantage against *Salmonella typhi*, or typhoid (Pier et al. 1998; Slatkin and Bertorelle 2001). Familial Mediterranean fever is caused by several defective alleles of the protein pyrin, which downregulates granulocyte-mediated inflammation. Homozygotes with mutant pyrin alleles can develop harmful fever and inflammation, while granulocytes produced in heterozygotes may protect against some pathogens (Harpending and Cochran 2006). Several different mutations of the gene connexin-26 cause congenital deafness in homozygotes but make heterozygotes' skin thicker and sweat saltier, potentially providing a better barrier against pathogens (Meyer et al. 2002).

Single locus changes therefore come in two flavors: advantageous genes and advantageous genotypes (i.e. balanced polymorphisms). The former replaces the latter if fitness reductions to homozygotes are severe enough (Cochran et al. 2011; Sellis et al. 2011; Venkataram et al. 2015). Alternative alleles conferring greater advantages than what heterozygotes enjoy would quickly proceed to fixation in the novel environment. However, alleles with such a significant advantage in the changed environment are exceedingly rare within the standing variation of the population and are also very unlikely to emerge via de novo mutation. Furthermore, as our LCT simulation illustrated, many potential advantageous genes introduced into the population require good luck to overcome genetic drift. Most of these advantageous genes would be lost by the accidents of history before they rose to fixation. Replacing balanced polymorphisms likely requires substantial amounts of time. Balanced polymorphisms should be more common following drastic environmental change: natural selection has less difficulty "finding" balanced polymorphisms than it does finding advantageous alleles. We would predict humans have numerous balanced polymorphisms but few simply advantageous genes.

Variants of this hypothesis have been hotly debated throughout the twentieth century (Morton et al. 1956; Ford 1964; Lewontin 1974; Gillespie 1991; Harpending and Rogers 2000; Orr, 2005; Crow 2008), which continues today to some degree (Andrés 2011; Hedrick 2012; Key et al. 2014; Fijarczyk and Babik 2015). Previous arguments suggested the fitness costs of many overdominant loci in a population must be compensated by implausibly large excess fertilities. However, Harpending and Rogers (2000) note this argument focuses attention on the fitness of a nonexistent optimal genotype. Instead, sufficiently large populations with ample coefficient of variation in fitness could maintain several hundred *million* overdominant loci (see Gillespie [1991] and Harpending and Rogers [2000] for a numerical example). Furthermore, recent genome-wide SNP surveys for candidate balanced polymorphisms (Andrés et al. 2009; Key et al. 2014; Fijarczyk and Babik 2015) suggest balanced polymorphisms are relatively rare in humans. Unfortunately, these scans are only capable of detecting old balanced polymorphisms extending to, or before, the human–chimpanzee split. More balanced polymorphisms should have developed following the advent of agriculture than in any other time in our evolutionary history. We therefore suggest most human balanced polymorphisms are young and are therefore outside the current scope of these scans.

Despite this controversy, several good candidate balanced polymorphisms have been given here (although there are likely more to be found). To our knowledge, however, only one allele has been proposed to be an advantageous variant that is replacing a balanced polymorphism. Bodmer and Cavalli-Sforza (1976) suggest HbC may eventually replace HbS in African malarial regions. HbS spread farther and faster than HbC, presumably because the selection pressures on the former were much greater than on the latter. However, HbS is near equilibrium, while HbC is not. Thus, over time, HbC may eventually reach fixation. Although HbC homozygotes suffer from anemia, it is much less severe than the sickling anemia in HbS homozygotes. Furthermore, the rate of HbC has been increasing in frequency at the rate that LCT increased in our simulations, suggesting it has a selective advantage (and the good luck) of an advantageous gene. Consequently, HbC may be a transient polymorphism on its way to becoming an advantageous allele that replaces a balanced polymorphism (HbS, in this case).

Quantitative Traits

Quantitative traits like stature, body weight, or the milk yield in cattle are traits that can be measured on a continuous scale. Fisher (1919) modeled quantitative traits as the sum of effects of a large number of genetic loci, with dominance, and random environmental effects. By the Central Limit Theorem, phenotypes of quantitative traits have Gaussian distributions within populations. Genotypic and environmental contributions to an individual's phenotypic value of a quantitative trait are stochastically independent. The phenotypic variance of a quantitative trait is therefore the sum of the fraction of variance given by additive genetic effects, or the narrow-sense heritability, and the remaining fraction of variance given by random environmental effects. Fisher's model further assumed no epistasis (nonadditive interactions between genes at different genomic locations) and a linear regression of genotypic values on phenotypes (Nagylaki 1992).

Under these assumptions, the *breeder's equation* relates the change in the average value of a quantitative trait responding to selection as:

$$r = h^2 s.$$

where the response to selection r is the difference between the average phenotype of parents and the average phenotype of offspring after a single generation of selection. It quantifies how far apart the parental and offspring distributions are after selection. The term h^2 is the narrow-sense heritability. Observe that the response to selection is proportional to the trait's narrow-sense heritability and completely independent of any environmental effects. Consequently, the narrow-sense heritability of a trait is the "raw material" that selection acts on. All else being equal, the effect of selection on a quantitative trait should be greater for traits with larger heritabilities. The selection differential S is the difference between the average phenotype in the population and the average phenotype of parents of the next generation. Again, all else being equal, the effect of a single generation of selection on parents with phenotypic values far from the population mean is larger than the effect of selection on parents with phenotypic values close to the population mean. Harsh selective breeding can thus produce substantial changes in several generations.

Any two terms of the breeder's equation estimate the third. For example, breeders or experimenters can fix the selection differential by choosing parents from the population under study. The response to selection r is obtained by measuring the offspring after selection. The heritability of the trait h^2 immediately follows from the breeder's equation. Historically, heritability of human traits was estimated from the similarity among relatives. Assessing the relationship between individuals by the number of shared SNPs is how heritability is currently estimated (Visscher et al. 2008; Yang et al. 2010; Davies et al. 2011; Lee and Chow 2014).

Crow (2010) suggests that, despite potential complications like epistasis, natural selection generally favors changes that lead to the simple model given by the breeder's equation. Effects of individual genes tend to be small, and hence additive (Hill et al. 2008). Thus, human quantitative traits come to evolve according to Fisher's model, as discussed in this chapter. Like the evolution of genes at a single locus, quantitative traits can undergo rapid and significant changes over the span of hundreds to tens of thousands of years *without losing genetic variation in the trait from selection* (Laurie et al. 2004; Hill 2005). Thus, small changes in allele frequencies induce substantial changes in phenotype without affecting overall genetic variation. The rate and magnitude of these changes can be quantified with the breeder's equation.

Biometrics is returning to prominence in anthropological genetics despite the twentieth-century takeover by Mendelians, given several methodical and theoretical reasons to consider quantitative traits when exploring problems in human evolution (Relethford 2007). Measuring how population structure and history affect quantitative traits hints at the relative influence of environmental and genetic effects (see, for example, McCullough [1982] for an example of this exercise on secular trends in stature among male Yucatec Maya). Measuring quantitative traits is relatively easy, and such data are frequently the best or only source of genetic information about some human populations. Inferences about the effects of population structure and history on extant populations can therefore be drawn from large available datasets (Chakraborty 1990). Quantitative traits may also give data about (pre-)historic populations that are otherwise unavailable (Relethford and Harpending 1994). This last methodical reason to consider quantitative traits is especially relevant for our re-evaluation of the tempo of human evolution. Quantitative traits may be the appropriate source of data for some questions about recent changes in humans: comparisons between the quantitative traits of humans today that are by-products of recent selection and those "[in] a world that no longer exists" (Relethford 2003, p. 33) would hint at the tempo and magnitude of recent evolutionary changes in our species.

Not all quantitative traits are equally useful for this last purpose. For example, craniometric (Relethford 2007) and dermatoglyphic variation (DiGangi and Jantz 2004) give the same F_{st} value for major continental groups as (presumably) neutral genetic markers. It's therefore unclear how to infer the history of selection from such traits. However, other quantitative traits like skin color show considerably more variation among populations than within due to natural selection (Relethford 2002). We suggest this may also be the case for some behavioral and psychometric traits given their (at least) moderate heritability (Bailey 1998; Turkheimer 2000; Davies et al. 2011). The social sciences have generally neglected quantitative genetic models. However, many quantitative traits like intelligence quotient (IQ; Herrnstein and Murray 1994; Jensen 1998), the propensity to develop mental illnesses like bipolar and schizophrenia (Owen and O'Donovan 2003; Barnett and Smoller 2009; Wray and Visscher 2010), aggression and antisociality (Barker et al. 2009), religiosity and the propensity to divorce (Bailey 1998), and other behavioral traits (Turkheimer 2011) have significant social and economic consequences. Even mild selection on these traits over dozens of generations can cause significant changes, which would themselves have dramatic social consequences. Larger populations and changes in social structure following the adoption of agriculture around the world introduced novel pressures on these behaviors that likely accelerated their evolution (Harpending and Cochran 2002; Cochran and Harpending 2009). Here, we outline assortative mating and genetic accommodation as two mechanisms of rapid evolution that have plausibly been important throughout our recent evolutionary past.

Assortative mating

Some couples pair according to their similarities along some trait. Although married couples are broadly similar to each other physically (Buss 1985; Silventoinen et al. 2003), they tend to be more similar along psychological and personality traits (Keller et al. 1996; Little et al. 2006). For example, a sample of Iowa newlyweds strongly assorted by age, religiousness, and political orientation (Watson et al. 2004). Assortment by education is well known (Mare 1991; Scwartz and Mare 2005) and likely has been increased by assortment by IQ after elite institutions became coeducational in the 1950s (Herrnstein and Murray 1994; Murray 2013). Since psychometric and personality traits are moderately heritable (as discussed here), we outline both the genetic effects of assortative mating along these traits and how rapidly differences among groups can evolve as a result.

Assortment for a quantitative trait generates either distinct groups or a continuum within populations, with those with more and less of the trait at their respective extremes. Such assortment induces correlations between causal loci (i.e. departures from linkage equilibrium) that increase the overall phenotypic variance of the trait as the groups or subpopulations grow more different (Fisher 1919; Crow and Felsenstein 1968; Nagylaki 1982). The trait's narrow-sense heritability increases as much as 10% after one generation of assortative mating as a consequence (Bulmer 1980; Falconer and MacKay 1996).

Note that assortative mating can be equivalent to directional selection. To illustrate this, consider a simple thought experiment about assortative mating for stature in a population. Assign an average stature of 69 in. for males and 64 in. for females, each with a standard deviation of 3 in. These figures are close to observed values in North America. Suddenly, assortative mating by height occurs such that those taller than the population mean mate only with others also above the mean, while those shorter than the mean mate only with others shorter than the mean.

First, we assume that the difference between male and female stature reflects fixed differences in the development of the two sexes. We can then normalize the data for each sex and measure stature in standard deviation units (SDs). The pooled population has a mean of zero and a SD of 1. The new units, called *z*-scores, are obtained by subtracting the population mean from everyone's stature (69 in. from each male and 64 in. from each female) and dividing by the standard deviation of 3 in.

After imposition of the mating rule, the population separates into two populations: talls and shorts. With our normal distribution assumption, we have split the overall normal distribution into two distributions. The mean of a left half-normal distribution is −0.8 SDs, and of the right half is +0.8 SDs. The difference in population means is therefore 1.6 SDs, or 4.8 in. in the original units.

The new subpopulations mate at random within themselves. The two sets of parents differ by 1.6 SDs. After reproduction, assuming an additive heritability of .80 for stature, the group difference will be $.8 \times 1.6 = 1.28$ SDs, corresponding to $1.28 \times 3 = 3.84$ in. After the first generation of reproduction, the genotypic group means differ by 3.84 in. without further movement between groups.

Before the new mating system, the overall variance of stature was 1. After a single generation of the new mating system, the between-subpopulation variance is 0.64^2 or 0.41 on the standard deviation scale. The variance of stature within each of two subpopulations is the complement, or 0.59. The fraction of stature variance that is between groups is now 41%. Back on the scale of inches, the SD within populations is the square root of 0.59 (0.773), corresponding to 2.3 in.

Richard Lewontin (1972) famously estimated the fraction of genic variance among continental human groups to be about 15% for a collection of presumably neutral genetic markers. That estimate of the ratio of between-group to total variance has remained constant as the number of loci available to estimate it grew from dozens to hundreds of thousands. After a single generation of our thought experiment, the stature variance between groups at genetic loci influencing stature is between two and three times as great as overall neutral genic variance among major human populations. Yet there is absolutely no selection in this thought experiment. Everything is selectively neutral. On the other hand, from the perspective from within subpopulations, there has been strong selection imposed by the mating rule allocating persons to groups according to stature. With this perspective, very strong selection has occurred.

This mechanism of selecting according to a threshold is called *truncation selection*. It is the most common selection rule in animal and plant breeding. Here, there is some ambiguity in whether or not we should speak of selection. From the overall population perspective, there has been no selection; while from the perspective from within groups, they are the product of effectively strong selection.

Agriculture offered permanent settlements, which in return meant accumulated wealth along with social and occupational stratification. Suppose these new specialized occupations required abilities that are quantitative traits, so an individual's value along the trait predicted their success within the occupational niche. Those with trait values above a certain threshold given by the demands of the job were generally successful, while those with trait values below this threshold were generally unsuccessful. Under these conditions, the social organization following the development of agriculture could have introduced novel social pressures, analogous to the ecological and dietary pressures discussed earlier in this chapter, favoring individuals above a certain threshold. Unsuccessful workers would be selected out of the niche through a process similar to truncation selection.

Conceivably, successful workers within these niches would have limited mating pools. Just as students are more likely to marry others with the same education level, individuals of a certain ability would likely pair up with others of comparable ability. These institutional pressures would then assort individuals into endogamous subgroups along the metric trait in the same way that the hypothetical mating rule (as discussed here) allocated individuals into groups. Assortative mating about these traits could therefore produce endogamous subgroups. The magnitude and rate of genetic differentiation along the trait between these subgroups would realistically be lower in real populations than what the numerical example implies since perfect assortative mating is the exception rather than the rule. However, even mild assortative mating can *eventually* induce significant differences between subgroups. More complete endogamy simply makes differentiation through assortative mating more rapid. In fact, assortative mating of this type is one of the most efficient mechanisms of selection (Crow and Kimura 1979).

History supplies several examples of this process. Beginning from at least 1250 CE, economically upper-class English left twice as many children as the poorest English (Clark 2007, 2009). Consequently, the modern English population largely descended from the Medieval upper economic classes. This substantial fitness advantage meant natural selection favored quantitative traits that increased the probability of entering, or staying in, the economic upper class. In other words, selection acted on individual economic success within the Medieval English economic niche. Given the English economy grew more capitalist around the time that fertility differentials between the economic classes developed, Clark (2007) suggests that individualism, low time preference rates, and the propensity to work long hours would be favored. If these individual characteristics were even moderately heritable, then they would be quantitative traits that would respond to selection according to the model given by the breeder's equation. The replacement of the lower economic class by the descendants of the upper economic class could then be modeled as truncation selection against individuals with low values of these traits.

The Medieval history of Ashkenazi, or Northern European, Jews is another example of possible recent selection through a process like truncation selection. Ashkenazi average verbal and mathematical IQ is three-quarters to a full SD greater than the Anglo average (i.e. 12–15 points greater). They also suffer from a suite of genetic diseases like Tay–Sachs, Niemann–Pick disease and other sphigolipid mutations, Gaucher's disease, and others at a disproportionately higher rate. Cochran et al. (2006) suggest these two peculiarities are consequences of recent and relatively rapid selection for IQ during the Middle Ages. The Ashkenazi were largely restricted to cognitively demanding occupations like finance, estate management, tax collection, and other managerial roles. Like in Medieval England, prosperous Ashkenazi individuals had more children than their nonprosperous counterparts. Since those with IQs below 100 essentially could not successfully perform within these jobs, selection could have acted on the fertility differentials between the prosperous and nonprosperous by favoring higher IQ. Prohibitions against intermarriage, rare conversions into Judaism, and prejudices against the Ashkenazim kept them almost completely endogamous and allowed selection to advance without interference from gene flow. Endogamy, the significant advantage to individuals with above-average IQ, and the significant heritability of IQ meant evolution within the Ashkenazi could have been very rapid. Genetic variants that boost IQ would have therefore risen in frequency very quickly. However, as discussed in this chapter, these are exactly the circumstances in which "broken" genes and balanced polymorphisms emerge. In particular, it seems heterozygotes for the mutations that cause any Ashkenazi sphigolipid disease and Tay–Sachs enjoy an increased IQ, while homozygotes for these mutations suffer from a fatal genetic disease (see Cochran et al. [2006] and Cochran and Harpending [2009] for both references and a more detailed discussion).

Genetic accommodation

Waddington (1942) was the first to outline how environmentally induced traits with no apparent genetic basis, like the sternal and alar callosities ostriches develop while crouching, are genetically inherited. Reasoning that an organism's genetic makeup constrains how they respond to environmental pressures, Waddington hypothesized that environmental pressures therefore act on the standing genetic variation among organisms. Developmental reactions conferring some adaptive advantage under the environmental pressures are favored by natural selection, implying the genes responsible for the development of the reaction increase in frequency. Thus, while development is fixed, certain developmental pathways will be favored according to their response to novel environmental stressors. If the environmental pressures are severe enough, then these reactions will be inherited even in the absence of the environmental effect. The acquired trait has been *genetically assimilated*. Two experiments, carried out by Waddington himself, corroborate this hypothesis. First, Waddington subjected *Drosphilia melanogaster* flies to a short, but intense, temperature shock (Waddington 1953). Several flies developed cross-veinless wings that are normally absent without the treatment. Breeding according to phenotype rapidly resulted in a subpopulation of cross-veinless flies generations after the initial temperature shock; cross-veinless wings were now a genetically assimilated characteristic.

Waddington (1956) then treated the same species of fly with ether. He found those flies subjected to the ether treatment developed a second thorax, although bithorax is absent without the ether treatment. Bithorax began appearing even in the absence of ether after 20 generations of selection for flies vulnerable to the ether treatment. Alleles that slightly increased the probability of bithorax, but not enough to bring out this phenotype without the environmental shock from the ether treatment, were apparently positively selected for during this experiment. Selection had then raised their frequencies high enough to trigger bithorax in the absence of the ether. Like cross-veinless wings, bithorax was a genetically assimilated trait.

Changing prey sizes in different parts of Australia have acted as environmental treatments on tiger snakes that favor larger heads (Aubret and Shine 2009). The width a tiger snake can open its jaw while consuming prey, and hence its overall head size, is highly plastic and variable among tiger snakes. Founding tiger snake populations can therefore adjust the width they open their mouths according to the size of the local prey populations. Over time, however, those snakes with alleles for the development of larger heads were favored in environments where local prey were larger, and thus these snakes increased in frequency. These changes were relatively rapid: comparisons between tiger snake populations in Australia show genetic assimilation can occur over a few thousand years. This result, coupled with others (Pigliucci et al. 2006) and several theoretical investigations (Pigliucci and Murren 2003; Lande 2009), reveals that genetic assimilation may be the mechanism behind some of the most rapid evolutionary changes.

Genetic accommodation is a generalization of genetic assimilation (West-Eberhard 2003; Braendle and Flatt 2006; Crispo 2007; Suzuki and Nijhout 2008). Like genetic assimilation, genetic accommodation begins with an environmental change that triggers the expression of a novel, heritable phenotypic variant. The new variant spreads through natural selection within a subpopulation, creating phenotypic variation within the overall population. Selection on the standing genetic variation within the population changes the regulation of the trait's expression, giving two possible outcomes. On one hand, the trait may become developmentally fixed, so it is expressed even without the environmental stimulus (i.e. genetic assimilation). On the other hand, selection may favor the trait to have a fixed, but heritable, expression only in certain environments. Thus, populations in highly variable environments will have adaptations in which their genomes produce different phenotypes in different environments (Suzuki and Nijhout 2008; Moczek et al. 2011).

We suggest genetic accommodation affects behaviors the same way it affects other phenotypes. For example, consider a hypothetical model of Scythian history, given only to illustrate how genetic accommodation could be used to model recent evolution. The Scythians were one of the earliest people to ride horses and were, by all accounts, unusually aggressive horseback warriors (Beckwith 2009; Wentz and de Grummond 2009). The ability of early Scythians to ride horses almost certainly did not have a genetic basis and therefore was not initially heritable. Riding horses was simply a cultural trait. However, skilled Scythian warriors likely enjoyed both mating advantages from their status as superior riders and a survivorship advantage over less skilled riders in battle (Anthony 2010). If this mating and survivorship differential was substantial enough, selection will then act on the standing genetic variation among poor and excellent riders and favor any alleles that improve "horsemanship." Horsemanship would then become more heritable within the Scythians, as initially rare genes present within the population became more common. In other words, over (our hypothetical reading of) Scythian history, horsemanship initially began as a nonheritable cultural trait and then became heritable through genetic accommodation because of the advantages it conferred. Once horsemanship's heritability increased, it would then evolve according to the breeder's equation.

The effects of this mechanism on the genome could be the same for other popular social traits that rapidly spread, like church attendance, certain political ideologies, or adherence to the law. Take heroin use as another example. Initially, there may not be a genetic component to who becomes a heroin user, nor any genetic advantage to being a heroin user. Instead, individuals likely begin using heroin as a response to social pressures from others. Over time, some users overdose and die (potentially before reproducing), while others survive. Selection then acts on the standing genetic variation among users and favors alleles that increase heroin tolerance because of the survivorship advantages they confer under these social circumstances. Tolerance to heroin would then become more heritable as these alleles increased in frequency.

Genetic accommodation is therefore a mechanism that plausibly increases the heritability of behavioral traits. As illustrated by the tiger snake example, genetic accommodation generally, and genetic assimilation specifically, may be the mechanism responsible for some of the most rapid evolutionary changes. Genetic accommodation may therefore be the mechanism driving some of the most substantial evolutionary changes in human behaviors. Like single locus changes, these evolutionary changes in behavior have likely been the greatest and most rapid following agriculture. The idea that cultural adaptation renders genetic adaptation irrelevant is widespread in anthropology. Instead, cultural pressures have increased overall rates of genetic adaptation (Hawks et al. 2007; Hughes et al. 2008; Cochran and Harpending 2009; Laland et al. 2010; Richerson et al. 2010). Genetic accommodation could be the mechanism behind many such genetic changes.

Conclusion

Our species has undergone rapid evolution during the past few millennia. In this chapter, we outlined the types and mechanisms that might account for these recent genetic changes and observed how pervasive they likely are. In particular, we discussed assortative mating and genetic accommodation as two mechanisms of rapid evolutionary changes in quantitative traits. Both of these mechanisms increase the heritability of quantitative traits, many of which have enormous social consequence. As the heritability of these traits increases over time, the difference between cultural and genetic transmission becomes blurred. Just as culture does not shield us from natural selection, we suggest cultural traits are also subject to genetic evolution.

CHAPTER 15

Natural Selection and Adaptation to Extreme Environments: High Latitudes and Altitudes

Abigail W. Bigham

Department of Anthropology, University of Michigan, Ann Arbor, MI, USA

Introduction

Modern human populations evolved in the tropical clines of equatorial Africa roughly 120 000 years ago. During the Late Pleistocene, we dispersed, migrating around the globe over the course of many thousands of years (Fu et al. 2013b). Humans eventually colonized nearly every inhabitable climatic zone across six continental landmasses. As people trekked across the biosphere into unchartered territory, they encountered novel environments, each with a unique set of ecological conditions. Variations in annual amount of rainfall, the presence of infectious diseases, the degree of solar radiation, as well as many other macro- and microvariables defined each habitat. Together, these niche-specific environmental pressures influenced the evolution of human populations and contributed to both their biological and cultural adaptation. Today, the allele frequencies of many genetic polymorphisms follow a clinal pattern of diversity, supporting the notion that climatic variables have been important selective pressures in human evolutionary

A Companion to Anthropological Genetics, First Edition. Edited by Dennis H. O'Rourke.

history (Cavalli-Sforza et al. 1994b). Furthermore, observed correlations between human traits such as skin color and environmental variables indicate that climate contributed to the geography of genetic variation (Jablonski and Chaplin 2000).

Many of the habitats experienced by territorially expanding human populations not only were novel, but also could be described as extreme. These demanding environments possessed conditions that were challenging for survival and reproduction, such as exceptional heat, cold, aridity, or ultraviolet light, or low ambient oxygen. Nonetheless, humans colonized countless regions with extreme climatic conditions, including locales with high altitude, subzero temperatures, and arid desert expanses. In so doing, human bodies responded biologically via both phenotypic adaptability and genetic adaptation. The latter is of particular interest to understanding evolution through natural selection, especially positive natural selection.

Positive natural selection is the route by which a beneficial, heritable trait increases in frequency over successive generations (Darwin and Wallace 1858). In extreme environments, geographically restricted positive selection has occurred wherein particular human groups have responded to unique sets of ecological conditions. These habitats provide researchers with a natural laboratory to study genetic adaptation. By understanding the process by which human populations underwent natural selection to extreme environments, we will be better suited to appreciate the effects of selection on modern biological variation and to identify the underlying genetic basis for adaptive phenotypes. Such knowledge can aid in identifying functionally important variants and uncovering attractive targets for producing beneficial phenotypes. Indeed, several studies of positive natural selection have uncovered functional roles of the loci under selection (Akey et al. 2004; Lamason et al. 2005). Additionally, recent work has indicated a link between selected genes and factors causing inherited complex disease (Fullerton et al. 2002; Rockman et al. 2004; Zhou et al. 2004) and resistance to infectious disease (Sabeti et al. 2005; Wood et al. 2005; Karlsson et al. 2013).

In this review, two unique environments are considered: extreme cold and extreme altitude. In what follows, I highlight population-specific physiological responses to particular environmental pressures to document the potential for natural selection to have acted. I then examine genomic evidence of natural selection and consider the data linking genotypes with adaptive phenotypes. I conclude with a discussion of the biomedical implications of characterizing the functional consequences of putatively adaptive alleles.

Cold Adaptation

Exceptionally cold temperatures harsh enough to threaten human survival characterize high-latitude regions of the earth. Cultural adaptations and technological innovations, such as fire, insulated clothing, protective shelters, and diets rich in high-caloric foods (i.e. meat and foods high in fat), enable human populations to overcome some of the stresses associated with severe cold. However, physiological mechanisms also are necessary to withstand continued subzero temperatures while maintaining an average core body temperature of 98.6° F. These biological responses to cold, including metabolic changes, thermogenic factors, and alterations in body morphology, are critical for prolonged sustainable human habitation in freezing climates such as those experienced by populations living in circumpolar regions.

One of the goals of early physiological studies was to disentangle the effects of technological innovations (e.g. shelter) from thermal acclimatization on cold survival. However, they failed to identify population-specific acclimatization. Research comparing Australian Aborigines with European-ancestry controls showed only that insulative cooling techniques differed between the two populations (Scholander et al. 1958). Furthermore, other studies documented elevated basal metabolic rates (BMRs) among Native Alaskan and Canadian groups, but suffered from sample size issues and a lack of control for confounding variables (Heinbecker 1928; Rabinowitch and Smith 1936; Rodahl 1952; Hart et al. 1962). As such, these analyses were unable to present convincing evidence for cold adaptation. Since these initial investigations, it has become clear that several circumpolar populations exhibit cold-adaptive phenotypes that distinguish them from acclimatized newcomers. In particular, Siberian populations have contributed extensively to our understanding of cold adaptation (for a review, see Leonard et al. 2005).

Physiological responses

Two classic ecogeographical rules in biological anthropology predict the distribution of body mass and limb proportions in humans (and other endothermic species) with respect to temperature, wherein the surface area–body mass ratio is maximized in colder climates and minimized in hot climates. Bergmann's (1847) rule specifies that species with greater body mass relative to surface area minimize heat loss through reduced body surface area. Allen's (1877) rule states that shorter appendages have higher mass to surface area and consequently possess better heat retention. Cold-adapted populations should thus be characterized by stocky stature and shorter appendages. Extensive empirical data from human populations and Neanderthals suggest that we and our close relatives follow these rules (Schreider 1950, 1964; Roberts 1953; Trinkaus 1981; Ruff 1991, 1994). In addition to body morphology, cranial phenotypes have been hypothesized to be functionally adaptive for cold stress (Thomson and Buxton 1923; Beals 1972). A narrow nasal aperture, an elongated nasal cavity, and an increase in nasal surface-to-volume ratio have been proposed to condition (i.e. heat) the inspired air in the upper airways of the respiratory tract (Franciscus and Long 1991). Furthermore, recent work exploring the evidence for midface adaptations to cold climate has shown that maxillary and nasal morphology associates with cold climatic variables (Evteev et al. 2014).

Morphological changes alone, however, are not sufficient mechanisms to prevent significant decreases in core body temperature. Human bodies must also rely on physiological processes to both conserve and produce heat. To this end, the human body has an arsenal of thermal conservation and production techniques to respond to environmental cold. With acute exposure, bodies will practice energetically inexpensive techniques to conserve heat such as vasoconstriction and vasodilation. Vasoconstriction in the subcutaneous blood vessels is a crucial mechanism for the retention of core body heat. Vasodilation increases warm blood flow near the skin's surface and is critical to protect tissues, especially those of the lower extremities, from cold injury. With sustained exposure to more severe cold, all homeotherms use energetically expensive techniques to produce heat instead of conserve heat.

Internal heat production is controlled by regulatory thermogenesis, which can be divided into shivering (ST) and nonshivering thermogenesis (NST). ST increases endogenous heat production by hydrolyzing adenosine triphosphate (ATP) through involuntary muscle contractions (Cannon and Nedergaard 2011). NST produces heat metabolically in brown adipose tissue (BAT) through the process of uncoupled oxidative respiration (Foster and Frydman 1979). To facilitate NST, uncoupling protein 1 (UCP1) expressed in the mitochondria of BAT (BAT is characterized by a high density of mitochondria) dissipates the proton gradient that drives ATP synthesis. Thus, BAT transfers energy from food into heat via NST.

BAT is found in all eutherians and likely arose very early on in mammalian evolution. In humans, BAT has been shown to persist into adulthood (Cypess et al. 2009), express UCP1 (Virtanen et al. 2009), exhibit seasonal variation (Saito et al. 2009), and contribute to cold-induced thermogenesis (Ouellet et al. 2012). BAT is particularly essential for maintaining homeostatic temperatures in infants who are unable to shiver as a result of weak musculature. It is noteworthy that European and East Asian infants possess greater BAT and white adipose tissue (WAT) than infants of African ancestry. This has led some to suggest that fat store distribution has evolutionary origins and a genetic basis (Merkestein et al. 2014). Combined, these observations suggest that cold-dwelling human populations may have adapted to subzero temperatures through genetic changes affecting their rates of NST.

It has been hypothesized that increased rates of NST among cold-adapted peoples lead to an increase in basal metabolic rate (BMR) (Leonard et al. 2005; Snodgrass et al. 2007). Seminal work by Roberts showed that BMR is inversely related to mean annual temperature (Roberts 1978), indicating that elevated metabolism is a physiological adaptation to cold stress. Recent research developed this observation further by demonstrating that arctic peoples exhibit the highest global BMRs. For example, Yakut (Sakha), Evenki, and Buryat peoples display high energy expenditures when compared to standard reference values (Galloway et al. 2000; Leonard et al. 2002; Snodgrass et al. 2005), with estimates based on fat-free mass (FFM) averaging 16.4% higher in men and 18.8% higher among women (Leonard et al. 2005). Furthermore, elevated BMR also has been observed among the Inuit (Heinbecker 1928).

A wide range of data suggest that changes in body morphology, regulatory thermogenesis, and basal metabolic rate all contribute to various populations' cold-adaptive techniques, and as such would be expected to undergo selective adaptation in populations living in cold environments. Together, the various physiological responses to cold exposure highlight the complex mechanisms involving many pathways that initiate and regulate the cold response process. Emerging evidence indicates that genetic changes are responsible for the observed cold-adaptive phenotypes, and thus provide compelling evidence that human populations have genetically adapted to sustained low temperatures. What remains to be demonstrated are the functional links between naturally selected mutations and particular cold-adaptive phenotypes.

Candidate genes for cold adaptation

Genome scans for natural selection show an enrichment of genic and nonsynonymous single nucleotide polymorphisms (SNPs) correlated with climatic variables in worldwide samples of human populations (Hancock et al. 2011b). In particular, genes involved in energy metabolism, an important component of the thermogenic response to cold, show spatially varying signals of selection on a global scale (Hancock et al. 2008). These include nonsynonymous SNPs in epoxide hydrolase 2 (*EPHX2*) and fatty acid binding protein 2 (*FABP2*) and noncoding SNPs in regulatory associated protein of MTOR (*RPTOR*) (Hancock et al. 2008). The strongest case for spatially varying selection in response to cold tolerance has been observed for the nonsynonymous 109R allele (rs1137100) in the leptin receptor (*LEPR*) (Hancock et al. 2008). This allele is part of the thermogenesis pathway (Dulloo et al. 2002), and it has been associated with an increased respiratory quotient consistent with a role in NST (Rosmond et al. 2000; Loos et al. 2006). Natural selection screens for cold adaptation loci also have been performed specifically among cold-adapted populations. Genome-wide analysis of indigenous Siberian populations indicated that genes involved in energy regulation and metabolism (mitochondrial carnitine palmitoyltransferase 1 alpha [*CPT1A*], low-density lipoprotein receptor-related protein 5 [*LRP5*], and thyroid adenoma associated [*THADA*]) as well as vascular smooth muscle contraction (protein kinase, cGMP-dependent, type I [*PRKG1*]) showed the strongest signals of selection in this group (Cardona et al. 2014). Together, these observations suggest the possibility for independent mechanisms for cold adaptation among modern human populations.

Interrogation of specific candidate loci implicated in genome-wide selection screens for cold adaptation or identified from animal models has revealed exciting results. One gene in particular that has emerged as a high-priority natural selection candidate gene is *UCP1*. Evidence suggests that variants at this locus increase thermogenic function in cold-exposed populations. Murine studies show that UCP1-knockout mice exhibit greater sensitivity to cold (Enerback et al. 1997), whereas UCP1 overexpression results in increased metabolic rate (Li et al. 2000). Among humans, the *UCP1* promoter region variant (rs1800592, −3826A/G) upregulates *UCP1* expression (Esterbauer et al. 1998) and has been associated with the level of cold-induced thermogenesis in healthy children (Nagai et al. 2007) as well as with winter climate variables (Hancock et al. 2011a). Another noteworthy feature of *UCP1* is that its expression is upregulated in WAT during winter months (Kern et al. 2014). DNA-sequencing efforts indicate that positive selection may have acted on standing variation to increase the frequency of the (−3826A) haplotype among Europeans and East Asians (Hancock et al. 2011a). Such a finding suggests that the common ancestor to modern-day Europeans and East Asians responded to lower temperatures experienced outside of Africa. It would be worthwhile to estimate the age of the selective event to clarify when in human evolutionary history such a selective sweep took place. The *UCP1* homolog, *UCP3*, also shows significant associations between the T allele at SNP rs1800849 (−55T/C) and winter climate variables, as well as statistical evidence of positive natural selection (Hancock et al. 2011a). This finding indicates a role for *UCP3* in NST, a point that has been debated in the past (Boss et al. 1997; Fleury et al. 1997; Krauss et al. 2005), further stressing the utility of population genetics to identify gene function.

One of the most compelling cases for natural selection to cold tolerance comes from the gene *CPT1A*, whose cellular role is to regulate mitochondrial long-chain fatty acid oxidation. A nonsynonymous variant (rs80356779, C1436T) among circum-Arctic populations shows one of the strongest selective sweeps reported in humans (Clemente et al. 2014). However, its phenotypic consequence in cold-adapted populations has not been explored. Furthermore, this gene's known role in hypoketotic hypoglycemia indicates that this variant could have swept to high frequency as a response to a diet rich in fats, not cold exposure alone. Several other genes have emerged as potential natural selection candidate loci for cold stress. Candidate-gene analysis of BAT metabolism loci among Europeans and East Asians again identified *LEPR* as a target of recent positive selection, but only among East Asians (Sazzini et al. 2014). Additionally, roles in adaptive thermogenesis for several other loci have been proposed, including adiponectin receptor 1 (*ADIPOR1*) and 2 (*ADIPOR2*) (Svensson et al. 2011), adrenoceptor 2 (*ADRB2*) and 3 (*ADRB3*) (Takenaka et al. 2012), and major facilitator superfamily domain containing protein 2a (*MFSD2A*) and 2b (*MFSD2B*) (Angers et al. 2008).

Health implications of cold adaptation

Both cold-adaptive phenotypes and particular cold-adaptive alleles influence susceptibility to complex disease. Cardiovascular disease, metabolic disease, and autoimmune thyroid disorders all have been linked to physiological correlates of cold adaptation, in particular changes in BMR. Elevated BMR likely results in increased rates of hypertension and lower serum lipid levels. Luke et al. (2004) hypothesized a positive association between BMR and blood pressure that is independent of body size. Data from indigenous Siberians support this hypothesis: Buryats, Evenki, and Yakuts (Sakha) possess elevated BMRs that positively associate with blood pressure and likely contribute to their high hypertension rates (Snodgrass et al. 2008). Increased BMR additionally may affect overall serum lipid levels in these populations, resulting in relatively low levels of low-density lipoproteins (LDLs) and triglycerides despite high dietary consumption of animal fats (Leonard et al. 1994; Rode and Shephard 1995). However, indigenous Siberians also are characterized by low levels of high-density lipoproteins (HDLs) (high levels are associated with a cardioprotective affect) (Snodgrass et al. 2010). Despite elevated blood pressure and low HDL cholesterol, metabolic syndrome, a cluster syndrome characterized by elevated blood pressure, high serum triglycerides, low HDL, impaired fasting glucose, and abdominal obesity, remains uncommon among indigenous Siberians (Snodgrass et al. 2010). In addition to effects on cardiovascular health, cold adaptation may play a role in increased autoimmune thyroid disorders. Thyroid hormones promote oxidative metabolism; thus, thyroid activity changes with seasonal variation in temperature to regulate metabolic processes involved in cold stress (Maes et al. 1997; Leonard et al. 1999). Increased levels of anti-thyroid peroxidase enzyme (TPOAb) and T4 (thyroxine) have been observed among Yakut and Evenki adult women, respectively (Leonard et al. 2005; Cepon et al. 2011).

Traditional lifeways such as herding and high-activity levels related to subsistence strategies likely have protected circumpolar populations from health risks associated with cold-adaptive phenotypes, but recent economic change in the post-Soviet era may have a deleterious effect on health, increasing the risk for metabolic disorders and cardiovascular disease among indigenous Siberians (Snodgrass et al. 2010). Identifying and functionally characterizing these genetic variants can potentially be helpful for devising intervention strategies in changing economic landscapes as well as devising novel approaches for treating chronic disease among global populations. To this end, the specific genes underlying cold-adaptive phenotypes among circumpolar populations are just beginning to be identified. *CPTA1* SNP rs80356779 (C1436T), discussed in this chapter, shows a cardioprotective role elevating plasma HDL cholesterol levels among Greenland Inuit (Rajakumar et al. 2009). However, mutations in this gene are known to cause the autosomal recessive metabolic disease carnitine-acylcarnitine translocase deficiency, a disorder involving long-chain fatty acid conversion into energy. Among Canadian Inuit and Siberian populations, the rs80356779 (C1436T) polymorphism is maintained at a high frequency (>70%) with few clinical effects (Greenberg et al. 2009; Cardona et al. 2014). Hence, some have suggested that this polymorphism confers a selective advantage and has been maintained in circumpolar populations despite its deleterious effects. Several cold-adaptive alleles have been identified in other populations. One noteworthy candidate gene is *UCP1*, as *UCP1* expression may contribute to cardiovascular health given its effect on the amount of energy stored as fat. Indeed, the A allele of SNP rs1800592 located upstream of *UCP1* is associated with increased expression of *UCP1* and therefore a decrease in the amount of energy stored as fat (Esterbauer et al. 1998). Likewise, the G allele at this locus is associated with an increased rate of weight gain among morbidly obese subjects (Clement et al. 1996).

High-Altitude Adaptation

High-altitude regions of the world, defined as areas lying 2500 m or more above sea level, are barren environments characterized by extreme temperatures, aridity, poor soil quality, high solar radiation, and low oxygen tension. These environmental conditions challenge the ability of humans to live and reproduce (i.e. adapt and/or acclimatize). Behavioral or cultural modifications such as clothing, shelter, and ecological-tier farming buffer many of these characteristics, but low ambient oxygen tension, caused by decreased barometric pressure and commonly referred to as high-altitude hypoxia, cannot be readily overcome by cultural buffers. Hypoxia is thus the fundamental challenge that high-altitude residents face, necessitating physiological acclimatization and/or genetic adaptation to overcome it.

Despite numerous harsh environmental conditions found at high altitude, human populations have inhabited three high-altitude zones for millennia: the Qinghai–Tibetan Plateau, the Andean Altiplano, and the Semien Plateau of Ethiopia. Archeological evidence indicates that these three geographic locations were first populated approximately 25 000, 11 000, and 8000 years ago, respectively, providing anthropologists with a natural laboratory to study the effects of hypobaric hypoxia on high-altitude populations (Baker and Little 1976; Schull 1990). Traditional attention to high-altitude-adaptive phenotypes largely focused on hematological and pulmonary factors, and much is known about these physiological adaptations. More recent research has pointed to the role of vascular factors as central components contributing to adaptation to chronic hypoxia. Overall, these efforts have led to a body of literature documenting the suite of human physiological responses to high-altitude habitation in the three human groups.

Physiological responses

Today, Andeans, Ethiopians, and Tibetans display a range of physiological traits that help them biologically respond to the hypoxic environment and that are distinct from those of low-altitude residents. They also exhibit distinct physiological adaptations compared to each other, highlighting the array of phenotypic diversity in altitude-response phenotypes. Tibetan and Andean physiological responses to high altitude have been studied extensively, whereas less is known regarding the Ethiopian pattern of adaptation. One widely studied, potentially altitude-adaptive trait is hemoglobin, the oxygen-carrying metalloprotein in erythrocytes. Tibetans display relatively low hemoglobin concentrations (Adams and Strang 1975; Beall and Reichsman 1984; Beall and Goldstein 1987), Andeans show elevated concentrations (Beall et al. 1990, 1998), and Ethiopians maintain concentrations within the sea-level range (Beall et al. 2002). A second trait influencing arterial oxygen content along with hemoglobin is arterial oxygen saturation (SaO_2). Tibetan and Andean highlanders exhibit higher levels of SaO_2 compared to low-altitude residents measured at the same altitude (Brutsaert et al. 2000; Wu and Kayser 2006). However, Tibetan average SaO_2 levels at rest are not significantly different from those of Han Chinese born and raised at high altitude, and furthermore they are lower than those observed among Andeans when measured using the same equipment and protocol (Beall et al. 1997b, 1999; Weitz and Garruto 2007). Emerging studies from Ethiopia suggest that high-altitude natives in that region have resting SaO_2s at 3530 m that are only moderately lower than sea-level values (Beall et al. 2002).

Pulmonary responses to high-altitude hypoxia also vary across populations. Tibetans maintain equal or higher resting ventilation compared to other Asian and European populations measured at the same altitude (Zhuang et al. 1993; Ge et al. 1994; Beall et al. 1997a). Moreover, the Tibetan hypoxic ventilatory response (HVR), the immediate rise in ventilation following acute exposure to increased hypoxia, is in line with that of acclimatized newcomers and low-altitude populations (Zhuang et al. 1993). This suggests that Tibetans have adapted to maintain the temporary hypoxia-induced increase in ventilation observed among low-altitude native populations. This increase in ventilation is not observed among Andeans, who exhibit no increase in resting ventilation levels over low-altitude values at rest or during exercise (Chiodi 1957; Brutsaert et al. 2000) and a blunted (low) HVR (Chiodi 1957; Beall et al. 1997a) that is commonly lower than sea-level values.

Perhaps some of the most compelling physiological adaptations to life at high altitude are the changes associated with pregnancy and fetal growth. High altitude presents numerous challenges to the developing fetus, including intrauterine growth restriction (IUGR), an important marker of morbidity and mortality in newborns that results in low birth weight. In general, birth weight declines an average of 102 g per 1000 m gain in altitude (Jensen and Moore 1997). However,

Tibetan and Andean residents of high altitude are protected from altitude-associated IUGR. Women from both of these regions give birth to larger babies than Han Chinese or European women, respectively, living at the same elevation regardless of socioeconomic status or gestational age (Zamudio et al. 1993; Moore et al. 2001; Soria et al. 2013).

The diverse biological responses to high-altitude hypoxia, including pulmonary, circulatory, and pregnancy-related traits, document the potential for natural selection to have acted on particular altitude-adaptive phenotypes. Emerging genetic data support an evolutionary origin for these observed physiological adaptations to high altitude. Research of this nature documents the role of local adaptation, not simply acclimatization, to the high-altitude environment. Early heritability studies identified traits with a likely genetic basis, although they did not isolate the genes responsible for the observed biological differences. For example, a major autosomal dominant allele for high SaO_2 among Tibetan women was identified, wherein women carrying the high-oxygen-saturation allele exhibited a greater offspring survival rate than women possessing the low-oxygen-saturation allele (Beall et al. 1994). Other investigations identified associations between the percentage of genomic Native American ancestry as a proxy for Andean ancestry and altitude-adaptive traits. Andeans' blunted ventilatory response, increased birth weight, and greater uterine artery diameter have all been shown to correlate with Andean ancestry (Brutsaert et al. 2005; Wilson et al. 2007; Bennett et al. 2008; Julian et al. 2009; Soria et al. 2013). More recently, genome scans have been a powerful tool to distinguish selection-nominated candidate genes or gene regions for high-altitude adaptation.

Based on their documented involvement in both oxygen regulation and altitude phenotypes, genes in three pathways have been hypothesized to help regulate the physiological responses to hypobaric hypoxia (Moore et al. 2006). Thus, they are a priori candidates for natural selection to hypoxia. They include the hypoxia inducible factor (HIF) pathway, the renin angiotensin system (RAS), and the globin family of genes. Genes that are members of the HIF pathway show the most compelling evidence of natural selection across all three long-term, high-altitude population groups. Nonetheless, the data also indicate that population-specific genetic changes have occurred wherein different genes have been acted upon by natural selection in each of the three populations. The mounting evidence for each highland population's genetic response to hypoxia is discussed in Sections below.

Ethiopian adaptation

Of the three high-altitude groups, Ethiopians have been the focus of the least research. Nonetheless, several genome scans have been performed in this population and have implicated HIF and non-HIF pathway genes in the Ethiopian pattern of adaptation to high altitude. Candidate genes include the basic helix–loop–helix family, member e41 (*BHLHE41*, also known as *DEC2* or *SHARP1*); mitochondrial calcium uptake 1 (*MICU*, also known as *CBARA1*); vav 3 guanine nucleotide exchange factor (*VAV3*); aryl-hydrocarbon receptor nuclear translocator 2 (*ARNT2*); thyroid hormone receptor, beta (*THRB*); as well as a gene-rich region on chromosome 19 containing several genes that are involved in vascular physiology, including chemokine (C-X-C motif) ligand 17 (*CXCL17*) and platelet-activating factor acetylhydrolase 1b, catalytic subunit 3 (*PAFAH1B3*), or that have been linked to hypoxia (lipase, hormone-sensitive [*LIPE*]) (Scheinfeldt et al. 2012; Huerta-Sanchez et al. 2013; Udpa et al. 2014). Two genes, *ARNT2* and *THRB*, show suggestive relationships with hemoglobin concentration. Ethiopian Amhara who are homozygous for one SNP in particular, the C allele of *THRB* rs826216, display higher hemoglobin levels than individuals who are homozygous for the T allele (Scheinfeldt et al. 2012).

Andean adaptation

Genomic analysis of Andean populations has revealed at least 40 HIF pathway genes as candidates for natural selection to high altitude (Bigham et al. 2009, 2010; Foll et al. 2014). To a lesser extent, it also has indicated the involvement of genes in the globin family and RAS (Bigham et al. 2008, 2010). One of the most compelling cases for genomic evidence of natural selection can be found in the gene *EGLN1* (Bigham et al. 2010; Foll et al. 2014). *EGLN1*, along with *EGLN2* and *EGLN3*, is a molecular oxygen sensor that regulates the *HIF* transcriptional pathway (Hirsila et al. 2003). In normoxia, *EGLN1* hydroxylates HIF1α's oxygen-dependent degradation domain, which targets this protein for breakdown by the E3 ubiquitin ligase complex (Maxwell et al. 1999; Ivan et al. 2001; Jaakkola et al. 2001). Under hypoxic stress, decreases in oxygen tension lead to a reduction in prolyl hydroxylation of HIF1α by *EGLN1*, thus increasing HIF levels and maintaining cellular oxygen homeostasis (Epstein et al. 2001). Standard population genetic metrics measuring evidence for natural selection showed extended regions of statistical significance within this gene as well as 500 kb upstream and downstream (Bigham et al. 2010). Despite the genetic signatures of positive directional selection, to date, no associations between putative functional alleles and Andean altitude-adaptive phenotypes have been identified for this locus (Bigham et al. 2013).

When considering other loci, several selection-nominated candidate genes have been linked to phenotypic outcomes among Andeans. The angiotensin-converting enzyme (ACE) insertion–deletion (I/D) polymorphism, associated with resting and exercise SaO_2 in a sample of Peruvian Quechua, explains ~4% of the variance in SaO_2 (Bigham et al. 2009). Polymorphisms near a second gene – protein kinase, AMP-activated, alpha 1 catalytic subunit (*PRKAA1*, also known as *AMPKα1*) – may be influential in achieving genetic adaptation to high altitude by affecting physiological responses to pregnancy that are instrumental for fetal growth (Bigham et al. 2014). Of note, the maternal genotype of a SNP upstream of *PRKAA1* shows associations both with advantageous phenotypes (heavier birth weight and larger uterine artery diameter) and with gene expression patterns of key pathways involved in the pathophysiology of fetal growth restriction. Furthermore, whole-genome sequencing of an Andean chronic mountain sickness (CMS) cohort implicated sentrin-specific peptidase 1 (SENP1), an enzyme implicated in erythropoiesis, and acidic leucine-rich nuclear phosphoprotein 32 family member D (*ANP32D*), an oncogene, in the development of CMS (Zhou et al. 2013).

Tibetan adaptation

Numerous genomic scans for positive selection have been performed among Tibetan highlanders (Beall et al. 2010; Bigham et al. 2010; Simonson et al. 2010; Yi et al. 2010; Peng et al. 2011; Wang et al. 2011; Xu et al. 2011). From this work, two genes in particular have emerged as the targets of natural selection, *EPAS1* and *EGLN1*. *EPAS1* shows evidence of positive directional selection across all of the genome-wide analyses (Beall et al. 2010; Bigham et al. 2010; Simonson et al. 2010; Yi et al. 2010; Peng et al. 2011; Wang et al. 2011; Xu et al. 2011), whereas *EGLN1* shows this across the majority (Simonson et al. 2010; Yi et al. 2010; Peng et al. 2011; Wang et al. 2011; Xu et al. 2011). This convergence and replication of results using distinct cohorts strongly support the roles of both of these genes in the Tibetan pattern of high-altitude adaptation. Furthermore, significant associations for *EPAS1* and *EGLN1* SNP genotypes and haplotypes with hemoglobin concentration have been identified in Tibetans (Beall et al. 2010; Simonson et al. 2010; Yi et al. 2010). These data now provide direct evidence that hypoxia has acted as a selective agent on HIF genes that influence hemoglobin concentration.

EPAS1 is a HIF regulatory gene encoding the transcription factor HIF2A. It induces downstream genes when cellular oxygen levels decrease. In addition, HIF2A is partly responsible for erythropoiesis and therefore can affect the concentration of hemoglobin in the bloodstream. Dating the emergence of the *EPAS1* allele has been controversial, with estimates as recent as 3000 years ago (Yi et al. 2010) and as early as 18 000 years ago (Peng et al. 2011). Intriguingly, a recent study suggests that introgression from Denisovans, a *Homo* species identified from remains found in the Altai Mountains of Siberia, or Denisovan-related individuals is responsible for the Tibetan pattern of variation observed for *EPAS1* (Huerta-Sanchez et al. 2014), thus suggesting ancient admixture may have contributed to functionally important adaptive changes in the Tibetan genome. Further research elucidating the molecular mechanisms underlying the low hemoglobin adaptation of Tibetans will be an important step in understanding the functional significance of *EPAS1* variation on hemoglobin concentration.

In comparison to *EPAS1*, experimental validation of *EGLN1*'s hypothesized effects on hemoglobin concentration has progressed to a greater extent. *EGLN1* DNA-sequencing studies have revealed two coding region SNPs, D4E (rs186996510) and C127S (rs12097901), that are highly enriched and are in strong linkage disequilibrium in the Tibetan population (Xiang et al. 2013; Lorenzo et al. 2014; Petousi et al. 2014). In fact, the D4E/C127S haplotype is enriched over 80-fold in Tibetans as compared to Han Chinese. Remarkably, *in vitro* studies indicate that the Tibetan D4E/C127S *EGLN1* is a hypomorphic allele diminished in its capacity to promote HIFα turnover, which is necessary to turn on genes in the HIF cascade (Song et al. 2014a). Age estimates for emergence of the *EGLN1* adaptive allele are placed at about 8000 years ago (Peng et al. 2011; Lorenzo et al. 2014). If the later age estimation for the adaptive *EPAS1* allele is accurate (i.e. 18 000 years ago), it is possible that the adaptive *EGLN1* and *EPAS1* polymorphisms work in concert, with the potentially deleterious effects of a proposed loss-of-function *EGLN1* allele blunted by a preexisting high-frequency protective *HIF2A* allele (Bigham and Lee 2014).

Convergent natural selection

The data discussed in the preceding sections suggest that the three high-altitude human populations have adapted to low ambient oxygen tension through population-specific genetic factors as a result of natural selection. We observe that the majority of the genes that are part of biochemical pathways known to be involved in oxygen homeostasis, such as the HIF pathway, do not exhibit overlap between the three high-altitude groups. However, a single HIF-regulatory candidate gene, *EGLN1*, shows evidence of natural selection in both Andeans and Tibetans. Intriguingly, both high-altitude groups show a single dominant haplotype, but the haplotype is population specific. Furthermore, *EGLN1* genotypes associate with hemoglobin concentration in Tibetans but not Andeans (Simonson et al. 2010; Bigham et al. 2013). These data suggest that natural selection may have operated on functionally different changes in the same gene.

It has been shown that the functional variation in the Tibetan *EGLN1* allele affects hemoglobin concentration, possibly as a loss-of-function allele (Song et al. 2014b). The yet-to-be-identified functional variation in the Andean *EGLN1* allele may also contribute to their hemoglobin phenotype (elevated concentration) or to an altogether different altitude-adaptive trait. Data from mouse models show that mice with heterozygous loss-of-function *Egln1* mutations display erythrocytosis and increased respiration under normoxic and hypoxic conditions (Arsenault et al. 2013; Bishop et al. 2013; Bigham and Lee 2014). This reveals that *Egln1* plays a central role in both erythropoietic and respiratory adaptation, and suggests that positively selected mutations in this gene have the capacity to affect different phenotypes in Andeans versus Tibetans. It seems likely that several interacting genes account for the overall adaptive phenotype observed among Andeans. Further research elucidating the genotype–phenotype relationship between polymorphisms in this gene and Andean phenotypes will be an important first step in characterizing the functional impact of Andean variation. Such analyses should be followed up by functional assays to gain insight into the molecular mechanism of the Andean adaptive allele.

Conclusion

A central objective of evolutionary biology is to elucidate the genetic basis of positive natural selection. Exploring genetic signatures of natural selection within the human genome offers a window into our recent evolutionary past and provides a deeper understanding of how this evolutionary force has shaped extant patterns of human variation. In particular, understanding natural selection to extreme environments is a productive avenue of research, as these environments are natural experimental designs and allow for the understanding of phenotypes in relatively controlled conditions. Continued investigation will (i) result in a more comprehensive understanding of how, when, and where natural selection has acted on the human genome to shape modern human diversity; (ii) reveal the extent of geographically restricted selective pressures; (iii) contribute to our understanding of the genetic architecture of adaptive human phenotypes; (iv) elucidate the molecular basis of adaptive change, including convergent human adaptations; and (v) ultimately render a more complete picture of the effects of adaptive genetic variation on phenotypic variation and disease susceptibility.

The majority of studies seeking to understand positive selection in the human lineage can be classified as genome-wide scans for positive selection. At this juncture, it is imperative to move forward with research linking genotype to phenotype, identify the causal mutation(s) responsible for the observed phenotypic adaptations, and functionally characterize the selected alleles. Until then, such loci remain only putatively adaptive. These endeavors are critical for answering evolutionary questions and will be important for identifying genetic factors that influence human phenotypes, including normal morphological variation and disease-related traits. Such knowledge is likely to profoundly impact our understanding of the genetic control of phenotypic traits as well as provide the basis for therapies for common diseases such as cardiovascular and metabolic diseases.

CHAPTER 16

Sensory Polymorphisms and Dietary Adaptation

Carrie C. Veilleux

Anthropology Department, University of Texas, Austin, TX, USA

Introduction

In humans, as in other animals, sensory systems mediate our relationships with the external environment. We use our five primary senses (vision, olfaction, taste, hearing, and touch) to find resources or predators, receive communicatory signals from other animals, and direct movement around our habitats. Consequently, studies across a wide range of primates and other mammals have repeatedly shown strong relationships between sensory adaptations and ecological variables (e.g. Dominy et al. 2001; Hayakawa et al. 2014; Veilleux and Kirk 2014). Because finding food is one of the most important activities for any animal, much research has focused on identifying links between sensory systems and an animal's diet or foraging strategy. For primates, all five senses can be critical in both detecting food resources and evaluating the quality of individual food items (Dominy et al. 2001). For example, vision (color vision and acuity), olfaction, and hearing can allow a primate to detect resources (e.g. fruit patches, insects) from a distance. Within food patches, color discrimination, olfaction, touch, and taste may all be involved in selecting preferred food items, such as through detecting fruit ripeness (touch, olfaction, color) or avoiding toxins (taste).

A Companion to Anthropological Genetics, First Edition. Edited by Dennis H. O'Rourke.

Differences in sensory perception within humans and between humans and other animals have long been a focus of interest (Fisher, Ford, and Huxley 1939; Post 1962). However, modern genetics has allowed researchers to begin identifying specific genes and gene families involved in sensory system function (Nathans, Thomas, and Hogness 1986b; Buck and Axel 1991; Hoon et al. 1999). Subsequently, our understanding of the evolution and ecology of primate and nonprimate mammal sensory genes has expanded dramatically over the last 20 years, especially for color vision and taste. While the genes for these two systems can vary structurally (e.g. in copy number variation, the number of exons and introns, and gene and protein length), they all encode G protein-coupled receptors (GPCRs), which are proteins with seven membrane-spanning domains (Figures 16.1a and 16.2) that interact with external stimuli (such as light or a chemical compound) and activate signal transduction pathways within the cell, resulting in the signal being sent to the brain for higher levels of sensory processing (Sharpe et al. 1999; Feeney 2011). Researchers have used a variety of investigative techniques to link variation in these genes to differences in sensitivity and perception of color and taste, including psychophysical experiments linking genotypes to behaviorally measured differences in sensitivity in humans and other mammals, experiments involving knockout genes in mice, and *in vitro* experiments using

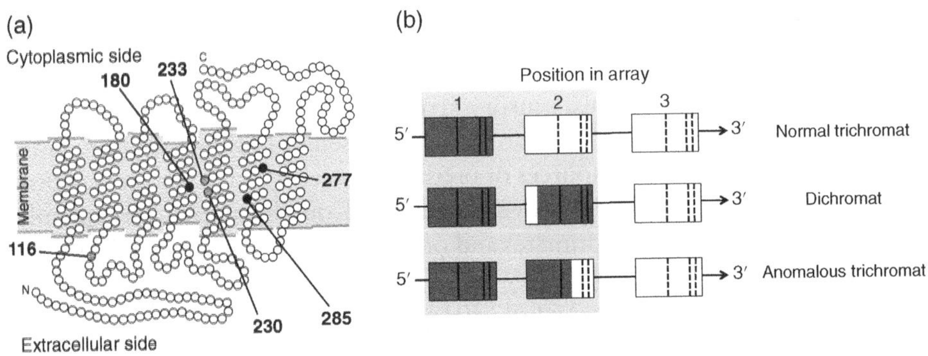

Figure 16.1 L and M opsin and L–M opsin array in humans. (a) The structure of the L and M opsin proteins, which are transmembrane GPCRs. Three critical spectral-tuning sites are responsible for the majority of the difference in spectral sensitivity between human L and M opsins (black circles), while three additional sites (gray circles) cause small shifts in sensitivity. (b) Three hypothetical L–M opsin arrays and their resulting color vision phenotypes. Most humans have a single L opsin gene (dark gray) with the specific SNPs for sensitivity to long wavelengths (black lines), and one or more copies of the M opsin gene (white) with specific SNPs for sensitivity to middle wavelengths (dashed lines). Due to gene conversion, some humans have L–M hybrid genes. When the hybrid genes occupy the first or second position of the array, they cause defective color vision. Dichromacy can result when the hybrid gene shares all of the spectral-tuning SNPs with the first gene in the array (typically, the L opsin gene), while anomalous trichromacy can result when the hybrid gene differs at some tuning sites from the first gene in the array.

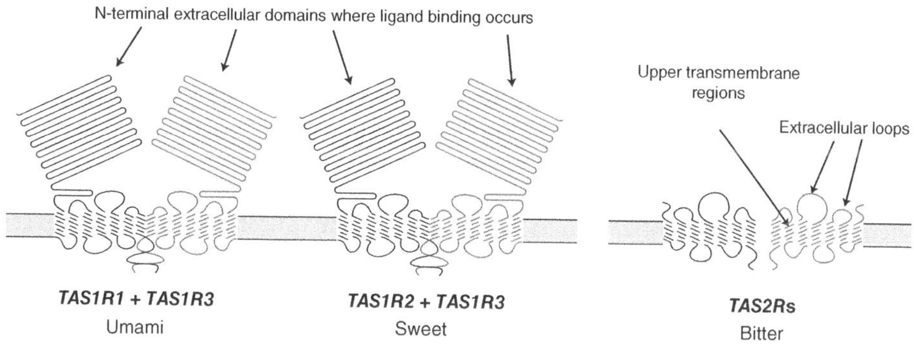

Figure 16.2 *TAS1Rs* and *TAS2Rs*. The structure of the *TAS* receptor proteins, which are transmembrane GPCRs (notice the same seven-helix structure to the opsin in Figure 16.1). *TAS1R1* and *TAS1R3* form a heterodimer that makes a taste receptor cell detect umami, while *TAS1R2* and *TAS2R3* form a heterodimer that makes a taste receptor cell detect sweet. Bitter-taste *TAS2Rs* do not form heterodimers, although more than one *TAS2R* gene can be expressed in a bitter-detecting taste receptor cell. In *TAS1Rs*, an extended N-terminal extracellular domain is involved in ligand recognition and binding (called the *Venus fly trap domain*). In *TAS2Rs*, ligand recognition and binding occur in upper transmembrane regions and extracellular loops. *Source*: Adapted by permission from Macmillan Publishers Ltd.: *Nature* (Chandrashekar et al. 2006), copyright 2006.

mutagenesis to explore how changes at critical positions in the coding sequence of the genes alter the sensitivities of sensory receptors. As a result of these types of studies, we know that there are often relationships between single nucleotide polymorphisms (SNPs) at critical positions in the coding regions and physiological differences in sensory receptor sensitivity, particularly for color vision and taste (Sharpe et al. 1999; Hayes, Feeney, and Allen 2013).

Although all five senses are frequently involved in some aspect of food detection and selection in nonhuman primates, in this chapter, I will focus on the two gene systems that are best understood and studied in relation to dietary adaptation in humans and nonhuman primates: opsin genes (color vision) and taste receptor genes (*TAS1R* and *TAS2R* genes, referred to collectively as *TAS* genes here). Substantial intraspecific variation exists in nucleotide diversity and gene functionality in these gene systems among humans (e.g. Verrelli and Tishkoff 2004; Kim et al. 2005), which suggests differences in sensory perception both between and within human populations. Moreover, humans also differ from chimpanzees and other primates in some aspects of the opsin and *TAS* gene repertoires (e.g. Go et al. 2005; Verrelli et al. 2008), suggesting that the selective pressures acting on these senses may have changed during hominin evolution. Some researchers have thus hypothesized that differences in diet and subsistence strategy may influence the evolution of sensory variation in and between human populations (Post 1962; Verrelli and Tishkoff 2004; Wang, Thomas, and Zhang 2004; Soranzo et al. 2005).

Several studies have identified adaptations in other genes between human populations in relation to diet. Perry et al. (2007) found that copy number variation in the salivary amylase gene (*AMY1*) differed between populations in relation to dietary starch consumption. In general, populations consuming low levels of starch (including rainforest hunter-gatherers, as well as pastoralists) exhibit a lower number of *AMY1* copies than high-starch-consuming populations. This difference in copy number variation corresponds to different levels of salivary amylase production (Perry et al. 2007). Humans also vary in the ability to digest lactose in milk upon reaching adulthood due to polymorphisms in the *LAC* gene (Tishkoff et al. 2007). Tishkoff et al. (2007) found evidence of convergent evolution of lactose tolerance in European and African populations that corresponds to population histories of pastoralism. The results of these studies emphasize the role of local adaptation, wherein positive selection favors traits within specific populations as adaptations to local ecological factors (Bamshad and Wooding 2003). Local adaptation may be particularly important in understanding the evolution of sensory genes in humans. In fact, a recent study observed signatures of local adaptation in the olfactory receptor genes of different African hunter-gatherer populations (Lachance et al. 2012).

To understand the ecological and evolutionary significance of human sensory polymorphisms relating to foraging and diet, it is necessary to view human variation within a broader framework of nonhuman primate sensory ecology. As such, each sensory system discussed in this chapter will begin with a brief overview of how the sensory system and sensory genes function, followed by a review of the variation and ecology of the sensory system in nonhuman primates. I will then summarize the polymorphisms observed in the sensory system in humans, how the sensory system in humans differs from that of closely related catarrhines, and hypotheses linking variation in human sensory genes to adaptive shifts in human diets.

Color Vision (Opsin Genes)

In vertebrates, color discrimination ability is determined by the types of cone photoreceptors with differing spectral sensitivities present in the retina. The spectral sensitivity (or color of light) to which each cone type is maximally responsive (λ_{max}) is determined by the visual pigment in the cone, a GPCR known as an *opsin protein* (Sharpe et al. 1999). These opsin proteins are encoded by different opsin genes (Nathans, Thomas, and Hogness 1986b). In mammals, interactions between certain critical amino acid residues in the protein (*spectral-tuning sites*) and an 11-*cis* retinal chromophore determine λ_{max} of the cone (Hunt et al. 2009). As a result, nonsynonymous SNPs in the opsin genes at these spectral-tuning sites can shift λ_{max} (Figure 16.1a), sometimes dramatically (Jacobs et al. 2002; Hunt et al. 2009). Most mammals possess two types of cone opsins (phenotypic dichromacy), one maximally responsive to shorter wavelengths (S opsin: blues and violets) encoded by the autosomal *SWS1* opsin gene (also known as *OPN1SW*), and one maximally responsive to medium and long wavelengths (M/L opsin: yellows, greens, and reds) encoded by the *LWS* opsin gene on the X chromosome.

Opsin gene variation and dietary adaptation in primates

As a clade, primates exhibit substantial variation in opsin genes and color vision compared to other mammals (Jacobs 2008; Hunt et al. 2009). Dichromacy is the ancestral condition for primates, and it is still found in many species (including tarsiers and several lemurs; Veilleux, Louis, and Bolnick 2013). However, several primate lineages have lost functional *SWS1* opsin genes due to the accumulation of deleterious mutations in coding regions (e.g. all lorises and galagos, all owl monkeys, and some cheirogaleid lemurs; Kawamura and Kubotera 2004; Jacobs 2013). These species exhibit monochromacy and, as such, lack functional color vision. Finally, three primate clades exhibit trichromacy, which is achieved through the evolution of a three-cone opsin protein system: the S opsin, and separate long- (L) and medium-wavelength sensitive (M) opsins. Consequently, while dichromat primates can only distinguish color on one dimension (blues/violets vs. yellows/greens/reds), trichromats can exploit two dimensions of color discrimination (blues/violets vs. yellows/greens/reds, and greens vs. reds).

The genetic mechanisms underlying trichromacy vary between the trichromat primate clades. Most species of New World monkeys (platyrrhines) and four lemur genera (*Eulemur*, *Varecia*, *Indri*, and *Propithecus*) achieve distinct L and M opsins through allelic variants at the *LWS* opsin gene locus, which is located on the X chromosome (polymorphic or allelic trichromacy; Jacobs 2008; Veilleux and Bolnick 2009; Jacobs et al. 2017). In these polymorphic trichromat species, SNPs at critical spectral-tuning sites result in alleles for opsins of differing λ_{max} (2–6 alleles in different species of platyrrhines, and 2–3 alleles in lemurs). Because the *LWS* gene is X-linked, only heterozygous females can exhibit trichromacy, while homozygous females and hemizygous males are dichromats (Surridge, Osorio, and Mundy 2003). By contrast, a duplication of the *LWS* gene in the ancestor of catarrhines (Old World monkeys, apes, and humans) has resulted in two separate *LWS*-type genes (L and M opsin genes, also known as *OPN1LW* and *OPN1MW*) adjacent to one another on the X chromosome. These genes differ at three spectral-tuning sites, resulting in one gene responsible for an L opsin and the other for an M opsin in both males and females (routine trichromacy; Jacobs 2008). Routine trichromacy due to an *LWS* opsin gene duplication also evolved independently in a single platyrrhine genus, *Alouatta* (Hunt et al. 2009).

Because of the importance of vision to hypotheses of primate evolution (Ankel-Simons and Rasmussen 2008), links between primate color vision and ecology have been the focus of considerable research effort in biological anthropology. Many hypotheses suggest that the wide variation in primate color vision reflects adaptations to different diets or foraging strategies. Color can be a particularly informative cue of nutritional quality and the presence of secondary compounds (e.g. toxins) for both fruit and young leaves (Dominy and Lucas 2004; Riba-Hernández, Stoner, and Lucas 2005). Consequently, diet is hypothesized to influence both the evolution of different color vision phenotypes and the λ_{max} of primate opsin proteins (i.e. spectral tuning).

The extra dimension of color vision provided by trichromacy is believed to offer a selective advantage to primates in foraging for yellow/orange/red food items (such as ripe fruit) in a green/brown foliage background. Experimental and modeling studies indicate that trichromat primates are better at detecting yellow/orange/red items, particularly at long distances, than dichromat conspecifics (Caine and Mundy 2000; Melin et al. 2014). The evolution of routine trichromacy and the spectral tuning of catarrhine opsins is suggested to reflect an adaptation for exploiting red protein-rich young leaves as fallback foods (Dominy and Lucas 2001; Dominy and Lucas 2004). By contrast, the ecological factors driving the maintenance of polymorphic trichromacy in platyrrhines and lemurs are subject to debate (Bunce 2015). Population analyses of the M/L opsin gene reveal signatures of balancing selection acting to maintain a diversity of opsin alleles at the locus in platyrrhines (Hiwatashi et al. 2010). However, it is unclear what form of balancing selection is acting in these populations. Field studies have not yet found strong evidence for a fitness advantage for trichromat females (e.g. heterozygous advantage), leading some researchers to hypothesize that balancing selection may be acting through frequency-dependent selection, such as niche divergence (Surridge, Osorio, and Mundy 2003; Bunce 2015). Because dichromacy is more effective in edge detection and breaking camouflage (Caine, Osorio, and Mundy 2010) and may offer an advantage in detecting cryptic insects (Melin et al. 2010), trichromats and dichromats may utilize different ecological niches to achieve similar fitness.

Diet and foraging strategy are also suggested to influence color vision phenotype and spectral tuning in nocturnal primates. A recent study by several colleagues and myself (Veilleux, Louis, and Bolnick 2013) found that, in general, the strength and direction of selection acting on the *SWS1* opsin gene to maintain dichromacy in nocturnal lemurs are related to habitat variation in nocturnal light environments. However, some genera deviated from these trends, implying an additional role of diet (Veilleux, Louis, and Bolnick 2013). For example, gummivorous lemurs from all habitats exhibit routine monochromacy, as indicated by a premature stop codon in the first *SWS1* exon in all fork-marked lemurs (*Phaner* spp.) and anatomical evidence of complete S cone loss in hairy-eared dwarf lemurs (*Allocebus trichotis*), which is in keeping with preliminary research that suggests color may not be a salient cue when foraging for gum (Moritz and Dominy 2010). In contrast, frugivorous-insectivorous mouse lemurs (*Microcebus* spp.) and folivorous woolly lemurs (*Avahi* spp.) from all habitats exhibit signatures of purifying selection to maintain dichromacy. Subsequent work has revealed that color cues are important for fruit foraging in *Microcebus* (Valenta et al. 2013), while the λ_{max} of *Avahi* S and M/L opsin proteins appears optimized for using color cues in detecting their preferred leaf foods in a foliage background at twilight and night (Veilleux et al. 2014).

Variation in opsin genes and color vision among humans

Humans, like other routine trichromats, typically express three cone opsin proteins: an S opsin (λ_{max} = 420 nm), an M opsin (λ_{max} = 530 nm), and an L opsin (λ_{max} = 558 nm). The human *SWS1* opsin gene is found on chromosome 7, while the L and M opsin genes are adjacent to one another in an array on the X chromosome, beginning with a single L opsin gene and followed by one or more copies of the M opsin gene (Nathans, Thomas, and Hogness 1986b). The L and M genes are very similar, sharing 98% sequence identity, and their ~30 nm difference in λ_{max} is primarily due to variation at three spectral-tuning sites (Figure 16.1a) – 180 (Ala/Ser), 277 (Tyr/Phe), and 285 (Ala/Thr) – although at least three other sites can influence small shifts (residues 116, 230, and 233; Neitz and Neitz 2011).

While the *SWS1* opsin gene is fairly invariant, humans exhibit substantial diversity in the structure, number, and λ_{max} of L and M opsin genes (for an excellent review, see Sharpe et al. 1999). For example, in addition to the normal-type L and M opsin genes, many humans exhibit hybrid L–M genes caused by gene conversion and unequal crossing over (Nathans et al. 1986a). These hybrid genes vary in the relative contributions from the M and L genes and can exhibit intermediate λ_{max}. Human populations also differ in the number of M and hybrid genes found in the L–M opsin array, with 78% of European men exhibiting more than one M or hybrid gene (average of two, range between one and five), compared to only 51.5% of Japanese men and 58% of African American men (Sharpe et al. 1999). Moreover, humans exhibit several nonsynonymous amino acid–changing SNPs in the L and M opsin genes, some of which are observed at spectral-tuning sites (Neitz and Neitz 2011). These SNPs can thus shift the cone opsin λ_{max}. One major SNP observed in both L and M opsin genes is found at the major spectral-tuning site 180 (Ala/Ser), and it can cause a shift in λ_{max} of 3–6 nm that is detectable in behavioral color-matching tests (Winderickx et al. 1992). The allele frequency of this SNP in the L opsin gene varies between geographic populations, with the Ser180 allele found in 62% of European men, 80% of African American men, and 84% of Japanese men (Deeb and Motulsky 1996). In general, the magnitude of the shift in spectral sensitivity caused by spectral-tuning site SNPs differs depending on whether it occurs in an L opsin or M opsin gene (Neitz and Neitz 2011).

This intraspecific variation in L and M opsin genes has important effects for the color vision phenotypes present in human populations. There are two general color vision phenotypes found in humans in addition to “normal” trichromacy:

anomalous trichromacy and dichromacy (monochromacy and loss of all cones can also occur, but these are extremely rare). Each phenotype can be achieved in a number of different ways genetically (Figure 16.1b). Dichromacy, for example, can result from the loss of a functional L or M opsin gene (or, more rarely, *SWS1* opsin gene) through deleterious mutations (Sharpe et al. 1999). However, because only the most proximal two genes on the L–M opsin gene array are expressed, it can also be achieved if the first and second genes on the array have the same λ_{max} due to the presence of a hybrid gene or SNPs at spectral-tuning sites (Figure 16.1; Nathans et al., 1986a; Hayashi, Motulsky, and Deeb 1999). The presence of hybrid genes can also result in anomalous trichromacy, wherein the two L–M opsins expressed have slightly different λ_{max}, permitting some degree of red-green discrimination but not as much as with normal trichromats. In men, the frequency of L and M opsin gene defects ranges between ~2 and 8% (but varies between populations, as discussed further in this chapter), but it is only between 0.03 and 0.63% for women (Sharpe et al. 1999). Defects leading to anomalous trichromacy are generally more common than dichromacy (5.7 vs. 2.3% of European men), and defects affecting the M opsin gene are more common than those affecting the L opsin gene (5.9 vs. 2.1% of European men; Sharpe et al. 1999).

Additionally, some evidence suggests that a fourth color vision phenotype may be present in human populations. Due to the occurrence of hybrid L–M genes and SNPs at spectral-tuning sites, some heterozygous women may express four cone opsins with different λ_{max} (tetrachromacy; Deeb and Motulsky 1996). These women are carriers of an anomalous trichromat L–M opsin array on one X chromosome. While most behavioral color-matching and discrimination tests have not found evidence of tetrachromacy, a few have offered limited support for increased color discrimination in a small subset of heterozygous women (Jordan and Mollon 1993; Jameson, Highnote, and Wasserman 2001; Neitz and Neitz 2011). For example, Jordan et al. (2010) used a combination of psychophysical color vision tests and molecular estimates of spectral sensitivity to evaluate color vision in 31 female carriers of anomalous alleles. They found that while most carriers did not exhibit results consistent with tetrachromacy, one exhibited tetrachromatic behavior on all tests and possessed widely separated L and M opsin pigments (λ_{max} at 530, 545, and 557 nm). Overall, however, if tetrachromacy is present in human populations, it is likely to be rare (Jordan et al. 2010).

Variation in opsin genes and color vision between humans and other catarrhines

In contrast to humans, L and M opsin genes appear to be strictly conserved in other catarrhine primates. Comparative studies of L–M opsin gene arrays in several cercopithecoid and hominoid species suggest that, while gene conversion between L and M opsin genes may also occur in nonhuman catarrhines, hybrid

L–M opsin genes with intermediate spectral sensitivity are very rare or not present at all (Ibbotson et al. 1992; Onishi et al. 2002; Terao et al. 2005; Verrelli et al. 2008; Hiwatashi et al. 2011). Moreover, nonhuman catarrhines exhibit much lower frequencies of abnormal color vision (dichromacy and anomalous trichromacy), with frequencies ranging from 0.4% in macaques (out of 3153 macaques; Onishi et al. 1999) to 1.9–2.2% in chimpanzees (out of 58 individuals, and possibly even lower if combined with previous studies: Terao et al. 2005). These studies suggest that purifying selection is acting on nonhuman catarrhines to remove nonsynonymous mutations that could shift L and M opsin λ_{max} (e.g. hybrid genes or SNPs at spectral-tuning sites).

The high levels of variation observed in human L and M opsin genes compared to those of nonhuman catarrhines suggest that the selective pressures acting on human opsin genes have changed at some point since our divergence with other hominoids (Jacobs and Williams 2001; Verrelli et al. 2008; Hiwatashi et al. 2011). Verrelli et al. (2008) explicitly tested this hypothesis by comparing signatures of selection acting at the L opsin gene in chimpanzees and humans. They found that patterns of variation differed substantially between chimpanzees and humans, supporting the hypothesis of divergent selective pressure (Verrelli et al. 2008). In particular, they identified signatures of purifying selection acting on the chimpanzee L opsin gene, including reduced levels of nonsynonymous nucleotide diversity relative to diversity at silent sites (synonymous coding sites and introns) and a very low number of nonsynonymous SNPs. These findings contrast with the results for humans, which exhibited high levels of nonsynonymous nucleotide diversity and a number of nonsynonymous SNPs occurring at higher frequencies in the population (Verrelli et al. 2008).

Dietary adaptation and evolution of human opsin genes

While evidence from L and M opsin genes suggests that there has been a shift in selective pressure for red-green color vision during the course of human evolution, there is debate regarding both the nature of this shift and the ecological factors driving it (Sharpe et al. 1999; Jacobs and Williams 2001; Jacobs 2008). Some researchers have interpreted the high frequency of color vision deficiency in humans as evidence of a relaxation of functional constraint for maintaining normal trichromacy (Post 1962; Salzano 1972; Barnicot and Woodburn 1975). In a large meta-analysis, Post (1962) observed that the lowest frequencies of color vision deficiencies were in hunter-gatherer men (0–2.5%) and the highest in European men (5–10.5%). He hypothesized that the adoption of agriculture led to relaxed selection for color vision. According to this hypothesis, hunter-gatherer populations experience purifying selection to maintain normal trichromatic color vision in order to best detect ripe fruits or other red food items. In contrast, the cultivation of crop species that can lack color cues (such as grains) may have reduced the importance of maintaining functional red-green discrimination.

While later psychophysical color discrimination tests offer some support for the relaxed selection hypothesis (Salzano 1972; Barnicot and Woodburn 1975), it has been subject to criticism. In particular, Adam (1969, 1986) found methodological issues with earlier work that tended to underestimate frequencies in non-Western populations. When data were collected using comparable techniques, populations did not vary in the frequency of the more severe color vision defects (i.e. those resulting in dichromacy), and frequencies of deficiencies varied dramatically within each subsistence strategy (Adam 1969, 1973, 1986).

Other researchers have instead interpreted the high level of diversity in human L and M opsin genes as signatures of balancing selection for variation in color vision (Verrelli and Tishkoff 2004; Bosten et al. 2005; Verrelli et al. 2008). In a large-scale study of human L opsin genes, Verrelli and Tishkoff (2004) sequenced the gene in 236 men from 19 populations around the world. For statistical analyses, they divided their sample into Sub-Saharan African and non-African subgroups. They found that nucleotide diversity at the L opsin locus was over three times greater than the average diversity seen in other human genes. They identified 9 nonsynonymous SNPs, 5 synonymous SNPs, and 69 intron SNPs across the global human sample. The nonsynonymous SNPs were polymorphic in both African and non-African subgroups, although the allele frequencies significantly differed between the subgroups for three SNPs (including residue 180, as mentioned in this chapter). Their results also identified significantly more nonsynonymous SNPs in both Africans and non-Africans, with nonsynonymous SNPs at higher frequencies in Africans than would be expected under neutral evolution. Verrelli and Tishkoff (2004) interpret their results as evidence of balancing selection to maintain a variety of high-frequency alleles. They hypothesize that selection may have favored increased L opsin gene variation in hunter-gatherer societies to increase spectral sensitivities for finding reddish food items during gathering, particularly for heterozygous women (Verrelli and Tishkoff 2004; Verrelli et al. 2008). At present, however, it is unclear whether fruit foraging among human women exerts different selective pressures than those seen in other catarrhines, which also forage for fruits and reddish plant parts but do not exhibit signatures of balancing selection. Bosten et al. (2005) have similarly hypothesized that there may be selection to maintain anomalous trichromacy in human men. They propose that anomalous trichromats may better detect camouflaged stimuli, such as predators or food items. This hypothesis is also intriguing and needs further attention, especially considering hominins might have faced increased predation pressure with obligate terrestriality and mosaic habitats (Treves and Palmqvist 2007).

Alternatively, the role of local adaptation in the evolution of diversity at human M and L opsin genes has not been explored. In their analysis, Verrelli and Tishkoff (2004) pooled all of their Sub-Saharan African populations into a single subgroup, which would thus contain populations who differed in subsistence strategy and habitat type. This type of large sampling regime may miss signatures of local adaptation (Tishkoff et al. 2007). In fact, some of the signatures of balancing selection found by Verrelli and Tishkoff (2004) could also result from a regime of weak purifying selection and adaptation to local ecology in some populations. For example, an excess of SNPs at intermediate frequencies, such as the nonsynonymous SNPs seen in the African subgroup, can also occur in the presence of population subdivision (Kreitman 2000). In Verrelli and Tishkoff's (2004) study, the African subgroup did not exhibit a similar excess of intermediate-frequency silent SNPs, which might be expected under a pattern of general population subdivision. However, if some of the populations included in the African subgroup exhibited adaptations to different local ecologies (such as differences in spectral tuning), population subdivision might be expected at nonsynonymous sites but not at silent sites. Future work should examine variation at the L opsin gene across a variety of populations from different subsistence strategies and habitat types to increase our understanding of the role of diet and other ecological variables in selection for color vision in humans.

Taste Discrimination (*TAS* Genes)

Taste is detected by specialized receptors in taste buds located in gustatory papillae on the tongue (as well as other parts of the body, as discussed further here; Behrens and Meyerhof 2011). Humans and many other mammals can perceive five major "qualities" of taste: sweet, umami (referring to a savory/meaty taste), bitter, salty, and sour. Each of these qualities plays an important role in food selection and rejection. For example, sweet and umami tastes are "calorie detectors" for carbohydrates and protein, respectively, while bitter discrimination detects toxins in foods (Meyerhof 2005). The molecular mechanisms underlying the first three taste qualities are fairly well understood. Sweet, umami, and bitter tastes are detected by GPCRs encoded by two major gene families (sweet and umami: *TAS1R*s, bitter: *TAS2R*s; Hoon et al. 1999; Nelson et al. 2001). By contrast, the genes involved in perception of salt, sour, and other proposed tastes or oral senses (e.g. fat) are less well understood. Here, I focus on the *TAS1R* and *TAS2R* genes. Those interested in the other taste senses are referred to recent excellent reviews (Hayes, Feeney, and Allen 2013; Bachmanov et al. 2014).

Three genes (*TAS1R1*, *TAS1R2*, and *TAS1R3*) are part of the *TAS1R* gene family and involved in the detection of sweet and umami. These genes are conserved across vertebrates (Shi and Zhang 2006). *TAS1R3* is involved in the detection of both tastes. When it is co-expressed as a heterodimer with *TAS1R1*, the resulting TAS1R1 + TAS1R3 cells detect umami, while when it is co-expressed as a heterodimer with *TAS1R2*, the resulting TAS1R2 + TAS1R3 cells detect sweetness (Nelson et al. 2001). All TAS1R proteins are characterized by long N-terminal extracellular domains (Figure 16.2) that are involved in recognizing and binding specific taste compounds (or ligands; Chandrashekar et al. 2006).

Taste receptor cells co-expressing the TAS1R2+TAS1R3 heterodimer bind to a large number of sweeteners (proteins that mimic sugars) and sugar ligands, while those co-expressing the TAS1R1+TAS1R3 heterodimer more narrowly bind to L-glutamate amino acids (such as monosodium glutamate; Chen et al. 2009).

In contrast to the *TAS1R* genes, the *TAS2R* gene family is highly diverse. The number of functional *TAS2R* genes differs between species (16–40 in mammals), with humans exhibiting ~26 functional *TAS2R* genes and 10 *TAS2R* pseudogenes. *TAS2R* genes vary in sequence similarity (17–90%), likely due to the large number of structurally diverse bitter compounds found in potential food items (Meyerhof 2005; Meyerhof et al. 2010). Mammalian lineages have experienced dramatic variation in *TAS2R* gene expansions (through gene duplication and diversification) and gene deaths (through pseudogenization or whole gene deletions) over the course of their evolutionary history in relation to dietary shifts (Hayakawa et al. 2014). Human *TAS2Rs* vary in the number of ligands to which they respond (i.e. tuning broadness). Some "specialist" *TAS2R* genes are narrowly tuned to only one or a few ligands, while others respond to many different ligands ("generalists"; Meyerhof et al. 2010). The same ligand can sometimes stimulate more than one *TAS2R* gene, particularly genes that share more recent ancestry. *TAS2R*s lack the extended N-terminal extracellular domain seen in *TAS1R*s; instead, the critical recognition and binding sites for bitter ligands are located in the extracellular loops and upper parts of the transmembrane regions (Figure 16.2).

TAS gene variation and dietary adaptation in primates and mammals

Because of the importance of sweet, umami, and bitter tastes in food selection and rejection, variation at *TAS1R* and *TAS2R* genes across primates and other mammals is hypothesized to reflect adaptive shifts in diet. For example, among nonprimate mammals, obligate meat eating is associated with pseudogenization of the *TAS1R2* gene and loss of sugar sensitivity across all felids (e.g. domestic cats, lions, etc.), as well as hyenas and seals (Jiang et al. 2012). Similarly, the timing of pseudogenization of the umami receptor gene *TAS1R1* in the panda is linked to its transition from meat eating to a diet of primarily bamboo (Zhao et al. 2010).

Traditionally, studies of taste discrimination in primates have employed behavioral assays to identify taste detection thresholds (i.e. the concentration at which a compound is discernable) and preferences, such as the *two-bottle test* where subjects are offered the opportunity to drink from two bottles with different concentrations of compounds. These studies highlight possible links between sensitivities to sugars and toxins and primate diets (Simmen and Hladik 1998). For example, the rejection threshold for a mixture of bitter tannins and fructose sugar was higher for the folivorous sifaka (*Propithecus verreauxi*), which normally consumes tannin-rich leaves, compared to the more frugivorous ring-tailed lemur

(*Lemur catta*) or mouse lemur (*Microcebus murinus*; Simmen and Hladik 1999), suggesting that sifaka are less sensitive to tannins. Other studies have suggested a relationship between species' sensitivity to sugars and the degree of frugivory in their diet (Simmen 1994; Laska et al. 1996), and a correlation between preference thresholds for monosodium glutamate and the proportion of animal matter in the diet (Laska and Hernandez Salazar 2004). The identification of *TAS* genes has permitted more detailed studies of dietary effects of variation in primate taste. Recent work using site-directed mutagenesis on the *TAS1R1* and *TAS1R3* genes, for example, suggests that humans, macaques (*Macaca mulatta*), and baboons (*Papio hamadryas*) share critical amino acid residues that result in umami receptors narrowly tuned to L-glutamates, while squirrel monkeys (*Saimiri boliviensis*) exhibit residues that encode receptors broadly tuned to different proteins (Toda et al. 2013).

The availability of genomic data has further allowed large-scale investigations into the evolution of *TAS* receptor genes across primate clades. Liu et al. (2014) evaluated selection acting on *TAS1R* genes for sweet and umami detection across 35 species of catarrhines and found substantial variation in the type and strength of selection acting on the genes in different clades. For example, colobine monkeys, which possess adaptations to folivory, exhibited signatures of a period of weakened purifying selection on *TAS1R2* in the ancestral colobine lineage for sugar sensitivity (but not a complete relaxation). Primates also exhibited signatures of positive selection at specific amino acid sites in *TAS1R1* and *TAS1R2*, most of which were found in the ligand-binding extracellular domains and are predicted to influence the sensitivity of the TAS1R proteins to different ligands (Liu et al. 2014). A phylogenetic analysis of the bitter *TAS2R*s found evidence of increases in the number of different *TAS2R* genes through gene duplication in the last common ancestors of anthropoids, hominoids, and cercopithecoids (Hayakawa et al. 2014). Because these lineages are hypothesized to have increased in body size and folivory, Hayakawa et al. (2014) hypothesize that *TAS2R* gene expansions in these clades reflect adaptations to detect novel leaf toxins in their expanded niches. Moreover, studies of *TAS2R* variation have detected evidence of local adaptation to species-specific or population-specific ecologies in chimpanzees, bonobos, and Japanese macaques (Parry, Erkner, and le Coutre 2004; Hayakawa et al. 2012; Suzuki-Hashido et al. 2015).

Variation in TAS genes and taste among humans

The genetic mechanisms underlying human variation in taste have been the focus of considerable research effort. A comparison of *TAS1R* genes in 88 individuals from eight populations found substantial genetic variation both globally and between populations (Kim et al. 2006). In particular, they identified 47 SNPs across all three genes, 64% of which were nonsynonymous and caused amino acid changes. Most of these nonsynonymous SNPs were observed in the first extracellular domain, where critical ligand-binding sites are found. Overall, global genetic diversity was highest for the sweet-specific *TAS1R2*, where nucleotide diversity was among the top 5–10% of human genes that Kim et al. surveyed. By contrast, *TAS1R3* (involved in both sweet and umami) was the most conserved of the *TAS1Rs*. Based on their results, Kim et al. (2006) predict that sweet perception may vary substantially between populations. Subsequent work has begun to link these and other nonsynonymous SNPs to behaviorally measured differences in sweet and umami sensitivity. Two recent studies (Eny et al. 2010; Dias et al. 2015) found relationships between *TAS1R2* SNPs, detection thresholds for sugar, and habitual sugar consumption, although these relationships are complicated by effects of body mass. Studies of *TAS1R1* and *TAS1R3* variation have also identified links between nonsynonymous SNPs and differences in sensitivity to glutamate amino acids (Chen et al. 2009; Shigemura et al. 2009). SNPs in *TAS* gene promoter regions and other genes in the taste transduction pathway, such as *GNAT3*, may also influence variation in taste perception (Hayes, Feeney, and Allen 2013).

Population genetic analyses have similarly identified substantial diversity within and between human populations in bitter *TAS* genes. In a study of global variation at 24 *TAS2R* genes, Kim et al. (2005) found high levels of allelic variation, with an average of 4.2 variant amino acid sites per gene and 2–11 haplotypes per gene. Furthermore, they found that mean population differentiation ($F_{ST} = 0.22$) was significantly higher in *TAS2Rs* than other genes, suggesting that human populations differ more in genes for bitter taste discrimination than they do other parts of the genome (Kim et al. 2005). Humans also exhibit intraspecific variation in the number of functional *TAS2R* genes. Of the 26 human *TAS2Rs*, at least 5 are segregating pseudogenes due to polymorphic frameshifts and premature stop codons, or whole gene deletions (Go et al. 2005; Kim et al. 2005; Hayakawa et al. 2014).

As with *TAS1Rs*, subsequent studies have sought to link SNPs at different *TAS2R* genes with perceptual differences of bitterness. The *TAS2R16* receptor, for example, responds to beta-glycopyranosides, which are a group of plant-derived compounds that includes salicin (an analgesic from willow bark) and cyanogenic compounds, which release cyanide when digested (Hayes, Feeney, and Allen 2013). Seven nonsynonymous SNPs have been identified in human populations at this locus, and one (amino acid residue 172) results in functional differences in the *in vitro* response to beta-glycopyranosides and in behaviorally measured salicin recognition thresholds (Soranzo et al. 2005; Campbell et al. 2014). Specifically, the derived allele (Lys172) is associated with a twofold increase in sensitivity compared to the ancestral allele (Asn172). The derived allele is almost fixed in non-African populations (99–100% frequency), but the frequency varies among African geographic regions, ranging between 75 and 94% in East Africa and 50–79% in Central Africa (Campbell et al. 2014). Other *TAS2R* loci that have received attention in genotype–phenotype studies are *TAS2R31* (related to bitter compounds in grapefruit) and *TAS2R43* and *TAS2R44* (related to aloins and bitterness in artificial sweeteners; Hayes, Feeney, and Allen 2013; Hayes et al. 2015).

One of the most famous *TAS2Rs* with known genotype–phenotype effects is *TAS2R38* (previously known as the *PTC* gene). *TAS2R38* is involved in discriminating ligands within the thiourea structural group, which includes phenylthiocarbamide (PTC) and goitrin, found in many cruciferous vegetables (e.g. Brussels sprouts, kale, and broccoli; Wooding et al. 2010; Feeney 2011). Early studies identified three nonsynonymous SNPs contributing to five haplotypes at the *TAS2R38* locus (Kim et al. 2003; Wooding et al. 2004; Bufe et al. 2005). The two most common haplotypes are PAV and AVI (named in the order of their three SNPs), which occur at intermediate to high frequencies across human populations. The frequencies of PAV are 61% in Africans, 55% in Asians, 46% in Europeans, and 95% in Native Americans, while AVI frequencies were 16, 44, 49, and 5%, respectively (Wooding et al. 2004). These *TAS2R38* haplotypes result in measurable differences in sensitivity to PTC and structurally related compounds (Kim et al. 2003; Bufe et al. 2005; Wooding et al. 2010; Behrens et al. 2013). The PAV haplotype is associated with high sensitivity to PTC (phenotypic "tasters"), while the AVI haplotype is associated with very low sensitivity (phenotypic "nontasters"). PAV–AVI heterozygotes (as well as the less common haplotypes, such as AAI) generally exhibit intermediate sensitivity to PTC. Differences in PTC sensitivity may influence food choices, as some studies have found that carriers of the PAV haplotype consume fewer cruciferous vegetables than AVI–AVI homozygotes (Duffy et al. 2010; Feeney 2011; Hayes, Feeney, and Allen 2013). Interestingly, more intensive sampling of African populations (57 populations) identified 19 additional nonsynonymous SNPs in *TAS2R38*, resulting in several novel subhaplotypes of the PAV, AVI, and AAI haplotypes that behavioral tests suggest have functional consequences (Campbell et al. 2012).

Dietary adaptation on the evolution of human TAS genes

Much of the work exploring the evolution of taste in humans has focused on *TAS2R* genes. Hominin evolution is characterized by major dietary shifts and new food-processing technologies that may have influenced the strength of selection acting on *TAS2R* diversity. Some researchers hypothesize that the use of fire as a detoxifying agent may have led to a relaxation of functional constraint on all bitter taste discrimination (Wang, Thomas, and Zhang 2004). Wang et al. (2004) analyzed *TAS2R* genes in 22 individuals from geographically diverse populations, and found that rates of nonsynonymous SNPs were similar to those of synonymous SNPs when averaged over all *TAS2R* genes, suggesting an equivalent selective constraint on neutral synonymous sites and nonsynonymous sites. They also found that the human lineage has experienced two fixed pseudogenes since the divergence with chimpanzees (Wang, Thomas, and Zhang 2004). Based on their analysis of pseudogenes, Wang et al. (2004) date the beginning of the complete relaxation of functional constraint on *TAS2Rs* in hominins to 750 kya (thousands of years ago), corresponding to when archeological evidence suggests hominins began to control fire.

The results of other studies, however, argue against a complete relaxation of functional constraint on all human *TAS2R* genes. Although the hominin lineage may have experienced a faster rate of pseudogenization than other primates (Go et al. 2005), humans do not significantly differ from other apes in the proportion of *TAS2R* pseudogenes (Fischer et al. 2005). Moreover, higher levels of nonsynonymous nucleotide variation in *TAS2R* extracellular domains compared to other domains may suggest positive selection in humans for increased *TAS2R* diversity in ligand recognition and binding (Fischer et al. 2005; Kim et al. 2005; Meyerhof 2005). It is likely that the adoption of new technology (e.g. control of fire) or subsistence strategies has relaxed selection on some human *TAS2R* genes. Two fixed human-specific pseudogenes (*TAS2R62* and *TAS2R64*), for example, are found in Neanderthals and Denisovans, suggesting loss of function at these loci by at least 550–590 kya (Perry et al. 2015). However, the history of selection acting on *TAS2Rs* in humans appears to be more complicated than simple relaxation. Signatures of selection vary between individual *TAS2R* genes, with different loci exhibiting evidence of positive selection, balancing selection, or relaxed selection in relation to dietary (and nondietary) factors. I will discuss two particular *TAS2R* genes that have been the focus of several recent evolutionary studies.

A popular focus of research has been *TAS2R38*, the gene involved in phenotypic differences in detecting PTC and related compounds. The maintenance of the PAV "taster" and AVI "nontaster" haplotypes at intermediate frequencies globally suggests that balancing selection is acting to maintain high diversity at *TAS2R38* (Wooding et al. 2004; Campbell et al. 2012). While early observations of PTC taster or nontaster phenotypes in chimpanzees initially suggested this balancing selection may predate the hominin–chimpanzee divergence (Fisher, Ford, and Huxley 1939), later work has revealed that the genetic mechanisms underlying these phenotypes differ between species (Wooding et al. 2006), suggesting that the derived nontaster phenotype independently evolved in humans and chimpanzees (and Japanese macaques: Suzuki-Hashido et al. 2015). Both timing of onset and the factors driving this balancing selection in humans are uncertain. Time estimates of the appearance of the three SNPs contributing to the derived haplotype (AVI) are 336 kya–1.3 mya (millions of year ago; Campbell et al. 2012), which corresponds to a study reporting the presence of one derived SNP in Neanderthals (Lalueza-Fox et al. 2009). Another ancient DNA study, however, found a novel SNP in a Neanderthal individual that would result in a truncated and nonfunctional *TAS2R38*, suggesting that different mechanisms may influence taster and nontaster phenotypes in humans and Neanderthals (Perry et al. 2015). Specific adaptive scenarios to explain the maintenance of the PAV and AVI haplotypes are currently lacking. Some researchers speculate that the AVI haplotype may respond to currently unknown ligands, and thus PAV–AVI heterozygotes would experience greater fitness by recognizing a greater number of bitter compounds (Kim and Drayna 2004; Wooding et al. 2004; Wooding 2006). A recent study testing the role of local adaptation found that the frequencies of major *TAS2R38* haplotypes do not differ with subsistence strategy or habitat in African populations (Campbell et al. 2012). Campbell et al. (2012) instead hypothesize that balancing selection at *TAS2R38* may be driven by extra-oral physiological functions of the gene related to human health (discussed further in this section).

Population genetic analyses indicate that positive selection may be acting at *TAS2R16* (involved in the detection of beta-glycopyranosides and cyanogenic compounds). In a large global study, Soranzo et al. (2005) found an excess of high-frequency derived SNPs (including the Lys172 high-sensitivity allele), high levels of population differentiation (F_{ST}) at the critical functional residue (172), and extended linkage disequilibrium around *TAS2R16.* These results suggest a selective sweep wherein the new mutation rises rapidly to high frequency due to strong positive selection (Wooding 2005). Soranzo et al. (2005) hypothesize that positive selection favoring the derived high-sensitivity allele may involve increased protection against cyanogenic foods during foraging in the Middle Pleistocene. They further speculate that higher frequencies of the less sensitive allele in African populations relative to the non-African populations may reflect an adaptation to malaria, as habitual low-level consumption of cyanogenic foods provides protection against malaria (Soranzo et al. 2005). A recent study with more intensive sampling across African populations similarly identified signatures of positive selection on *TAS2R16*, as well as novel haplotypes of the derived allele and evidence of geographic variation in the frequency of the high-sensitivity haplotypes between East and Central Africa (Campbell et al. 2014). The results of this new study suggest that the derived Lys172 SNP originated ~1 mya and arose to high frequencies in East Africa on multiple derived haplotype backgrounds, consistent with positive selection from standing variation (Campbell et al. 2014). However, the frequency of the less sensitive haplotypes in African populations did not significantly correlate with malaria endemicity. Moreover, while unusually high F_{ST} at the critical residue suggests local adaptation, Campbell et al. (2014) found no effect of diet or subsistence strategy on allele frequencies. As such, they hypothesize that positive selection and local adaptation acting on *TAS2R16* may be related to extraoral physiological functions rather than diet.

Extraoral expression of TAS genes: A caveat

For both *TAS2R38* and *TAS2R16*, Campbell et al. (2012, 2014) speculate that signatures of selection are not related to differences in diet, but instead extraoral physiological functions. In doing so, they highlight a recent and important concern regarding studies of selection, ligand tuning, and *TAS* genes: many of these genes are expressed in nongustatory tissues. *TAS1R* and *TAS2R* genes have been found in the gastrointestinal tract, thyroid, pancreas, brain, respiratory tissues, and testis (Behrens and Meyerhof 2011). These extraorally expressed *TAS* genes are hypothesized to mediate physiological responses to toxins (such as in the gut or airway), regulate circulating hormone levels, and aid in hormonal metabolic control (Jang et al. 2007; Clark et al. 2015). As such, functional polymorphisms in these genes can significantly impact human health. For example, *TAS2R38* is expressed in respiratory epithelial cells in the sinuses, and PAV–PAV homozygotes exhibit higher immune responses to bacteria and have significantly lower incidences of gram-negative respiratory infections compared to PAV–AVI heterozygotes and AVI–AVI homozygotes (Lee and Cohen 2013). Interspecific differences in the ligand specificity of individual *TAS* genes (such as the narrow specificity of human umami receptors vs. the broad specificity in rodents) may be important in extraoral function rather than specific taste adaptation (Toda et al. 2013). Thus, it is likely that both taste discrimination and extraoral functions contribute to selection pressure acting on many or all *TAS* genes.

Conclusions

Sensory ecology research on primates and other mammals has identified strong relationships between sensory systems (especially color vision and taste) and species feeding ecology. The field of human sensory ecology is relatively young, although it builds upon a substantial foundation of biological and psychological research of human sensory physiology and function. Some aspects of human feeding ecology are dramatically different compared to those of our closest hominoid relatives, such as the control of fire, use of hunting tools, or adoption of agriculture. It is reasonable to suggest that aspects of human sensory systems have been shaped by our unique hominin history. Additionally, however, humans are an ecologically flexible species, spreading into a variety of habitat types and utilizing a number of different subsistence strategies and diets. Thus, it is also reasonable to suggest that these different ecologies may exert different selective pressures on human senses, leading to local adaptation (e.g. Soranzo et al. 2005; Lachance et al. 2012; Campbell et al. 2014). The results of the recent studies of human sensory genetics reviewed here highlight the importance of sampling strategy when designing population studies of selection on sensory genes. Studies must include ecologically and geographically diverse populations and avoid treating ecologically heterogeneous populations as monolithic groups ("African"), or they risk missing possible signatures of local adaptation. The increased availability of genomic-sequencing techniques offers new opportunities for exploring variation in human sensory genes and promises to deepen our understanding of the ecology and evolution of human sensory systems.

CHAPTER 17

The Evolution of the Human Pigmentation Phenotype

Heather L. Norton

Department of Anthropology, University of Cincinnati, Cincinnati, OH, USA

Skin pigmentation is a complex phenotype that exhibits extensive variation across the human species. Perhaps because it is such an easily visible trait, skin color has often been misused as a tool to place humans into seemingly discrete "racial" categories. With the increasing realization that such categories cannot possibly reflect the continuous nature of human phenotypic and genetic variation in a biologically meaningful way, a re-examination of pigmentation diversity and its evolutionary and genetic causes has taken place. The result of these studies is a better understanding of the extent of pigmentation variation across the species, as well as an improved characterization of the global patterning of this variation and the extent to which it has been influenced by adaptation to a specific environmental pressure: ultraviolet radiation (UVR). Recent studies of genetic changes in pigmentation genes indicate that some mutations with large effects on pigmentation phenotypes may be relatively young, implying that rather than reflecting long-term and essentially stable population distinctions, skin color has changed several times over the course of human evolution in response to shifting environmental pressures. In this chapter, I describe basic elements of pigmentation biology, methods of characterizing pigmentation variation, and the geographic structure of human skin pigmentation variation, and review genetic evidence supporting the role of natural selection in shaping global patterns of pigmentation. Finally, I identify

A Companion to Anthropological Genetics, First Edition. Edited by Dennis H. O'Rourke.

areas where further study is required to gain a more complete picture of the evolution of the human pigmentation phenotype.

Pigmentation Biology

Pigmentation of the skin, hair, and eyes is governed by melanin, a complex biopolymer synthesized in specialized cells in the basal layer of the epidermis known as melanocytes. Melanin is produced in two basic forms: red/yellow pheomelanin and brown/black eumelanin (Meredith and Sarna 2006). After maturation, melanosomes are transferred to keratinocyte cells, where over time the melanosomal membrane degrades, scattering melanin particles throughout the cell (Szabo et al. 1969; Montagna and Carlisle 1991; Minwalla et al. 2001; Thong et al. 2003). Skin color is determined by a number of factors, including the ratio of eumelanin to pheomelanin, the size and shape of the melanosome, and how the melanosomes are packaged and distributed to the surrounding keratinocyte cells. In general, the skin of more darkly pigmented individuals can be characterized as having a higher eumelanin-to-pheomelanin ratio; larger, more ellipsoidal-shaped melanosomes; and melanosomes that are transferred to keratinocyte cells in single units (Szabo et al. 1969; Montagna and Carlisle 1991; Alaluf et al., 2002a, 2002b; Thong et al. 2003; Wakamatsu et al. 2006). In contrast, the skin of more lightly pigmented individuals tends to have greater amounts of pheomelanin, and more spherical melanosomes that are transferred to the keratinocyte cells in aggregations of 2–3 at a time.

The initial steps in the synthesis of both eumelanin and pheomelanin are mediated by the enzyme tyrosinase, which hydroxylizes tyrosine to dopa and then is involved in the oxidation of dopa to dopaquinone. At this point, the melanin synthesis pathway diverges – in the presence of the amino acid cysteine, dopaquinone is transformed into the lighter pheomelanin, while in its absence eumelanin is produced. Melanin acts as a barrier to UVR, limiting its penetration through the epidermis and its ability to induce DNA damage, which, if not repaired, can lead to skin cancer. Melanin's role as a photoprotective agent has other implications as well – by limiting the ability of UVR to penetrate the epidermis, it mitigates potential photodestruction of folic acid and limits the capacity for endogenous vitamin D synthesis (Loomis 1967; Branda and Eaton 1978; Jablonski and Chaplin 2000).

Assessing Phenotype

Quantitative assessments of human skin pigmentation phenotypes typically rely on skin reflectance spectrophotometry. While available instruments differ somewhat in their design, they all assess pigmentation levels by shining a controlled amount of light (sometimes at specific wavelengths) on the surface of the skin and measuring the percentage of light that is reflected back. Lighter skin pigmentation yields higher reflectance values, while darker skin color corresponds with lower levels of reflectance. Pigmentation can be expressed in terms of various standardized color systems, such as the CIELab color system (developed by the Commission International d'Eclairage), which measures a color in terms of its position along three axes: light–dark (the L* parameter), red-green (the a* parameter), and blue-yellow (the b* parameter) (Weatherall and Coombs 1992). Other reflectometers, such as narrow-band reflectometers, are designed to specifically assess differences in the absorbance spectra between the two primary chromophores of the skin, hemoglobin and melanin (Shriver and Parra 2000). However, regardless of the measurement method used, studies consistently demonstrate that (i) variation in pigmentation exists within and between populations, (ii) different populations may overlap considerably in their pigmentation levels, and (iii) overarching patterns of human pigmentation diversity reflect broad geographic substructure, but variation may still exist at local levels (Akey et al. 2001; Shriver et al. 2003; Norton et al. 2006; Edwards et al. 2010; Valenzuela et al. 2010; Leite et al. 2011; Ang et al. 2012; Candille et al. 2012; Beleza et al. 2013a).

Geographic Structure of Human Skin Pigmentation Variation

Human skin pigmentation varies in predictable ways across broad geographic regions. Specifically, darker pigmentation is more commonly observed in populations indigenous to tropical regions, while lighter skin color is more common as one moves away from the equator and into the upper latitudes (Jablonski and Chaplin 2000; Relethford 2000). This strong geographic structuring suggests that pigmentation phenotype reflects adaptation to local environmental conditions (Jablonski and Chaplin 2000). The majority of evolutionary explanations for variation in phenotype (as discussed in this chapter) argue for UVR as the driving force behind this localized adaption, although other factors (e.g. sexual selection and pathogen resistance) have also been proposed.

In 1997, Relethford examined a large number of skin reflectance datasets from Old World populations in order to formally test the hypothesis that skin color was darkest in equatorial regions (where UVR is presumed to be the strongest) and to test for differences in pigmentation levels between the Northern and Southern Hemispheres (Relethford 1997). This work confirmed the association between pigmentation and latitude, supporting UVR-based explanations of pigmentation variation. Rapidly following this came a seminal paper by Jablonski and Chaplin (2000), in which the authors more formally characterized the relationship between skin color and UVR by comparing published pigmentation measurements to direct estimates of UVR levels from satellite data. Using these data, they reported a strong correlation between UVR and skin reflectance.

While UVR has likely shaped broad patterns of pigmentation diversity, pigmentation also varies over smaller geographic scales where variation in UVR intensity seems unlikely to be the primary cause. For example, Norton et al. (2006) observed extensive variation in skin pigmentation in populations sampled throughout Island Melanesia, where UVR is high and pigmentation is predicted to be relatively dark. Similarly, high levels of variation are observed among the Orang Asli tribes of Malaysia (Ang et al. 2012), and among populations of South Asia (Basu Mallick et al. 2013). It may be that the selective constraints imposed by high UVR (discussed further in this chapter) generally limit pigmentation variation in these regions, but that, as long as pigmentation levels remain above a protective melanin threshold, the phenotype is free to vary, perhaps driven by factors such as genetic drift and sexual selection (Chaplin 2004; Norton et al. 2006).

Variation in Human Hair and Eye Pigmentation

In addition to skin color, pigmentation of the hair and eyes also varies within and between human populations. Quantitative methods to assess hair and iris pigmentation lag behind those used in skin pigmentation studies. Recently, more refined methods of quantitative measurement have demonstrated that both traits exhibit greater variation outside of Europe than had been previously appreciated (Norton et al. 2006; Edwards et al. 2012; Kenny et al. 2012; Ulivi et al. 2013), although variation of these traits observed in non-European populations is not as extreme as that observed within Europe. A well-known example of variation in hair color outside of Europe is the blond hair common to some populations from the South Pacific. These individuals exhibit a decoupling of hair and skin phenotype: as expected, given the high-UVR regions they inhabit, their skin color is dark, but they also exhibit strikingly blond hair (Norton et al. 2006; Kenny et al. 2012).

Recent refinements in iris-phenotyping methods have been developed that rely on the digital scoring of color values (using CIELab color space) from close-up, high-resolution photographs of the iris (Liu et al. 2010; Edwards et al. 2012). These methods have demonstrated that even in populations where eyes have traditionally been described using the categorical term *brown*, variation in eye color exists (Edwards et al. 2012), and different categories of brown eyes can be distinguished. These increasingly refined methods for characterizing hair and iris pigmentation are expected to increase the power to identify genetic loci associated with more subtle gradations in phenotype (Liu et al. 2010).

Evolutionary Hypotheses for Global Pigmentation Variation

Several hypotheses have been put forth to explain the evolution of human skin pigmentation variation, the majority of which focus on the relationship between UVR and melanin. UVR comes in three forms – ultraviolet-A (UVA), -B (UVB), and -C (UVC) – but only the long (UVA: 320–400 nm) and middle (UVB: 280–320 nm) wavelengths are thought to have a significant biological impact on humans (Jablonski and Chaplin 2000). UVA and UVB are biologically relevant to different physiological systems, and the global structure of human pigmentation variation may reflect adaptation to UVA and UVB intensity at different times of the year (Jablonski and Chaplin 2010).

Photoprotection hypothesis. As discussed here, the photoprotective properties of melanin enable it to function much like a natural sunscreen, blocking and scattering UVR. UVR has the potential to cause significant damage to DNA, as it may produce cyclobutane pyrimidine dimers (CPDs) and pyrimidine (6–4) pyrimidone photoproducts, both of which are implicated in photocarcinogenesis (Sage 1993; Kvam and Tyrrell 1997; Sander et al. 2004), and single-strand DNA breaks caused by reactive oxygen species (ROS) (Kvam and Tyrrell 1997; Sander et al. 2004). Over time, such damage can lead to the development of various skin cancers, including cutaneous malignant melanoma. Compared to the more soluble and reactive pheomelanin, eumelanin may provide increased protection from UVR as it is more resistant to photodegradation and can reduce ROS produced by UV damage (Abdel-Maleket al. 2010; Kadekaro et al. 2010). This is consistent with numerous studies demonstrating that rates of skin cancers are significantly higher in lightly pigmented individuals (with low levels of eumelanin) compared to darkly pigmented individuals (with higher levels of eumelanin) (e.g. Kollias et al. 1991).

According to the photoprotection hypothesis, darker skin pigmentation likely evolved soon after the ancestors of modern humans lost their fur, as lighter skin color would have been disadvantageous in the presumably hot, high-UVR environment (Robbins 1991). Purifying selection then maintained this phenotype in populations who inhabited high-UVR regions. A key criticism of this hypothesis is the late age of onset and comparatively low fatality rates of most skin cancers, which suggest that skin cancer may only be a weak selective force for maintaining darker skin color in high-UVR regions (Blum 1961; Jablonski and Chaplin 2000, 2010, 2014).

The folic acid hypothesis. While sunburn and skin cancer are the deleterious effects most commonly associated with sun exposure, UVR also influences health in other, potentially more evolutionary significant ways. Of particular concern from an evolutionary standpoint is the ability of UVA radiation to degrade the nutrient folate, a B vitamin that plays important roles in DNA synthesis and repair, as well as spermatogenesis (Mathur et al. 1977; Fenech 2012). Pregnant women deficient in folate have a higher risk of the fetus developing neural tube birth defects, including spina bifida and anencephalus (Lucock et al. 2000; Off et al. 2005). Branda and Eaton first reported on the link between UVR exposure and decreased folate levels (Branda and Eaton 1978), and argued that heavily melanized skin could provide protection against UV-induced breakdown of key nutrients. This argument was supported by the UVR and skin reflectance dataset analyzed by Jablonski and Chaplin (2000), in which the highest correlations between UVR intensity and skin reflectance were observed at 545 nm – the peak absorbance of hemoglobin. Jablonski and Chaplin argued that this correlation supported a model in which the primary function of melanin was to protect the contents of cutaneous blood vessels in the dermis from photodegradation. The folic acid hypothesis predicts that darker skin should be conserved in regions of high UVR, specifically high UVA (Jablonski and Chaplin 2010), and natural selection should eliminate variants associated with lighter skin color in these regions, given the specific fitness implications of folic acid deficiencies for women of reproductive age (and males, given the role of folate in spermatogenesis). An attractive feature of this hypothesis is the clear effect that folic acid degradation may have on reproductive fitness.

The vitamin D hypothesis. The photoprotection and folic acid hypotheses can both be invoked to explain the prevalence of darker skin color in high-UVR regions. However, neither can speak directly to the evolution of lighter skin color and its prevalence in lower UVR environments. One hypothesis is that lighter skin color in these regions simply reflects a relaxation of functional constraint: because the lower UVR levels decrease the risk of photocarcinogenesis or folic acid photolysis, mutations that lead to a decrease in pigmentation may have increased in frequency due to a combination of genetic drift, founder effects, and/or sexual selection. However, an alternative hypothesis is that lighter skin color evolved specifically as an adaptation to low-UVR conditions and that natural selection has favored genetic variants that lead to a decrease in eumelanin synthesis (Loomis 1967). In 1967, Loomis argued that light skin evolved as a way to increase the potential for vitamin D synthesis in regions of low UVR. Vitamin D is a prosteroid hormone that is involved in a number of physiological systems, including immune response, cell growth and differentiation, and calcium metabolism (Holick 2004b). Vitamin D deficiencies have been linked to increased risk of infection from certain pathogens, and the skeletal disorders of osteomalacia and rickets (Holick 2004a, 2004b; Lang et al. 2013).

Although vitamin D is naturally occurring in some foods (most notably, fatty fish and organ meats), humans are also capable of endogenous vitamin D synthesis (Holick 2003). When UVR (specifically, the midrange UVB wavelengths) penetrates the skin, it stimulates a series of chemical pathways that ultimately culminate in the liver and kidneys, resulting in the production of vitamin D (for a review, see Jablonski 2004). Loomis argued that in regions where UVR intensity is weak, selection should favor lower levels of melanin in order to allow more UVR to penetrate the skin, thus increasing the potential for vitamin D synthesis. This has been supported by estimates showing that large portions of the Northern Hemisphere do not receive enough annual UVR for even lightly pigmented individuals to synthesize sufficient quantities of vitamin D (Jablonski and Chaplin 2000).

It is worth noting that in some modern human populations, constitutive skin color appears to be mismatched to UVR levels given the predictions made by the vitamin D hypothesis. Oft-cited examples of this phenomenon are the

Eskimo–Aleut populations of the North American Arctic (Murray 1934; Jablonski and Chaplin 2000). However, the potential vitamin D deficiency caused by highly melanized skin in these regions may be countered by the vitamin D–rich diets consumed by individuals resident to these low-UVR environments (Jablonski and Chaplin 2000; Jablonski 2004).

Sexual selection. Finally, while not a specific explanation for global patterns of pigmentation diversity, it is necessary to briefly discuss the potential role of sexual selection in shaping pigmentation variation. A role for sexual selection in shaping human pigmentation was proposed by Darwin himself, who argued that mate preference for different skin tones could have led to divergence among populations (Darwin 1871). Many (but not all) quantitative studies of skin pigmentation have demonstrated that females are slightly, but significantly, lighter than males (see references in Madrigal and Kelly 2007), possibly indicating a universal male preference for slightly lighter females (van den Berghe and Frost 1986; Aoki 2002). Such a preference has been reported in 47 out of 51 societies surveyed, and it is explained in terms of fertility cues: human females experience a slight decrease in pigmentation around the age of menarche, suggesting that lighter skin color may be used by males as a proxy for female fecundity (van den Berghe and Frost 1986). Combining natural selection and sexual selection models, Ihara and Aoki (1999) proposed that sexual selection for lighter females will be overridden in high-UVR environments (where the evolutionary imperative for protection from UVR trumps mate preference), but that in low-UVR regions (where photoprotection has less of an impact on fitness), the sexes may diverge in pigmentation phenotype. However, an analysis of published reflectance data failed to support a relationship between increasing latitude and increasing levels of sexual dimorphism in skin pigmentation (Madrigal and Kelly 2007), suggesting that natural selection may be sufficient to explain global patterns of skin pigmentation variation. While natural selection may be a more parsimonious explanation for global patterns of pigmentation variation (Parra 2007), sexual selection may still play a role in local variation in skin, hair, and eye color (Frost 2014).

As the reader has no doubt surmised, these hypotheses are not mutually exclusive. In fact, it is likely that global patterns of human skin pigmentation variation reflect a balance between the need for protection from UVA-induced DNA damage and folic acid degradation in high-UVR regions and the need for adequate UVB-mediated vitamin D synthesis in low-UVR regions (Jablonski and Chaplin 2000). Genetic drift and sexual selection may play important roles in determining more localized patterns of variation in pigmentation phenotype, although such variation should still be broadly constrained by natural selection. Identifying genes targeted by selection, and in particular dating the origin of mutations associated with phenotypic change, may help us to assess the contribution of these forces to the global evolution of human pigmentation variation.

Role of Natural Selection in Shaping Human Pigmentation Variation: Genetic Evidence

The pigmentation phenotype of our ancestors (and the extent of variation in that phenotype) is unknown. The skin of our closest living relatives, chimpanzees and bonobos, suggests that our ancestors' skin was relatively lightly pigmented beneath a covering of fur (Jablonski and Chaplin 2000). Sun-exposed regions of the skin may have had the potential to darken (tan) in response to UVR, similar to the facial skin of modern chimpanzees (Jablonski and Chaplin 2000). However, the evolution of hairlessness, possibly as a response to new thermoregulatory demands brought about by the transition to bipedalism, suggests that darker skin color and its photoprotective benefits likely characterized hominins living in tropical regions early in the evolution of the genus *Homo* (Jablonski 2004). An analysis of sequence diversity at the pigmentation gene *MC1R* supports a model in which alleles associated with eumelanogenesis at this gene likely swept to fixation in populations of *Homo* around 1.2 million years ago (mya) (Rogers, Iltis, and Wooding 2004), indicating that the common ancestor of modern humans likely exhibited darkly pigmented skin.

The evolutionary hypotheses discussed here make specific predictions about expected patterns of DNA sequence variation in and around pigmentation genes. Both the photoprotection and folic acid hypotheses predict that variation in these genes should be highly conserved in populations from high-UVR regions. This conservation would manifest itself in low levels of diversity, and particularly in a low ratio of nonsynonymous to synonymous polymorphisms. Conversely, the vitamin D hypothesis predicts that natural selection will have favored derived alleles that decrease skin pigmentation, specifically in populations resident to low-UVR regions. A great deal of recent attention has focused on identifying genes targeted by natural selection, leading to the development of a variety of statistical tests that may be used to assess this. These include tests based on extended blocks of linkage disequilibrium (LD) around the selected site, population differentiation (F_{ST}), and distortions in the site frequency spectrum. *For a more in-depth review of tests of selection, please see Chapter 10.* In the remainder of this section, I outline some of the major pigmentation loci that have been targeted by natural selection. These examples are presented here with a few notable caveats: first, this is not an exhaustive list of loci affecting human pigmentation phenotype; second, due to space constraints, only select examples of pigmentation loci affected by selection are discussed (for a more complete listing, see Table 17.1); and, third, because of biases in populations studied (a heavy emphasis on populations of European descent), it is likely that in the near future, additional genes (or novel mutations within these known candidate genes) will be identified that have played an important role in shaping pigmentation in populations outside of Europe.

Table 17.1 A (non-exhaustive) list of pigmentation candidate genes targeted by selection in different human populations.

Gene	*Function*	*Phenotypic association*	*Evidence for Selection?*	*Population(s) in which selection is acting*	*Selection-based references*
ASIP	Eumelanin/ pheomelanin synthesis	Lighter skin color	F_{ST}	EAS, EUR	Norton et al. (2007)
ATRN	Eumelanin/ pheomelanin synthesis		SFS	EAS	Hider et al. (2013)
DCT	Melanogenesis stabilizing enzyme		F_{ST}, I_n, LRH, LSBL, lnRH, SFS	EAS, AFR	Myles et al. (2006), Edwards et al. (2010), Lao et al. (2007), Alonso et al. (2008), Hider et al. (2013)
DRD2	Dopamine receptor 2		LRH	EAS, AFR	Lao et al. (2005)
EDAR	Ectodysplasin-A receptor		LRH, SFS	EAS	Sabeti et al. (2007), Hider et al. (2013)
EGFR	Epidermal growth factor receptor	Lighter skin color	F_{ST}, I_n	EAS	Lao et al. (2005), Hider et al. (2013)
KITLG	Melanoblast growth and differentiation	Lighter skin color	F_{ST}, I_n, SFS	EAS, EUR	Miller et al. (2007), Lao et al. (2005)
LYST	Organelle biogenesis and size	Chediak Higashi Syndrome	SFS	EAS	Hider et al. (2013)
MC1R	Eumelanin/ pheomelanin synthesis	Red hair, fair skin, freckles, increased melanoma risk	SFS, LRH	AFR (purifying); EAS, EUR (positive)	Rana et al. (1999), Harding et al. (2000), Rogers et al. (2004), Hider et al. (2013)

(Continued)

Table 17.1 (Continued)

Gene	*Function*	*Phenotypic association*	*Evidence for Selection?*	*Population(s) in which selection is acting*	*Selection-based references*
OCA2	Melanosomal membrane protein	OCA type 2; lighter skin color	F_{ST}, LSBL lnRH Tajima's D, EHH, iHS	EAS, EUR	Norton et al. (2007), Hider et al. (2013), Edwards et al. (2010), Izagirre et al. (2006), Lao et al. (2005), Myles et al. (2006), Voight et al. (2006b)
SLC24A5	Calcium ion transporter	Lighter skin color	F_{ST}, LRH	EUR	Lamason et al. (2005), Izagirre et al. (2006), Myles et al. (2007), Norton et al. (2007), Sabeti et al. (2007)
SLC45A2	Membrane associated transporter protein	Lighter skin color; OCA type 4	F_{ST}, I_n, LRH, SFS	EUR	Soejima et al. (2007), Lao et al. (2007), Norton et al. (2007)
TYR	Melanogenesis enzyme	Lighter skin color; OCA type I	F_{ST}, LRH	EUR, EAS	Norton et al. (2007), Myles et al. (2006)
TYRP1	Melanogenesis stabilizing enzyme	OCA type 3	F_{ST}, LRH	EUR, EAS, AFR	Myles et al. (2007), Lao et al. (2005), Izagirre et al. (2006), Hider et al. (2013), Alonso et al. (2008)

MC1R. One of the first loci to be extensively studied because of its association with normal pigmentation variation, the *MC1R* gene has a large impact on pigmentation phenotype despite its small size (~1 kb) (Valverde et al. 1995; Cone et al. 1996; Box et al. 1997). The MC1R gene product is a 7-pass transmembrane G-protein coupled receptor that is expressed on the surface of melanocyte cells. When bound by its agonists, α-MSH or adrenocorticotropic hormone (ACTH), the synthesis of cyclic adenosine monophosphate (cAMP) increases, leading to the increased expression of several enzymes involved in eumelanin production, most notably tyrosinase and tyrosinase-related protein 1. The result is an increase in eumelanogenesis. However, several mutations in the *MC1R* gene that limit its binding efficiency may result in lower levels of eumelanin production or, in some cases, increased pheomelanin production. Early sequencing studies of *MC1R* variation demonstrated that the gene was highly conserved in high-UVR populations from Africa (Rana et al. 1999; Harding et al. 2000), consistent with predictions made by the photoprotection and folic acid hypotheses. However, lower UVR populations in both Europe and East Asia exhibited a high rate of nonsynonymous polymorphisms, which was alternately interpreted as a signal of either relaxed selection (Harding et al. 2000) or directional selection favoring lighter skin color in these regions (Rana et al. 1999). The pattern of sequence variation in *MC1R* fits well with expectations under a model in which variation is constrained by high UVR.

From a phenotypic perspective, *MC1R* is best known for its association with a red-haired, fair-skinned, and freckled phenotype in Europeans (Valverde et al. 1995; Box et al. 1997; Flanagan et al. 2000; Rees 2004). Mutations in *MC1R* are also associated with increased melanoma risk (Fernandez et al. 2007; Martinez-Cadenas et al. 2013), which is consistent with the lower rates of DNA repair and higher rates of oxidative damage observed in *MC1R* mutants compared to individuals carrying the ancestral *MC1R* sequence (Kadekaro et al. 2005; Song et al. 2009). Other *MC1R* variants are common in East Asian populations and are significantly associated with normal pigmentation variation, including freckling status (Akey et al. 2001; Yamaguchi et al. 2012). However, there is no evidence that these specific variants have been targeted by selection in East Asian populations.

SLC24A5. *SLC24A5* is a cation exchanger that plays a role in melanosome morphogenesis and melanogenesis. This gene was first investigated for a role in human pigmentation variation when it was discovered that a premature stop codon in the gene was responsible for the *golden* phenotype in zebrafish (*Danio rerio*) (Lamason et al. 2005). While *golden* mutants exhibit smaller, more transparent, and paler melanophores than their wild-type counterparts, the wild-type pigmentation could be rescued using human messenger RNA (mRNA). When human populations were examined for variation in *SLC24A5*, Lamason and colleagues observed a striking pattern of population differentiation: nearly all European samples carried a nonsynonymous mutation (rs1426654) in exon 3 of the gene, resulting in the production of threonine instead of alanine; this allele was nearly absent or at very low frequencies in the East Asian and African populations. Further study demonstrated that the derived allele in Europeans was surrounded by an extensive block (~150kb) of extremely low heterozygosity, interpreted as evidence of a selective sweep. Using an admixture mapping-based method (Shriver et al. 2003), it was shown that the derived allele at rs1426654 could explain ~30% of variation in skin pigmentation among admixed African American and African Caribbean individuals (Lamason et al. 2005).

Several lines of evidence, including measures of extended haplotype homozygosity (Sabeti et al. 2007) and population divergence estimates using a larger sample of populations from the Human Genome Diversity Panel (HGDP) (Norton et al. 2007), indicated that *SLC24A5* had been affected by a recent and extremely strong selective sweep. Using methods that examine the decay of haplotypes around the selected allele to jointly estimate time to the most recent common ancestor (TMRCA) and selection coefficient, Beleza et al. estimated that modern haplotypes carrying the derived allele arose from a common ancestral haplotype ~11 thousand years ago (kya) and that this new, derived haplotype was favored by very strong selection in Europeans (selection coefficient [s], ~0.08) (Beleza et al. 2013b).

SLC24A5 was an exciting addition to the story of pigmentation change in humans for several reasons. First, the high F_{ST} between Europeans and East Asians strongly suggested that adaptation to low UVR was due to independent genetic mechanisms in each population – a case of evolutionary convergence (Lamason et al. 2005; Norton et al. 2007). Second, the relatively recent TMRCA and strong selection coefficient suggest that the transition to lighter skin color in European populations occurred far later than was originally thought, significantly postdating the arrival of modern humans to Europe 40 000–45 000 years ago (Beleza et al. 2013b). This model of the convergent evolution of light skin pigmentation would soon be supported by data from several other pigmentation candidate genes (Izagirre et al. 2006; Soejima et al. 2006; Lao et al. 2007; Myles et al. 2007; Norton et al. 2007).

SLC45A2. SLC45A2 was first identified for its association with oculocutaneous albinism type IV (Newton et al. 2001). However, two nonsynonymous mutations in this gene are significantly associated with normal variation in skin, hair, and eye pigmentation in a diverse sample of populations (Graf, Hodgson, and van Daal 2005). The derived allele at one of these, rs16891982, occurs at intermediate frequencies in European populations, but it is rare or absent elsewhere, including in many East Asian populations (Graf, Hodgson, and van Daal 2005). Like *SLC24A5*, F_{ST} values at this locus also support a model of convergence in the evolution of lighter skin color, and admixture-mapping methods confirm its association with skin pigmentation variation in African Americans and African Caribbeans (Norton et al. 2007). Using haplotype and site frequency spectrum-based statistics, Soejima and colleagues demonstrated that the derived rs16891982 allele had been affected by selection in European (but not East Asian) populations (Soejima et al. 2006), with an estimated TMRCA of ~40 kya. Using methods similar to those used to date the TMRCA of rs1426654 in *SLC24A5*, Beleza et al. obtained slightly more recent TMRCAs (6–26 kya) (Beleza et al. 2013b).

OCA2. *OCA2*, sometimes known as the P-protein, was originally studied because of its association with oculocutaneous albinism type II (see references in Oetting and King 1999). However, mutations in this gene have also been associated with normal phenotypic variation (Akey et al. 2001; Shriver et al. 2003; Edwards et al. 2010), and mutations in *HERC2*, a gene upstream of *OCA2*, are associated with blue eye color (Eiberg et al. 2008; Kayser et al. 2008; Sturm et al. 2008). Multiple studies have identified different regions of *OCA2* as being targeted by selection in Europeans (Voight et al. 2006b; Lao et al. 2007; Myles et al. 2007) and East Asians (Lao et al. 2007; Edwards et al. 2010; Hider et al. 2013). Interestingly, distinct regions of *OCA2* seem to have been favored in each population, implying that selection may have caused unique haplotypes to increase in frequency in each geographic region.

KITLG. *KITLG* encodes the ligand of the KIT-locus coded tyrosine-kinase receptor, which plays a role in melanocyte differentiation and melanosome distribution. A large European-based genome-wide association study (GWAS) reported associations between *KITLG* and red hair color (Sulem et al. 2007), and a mutation upstream of this gene (rs1281256) alters a regulatory enhancer that is expressed in developing hair follicles and is associated with blond hair color in European populations (Guenther et al. 2014). Miller et al. (2007) identified a derived regulatory mutation (rs642742) of *KITLG* located 326 kb upstream of the transcription start site that is significantly associated with lighter skin pigmentation in an admixed African American population. Values of Fay and Wu's H (a site frequency spectrum summary statistic associated with positive directional selection) calculated from HapMap data indicate that *KITLG* exhibits significantly high frequencies of derived alleles in European and East Asian populations compared to the rest of the genome (Miller et al. 2007). HapMap data also indicate that rs642742 exhibits significantly high allele frequency differences between African and non-African populations (Miller et al. 2007), further evidence that the allele has been impacted by positive selection. *KITLG* differs from the loci discussed here in that the derived allele at rs642742 is present at similar frequencies in both Europeans and East Asians. Dating estimates indicate that the selective sweep that led to this increase in the derived allele frequency began before the divergence of modern Eurasian populations (Beleza et al. 2013b). This relatively early date is notable, since it indicates that the derived allele arose and began to increase in frequency before the European-specific alleles at rs1426654 and rs16891982 – this may make rs642742 one of the earliest examples of evolutionary change in response to lower UVR conditions observed in modern humans.

TYR, *TYRP1*, and *DCT*. *TYR*, *TYRP1*, and *DCT* produce enzymes involved in melanin synthesis. Tyrosinase, the product of *TYR*, is often described as the rate-limiting enzyme of melanogenesis, due to its role in the very early stages of melanin synthesis. Mutations in *TYR* can lead to oculocutaneous albinism type I, also known as tyrosinase-negative albinism, the most severe form of albinism. Individuals with these mutations produce no melanin (as reviewed in Oetting and King 1999). The enzymes produced by *TYRP1* and *DCT* are both stabilizing enzymes in the melanogenesis process, and are less critical to melanin production than *TYR*. While *TYRP1* mutations have been linked with a third type of albinism, OCA type III (Boissy et al. 1996), *DCT* is not known to be associated with pigmentary pathologies in humans. Given the role that all three loci play in melanogenesis, it is reasonable to assume that one or more may play roles in regulating normal pigmentation variation and may have been targeted by natural selection.

A nonsynonymous mutation in *TYR* (rs1042602) has been linked to normal variation in skin pigmentation using admixture-mapping techniques (Shriver et al. 2003), and the derived allele at this locus is associated with lighter skin pigmentation in a heterogeneous sample of South Asians (Stokowski et al. 2007). F_{ST} evidence for selection acting on this allele in Europeans (Norton et al. 2007) has also been reported, although the signal is far weaker than that observed for *SLC24A5* or *SLC45A2*. Selective scans using F_{ST} and/or haplotype data suggest that mutations in *TYRP1* and *DCT* have increased in response to directional selection in European and/or East Asian populations (Izagirre et al. 2006; Lao et al. 2007; Myles et al. 2007; Beleza et al. 2013b; Hider et al. 2013). However, it has proven difficult to directly associate mutations in these favored haplotypes to normal phenotypic variation in skin pigmentation. Surprisingly, a novel nonsynonymous allele in *TYRP1* is associated with blond hair color in Melanesian populations (Kenny et al. 2012; Norton et al. 2014). This allele exhibits sharp geographic localization, extending only from the island of New Britain in the east to the Temotu region of the Solomon Islands in the west – it has not been detected outside of this region, indicating that blond hair color has evolved multiple times in human populations. This result emphasizes the possibility that locally restricted mutations may have significant impact on pigmentation phenotype within small geographic regions and demonstrates the need for resequencing studies to identify phenotypically relevant alleles.

Conclusions and Future Directions

Skin pigmentation variation in humans is extensive, and there is good evidence that global patterns of variation have been strongly influenced by natural selection. The development of quantitative methods to assess skin, hair, and iris pigmentation variation has improved our knowledge of the extensive variation in these traits within and between populations. With the increasing use of genetic and genomic data, our understanding of the genetic and evolutionary causes underlying modern patterns of variation has also improved tremendously, but additional work is clearly needed.

First, genetic studies point overwhelmingly to the convergent evolution of light skin color in European and East Asian populations. However, the majority of selected loci identified to date impact European, but not East Asian, pigmentation. While this may in part reflect historical biases in the populations who were studied and/or ascertainment bias in early SNP genotyping assays, with the advent of the 1000 Genomes Project and the decreasing costs of genome sequencing, a new focus on both phenotypic and genetic variation in East Asian populations should help us to identify relevant loci in this part of the world.

Second, populations in high-UVR regions have largely been ignored from the discussion, possibly due to the assumption that darker skin color represents the retention of an ancestral phenotype maintained by purifying selection. However, the presence of extensive variation in skin pigmentation phenotype within high-UVR populations suggests that phenotypic variation may not be as constrained by UVR as was once thought (Norton et al. 2007; Ang et al. 2012; Basu Mallick et al. 2013). In addition, comparisons of Melanesian and African populations at known pigmentation loci using measures of interpopulation divergence, which should be low if variation is constrained by selection, indicate that for at least some loci, the two regions have diverged (Lao et al. 2007). This suggests that dark skin color in each region may be attributed (in part) to distinctive genetic architecture.

Finally, while loci with moderate to large effects on variation in skin, hair, and iris pigmentation have been identified, we know far less about loci with minor effects on phenotype. Future studies that include a more refined assessment of phenotype (e.g. digital scoring of iris color) may increase our understanding of the fine-scale genetic architecture that underlies pigmentation variation. Such studies may also be useful in the development of more sensitive forensic tests designed to predict phenotypic variation (Kayser and de Kniff 2011).

Historically, skin pigmentation was one of several phenotypic traits that anthropologists and biologists attempted to use to place humans into discrete taxonomic (racial) categories. Today, most would agree that such classification schemes are problematic for a number of reasons, including the continuous nature of these traits, the fact that these traits do not covary with each other or with geographic region, and because an overwhelming amount of genetic evidence does not support the existence of a limited number of clear biological clusters of human populations. However, this does not mean that pigmentation should be ignored in studies of complex phenotypic variation. On the contrary, the study of pigmentation variation and its underlying genetic causes can teach us a good deal about the role of natural selection in shaping global patterns of human diversity, and it may provide a model to assist in the understanding of the evolutionary processes that have shaped other complex traits in the human species.

CHAPTER 18

Anthropological Genetics and Growth and Development

Bradford Towne[1], Ellen W. Demerath[2], and Stefan A. Czerwinski[3]

[1]Boonshoft School of Medicine, Wright State University, Dayton, OH, USA
[2]University of Minnesota School of Public Health, Minneapolis, MN, USA
[3]University of Texas School of Public Health, Brownsville, TX, USA

In spite of the predominant role that genetic variation plays in the observed variability among children in their growth and development, studies of genetic influences on growth and development are few in comparison to the plethora of descriptive studies, population comparisons, and studies of the impact of specific environmental factors. There are two main reasons for this. First, the courses of study that many investigators of growth and development are trained under (e.g. physical anthropology or human biology) usually provide little formal training in human genetics and statistical genetic analysis. And, second, in order to best study genetic influences on growth and development, data from related children are needed. Preferably those data from related children are longitudinal, and ideally they are longitudinal data from large numbers of related children reared under different household environments. Unfortunately, such data are very rare.

A Companion to Anthropological Genetics, First Edition. Edited by Dennis H. O'Rourke.

The purpose of this chapter is to provide an overview of the genetic epidemiology of normal human growth and development. A full treatise on quantitative genetic approaches to the study of growth and development and a complete review of the existing literature on the genetics of growth and development are beyond the scope of this chapter, which provides a starting point for further pursuit.

Almost half a century ago, Neel and Schull (1954) proposed that the epidemiological approach can be extended to the study of nondiseased states, and argued that "genetic concepts must be an integral part of the armamentarium of the modern epidemiologist" (p. 302). The "epidemiological genetics" that Neel and Schull envisioned has become known as *genetic epidemiology*. Upon the establishment of the International Genetic Epidemiology Society (IGES) in 1992, its founding president, James V. Neel, succinctly defined genetic epidemiology as "The study of genetic components in complex biological phenomena" (Garn, Bailey, and Cole 1976).

From this perspective, the genetic epidemiology of growth and development may be considered as the study of the genetic underpinnings of the size, conformation, and maturity status of individuals over the course of childhood. This includes characterizing the magnitude of genetic influences on growth and development phenotypes, examining how those genetic influences operate over time, identifying and localizing genes and specific genetic polymorphisms in those genes that contribute to variation in growth and development, and elucidating how genetic and environmental factors interact during growth and development. The advances made over the past few decades in both molecular and statistical genetics have led to highly sophisticated analyses used to increasingly better elucidate the roles of genes and environment in the complex biological phenomena that comprise growth and development.

Anthropological genetics arose during the 1960s, with many of the early practitioners of genetic epidemiology being physical anthropologists or geneticists with an appreciation of what an anthropological perspective and the study of traditional anthropological populations can bring to "the study of genetic components in complex biological phenomena." In that context, anthropological genetics explicitly broadens the genetic epidemiology of growth and development to include the diversity of Western and non-Western populations, thereby potentially taking into account the diversity of genetic polymorphisms influencing (to one degree or another) growth and development, as well as the diversity of environmental factors that may contribute to phenotypic variation in measures of growth and development.

This chapter is divided into four sections. Section one provides an introduction to basic statistical genetic terminology and methods. The next section discusses different study designs used to examine genetic influences on quantitative traits. The following section presents a sampling of published findings from various studies of the genetics of growth and development. The final section presents example findings from current genetic epidemiological studies of the growth and development of US children in the Fels Longitudinal Study.

Statistical Genetic Terms and Methods

Statistical genetics refers to various methods for analyzing phenotypic variation among individuals, with particular emphasis on methods for analyzing phenotypic variation among related individuals. Most growth and development phenotypes exhibit a continuous distribution over a delimited range, and because the growth and development status of a child can usually be measured in some way, most growth and development phenotypes are quantitative traits. Growth and development phenotypes also are referred to as being *complex traits*, meaning that genes at a few and perhaps several loci contribute to the variation observed in the trait, as do environmental factors, possibly through interaction with those genes.

Relatedness of individuals

Because related individuals are not independent, but share some of their genes by virtue of sharing common ancestry, it is necessary to consider their degree of relatedness in assessing the extent of their resemblance for a trait. The coefficient of kinship between two individuals is the probability that an allele taken at random from the two alleles at a locus in one individual is identical to an allele taken at random from the two alleles at the same locus in another individual. The coefficient of kinship between first-degree relatives is 0.25, meaning that, for example, between a pair of full siblings there is a 25% chance that they each have at a locus the very same allele that they each inherited from a common ancestor.

Most of what we know about the genetic control of growth and development comes from family-based studies in which the correlations between relatives and between unrelated individuals for a trait, such as stature or weight, are calculated. The basic premise underlying these investigations is straightforward – if the variation in a trait is largely under genetic control, then related individuals will be more similar for the trait than will unrelated individuals (i.e. the intrafamily variance of the trait is low compared to the interfamily variance). Conversely, if the variation in a trait is only partly determined by genes, then related individuals may only resemble each other a little bit more so than do unrelated individuals (i.e. the intrafamily variance of the trait is a little smaller than the interfamily variance).

Heritability

Through examination of correlations between different pairs of relatives, heritabilities can be calculated. The concept of heritability (h^2) is central to understanding the nature of genetic control for a trait. The h^2 of a trait is a measure of the degree of genetic control of a phenotype, ranging from 0 (no genetic control) to 100% (complete genetic control). Heritabilities are population-level estimates, specific to a particular population in a given environment, and this can sometimes be an important consideration when comparing h^2 estimates across populations.

According to classical quantitative genetics theory (e.g. see texts Falconer and Mackay 1996; Lynch and Walsh 1998), the observed phenotypic variation (σ^2_P) in a trait can be expressed as the sum of both genetic (σ^2_G) and random environmental effects (σ^2_E). This is written as:

$$\sigma^2_P = \sigma^2_G + \sigma^2_E \tag{18.1}$$

In its simplest form, this model provides a starting point for understanding the quantitative genetics of complex traits. For example, σ^2_P can be decomposed further into components representing the variance due to additive effects of genes at several loci (σ^2_A), dominance effects (σ^2_D), and epistasis (σ^2_I), while σ^2_E can be decomposed into the variance due to specific measured environmental factors ($\sigma^2_{E\ factor\ \#1}$) and that due to random, unmeasured environmental factors ($\sigma^2_{E\ random}$). *Broad sense heritability* refers to the proportion of the phenotypic variance attributable to all sources of genetic variance, and it is written as:

$$h^2 = \sigma^2_G / \sigma^2_P \tag{18.2}$$

Narrow-sense heritability refers to the proportion of the phenotypic variance attributable only to the additive genetic variance, and it is written as:

$$h^2 = \sigma^2_A / \sigma^2_P \tag{18.3}$$

Generally speaking, at least initially, narrow-sense heritability is the most useful in characterizing the genetic effects of continuously distributed traits, such as stature or weight. Inheritance of such quantitative traits is likely to be influenced by a number of genes exerting mostly small to moderate effects. For that reason, quantitative traits are often also referred to as being *polygenic traits*. However, not all genes influencing a trait likely make the same contribution to the phenotypic variance of the trait. Also, since it is typically very difficult (e.g. because of sample size constraints) to identify genes explaining only a small proportion of the phenotypic variance of a trait (e.g. 5% or less), it is perhaps more practical to refer to most quantitative traits as being *oligogenic traits*, meaning that it is likely that a few genes with pronounced and identifiable effects of varying degrees are together responsible for most of the genetic contribution to the phenotypic variance of a trait. In most instances, h^2 estimates refer to narrow-sense heritabilities. The variance components approach to decomposing the phenotypic variation exhibited in a quantitative trait briefly described here has its roots in the seminal work by Fisher (1918), and it is an elegant and powerful method for evaluating the different sources of variation contributing to the overall variance of a complex trait.

Genetic and Environmental Correlations

Quantitative genetics is much more than simply calculating h^2 estimates. Since it is well established that measures of growth and development have substantial and significant heritable components, intellectual focus turns to the nature of the genetic regulation of growth and development. For example, significant phenotypic correlations often exist between different measures of growth and development. These phenotypic correlations may be due to pleiotropy, the joint effects of a gene or genes on different traits, or shared environmental factors. In most cases, significant phenotypic correlations between two traits are due to both pleiotropy and shared environmental effects.

Just as the phenotypic variance of one trait can be decomposed into genetic and environmental variance components, so too the phenotypic correlation between two traits can be decomposed into genetic and environmental covariance components. Thus, the phenotypic correlation between two traits is a function of the h^2 of each trait and the genetic and environmental correlations between them. This is written as:

$$\rho_P = \sqrt{h_1^2}\sqrt{h_2^2}\,\rho_G + \sqrt{(1-h_1^2)}\sqrt{(1-h_2^2)}\,\rho_E \qquad (18.4)$$

where ρ_p is the phenotypic correlation, ρ_G is the genetic correlation, ρ_E is the environmental correlation, h^2_1 is the heritability of trait 1, and h^2_2 is the heritability of trait 2. If both traits have low heritabilities, the phenotypic correlation between them is largely due to the environmental correlation, whereas if both traits have high heritabilities, the phenotypic correlation between them is due largely to the genetic correlation.

As with phenotypic correlations, additive genetic and random environmental correlations range from −1.0 to 1.0. A genetic correlation of 1.0, for example, indicates complete positive pleiotropy between two traits. That is, there are genes that affect in the same manner both of the traits being examined. A genetic correlation significantly less than one indicates incomplete pleiotropy, meaning that the two traits are influenced to some extent by the same set of genes, but that other genes also are influencing the value of one or the other of the two traits. A genetic correlation of zero between two traits indicates that the two traits have different genes controlling them. Finally, a negative genetic correlation indicates that the same set of genes operates in an opposite manner on the two traits. Similarly, the random environmental correlation is a measure of the direction and strength of the correlated response of two traits to nongenetic factors. If specific nongenetic factors have been identified that influence the covariance of the two traits, the environmental correlation can be decomposed into nonrandom and random components.

Applications of genetic and environmental correlations to longitudinal data

Of particular interest in the field of growth and development is the nature of the genetic control of a trait over time. For these types of analyses, it is necessary to have serial measurements of the trait. Serial measurements of traits separated by time are normally correlated to some degree, with higher phenotypic correlations often found over short intervals and lower phenotypic correlations found over longer intervals. *Canalization* is a familiar term to auxologists, referring to the tendency of a trait to follow a certain course or trajectory over time. The more highly canalized a trait, the higher the phenotypic correlations between repeated measurements. From a genetic perspective, traits that are highly canalized, and that are relatively insensitive to changes in environmental conditions, are likely to have relatively high heritabilities. The same genes, however, may or may not be influencing the trait to the same extent over the entire course of growth and development.

To test hypotheses concerning the genetic control of growth at different ages, the same approach discussed here for the examination of two traits at one point in time is taken. Now, however, the two "traits" are actually the same trait measured at two points in time. The genetic and environmental correlations between repeated measures of the trait at different ages are then calculated. This approach allows for disentangling shared genetic effects from shared environmental effects on a trait measured over the course of childhood.

The strength of a genetic correlation for a single trait with repeated measures is indicative of the degree of consistency or uniformity in the genetic control of the trait over time. For example, if a genetic correlation of 1.0 is found between stature measured at age 8 years and measured again at age 18, then it can be inferred that the genes influencing stature during the middle of childhood are the same as those that influence height in early adulthood. If a genetic correlation is obtained that is significantly lower than 1.0, however, then there is evidence that different suites of genes control stature at ages 8 and 18 years. Similarly, the environmental correlation is a measure of the consistency or uniformity of the response of the trait to nongenetic factors over time. We will return to genotype-by-age interaction after discussing genotype-by-environment and genotype-by-sex interactions.

Genotype-by-environment interaction

Understanding how genes interact with the physical and internal biological environments is essential for better understanding the genetic architecture of complex traits. In studies where relatives live in different environments, genotype-by-environment (G×E) interactions can be examined using extensions of variance components methods for studying quantitative trait variation.

G×E interaction is likely an important influence on the variation observed among children in their growth and development, particularly in populations with high prevalences of environmental factors known to negatively impact growth and development. The key to G×E interaction is that not all children may respond the same to such environmental factors, and a portion of that differential response or susceptibility at the phenotypic level may be due to genetic variation among individuals.

The simplest approach to modeling G×E interaction is to make the genetic variance in a trait a function of a dichotomous environmental variable. Examples of

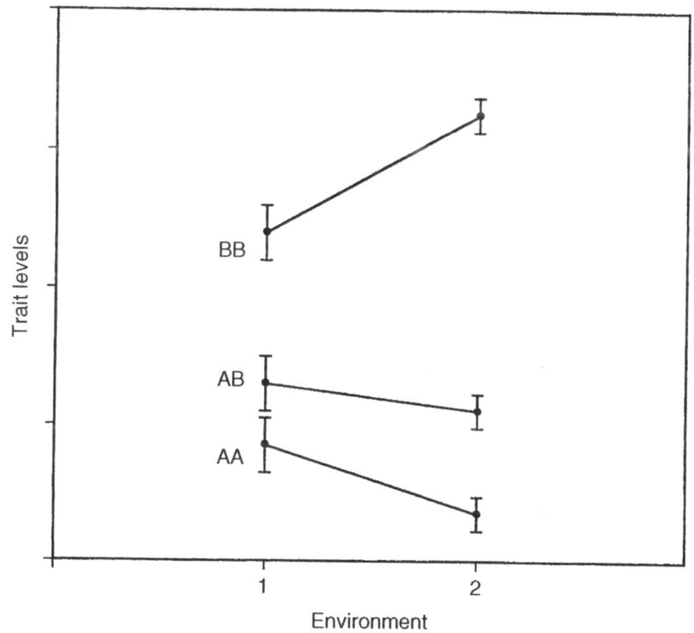

Figure 18.1 Hypothetical depiction of gene-by-environment interaction with the response of three genotypes at a locus to two different environments. *Source*: Towne, Demerath, and Czerwinski (2012). Reproduced with permission of Elsevier.

this could be the presence or absence of a particular disease in a child, high or low protein intake, and so on. Figure 18.1 shows a simple hypothetical depiction of the response of three genotypes at a locus to two different environments. In the presence of G×E interaction, the relationship between trait levels and specific genotypes will vary as a function of the environment. In this case, trait levels in Environment 1 are substantially less variable compared to trait levels in Environment 2. For genotypes AA and AB, trait levels remain stable or decrease from Environment 1 to Environment 2. For genotype BB, trait levels increase from Environment 1 to 2. This example demonstrates how gene expression may vary under different environmental conditions.

In G×E analyses of the response of a quantitative trait, the variance components method is expanded to include environment-specific additive genetic variances that are then estimated. For example, a large number of related children might be measured for a trait (e.g., stature) at a specific age, and also tested for the presence of a particular infection at that age. If the additive genetic variances of the measured trait are not significantly different between infected and non-infected children, then that would be an indication that there is no G×E interaction between that trait and that infection at that age. If, on the other hand, the additive genetic variances of the measured trait are significantly different between infected and non-infected children, then that would indicate a genetic basis to the differential response of the growth status of children to infection at that age. G×E interaction is also tested by examining the genetic correlation between the trait measured in different environments. A genetic correlation significantly different from 1.0 is another indication of G×E interaction. In the example here, a genetic correlation significantly less than 1.0 would indicate that the G×E interaction is due to an incompletely correlated genetic response of the trait in infected and non-infected children.

Genotype-by-sex interaction

Sexual dimorphism in the growth and development of children is well known, but the genetic basis of this sexual dimorphism is poorly understood. The approach for studying G×E interaction using related individuals living in different environments, as described in the previous section, can be used to study genotype-by-sex (G×S) interactions. The rationale here is that the hormonal environments of males and females differ considerably, and the expression of autosomal genes controlling a quantitative trait may be influenced by the sex environment encountered.

In analyses of G×S interaction, additional parameters are estimated, the most important being sex-specific variance components and the genetic correlation between the sexes for the trait. G×S interaction is indicated by significantly different additive genetic variances for males and females and/or a genetic correlation between the sexes significantly less than 1.0.

G×S interaction analyses can be used to examine the genetic basis to the sexual dimorphism in measures of growth and development. The aim of G×S interaction analyses is to determine if the sexual dimorphism evidenced in a trait during childhood age is itself a heritable trait. In some families, for example, male and female children might not be very different in a measure of growth or development at a particular age, while in other families there might be significant differences between male and female relatives in that measure of growth or development at that age.

Genotype-by-age interaction

The nature of genetic influences on measures of growth and development may change over the course of childhood. If extensive longitudinal data from related children are available, genotype-by-age (G×A) interactions can be more rigorously examined. Like G×S interactions, G×A interactions are a type of G×E interaction. In this case, the "environment" is the age of the child at the time of the measurement of a trait. In these analyses, the additive genetic variance of a trait is modeled as a function of age. From these age-specific additive genetic variances, age-specific heritabilities of the trait can be determined, as can additive genetic and environmental correlations between the trait measured across time.

G×A interaction is indicated by an additive genetic variance of a trait changing over a span of ages. This suggests that the genetic expression of a trait depends on the age of the child. G×A interaction also is indicated by a change in the genetic correlation between a trait measured over time. For example, a genetic correlation between time points of a serially measured trait that decreases significantly from 1.0 over a span of ages indicates G×A interaction. And, if sufficient serial data are available, the function or shape of a genetic correlation curve can provide further insights into dynamic genetically mediated biological processes underlying such G×A interactions.

Identifying genes influencing growth and development

Unlike monogenic traits that are influenced by a single gene with large effects, most complex traits are influenced by genes at a number of loci whose individual effects can be of small to moderate size. While understanding of monogenic growth disorders has significantly increased over the last several decades, understanding the genetics of normal variation in quantitative measures of growth and development has continued to be a daunting task. Technological advances in molecular biology, however, including high-throughput genotyping of millions of single nucleotide polymorphisms (SNPs) and whole-exome and whole-genome sequencing, along with attendant methodological advances in statistical genetics, have made it possible to identify genes exerting small or moderate effects, and to even identify rare polymorphisms influencing a trait in only some populations or pedigrees. The two strategies used most often in searching for genes involved in the regulation of growth and development are population-based association studies and family-based quantitative trait linkage studies.

Candidate gene association studies

In the candidate gene association approach, genes suspected to be physiologically involved in the trait are examined. For example, a sample of unrelated individuals drawn from a population is selected and genotyped for a specific polymorphism in or near the candidate gene. Simple statistical tests are then used to evaluate associations between marker genotype status and the value of a trait. Carriers of a particular allele, for example, may have a mean value for the trait that is significantly different than the mean value of the trait in those who do not have a copy of that allele.

Population-based association studies such as the candidate gene approach have obvious appeal – they are computationally straightforward. There is a significant problem with population-based association studies, however, that has become evident as greater knowledge has been gained regarding linkage disequilibrium. Two loci are in equilibrium when alleles at the two loci are randomly associated with each other. If the relationship between the loci is not random, then linkage disequilibrium is present. Linkage disequilibrium can occur for a number of reasons, including new mutations, genetic drift, and the presence of selection.

Unfortunately, disequilibrium cannot be predicted. Two loci may be very close to each other and yet be in equilibrium. Conversely, two loci may be relatively far apart from each other and yet be in disequilibrium. There is no sure way to know that the marker that has been typed is in disequilibrium with the trait that has been measured. If it is known a priori, however, that the typed marker is in fact a functional polymorphism (i.e. there is a measurable difference among marker genotypes in gene expression; e.g. one genotype results in much lower levels of a particular protein compared to the other genotypes), then association studies become a more viable strategy to pursue.

Genome-wide association studies

In recent years, genome-wide association studies (GWAS) of complex traits have proliferated. Genome-wide association (GWA) analysis is made possible by relatively low-cost genotyping of typically from 500 000 to 1 000 000 SNPs in each subject in a study sample. SNPs are bi-allelic genetic markers that are coded either "0" or "1", and in most standard SNP panels are typically separated by only fairly small intervals across the entire genome. Associations between genotypes and phenotypes are evaluated over every marker. Of course, given the large number of markers, these analyses are computationally intensive, and stringent strategies to control for multiple testing must be followed.

GWAS designs typically take one of two forms. One is the case–control approach where individuals with a certain disease or condition are compared to unaffected individuals with regard to genotype status at every genotyped SNP across the genome. SNPs significantly more prevalent in either cases or controls are identified for follow-up to assess possible causative or protective roles of nearby functional polymorphisms. A second GWAS design focuses on quantitative traits where different genotypes at a single locus are examined for differences in trait levels. Associations are denoted when individuals with certain SNP genotypes have consistently and significantly higher or lower trait values than individuals with the other SNP genotypes.

There are several strengths to the GWA approach. First, the high density of SNP markers helps to better localize association signals. GWA signals can typically be reduced to approximately a 500 kb interval, compared to a much broader interval obtained from quantitative trait linkage (discussed further in this chapter). Another strength, alluded to here, is that data can be obtained from unrelated individuals, potentially making data collection more efficient. Indeed, numerous GWAS have been based on already existing epidemiological study samples.

Despite the popularity of GWAS over the last two decades, important criticisms of the approach have emerged in recent years. To date, GWAS have collectively had somewhat limited success, despite numerous published studies and several large meta-analyses of various complex traits that have sometimes included samples sizes of over 100 000 subjects. In most cases, reported associations have accounted for only a very small proportion of the overall heritability of the traits examined. Some researchers speculate that the reason for this is that GWAS are only useful for identifying common disease variants. Unfortunately, it now appears that many common disease or quantitative risk factor variants explain only a relatively small proportion of the total phenotypic variance in the disease or trait examined. The population-based association approach is therefore not likely to be able to identify rare disease variants that most likely have larger effect sizes.

Additionally, significant associations can be due to heterogeneity in the population sampled. This occurs when population subgroups differ systematically in both allele frequencies and levels of the quantitative trait of interest. Even among seemingly fairly homogeneous families (e.g. similar ethnic background), there can be significant differences in specific allele frequencies, and all families carry unique "private" polymorphisms, some of which may impact a trait in some families but not in others. Replication of findings is critically important for GWAS, providing confirmation of the role that at least common polymorphisms play in contributing to phenotypic variation.

Quantitative trait linkage analysis

The second predominant approach for discovering genes influencing the types of measurable traits of most interest to auxologists is quantitative trait linkage mapping. The premise behind linkage analysis is that if two loci are physically located close to each other, then alleles at these loci will be more likely to be inherited together. In this case, the loci are said to be linked. As the distance between loci increases, the probability that alleles at these loci will cross over or recombine during meiosis increases. Through investigation of the frequency of recombination events among genetic markers, one can identify chromosomal regions harboring genes that influence variation observed in a trait. Once a region has been identified, molecular-mapping techniques such as high-density SNP typing or sequencing can be used to better delineate chromosomal regions of interest and to identify functional polymorphisms.

Over the last three decades, there have been many advances in quantitative trait linkage analysis as applied to complex traits. Over that time, allele-sharing methods have gained prominence for the analysis of quantitative traits. The key premise behind allele-sharing methods is the concept of identity by descent (IBD). In comparisons between relatives, two alleles that are structurally identical are said to be identical by state (IBS); alleles that are structurally identical and inherited from a common ancestor (e.g. two siblings getting the same allele from their mother) are further classified as IBD. A pair of relatives can share 0, 1, or 2 alleles IBD at any given marker locus. The likelihood of their sharing 0, 1, or 2 alleles IBD is contingent upon their coefficient of kinship. Linkage between a quantitative trait locus (QTL) and a marker exists in chromosomal regions when pairs of relatives who are more phenotypically similar share more alleles at a marker locus than pairs of relatives who are less phenotypically similar.

The power to detect and localize QTLs is a function of several factors, the most important being the strength of the genetic effect. Traits that are highly heritable will tend to have a higher probability of being mapped compared to those with low to modest heritability, but this is not always the case. Also, as in any statistical analysis, sample size is of importance, but in linkage studies other aspects of the study sample are also important, most especially the family structure of the study sample. Having many families is good, but having fewer more complex extended pedigrees, preferably with several generations represented, will yield increased statistical power because of the greater number and variety of relationships between relatives.

Linkage analysis has several strengths and some weaknesses. One strength of linkage analysis is the ability to identify rare genetic variants in family-based samples. Because genes segregate in families, the ability to identify rare genes of moderate effect is possible. Identification of rare variants may help to explain what has been termed the *missing heritability* observed from population-based GWAS of common traits (i.e. the portion of the heritability not explained by the common variants; Manolio et al. 2009). Rare variants are likely to have larger effect sizes and could contribute to the unexplained heritability. Furthermore, some researchers suggest that these rare variants are likely to have obvious functional consequences (Manolio et al. 2009; Cirulli and Goldstein 2010). A benefit of pedigree-based studies for the identification of rare variation is that rarer variants, if present, will be present at a much higher frequency than in the general population. Thus, pedigree-based studies inherently have greater power to detect the effects of such rare variants. However, traditional linkage-based studies have limited resolution (due to typically having fewer genetic markers typed, although this is somewhat ameliorated with SNP-based linkage analysis vs. earlier STR-based linkage analysis; see further in this chapter) and are only able to localize QTL to approximately 10–15 Mb of sequence, a much broader region compared with GWAS.

Quantitative trait linkage and association

Combined linkage and association analysis effectively utilizes the strength of both genetic paradigms, but it can only be accomplished with family-based data. As a first step, a family-based approach to association analysis (e.g. measured genotype) can be implemented to test for associations assuming additive genetic effects on each available SNP in the panel (Almasy and Blangero 2004; Boerwinkle, Chakraborty, and Sing 1986). Since data from family members cannot be treated as independent observations, family-based methods such as variance-component analysis are able to use a polygenic component to absorb any non-independence among individuals by incorporating a residual heritability parameter. In this context, quantitative trait linkage provides an additional, independent source of information that, when used in conjunction with GWA, can augment power to detect loci influencing growth-related traits. SNPs used for linkage can be selected from among the typed SNP panel to maximize heterozygosity and minimize linkage disequilibrium among the selected markers.

Approximately 10 000 SNPs are required for adequate genomic coverage in SNP-based linkage analysis. A joint test of linkage and association can be performed by comparing the likelihood of a model in which both the SNP-specific association parameter and the linkage variance component are estimated to the likelihood of a model in which both are constrained to be zero. The power of this test depends on the underlying trait model and on how many functional variants there are within a gene or region; however, under certain circumstances, combined linkage and association can be more powerful than association alone.

Study Designs

Various family-based study designs can be used to examine the genetics of complex traits. Each study design has certain advantages and disadvantages. This section describes some of the major types of family study designs used to study complex quantitative traits.

Twin studies

Over the years, studies of twins have been useful in establishing the familial aggregation of many complex traits. In its basic form, the twin model compares phenotypic differences between two classes of twins, monozygotic (MZ) and dizygotic (DZ). MZ twins share 100% of their genetic makeup, while DZ twins share on average only half of their genetic makeup (i.e. on the genetic level, they are the same as any other pair of full siblings). Because of this, phenotypic differences observed between MZ twins are assumed to be the result of environmental factors only, while phenotypic differences between DZ twins are considered to be due to differences in both genes and environmental exposure. Thus, by calculating phenotypic correlations in groups of MZ and DZ twins and comparing them, assumptions can be made about the degree of genetic control of different traits.

One important assumption in the classical twin study design is that both MZ and DZ twin pairs are equally likely to share a common environment. This assumption may not necessarily be valid, however, because MZ twins are often more likely to share common activities, foods, and other aspects of the environment to a greater extent than DZ twins. Because there is no fully satisfactory way to separate shared genetic and environmental effects, studies of twins often yield inflated h^2 estimates.

The twin study design is especially problematic if the focus of the study is a growth-related outcome. Twin births are physiologically different from singleton births due to competition over maternal resources during pregnancy. Fetal growth rates among twins may therefore be considerably discordant, and the postnatal growth of twins is often different from that of siblings from singleton births (e.g. early catch-up growth in twins).

Nuclear families

Another commonly used study design is that of nuclear families. In this study design, correlations between the various classes of first-degree relatives in a nuclear family are estimated. These include parent–offspring, sibling–sibling, and spouse–spouse correlations. Heritabilities can be estimated from these different familial correlations. Heritability estimates calculated from nuclear family data, however, are subject to inflation due to the effects of shared environmental factors such as diet and lifestyle among family members living in a single residence. Given this, heritabilities are often adjusted by taking into account the degree of spousal correlation in the family. It is assumed that any correlation found between spouses is the result of shared environmental factors. Such spousal correlations may depend upon the length of time that the couple has been married. But, such spousal correlations may also be the result of initial assortative mating.

There are also practical considerations to be taken into account in studies of nuclear family members. For example, it is sometimes difficult to obtain information about certain life events because they are often separated in time by a generation – it may take 20–30 years of waiting to collect growth measures of the children of parents who were measured when they were children. Also, generational differences in growth may be due to secular trends. This may effectively reduce the heritability of certain traits by diminishing the degree of phenotypic correlation observed. These two problems can be eliminated by examining only sibling correlations, but the problem of shared environment remains.

Extended pedigrees

The study design that offers considerable advantages for elucidating the genetic architecture of complex traits is the extended family approach. This approach involves collecting information from all available family members and estimating phenotypic correlations between all relatives of varying degrees of relationship. By sampling members outside of the immediate nuclear family, many of the problems encountered with immediate shared environmental effects in other study designs are minimized because family members come from a number of different households. This results in more accurate and reliable h^2 estimates. In addition, by sampling family members in different households (who thereby live in potentially different environmental circumstances), there is the opportunity to investigate

G×E interactions. With regard to the study of growth and development, within large extended pedigrees there will be several related children of approximately the same age. This will enable analyses to proceed very quickly after the initiation of data collection.

There are a few practical drawbacks to this approach, however. The single most important consideration is that the methods involved in calculating statistical genetic parameters can be computationally intensive. This is much less of an obstacle now as computer technologies have continued to progress. Indeed, advances in computer technology over the last three decades have made statistical genetic analyses of data from large pedigrees tractable.

Studies of the Genetics of Growth and Development

This section provides a brief overview of numerous studies of genetic influences on growth and development that have been conducted over the last century. Most of these studies fall into two general categories – those that infer genetic determination of growth and development through comparison of different human populations, and those that examine growth and development traits in families. The review presented here provides just a sampling of this literature, focusing on studies of height, birth weight, menarche, and skeletal development.[1]

Population differences in growth and development

There is considerable variation across populations in growth in height, weight, and other body dimensions, as well as in the tempo and timing of maturation (Eveleth and Tanner 1990). For example, mean adult height varies from approximately 150 cm for males in the shortest populations on earth (e.g. Mbuti pygmies of Central Africa) to 180 cm for males in Northern European populations. These long-standing observations of racial or ethnic differences in growth and development lent support for the notion that genetic factors are likely involved.

The degree to which genetic factors influence growth and development cannot be addressed, however, by the simple comparison of measures of growth and development traits across populations. The populations compared often are exposed to vastly different environments, and the shortest and smallest populations also tend to have the poorest economic status, while the tallest populations tend to be from industrialized nations. Between-population differences may be due to differences in both genetic and environmental factors, whose relative importance are often confounded. For example, evidence of secular trends in

[1] Recently compiled tables of results from numerous genetic studies of various growth and development traits can be found in Towne, Demerath, and Czerwinski (2012), from which this current chapter was derived.

stature and pubertal maturation (Eveleth and Tanner 1990), and the degree of similarity in stature in high-socioeconomic-status groups from various parts of the world (e.g. Martorell, Mendoza, and Caastillo 1988), argue that a significant part of interpopulation variation in growth and development is due to environmental factors.

Family studies of growth and development

Some general comments can be made regarding family-based studies of growth and development. First, most studies have been based on first-degree familial correlations. That is, they are based on either nuclear family or twin pair designs. And, as discussed in this chapter, there are important concerns when studying only first-degree relatives – particularly when studying growth and development. These concerns include secular trends that may reduce correlations between parents and offspring, and the shared environments of siblings, especially twins, that may inflate correlations between them. Second, specific environmental sources of variation such as diet and disease usually have not been incorporated into the analyses. Not accounting for the variance in a trait attributable to such environmental factors can lead to underestimation of the h^2 of the trait. Third, the majority of studies have focused solely on height at a given point in time (mostly adult height). A much smaller number of studies have examined other anthropometrics. Fourth, the majority of studies are based on cross-sectional data. Only a few studies have longitudinal growth and development data from related individuals that permit examination of genetic influences on patterns of change in height, weight, and other measures over time. And, fifth, almost all of the studies have focused solely on heritability estimation. There are very few multivariate quantitative genetic analyses of measures of growth and development, or analyses of genotype-by-environment, sex, or age interactions.

Although comparatively few GWA or linkage analyses of growth and development traits (e.g. birth weight, height, body mass, and age at menarche) have been conducted, this is an expanding area of auxological genetics research (indeed, the most relevant area of auxological genetics research), as primary interest has shifted from heritability estimation to identifying specific genes and genetic polymorphisms that influence such measures.

Birth weight The genetics of prenatal growth has almost entirely been approached by examining the heritability of birth weight. Initially, genetic influences on birth weight were deduced from the known effects of quantitative changes in chromosomes. For example, supernumerary autosomes (trisomy 21, 18, and 13) and abnormal numbers of X chromosomes (as in Turner syndrome) all result in growth retardation. Formal quantitative genetic analyses of birth weight find somewhat lower heritability estimates than for body weight and length in postnatal life, which are both highly heritable.

Assessment of genetic influences on birth weight is complicated, however, by the fact that prenatal growth (at least as measured by birth weight) is influenced by both the genetic makeup of the fetus and the maternal intrauterine environment, and there is no fully satisfactory way to partition these two sources of variation. Therefore, not surprisingly, estimates of the influences of fetal genes, maternal genes, nongenetic maternal factors, and random environmental effects on fetal growth vary considerably across studies. The role of fetal genes varies from 0 to 50%, maternal factors from 27 to 50%, and random environmental factors from 8 to 43% in the variation in birth weight (Mueller 1998). For example, a classic study by Penrose (1954) attempted to partition the variance in birth weight among fetal genes, maternal genes, nongenetic maternal factors, and random environmental effects. He concluded that fetal genes accounted for approximately 18% of the phenotypic variance, while "maternal factors" (a combination of both genetic and uterine environment) explained approximately 40% of the phenotypic variance.

The importance of uterine environment in the control of prenatal growth is also demonstrated by the changes in twin correlations from birth onward (e.g. Wilson 1976). Intrapair differences in the birth weight of MZ twins are often significant at birth (tending to be larger than differences between DZ twins) because MZ twins compete for placental resources. Differences in weight between MZ twins decrease quickly during infancy. By three years of age, the MZ twin correlation is about 0.80–0.90, whereas the DZ twin correlation is about 0.40–0.50.

Another issue with the use of birth weight as a measure of prenatal growth is that it represents growth status at a variety of maturational ages depending on gestational age. Most studies of the genetics of birth weight have not controlled for gestational age. This flaw has likely led to underestimates of genetic influences. Indeed, using a variance components method for pedigree data, and modeling a gestational age covariate effect, we have found a high heritability of birth weight in the Fels Longitudinal Study population (h^2 = 0.80; Demerath et al. unpublished results). Continued work along these lines will help to identify specific factors influencing fetal growth and development. However, progress depends on measurement strategies that better capture the process of fetal development (e.g. serial ultrasound biometry).

Height Data from nearly 4000 individuals in 1100 nuclear families in England analyzed by Pearson and Lee (1903) provide perhaps the earliest evidence for the inheritance of height. This landmark study found a significant correlation between spouses of 0.28 (showing positive assortative mating for height), but higher correlations between siblings (0.54) and between parents and offspring (0.50). Since the expected correlation between full-siblings and between parents and offspring would be 0.50 if the h^2 of the trait was 1.0, they concluded that the population variation in height was highly determined by genetic factors.

These early results have been corroborated by hundreds of subsequent family studies. In populations around the world, the estimates of the h^2 of height range from 0.60 to above 0.90, clearly showing that height is a highly heritable trait.

In a review of 24 studies of parent–child correlations of height and weight, however, Mueller (1976) indicated that population estimates of heritability tend to be systematically lower in developing countries than in affluent countries. There are a number of reasons why this might be so. As mentioned in this chapter, according to classic quantitative genetic theory, the heritability of height or any trait is a function of the population in which the estimate is made, as well as of the trait itself. Heritability estimates will tend to be higher if there is positive assortative mating (i.e. a significant phenotypic correlation between parents). And, indeed, assortative mating for height has been found in European or European-derived populations more frequently than in non-European populations. Also, non-European populations in the developing world tend to live under more nutritional and disease stress than European populations. In these populations, such environmental factors have the potential to impact a given trait more than in affluent populations. Since heritability is the proportion of variance due to genetic influences, a larger proportion of environmentally induced variation will reduce the heritability. Additionally, many non-European populations are experiencing rapid economic change, which results in the growth environments of children differing quite markedly from those of their parents, thus decreasing parent–offspring correlations and the estimate of total variation attributable to genes.

Weight, circumferences, and skinfolds Whereas the heritability of skeletal lengths (e.g. height and sitting height) tends to be high, the h^2 of skeletal breadths (e.g. bi-iliac and bi-acromial diameters) tends to be somewhat lower, averaging between 0.40 and 0.80. In turn, skeletal breadths tend to have higher heritabilities than circumferences and skinfolds. It has been assumed that soft-tissue traits are more easily altered by the changing nutritional environment of individuals than are skeletal tissues that respond less quickly to changes in nutritional status, and as a result have a greater proportion of their variance explained by environmental, rather than genetic, factors.

Longitudinal studies As mentioned in this chapter, the vast majority of family studies of growth and development are cross-sectional. Only a few studies have longitudinal growth and development data from related children that permit genetic analyses of the processes of growth and development. Some of these longitudinal studies of the genetics of growth, for example, examined changes in parent–child or sibling correlations from age to age. Reports from the Fels Longitudinal Study (Garn and Rohmann 1966), Poland (Welon and Bielicki 1971), and elsewhere (Furusho 1968; Tanner, Goldstein, and Whitehouse 1970) found that parent–child correlations for height increased during the first four years of life, decreased during adolescence (when heterogeneity of maturational tempo disrupted familial similarity), and subsequently rose above the prepubertal level.

Modern longitudinal genetic epidemiological studies of growth and development use growth curve–fitting methods to pinpoint growth and maturational events, particularly of changes in the tempo of growth in a measure, and then examine growth curve parameters in genetic analyses. For example, Beunen et al. (2000) reports high h^2 estimates for the ages at take-off and at peak height velocities, and the heights at those ages. And van Dommelen et al. (2004) fit curves to serial infant height (length) and weight data from a large sample of Dutch twin infants and found significant heritabilities of various growth curve parameters. Similar analyses of Fels Longitudinal Study data are discussed in more detail below.

Maturation Not only is physical size heritable, but also the timing and tempo of maturation are significantly controlled by genes. A number of early studies of dental development found that radiographic measures of the timing of tooth formation (calcification) and dental emergence were more highly correlated within MZ twin pairs than DZ twin pairs, suggesting a heritability of 0.85–0.90 (Garn, Lewis, and Polacheck 1960). Also, the number and pattern of dental cusps were found to be under genetic control. The rate of skeletal maturation has been compared in siblings over time in several reports, with the general finding being that there is a great deal of similarity between siblings in the age of ossification onset of bones in the hand and foot. The general pattern of skeletal maturation (i.e. the tendency to be an "early" or "late" maturing individual) also suggests that the tempo of development is highly heritable, with sib–sib correlations of 0.45 (Hewitt 1957).

The process of maturation is commonly believed to be controlled, at least partially, by genes independent from those controlling final size. This conjecture stems from the observation that siblings may reach identical height, even though they differed in the timing of maturational events (Tanner 1962). Further and more widespread use of the multivariate quantitative approaches is discussed in a later section, in which genetic and environmental correlations between different traits may be calculated, will allow for greater understanding of the extent of shared genetic and nongenetic factors underlying growth and development traits.

Age at menarche is one of the most studied developmental traits. A number of early studies suggested that age at menarche has a genetic basis (e.g. Boas 1932). The mother–daughter and sister–sister correlations in the age at menarche were close to 0.50, indicating a high degree of genetic determination of age at menarche. These and later studies, however, have relied primarily on recalled ages at menarche, and thus recall bias (greater in mothers than in daughters) is introduced into these estimates. Later studies have confirmed a strong genetic influence on age at menarche (Damon et al. 1969; Zacharias, Rand, and Wurtman 1976), although the familial correlations were lower than in the early studies (~0.25–0.45). In a sample of 371 female Fels Longitudinal Study subjects of varying degrees of relationship to each other, and from whom age at menarche data had been collected during their participation in the growth and development aspect of the study, Towne et al. (2005) found a substantial and significant heritability of 0.49 for age at menarche.

Examples from the Fels Longitudinal Study

This final section highlights some of the topics discussed in the preceding sections through examples of published and ongoing genetic analyses conducted over the years in the Fels Longitudinal Study.

Fels longitudinal study

The Fels Longitudinal Study began in 1929 in Yellow Springs, Ohio. It was one of several longitudinal studies of child growth and development initiated in the USA between the end of World War I and the start of the Great Depression, and it is the only one that has survived to today. Although the Fels Longitudinal Study did not begin with an interest in genetics, familial data began to be collected soon after the study began. Most of the mothers who enrolled their children in the early years of the study had more children later, and many of those children were subsequently enrolled. Also, a set of monozygotic, dichorionic triplets was recruited early in the study specifically to examine their similarities in growth and development. Another set of triplets and a few twin pairs also were recruited in later years. Over time, other relatives were incorporated into the study, the first of these being siblings of original subjects as mentioned above, and then offspring of original study subjects. The Fels Longitudinal Study today has more than 1000 active research subjects with various serial data from infancy, and mixed cross-sectional and longitudinal data from more than 1000 of their relatives. These individuals represent about 200 kindreds consisting of both nuclear and extended families.

The description of the "Genetics Program" of the Fels Longitudinal Study written by its first director, Lester W. Sontag (1946), is remarkable for its modern-sounding tone. Sontag noted that many aspects of growth and development were likely to have significant genetic determination, but are influenced by environmental factors as well. He noted that the study included many families with two or more children, and that these "constitute the material for the study of inheritance of growth patterns as well as of metabolic characteristics."

For example, the set of monozygotic, dichorionic triplets mentioned here were the subject of three early reports that described their similarities in physical and mental traits as young children, striae in their bones, and the onset of ossification from infancy through pubescence (Sontag and Nelson 1933; Sontag and

Comstock 1938; Sontag and Reynolds 1944). Soon after the triplets' 18th birthday, Reynolds and Schoen (1947) published a description of their growth patterns. A paper by Reynolds (1943) is especially noteworthy because it used familial data from different types of relatives to examine the effects of degree of kinship on patterns of ossification. Included in this analysis were the set of identical triplets, as well as 3 pairs of identical twins, 22 pairs of siblings, 8 pairs of first cousins, and 18 unrelated children. Reynolds found that close relatives were very similar in their pattern of ossification, distant relatives less so, and unrelated people even less similar.

A series of studies from the late 1950s to the late 1960s by Garn and colleagues used data from siblings, parents, and offspring to examine patterns of familial correlations in traits pertaining especially to dental and skeletal maturation. An example of the analyses and sample sizes from this period is provided by Garn, Lewis, and Kerewsky (1963), who examined ossification data from radiographs of the hand–wrist and chest for 72 parent–child pairs, 318 sibling pairs, four pairs of dizygotic twins, and four pairs of monozygotic twins. Since these were serial data taken at half-yearly intervals from ages one to seven years, there were 1211 pairings of parent–child data, 6690 pairings of sibling data, 102 pairings of data from dizygotic twins, and 176 pairings of data from monozygotic twins. Garn, Lewis, and Kerewsky (1963) concluded, "In these well-nourished ... Ohio-born white children, genes appear to account for a major proportion of ossification variance during growth." These investigators also examined the genetics of various dental traits, including the timing of stages of dental development (Garn, Lewis, and Polacheck 1960), tooth morphology (Garn, Lewis, and Walenga 1968), and the appearance of discrete dental traits (Garn et al. 1966). The influence of familial factors on growth in body size also was examined (Garn, Lewis, and Polacheck 1960; Garn and Rohmann 1966).

Genetic analyses of growth and development data from the Fels Longitudinal Study data have had a resurgence in the last 20 years. This is largely due to advances in statistical genetic methods that maximize the amount of information available in longitudinal data from large numbers of relatives of varying degrees of relationship to one another, as well as advances in molecular genetic methodology that allow for relatively low-cost genotyping. For example, Towne et al. (1993) fitted a three-parameter function to serial recumbent lengths from 569 infants in order to characterize each individual's unique pattern of growth during infancy. Figure 18.2 shows the growth curves of two infant boys who differ in their patterns of growth. Boy #1 started out in life shorter than Boy #2, but had a rate of increase in recumbent length that was much greater than that of Boy #2. Both boys, however, experienced about the same amount of growth (~42 cm) from birth to age two years. In this study, substantial h^2 estimates of 0.83 for recumbent length at birth, 0.67 for rate of increase in length, and 0.78 for a parameter describing the curvilinear shape of growth in recumbent length from birth to two years were found.

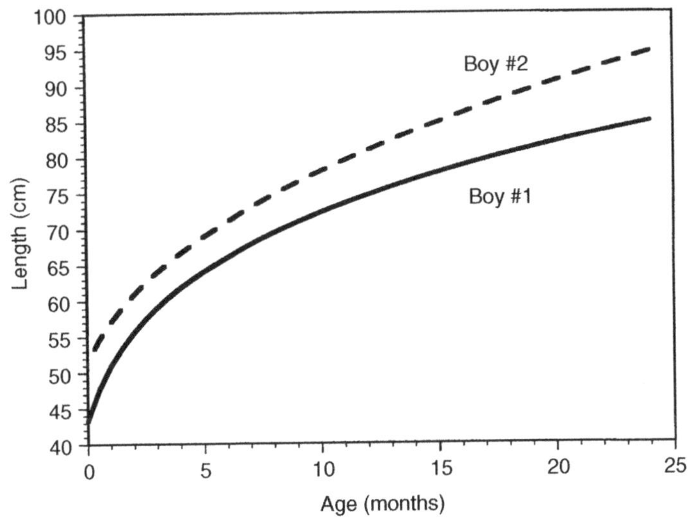

Figure 18.2 Height distance curves for two boys with differing growth patterns between birth and 24 months old. *Source*: Towne, Demerath, and Czerwinski (2012). Reproduced with permission of Elsevier.

Fels Longitudinal Study investigators have used the triple logistic model of Bock, du Toit, and Thissen (1994) as implemented in the AUXAL program (Bock, du Toit, and Thissen 2003) to fit growth curves to extensive serial stature data from some 600 study subjects in order to examine individual differences in patterns of growth and to conduct multivariate quantitative genetic analyses of different parameters of the pubertal growth spurt. For example, Figure 18.3 shows the growth and velocity curves of two girls with visibly different growth patterns. Girl #1 was only 9.37 years old when she was at the peak of her pubertal growth spurt, whereas the age at peak height velocity of Girl #2 was 13.09 years. At the time of peak height velocity, Girl #1 was shorter than Girl #2 (141.1 vs. 156.5 cm), which is expected given her younger age at peak height velocity, but at the age at peak height velocity, Girl #1 had a higher rate of growth than Girl #2 (8.8 vs. 6.4 cm/year^{-1}). By the end of their growth, Girl #1 was a petite woman (158.5 cm), while Girl #2 was somewhat taller than average (170.6 cm). Highly significant h^2 estimates on the order of 0.85 for age at peak height velocity, 0.61 for growth rate at peak height velocity, and 0.96 for stature at the age of peak height velocity were found. Especially interesting was the finding of additive genetic correlations between these pubertal growth spurt parameters that were significantly lower than 1.0, suggesting incomplete pleiotropic effects of genes on different aspects of growth. That is, these three different growth curve parameters may have, to some extent, unique genetic underpinnings.

In an association study, Towne et al. (2000) found evidence of the effects of a functional polymorphism in the β-subunit of the luteinizing hormone gene

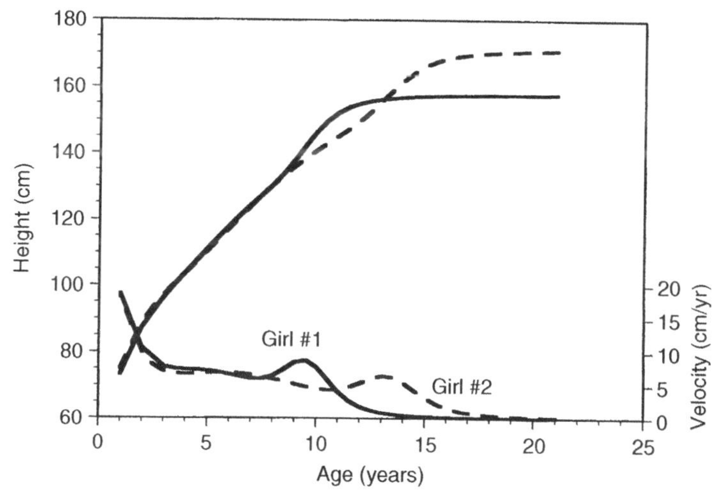

Figure 18.3 Height distance and velocity curves for two girls with different growth patterns between 12 months and 20 years old. *Source*: Towne, Demerath, and Czerwinski (2012). Reproduced with permission of Elsevier.

(LH-β) on stature during childhood. A total of 736 individuals, from 137 nuclear and extended families, measured a total of 13 300 times between the ages of 2 and 18 years, were genotyped for the LH-β polymorphism. Individuals with the less common LH-β allele were shorter than those homozygous for the common LH-β allele at all ages, with this difference steadily increasing with age (e.g. on average, heterozygotes were 0.5 cm shorter than homozygotes at age 2, and they were 2.0 cm shorter at age 18). These results suggest that the LH-β polymorphism is associated with at least modest differences in the stature of children at all childhood ages.

Towne et al. (2008) examined the mid-childhood growth spurt in Fels Longitudinal Study subjects. Presence of a mid-childhood growth spurt was found to have a significant heritability of 0.37 ± 0.14 ($p = 0.003$). From linkage analysis, two QTLs with suggestive LOD scores were found – one at 12 cM on chromosome 17p13.2 (LOD = 2.13), and one at 85 cM on chromosome 12q14 (LOD = 2.06). While the chromosome 17 finding was novel, other investigators have found evidence of linkage or association of growth measures to markers in the same region of chromosome 12.

Fels Longitudinal Study investigators have used multivariate variance components methods incorporating parametric correlation functions to model the heritability and genetic correlational structures of skeletal maturity throughout childhood. For example, a total of 6893 annual hand–wrist radiograph skeletal age assessments, from a sample of 807 children aged 3–15 years representing 192 nuclear and extended families, were simultaneously analyzed. The best fitting model had 65 parameters and allowed for an exponential decay in genetic and environmental correlations as a function of chronological age differences. From this model, the h^2 estimates of skeletal age at each chronological age were: 3 = 0.71, 4 = 0.73, 5 = 0.77, 6 = 0.93, 7 = 0.78, 8 = 0.77, 9 = 0.73, 10 = 0.63, 11 = 0.45, 12 = 0.39, 13 = 0.34, 14 = 0.23, and 15 = 0.11. The genetic correlation matrix showed a pattern of decreasing correlations between skeletal age at different chronological ages as age differences increased (e.g. the ρ_G between skeletal age at age 3 and skeletal age at age 4 was 0.96, but between skeletal age at age 3 and skeletal age at age 15 ρ_G was 0.56). The random environmental correlation matrix showed an even more pronounced pattern of decreasing correlations between skeletal ages at different chronological ages as age differences increased (e.g. the ρ_E between skeletal age at age 3 and skeletal age at age 4 was 0.77, but the ρ_E between skeletal age at age 3 and skeletal age at age 15 was only 0.12). These results show a high heritability of skeletal age through early puberty, and suggest that skeletal maturation at different stages of development is influenced by different sets of genes and environmental factors.

Summary

For over a century, there has been scientific interest in the genetic underpinnings of growth and development. But, as with any area of scientific inquiry, to one degree or another all of these studies were limited by the methods and technologies available to them at the time. For that reason, most of the literature on the genetics of growth and development until relatively recently is limited to h^2 estimates of measures of growth and development gathered mostly from first-degree relatives. The opportunities exist today, however, for far more sophisticated genetic epidemiological studies of growth and development.

One major problem, though, is that modern genetic epidemiological studies of growth and development can be expensive undertakings. Such studies are readily justified, however, on very practical and applied grounds. Foremost among these is that the growth and development of children can have health consequences later in life. Thus, to a large extent, genetic epidemiological studies of growth and development are inherently of biomedical interest. Indeed, much of the current research emphasis in the Fels Longitudinal Study pertains to studies of the relationships between age-related changes in body composition (including those that occur during childhood) and the development and progression of cardiovascular disease (CVD) and type II diabetes mellitus (T2DM) risks in later life, an area of active research today. For example, work by Barker and colleagues in the UK (Barker, 1988), as well as others, suggests that early growth variation in size and body composition (during prenatal as well as postnatal periods) influences the risk of a number of disorders, including hypertension, obesity, heart disease, and diabetes. Longitudinal family studies of growth and development are needed to fully evaluate these hypotheses.

A current Fels Longitudinal Study project is aimed at evaluating the role of birth weight and infant growth in predisposing to adult-onset disease risk factors, taking into account the significant heritable components of both early growth and various adult disease risk factors. Demerath et al. (2000), for example, found that birth weight was negatively associated with fasting insulin concentration in adulthood after adjusting for body mass index (BMI) and age; but, after taking into account the significant heritability of insulin concentration, birth weight accounted for only 1–2% of the phenotypic variance of fasting insulin concentration. Demerath et al. (2007) found significant heritabilities of measures of infant weight and weight change, and Demerath et al. (2009) found significant associations between measures of postnatal weight gain (which had earlier been found to be significantly heritable) and direct measures of adiposity in adulthood (which also are significantly heritable).

In a similar vein, another Fels Longitudinal Study project is examining changes in traditional CVD risk factors during growth and development, and associations between growth-related events and CVD and T2DM risk in adulthood. For example, although serum lipid and lipoprotein levels track from childhood to adulthood, Czerwinski et al. (2000) found differences in the heritabilities of lipid and lipoprotein measures in children sampled pre- and post-puberty. Generally, heritability estimates were higher after puberty, suggesting that the genetic control of lipid and lipoprotein levels may be influenced by maturational factors. Czerwinski et al. (2007) found highly heritable measures of patterns of growth (e.g. timing of the pubertal growth spurt) to be associated, at least on the phenotypic level, with heritable measures of adult disease risk such as blood pressure and BMI. Future studies based on these findings will explicitly explore pleiotropic effects of genes on measures of growth and measures of disease risk in adulthood.

Placing studies of growth and development more squarely in the context of biomedical research will allow auxological investigations to move beyond being descriptive studies, and it will help open the door to the sorts of resources needed to conduct modern genetic epidemiological studies of growth and development in various populations worldwide.

CHAPTER 19

Epigenetics and Human Variation

Amy L. Non[1] *and Zaneta M. Thayer*[2]

[1] *Department of Anthropology, University of California, San Diego, CA, USA*
[2] *Department of Anthropology, Dartmouth College, Hanover, NH, USA*

Introduction

The study of epigenetics has revolutionized our knowledge of human variation, and thus has major implications for the field of genetic anthropology. Epigenetics was first defined by Conrad Waddington in 1942 as the whole set of developmental processes that link the genotype with the phenotype (reprinted in Waddington 2011). Today, epigenetics is commonly defined more narrowly as stable and heritable modifications to the genome that can alter gene expression but do not alter the underlying DNA sequence (Russo et al. 1996). Epigenetic marks have long been recognized as a mechanism responsible for differentiating cell types (Meissner 2010), but only recently has the sensitivity to environmental stimuli been explored. It is now believed that epigenetic changes are influenced both by environmental stimuli and by the variation in the underlying genome. Because epigenetic mechanisms can serve as a bridge between the environment and the genome, study of these processes has helped to break down the false opposition of nature versus nurture and reframe it as "nature via nurture"

A Companion to Anthropological Genetics, First Edition. Edited by Dennis H. O'Rourke.

(Lynch and Kemp 2014). Epigenetic marks on the genome thus represent a new layer of biological complexity of increasing interest to genetic anthropologists.

Types of Epigenetic Modifications

There are three main epigenetic mechanisms: DNA methylation, histone modifications, and noncoding (nc) RNAs, all of which are illustrated in Figure 19.1. *DNA methylation* refers to the attachment of a methyl group (CH_3) to the 5′ carbon of a cytosine in DNA. This occurs via the action of a methyltransferase enzyme, which usually attaches the methyl group via a covalent bond at a cytosine–phosphate–guanine dinuclueotide, termed a *CpG site* (Goll and Bestor 2005). While DNA methylation is a binary state (e.g. each nucleotide can be methylated or not), it is usually measured as an average proportion across DNA

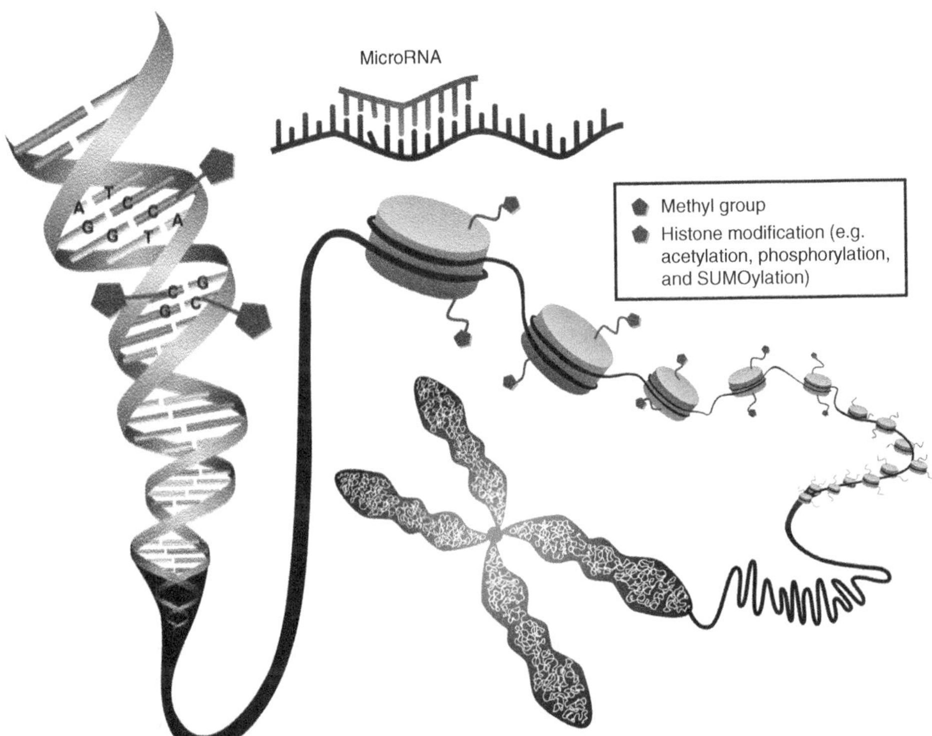

Figure 19.1 Three main types of epigenetic mechanisms. There are many types of epigenetic modifications. Methylation can occur at CG dinucleotides as well as within histones (red hexagons). Histones can also undergo different types of epigenetic modifications (green hexagons). Lastly, microRNA is shown, which is one type of ncRNA that plays a role in posttranslational regulation. Note that the imperfect complement of the miRNA allows each miRNA to bind to more than one target mRNA.

strands and cells in a sample, and is thus represented as a percentage ranging from 0 to 100%. Most CpG sites in the human genome are methylated (70–80%), unless they occur at CpG islands (regions of DNA rich in CpG sites), which are generally unmethylated (Bird 2002).

Chromatin is made up of nucleosomes, each of which contains 147 base pairs of DNA wrapped around a set of proteins called *histones.* Histones are found in four main types: H2A, H2B, H3, and H4. DNA wraps 1.65 times around an octamer of two of each of these histone types (Luger et al. 1997). Each histone has an amino-terminal tail of 25–40 amino acids. Histone modifications include a range of somatically heritable posttranslational changes in which molecules bind to various amino acids in these amino-terminal tails, or on the globular core region (Cosgrove et al. 2004). Among those recognized to date, these modifications include acetylation, methylation, phosphorylation, ubiquitination, SUMOylation, adenosine diphosphate (ADP)-ribosylation, and proline isomerization (Vaquero et al. 2003; Kouzarides 2007; Bannister and Kouzarides 2011). Histone acetylation is the transfer of an acetyl group ($COCH_3$) from acetyl-CoA to the lysine residue on the amino-terminal tails of histones, catalyzed by histone acetyltransferases (HATs). Histone methylation is the addition of a methyl group to arginine or lysine residues from the donor S-adenosylmethionine (SAM) (Kouzarides 2007). Histone phosphorylation is the addition of a phosphate group (PO_4) to a histone tail, catalyzed by protein kinases (Herceg and Murr 2011). Ubiquitination is the attachment of ubiquitin monomers (a 76-amino-acid protein) to a lysine residue on a histone core (Geng and Tansey 2008; Zhou et al. 2009). SUMOylation is the addition of a small (100-amino-acid) ubiquitin-related modifier protein (SUMO) that can attach to histone lysines on all four histone cores (Shiio and Eisenman 2003). ADP-ribosylation is the addition of an ADP-ribose molecule onto a protein, and it can occur on an amino acid acceptor or on an acetyl group (Schreiber et al. 2006). Proline isomerization is the transformation of a proline molecule that can alter the conformational status of a histone (Nelson et al. 2006). All of these histone modifications work together to comprise a "histone code" that can either enhance or repress transcription (Jenuwein and Allis 2001).

The ncRNAs are RNA molecules that do not encode a protein, but act to regulate gene expression (Kaikkonen et al. 2011). These can be divided into two types: long ncRNAS and short ncRNAs. Short ncRNAs include microRNAs (miRNAs), small interfering RNAs (siRNAs), and Piwi-interacting RNAs (piRNAs) (Shrey et al. 2009). miRNAs are small (20–40 nucleotides) single-stranded ncRNA molecules that are responsible for posttranscriptional regulation of gene expression (Figure 19.1). siRNAs, aka silencing RNAs, are double-stranded RNA molecules (20–25 nucleotides) that silence expression of specific genes that carry a complementary sequence. Long ncRNAs (>200 nucleotides) perform important regulatory functions, such as recruiting chromatin remodeling complexes, which help to mediate other epigenetic changes (Mercer et al. 2009). More detail on ncRNAs and their role in epigenetic processes can be found elsewhere (Costa 2008; Kurokawa et al. 2009; Collins et al. 2011).

In addition to the three main mechanisms of DNA methylation, histone modifications, and ncRNAs, a number of other processes are also considered epigenetic. These include DNA-binding proteins, such as transcription factors and methyl CpG-binding proteins, which can regulate gene expression (Ballestar and Wolffe 2001); prions; chromosome position effects; and polycomb mechanisms (Tollefsbol 2011). These processes are less well studied and will not be discussed further in this chapter.

Role in Biological Processes

Epigenetic processes perform complex regulatory functions for the DNA. This primarily occurs by changing the chromatin structure; making it more tightly closed reduces DNA transcription, while loosening the DNA increases transcription. The effect of DNA methylation on gene expression varies based on where in a gene it is located. DNA methylation can reduce expression of a gene when it occurs in the promoter region, where it interferes with binding of transcription factors or by recruiting repressive protein complexes. In contrast, within the gene body, DNA methylation can enhance expression (Jones 2012; Liyanage et al. 2014). It may be that the initiation of transcription, but not elongation, is sensitive to DNA methylation silencing. However, this relationship between methylation in the gene body and levels of expression seems to vary across tissue types (Aran et al. 2011; Zeng et al. 2012). In addition to regulating levels of expression, intragenic DNA methylation may also play a role in alternative RNA splicing (Maunakea et al. 2013).

Histone modifications may change the expression of a gene by altering the electrostatic charge of the histone, or by creating binding sites for protein recognition modules (Bannister and Kouzarides 2011). For example, acetylation of lysine residues is associated with gene activation, potentially because it loosens the bond between DNA and histones by removing the positive charge on the histones, decreasing their interaction with the negatively charged phosphate group of the DNA. Other histone modifications can have multiple effects on expression: for example, ubiquitination has been linked with both activation and inhibition of transcription (Zhang 2003; Laribee et al. 2007).

ncRNAs, such as miRNAs and siRNAs, can regulate gene expression by binding to complementary stretches of DNA and blocking their translation into protein or triggering their degradation (Lai 2005). Because miRNAs only bind by incomplete complementary base-pairing, this allows for one miRNA transcript to regulate hundreds of different genes (Lim et al. 2005). In contrast to the flexibility of miRNAs, siRNAs create a perfect match to a single messenger RNA (mRNA) target, and then cleave it. Thus, siRNAs can only knock down specific genes (the same ones that express them), with rare exceptions (Mack 2007).

Since epigenetic changes help to differentiate cell types, they play a critical role in embryonic development. DNA methylation gets established during embryogenesis, after parental methylation patterns are erased. Specifically, demethylation occurs in both maternal and paternal genomes once upon fertilization and again after specification of primordial germ cells (Meissner 2010). This widespread demethylation causes the vast majority of methylation patterns in our genes to reset at each generation.

An exception to the initial genome-wide demethylation after fertilization occurs at imprinted genes (Morgan et al. 2005). Imprinting is the process in which maternal or paternal alleles are distinguished by epigenetic marks, resulting in expression of only one parent's allele. There are approximately 100 known imprinted genes in the human genome (Bartolomei 2009). These genes have many important effects on growth, development, and behavior. Incorrect silencing of the paternal or maternal imprinted gene can result in severe disorders, as discussed in greater detail in this chapter.

Epigenetic marks can also respond to environmental stimuli, including diet, toxins, stress, and variation in parental care (Thayer and Kuzawa 2011). The most well-known example of epigenetic sensitivity to parental care is a reduction in DNA methylation at the promoter region of the glucocorticoid receptor in the hippocampus in response to licking and grooming of a rat pup during the first week after birth (Weaver et al. 2004). This decrease in DNA methylation is associated with an increase in the expression of the glucocorticoid receptor, thereby enhancing the stress response and reducing lifetime anxiety. In another example, macaques who experience early maternal separation have shown increased methylation of *SLC6A4* in blood cells, lower serotonin transporter expression, and increased stress reactivity (Kinnally et al. 2010). In humans, a twin study of bullying at age 10 demonstrated increased buccal cell DNA methylation of the promoter region of this same gene in bullied twins relative to nonbullied twins (Ouellet-Morin et al. 2013). These studies demonstrate that the epigenome is sensitive to, and may in some cases facilitate, adaptive responses to the environment.

One important complication of epigenetic research is that epigenetic marks are cell and tissue specific. This means that unlike the genome, which is the same across cells, epigenetic profiles vary across cell and tissue types, and in part are responsible for differences in biological functioning across tissues (Ohgane et al. 2008). Practically, this means that measuring the epigenetic profile in one tissue does not necessarily correlate with the epigenetic profile in a different tissue. This is an important limitation when the tissue of interest, for example the brain or liver, is not easily sampled. It is important to remember that the majority of what we know about the human epigenome comes from peripheral tissues (e.g. epithelial cells and blood leukocytes).

Epigenetics and disease

Epigenetic modifications have been implicated in the development of nearly all studied diseases (Egger et al. 2004; Ptak and Petronis 2008). However, the specific role of epigenetic changes in most disease phenotypes is underexplored (Mill and Heijmans 2013). The role of epigenetics is best understood in imprinting disorders, such as Prader–Willi and Angelman syndromes, which occur when one copy of a gene is lost and the second copy is epigenetically silenced (Nicholls 1993). Imprinting disorders are a notable concern for individuals conceived through assisted reproductive technologies due to the difficulties in precisely replicating conditions that facilitate genomic imprinting (Hiura et al. 2014). However, no functionally relevant methylation differences have yet been identified at important imprinting control regions with ovulation induction or *in vitro* fertilization (Rancourt et al. 2012). Given the increase in utilization of assisted reproduction, further research on the epigenetic consequences of these technologies is needed.

The role of epigenetic modifications in cancer development is also fairly well studied. Tumors demonstrate global DNA hypomethylation, which promotes genomic instability and activation of oncogenes (Eden et al. 2003). In contrast, hypermethylation is typically found at tumor suppressor genes and DNA repair genes (Esteller 2003). Epigenetic markers, including DNA methylation and histone modifications, are therefore being explored as biomarkers of cancer risk (Rodriguez-Paredes and Esteller 2011).

Although less is known about the role of the epigenome in complex and chronic diseases, such as obesity and diabetes, there is justified interest in exploring the potential causal role of epigenetic changes, since many environmental exposures that affect these types of disease outcomes also affect methylation. For example, exposure to tobacco smoke and high-fat diets, even prenatally, can influence DNA methylation (Cortessis et al. 2012) and has been associated with cardiometabolic disease risk (Sarrafzadegan and Gotay 2015).

Epigenetic changes may also contribute to the development of psychiatric disorders, including depression and posttraumatic stress disorder, which have both been associated with DNA methylation changes at genes in the hypothalamic-pituitary-adrenal (HPA) axis and the serotonergic system (Booij et al. 2013). In addition, bioaccumulation of persistent organic pollutants and low folic acid levels in pregnancy may contribute to autism spectrum disorders through epigenetic changes (LaSalle 2013).

Since epigenetic mechanisms, such as DNA methylation, are sensitive to the environment, they have been interpreted as a literal biological mechanism of embodiment. This makes epigenetic marks a key target of study for researchers interested in understanding how environments "get under the skin" to influence human biology and patterns of health and disease within and across populations (Hertzman and Boyce 2010). Socially disadvantaged individuals are more likely to be exposed to psychosocial stress, nutritional stress, and toxicants, all of which are linked to the development of altered epigenetic patterns (Thayer and Kuzawa 2011). Epigenetic mechanisms may therefore represent a pathway whereby social inequalities create health inequalities.

As an example, exposure to the toxicant arsenic, which is greater among the rural poor, is associated with changes in DNA methylation measured in maternal blood and umbilical cord blood at birth in studies in Bangladesh (Broberg et al. 2014; Kile et al. 2014). In another example, maternal exposure to warfare and acute violence in the Democratic Republic of the Congo has been associated with epigenetic changes in newborn children (Mulligan et al. 2012). These studies suggest that stressors associated with social disadvantage may be transmitted across generations by altering the quality of the fetal environment and, subsequently, the establishment of epigenetic markers.

Epigenetics in Anthropology and Related Fields

The discovery of epigenetics has had a profound influence on a wide range of anthropological questions about human variation, human evolution, and the influence of the environment on human biology. In the remainder of this section, we discuss how epigenetic concepts have expanded the ability of anthropologists to address these important questions. More discussion on the value of epigenetics in anthropology can be found in Thayer and Non (2015).

Patterns of methylation across human populations

While we are working toward a relatively deep understanding of genetic variation across human populations (Abecasis et al. 2010; Willems et al. 2014), relatively little is known about epigenetic variation. In addition to how much epigenetic variation exists across populations, one important question is: how much of the epigenetic variation is explained by underlying genetic differences, and how much is driven by the environment, or gene-by-environment (G×E) interactions? One study has examined patterns of DNA methylation across lymphoblastoid cell lines in Yorubans and a Northern European population in HapMap (Fraser et al. 2012). They identified population-specific differences at 8475 CpG sites, located within over a third of the 14000 genes assayed. Most of these population differences were very small in magnitude (between 5 and 10%). By examining the correlation between parents and offspring, they determined "heritable" methylation at less than 1000 sites in each population. The authors concluded that these population-specific patterns were likely due to a combination of underlying genetic differences as well as gene-by-gene and/or G×E interactions (Fraser et al. 2012). For example, some genetic variants may affect DNA methylation only in the presence of certain diet-related production of methyl donors (Friso and Choi 2005).

Another study investigated DNA methylation differences in lymphoblastoid cell lines and blood samples between European, African, and Asian Americans using samples from the Coriell Cell Repositories in the Human Variation Panel. The study reported 439 differences out of ~450 000 CpG sites assayed. Among these sites, 178 were located in gene promoter regions, 147 in gene bodies, and 114 in intergenic regions (Heyn et al. 2013). Additionally, enrichment of population-specific CpG sites was found within intragenic enhancer regions and insulator sites, emphasizing their role in regulatory function. Many of these differences were located in genes associated with important biological functions, including drug metabolism, environmental information processing, immune response, growth factors, and appearance (e.g. genes involved in keratinocytes and melanogenesis) (Heyn et al. 2013). One-third of these population-specific CpG sites were not associated with underlying genetic variation, suggesting that these sites may have evolved rapidly in response to selective pressures from varying environmental stimuli. Thus, changes in methylation may facilitate adaptation to local environments, which will be discussed further in this chapter.

In another study comparing genome-wide DNA methylation data across ethnicities, umbilical cord tissue DNA methylation was compared with exome genotyping data across a total of 237 Chinese, Malay, and Indian individuals (Teh et al. 2014). The exome is the part of the genome formed by exons, and therefore it provides an interesting comparison to DNA methylation since the latter is hypothesized to modify transcription. Unlike the exome genotype data, DNA methylation data were not sufficient to separate ethnic groups via principal component analysis. However, 1423 CpG sites varied significantly across the individuals in the study. Many of these methylation differences were linked with underlying genotype differences. Specifically, among the 1423 variable sites, 12 CpG sites contained a genetic mutation within them that directly created or eliminated a CpG site (i.e. disrupting SNPs), while the majority (58%) of the variable CpG sites were located near a SNP on the same chromosome (i.e. *cis*), and 41% were located near a SNP on a different chromosome from the CpG site (i.e. *trans*). Disrupting SNPs and *cis*-SNPs are likely important methQTLs, which are genetic variants associated with methylation levels of a neighboring cytosine. The term *methQTL* is adopted from the term *quantitative trait locus* (QTL), which is a segment of DNA that is associated with a quantitative trait, or a trait that shows continuous variation. MethQTLs are likely important contributors to variation in methylation both between individuals and across human populations (Teh et al. 2014).

In addition to ethnicity, other demographic and environmental factors have been linked with interindividual differences in DNA methylation across humans. Sex differences in DNA methylation have been identified across a range of studies, although most differences are small in magnitude (<5%) (Lam et al. 2012). Additionally, age is associated with DNA methylation (Boks et al. 2009; Lam et al. 2012), and methylation may even be used as a molecular clock to predict age with high accuracy (±3.6 years, when measured across multiple tissues) (Horvath 2013). However, age predictions may be complicated by the fact that DNA methylation is also affected by stress and other environmental factors. The Teh et al. study demonstrated that the variation within the majority (75%) of the 1423 variable methylation sites was best explained by models incorporating G×E interactions, including environmental variables such as maternal age, depression, smoking, folate, and other vitamin levels; no models were best explained by environment alone. In sum, these results suggest that both the genotype and the environmental factors must be considered together as important determinants of population-level differences in DNA methylation levels.

Evolution and the epigenome

Epigenetic mechanisms can influence evolutionary change in many ways. For one, epigenetic marks can change much more rapidly than genetic mutations since they are sensitive to such a wide range of environmental exposures. This increases the chance that a favorable phenotypic trait may arise that enhances organismal adaptation. Second, epigenetic changes can facilitate evolutionary change more quickly because unlike genetic mutations, which may arise in only a few individuals within a population at any given time, epigenetic changes in response to environmental conditions can arise simultaneously among many individuals in a population. This creates the potential for a trait to reach fixation at a much faster rate than is possible through genetic change via natural selection (West-Eberhard 2003). Additionally, DNA methylation can directly increase the rate of genetic change via evolution. This is because methylated cytosine has a higher deamination rate and thus a higher base mutation rate than unmethylated cytosine (Xia et al. 2012). Notably, bacteria exposed to oxidative stress can increase their rate of mutagenesis (Foster 2007), and this may be driven by epigenetic changes. Thus, DNA methylation may in fact be a mechanism to increase mutations in a time of environmental instability.

Recent research on the link between the underlying genome and the epigenome has revealed insights into the phenomena driving the evolution of the epigenome itself. Changes to DNA methylation likely occur as a combination of spontaneous changes independent of the underlying genome, as well as changes that are affected by particular genetic variants (Gibbs et al. 2010). Additionally, repetitive elements in the human genome, such as *Alu* elements, may have an important role in shaping the evolution of the human epigenome (Prendergast et al. 2014). For example, *Alu* elements may be responsible for remodeling local methylation patterns such that when they get inserted into a region, they may act as seeds spreading new methylation to specific CpG sites in the neighboring region (Zhang et al. 2012). Genetic rearrangements may also influence DNA methylation changes (Foerster et al. 2011). However, Prendergast et al. (2014) demonstrated that DNA methylation and chromatin levels are unexpectedly well preserved following segmental duplication events. They also observed that the variation in methylation that does occur between duplicated and paralogous regions was mostly independent of the average rate of divergence of the surrounding DNA sequence, suggesting that the evolution of the methylome may be more complex and not entirely linked to the evolution of the genome.

The evolution of the epigenome can also provide insight into the divergence of hominoid species over time. Given the small genetic differences between humans and chimpanzees, epigenetic changes in regulatory factors may help explain the large phenotype differences between these closely related species. Recent studies have indicated that the evolution of the epigenome closely parallels the evolution of the underlying genome, with 800 genes in whole-blood samples displaying significantly different methylation patterns across wild-born great apes (Hernando-Herraez et al. 2013). Notably, 184 DNA methylation differences were discovered in genes that show no difference at the protein level between humans and chimpanzees. Many of these differences between species are related to genes that influence cognitive ability, bone growth, and human diseases (Fukuda et al. 2013; Schneider et al. 2014). While intriguing, the examination of methylation in DNA from whole blood can only provide an average estimate across mixed cell types that can vary in levels of methylation. In a related study, Zeng et al. (2012) compared DNA methylation in the prefrontal cortex of chimpanzees and humans. They identified very heavy methylation generally across both species' brains, but also in hundreds of sites with significantly lower levels of promoter methylation in the brains of humans relative to chimpanzees (23% lower methylation). Again, many of these differences were located in regions enriched with loci associated with diseases, such as cancer, and neurological and psychological disorders, including autism. Incidentally, the high levels of methylation in the brain may imply that DNA methylation plays a role in constraining evolution of gene expression in this important tissue by reducing *transcriptional noise*, or spurious gene expression (Bird 1995; Suzuki and Bird 2008).

Developmental plasticity

Epigenetics may also represent a mechanism for developmental plasticity. *Developmental plasticity* refers to the evolved capacity of an organism to modify its developmental trajectory in response to environmental exposures during a sensitive period of growth and development. Anthropological evidence demonstrating developmental plasticity was recorded as far back as the early twentieth century, when anthropologist Franz Boas investigated anatomical changes in immigrant children's growth and development after migration to the USA (Boas 1912). Because the US-born immigrant children's cranial form was less similar to their immigrant parents' than that of foreign-born children, his data supported the idea that a phenotype is determined by more than just genes. Instead, the phenotype results from a particular genotype expressed in a particular environmental context. Biological anthropologists have recently advanced this idea of developmental plasticity by investigating biological mechanisms through which environments can change biology. Since epigenetic marks are environmentally sensitive and have the capacity to influence biological form, epigenetic mechanisms are likely important for understanding organismal adaptation. For example, data from the Dutch famine cohort have documented changes in methylation at the *IGF2* gene that are still present 60 years after prenatal exposure to undernutrition (Heijmans et al. 2008). Changes in methylation at this gene could serve to modify growth trajectory, potentially downregulating growth in order to avoid growth faltering or stunting in a low-nutrition environment. As another example, methylation at *NR3C1*, the gene that encodes the glucocorticoid receptor, is modified in adolescence in response to maternal experience of domestic abuse (Radtke et al. 2011). These modifications may lead to durable changes in HPA axis function in order to allow offspring to adjust their stress physiology response in anticipation of how stressful they predict their environment to be. More research is needed to identify additional examples of methylation underlying plasticity. For example, one possibility is to explore the potential role of epigenetic changes in growth rate and age of sexual maturity, which are sensitive to environmental conditions.

Epigenetic mechanisms may have been important for facilitating human adaptation to different environments. In addition to developmental plasticity, epigenetics may actually influence the rate of mutation, as discussed earlier in this chapter, and therefore the potential for genetic adaptation. Epigenetic mechanisms could thus be a useful mechanism for facilitating adaptation to local environments, and they may have facilitated survival of our human ancestors who colonized extremely diverse and extreme environments.

Ancient DNA methylation

With recent advances in ancient DNA (aDNA) techniques, opportunities for exploring aDNA methylation are becoming readily available. Because DNA methylation can be influenced by various types of environmental exposures, studying aDNA methylation may provide clues to the evolution of our hominin ancestors; exposure to disease, violence, or other types of stress; and details about ancient diets. Using techniques of bisulfite sequencing, aDNA methylation data have thus far been recovered from the bones of Late Pleistocene bison (Llamas et al. 2012b), a 38 000-year-old Neanderthal, and a 43 000-year-old Siberian mammoth (Briggs et al. 2010), and even from archeological samples of >2000-year-old barley (Smith et al. 2014). Despite some evidence of cytosine degradation, these samples established the stability of DNA methylation patterns over time. Furthermore, the barley samples demonstrate hypermethylation in response to viral infection, indicating that signatures of disease are identifiable in ancient samples.

Whole-genome aDNA methylation data have also been obtained from limb bones from Neanderthals and Denisovans (Gokhman et al. 2014), and hair from Inuk, a Paleo-Eskimo (Pedersen et al. 2014). These studies have identified ~2000 key regions of difference between humans and archaic hominins that may explain some of the phenotypic differences we see between species, such as at the HOXD cluster, a regulator of limb development. Rather than measure methylation directly, as in bisulfite sequencing, these studies have taken advantage of the natural degradation process of cytosines in the DNA to infer ancient methylation status from standard aDNA data. While unmethylated cytosines degrade into uracils, methylated cytosines degrade into thymines. When the DNA is amplified for aDNA analysis, the uracils get replicated as thymines, introducing C→T mutations into the DNA. If certain DNA polymerases (e.g. Phusion) are used, which cannot replicate uracils into thymines, then the only sites at which these C→T misincorporations may occur is at methylated cytosines (Pedersen et al. 2014). Thus, one can infer aDNA methylation status by measuring the degree of C→T misincorporation events. Additionally, other epigenetic information can be obtained, such as nucleosome occupancy, as packaging of the DNA into nucleosomes can shelter the DNA even after an organism dies. Nucleosome occupancy maps have been created from the hair of Inuk's genome, and they provide insight into gene expression.

These studies demonstrate that epigenetic data are readily recoverable from ancient samples with good preservation. However, there are still many challenges to generating and interpreting ancient epigenetic data, for which further research will be needed. For example, it is difficult to draw conclusions about species differences when only one individual from a hominin species has been studied, particularly given how much the environment can influence methylation patterns. Furthermore, technical problems such as overrepresentation of GC-rich regions during library creation can bias coverage of ancient genomes (Orlando and Willerslev 2014).

Methods Overview

There are many available technologies for assay of epigenetic markers. While a detailed description of methodological approaches is outside the scope of this chapter, a brief overview of the primary methods is presented. For greater detail on methodological approaches to studying epigenetic mechanisms, see Barault and Rancourt (2012), Erkmann (2011), and Non and Thayer (2015).

The first consideration for any epigenetic study is the choice of tissue to analyze, given that large differences exist across tissue and cell types. As anthropologists often strive to collect minimally invasive samples from populations in the field, buccal cells, saliva, and dried blood spots tend to be most appropriate. While all three sample types provide sufficient DNA for assays of DNA methylation, a number of additional factors should also be considered when selecting a tissue. Choice of tissue should ultimately be a balance between the biological relevance of the tissue for the environmental exposure or disease pathway of interest (e.g. blood is preferable for investigation of cardiovascular traits), the heterogeneity of the cell population within the sample (e.g. buccal swabs are primarily epithelial cells, while blood samples provide a mix of white blood cell types), and the quality and quantity of the DNA that can be obtained from the sample (e.g. blood provides much higher quantity DNA than buccal cells, which is preferable if whole-genome assays are planned) (Non and Thayer 2015).

Measuring DNA methylation

The primary methodological approaches for DNA methylation can be divided into locus-specific versus genome-wide approaches. Locus-specific assays target a small genic or intergenic region, while genome-wide approaches analyze thousands to millions of sites throughout the genome. Locus-specific approaches are preferred for those with clear hypotheses about gene regions of interest, and minimal available input DNA. Locus-specific methods include bisulfite pyrosequencing, methylation-specific polymerase chain reaction (MSP), and matrix-assisted laser desorption/ionization time-of-flight (MALDI-TOF). Genome-wide approaches can be performed using microarrays or high-throughput sequencing following a methylation-dependent treatment of DNA, such as bisulfite treatment or immunoprecipitation. The most commonly used methods include the Illumina Infinium HumanMethylation (27K or 450K) microarrays, reduced-representation bisulfite sequencing (RRBS), and methylation DNA immunoprecipitation, combined with sequencing or a microarray (meDIP-seq or meDIP-chip), and whole-genome bisulfite sequencing (WGBS), which is the most comprehensive method. For a detailed comparison of these methods, see Michels et al. (2013) and Non and Thayer (2015).

Measuring histone modifications

There are three primary categories of experiments for measuring histone modifications. The first consists of experiments to measure nucleosome occupancy, which is a measure of the condensation of the chromatin structure. To measure nucleosome occupancy, cells can be treated with a DNA methylase in order to add a methyl group to cytosines that are not within the histone core. Thus, one can distinguish the presence of histones based on differences in methylation patterns (Fatemi et al. 2005). For a review of methods on genome-wide mapping of the chromatin and nucleosome structures, see Zhang and Pugh (2011).

The second category consists of experiments to measure direct modifications to the histones themselves. The most commonly used method to measure histone modifications is chromatin immunoprecipitation (ChIP). Immunoprecipitation is a process by which cells are incubated with antibodies that target a specific modification after crosslinking the histones to the DNA (Collas 2010). After enrichment for DNA associated with histone modifications, the DNA can be paired with deep sequencing (Chip-seq) or with microarrays (Chip-chip) (Pellegrini and Ferrari 2012; Mundade et al. 2014). Chip-seq allows for genome-wide mapping of transcription factor binding sites, the distribution of histone methylation, acetylation, SUMOylation, and phosphorylation (Meyer et al. 2015).

The third category of experiments can measure chromatin conformation. The chromosome conformation capture assay can be used to quantify physical interactions between genomic loci, even when located over hundreds of kilobases apart (Dekker et al. 2002). More recently, this technique has been adapted for genome-wide analysis of chromatin conformation (Lieberman-Aiden et al. 2009).

Measuring noncoding RNAs

Global RNA profiling can be performed using quantitative real-time polymerase chain reaction (PCR) or with microarrays. Comparison of these techniques can be found in Ach et al. (2008). Additionally, high-throughput sequencing of RNA libraries can now be performed via RNAseq, aka whole transcriptome shotgun sequencing (WTTS). Library preparation for RNAseq usually entails converting RNA into complementary DNA with reverse transcriptase. For an overview of methods and platforms available for RNA sequencing, see Chu and Corey (2012) and Wang et al. (2009).

Current Debates in Epigenetics Research

With the growing interest in epigenetics across the biological and social sciences, a number of interesting theoretical debates have arisen. Some researchers view the potential for transgenerational inheritance of epigenetic marks as a reversion to a Lamarckian model of the inheritance of acquired characteristics (Burggren and Crews 2014). Some even argue that the evolutionary synthesis must be extended to account for these new forms of evolutionary change (Jablonka and Lamb 2005; Danchin et al. 2011). Others argue that epigenetics relies on inheritance of underlying genetics, and thus no extension of the evolutionary synthesis is necessary (Dickins and Rahman 2012).

There are now real examples of epigenetic inheritance through the germline in plants, nematodes, and fruit flies (Cavalli and Paro 1998; Greer et al. 2011; Hauser et al. 2011; Seong et al. 2011). There have even been reports of mammalian examples of transgenerational inheritance, such as transmission of natural methylation states at the *Agouti* locus, which changes coat color in mice (Morgan et al. 1999), although intriguingly the diet-induced changes in coat color do not appear to be transgenerationally inherited (Waterland et al. 2007). Regardless, this type of inheritance is much more difficult to demonstrate in humans due to our long generation time. Some studies claim to demonstrate transgenerational inheritance of epimutations in humans at specific loci (e.g. *MLH1*) (Morak et al. 2008). However, there is a counterargument that many of the epigenetic marks that appear to be transgenerationally inherited may in fact be caused by non-epigenetic forces, such as undetected mutations in the DNA sequence, behavioral changes (e.g. parental programming), alterations in the microbiome, or intergenerational transmission of metabolites (Heard and Martienssen 2014). Furthermore, some of the strongest evidence for transgenerational inheritance in mammals has been called into question, with many experiments failing to replicate (Kaiser 2014). We suggest there is not yet evidence to support a Lamarckian model of transgenerational epigenetic inheritance in mammals. Longitudinal studies across generations are necessary to fully disentangle the heritability of genetic, epigenetic, and environmental effects.

Another open question regards the interpretation of the epigenome as flexible versus stable. Epigenetics was initially lauded as a mechanism to counter traditional genetic determinism because the epigenome can be modified throughout one's lifespan. This idea was popularized in the *Time* magazine cover, "Why Your DNA Is Not Your Destiny" (Cloud 2010). Despite this view, there are others who fear that epigenetic data may become viewed as deterministic. In one example, there is a danger of reifying biological notions of race based on a new nongenetic system, in which racial groups could be classified based on epigenetic differences (Meloni 2014a). While epigenetic marks demonstrate both stability and modifiability to varying extents, under which conditions and how these play out across the lifespan are not yet known. Furthermore, there exists an inherent contradiction in the interpretation of the epigenome as a passive receiver of environmental insults versus a set of actively modifiable marks that can be manipulated with social interventions, with altered diet and exercise, or even pharmaceutically (Meloni 2014b). These oppositional interpretations are each extreme views that do not take into account the complex interplay of the environment, the genome, and the epigenome, which all act together to produce a phenotype that can change across time.

One unintended consequence of epigenetics research is an intense focus on how maternal biology and behavior affect the health of children. Consequently, the entire burden of creating a healthy prenatal (and postnatal) environment gets placed on mothers, who in turn get blamed for any adverse health outcomes of their children. In light of this tendency to hold women responsible for all potential and actual biological and behavioral decisions that may influence their children, it is important to recognize that broader structural factors shape and constrain a woman's choices and opportunities. For example, women struggling with poverty may not have the ability to maintain a healthy diet or exercise habits during pregnancy, or may be exposed to high levels of stress or toxins in the domestic environment that are outside of their control. Thus, anthropological researchers have an obligation to emphasize these broader political and socioeconomic factors that influence the larger environment of the developing child. Furthermore, the behaviors of the father should not be neglected, as paternal interactions with the child can influence development, and the epigenome of the sperm may also have a direct effect on the health and development of children (Rando 2012). Finally, in addition to blaming mothers, there is also the risk that entire racial groups or social classes may be blamed for behaviors that expose their children to harmful environments (Meloni 2014a). Given these numerous concerns for misinterpretation, anthropologists interested in working with epigenetic data should be mindful of how they communicate their results with the public.

Conclusions

In conclusion, epigenetic research has the potential to transform our understanding of human diversity, as well as provide new insight into our knowledge of human evolution. Genetic anthropology is uniquely positioned to be at the forefront of this exciting new discipline, given its position at the intersection of the genome and the environment. We expect to see an explosion of epigenetics research among the next generation of genetic anthropologists, and are excited for the discoveries to come.

PART V Anthropology, Genetics, and the Evolution of Health and Disease

CHAPTER 20

The Primate Immune System: A Survey of Innate and Acquired Immune Genes

Leslie A. Knapp and Nicole S. Torosin

Department of Anthropology, University of Utah, Salt Lake City, UT, USA

Introduction

Common ancestry and overall genome similarity suggest that the human and nonhuman primate immune systems should exhibit a high degree of homology with respect to shared genes and immune response. While we do share the majority of our genetic makeup with our primate relatives, studies have shown that humans are more susceptible to HIV, cancer, hepatitis, and other viral infections when compared to apes and monkeys. Countering this, however, are findings that nonhuman primates are highly susceptible to some diseases that are not deadly to humans, such as respiratory viruses (Williams et al. 2008), metapneumovirus (Kaur et al. 2008), herpes virus (Eberle and Hilliard 1989), and infections like *Staphylococcus aureus* (Nagel et al. 2012) and *Mycobacterium tuberculosis* (Walsh et al. 1996). Thus, disease susceptibility and pathogenesis of some infections may be very different, even when nonhuman primates share homologous immune genes with humans.

A Companion to Anthropological Genetics, First Edition. Edited by Dennis H. O'Rourke.

After several decades of research, it is clear that gene activity and expression are two major factors that explain important differences in disease susceptibility and response. Other factors such as evolutionary history and exposure to diseases have also shaped the way the immune system works in primates. By studying molecular, genetic, and functional processes using DNA, RNA, and immune cells of model nonhuman primate species, we are gaining a richer appreciation of the primate immune system. Nevertheless, there is much more to learn about other primates besides those that are typically used as model organisms (i.e. chimpanzees, rhesus and crab-eating macaques, and marmosets).

Like other mammals, humans and nonhuman primates have two kinds of immune responses that are induced by pathogens. The innate immune response is nonspecific and is activated almost immediately after infection. Innate immune mechanisms include physical boundaries such as skin, chemicals in the blood, and immune cells that are activated by the chemical characteristics of antigens. The adaptive immune responses are more complicated and are antigen specific. An adaptive response requires processing and recognition of an antigen, followed by the production of immune cells that attack the foreign antigen. Typically, adaptive responses retain memory of the foreign antigens in preparation for future re-exposure to specific antigens.

The Innate Immune System

The innate immune system is the initial line of defense against pathogens. It relies on mechanisms of pathogen recognition, which include identification of microbial nonself, recognition of missing self, and recognition of modified self (Medzhitov and Janeway 2002). Interferons and interleukins signal proteins that play a key role in stimulating the immune response. Several types of interferons are known to recognize viruses and upregulate adaptive immune responses. Similarly, interleukins play an immunomodulatory function in the presence of virus. Macrophages and neutrophils are inflammatory cells and are also part of the innate immune system.

Interferons and Interleukins

An interferon is an important modulator of the immune response. It is one of several proteins produced as a defense against viruses, impacting the quality of the host's cellular immune response and amplifying antigen presentation to specific immune cells. There are at least two basic types of interferon. Type I interferons – IFN-alpha (IFN-α), IFN-beta (IFN-β), and IFN-omega (IFN-ω) – bind to specific cell surface receptors. The type II interferon, IFN-gamma (IFN-γ), is mainly secreted by (NK) cells, macrophages, and T cells. Molecular genetic studies of most human IFN-α subtypes and IFN-γ reveal that these molecules evolved under strong purifying selection, probably due to their essential function in immunity (Manry et al. 2011).

Comparative studies of IFN-α in humans and other mammals suggest species-specific sequence patterns, except when nonhuman primates are included in the analyses. When the IFN-α (e.g. IFN-α1 and IFN-α13, or IFN-α4 and IFN-α10) genes of humans and chimpanzees are compared, a pattern of gene duplication prior to divergence from a common ancestor is observed. When rhesus macaques are included in the analyses, there is also evidence for gene conversion following gene duplication (Woelk et al. 2007). Due to extreme similarity, it is difficult to make direct comparisons between the human and most nonhuman primate interferons (Steinmann, Rosenkaimer, and Leitz 2013). Since humans and nonhuman primates are more similar to each other overall than they are to other mammals studied thus far, researchers in this field are calling for more comparative studies on nonhuman primate IFN genes (Woelk et al. 2007) to gain a greater understanding of interferons and the way they affect primate immune response.

Over the course of many decades, nearly three dozen interleukins (ILs) have been identified in humans. ILs are molecules produced by leukocytes, which are generic white blood cells in the body. Interferons are also produced by other cells to serve as mediators between leucocytes. Classification of the IL gene family can be made on the basis of characteristics that include biological function, receptor-binding properties, sequence homology, common structural motifs, and/or the cell types responsible for IL production. ILs have special roles in modulating immune response (Akdis et al. 2016). Interestingly, a large number of ILs share common features, in terms of either cross-reactivity to human antibodies and/or DNA sequence, in humans and nonhuman primates (e.g. macaques, baboons, guenons, and New World monkeys like marmosets and night monkeys) (Höglind et al. 2017). In recent years, research has suggested that IL3, or a nearby gene, in particular has undergone positive selection in humans and other primates (Li et al. 2016).

Macrophages, neutrophils, and dendritic cells

Macrophages and neutrophils are inflammatory cells and are also part of the innate immune system. These cells initiate the adaptive immune response. Immune cell subsets such as natural killer (NK) cells, dendritic cells (DCs), and Toll-like receptors (TLRs) also work as part of the innate immune response and play critical roles in mediating other innate immune responses. Recognition of microbial entities relies on the detection of conserved molecular patterns referred to as *pathogen-associated molecular patterns* (PAMPs).

Macrophages, neutrophils, and DCs are capable of engulfing, or phagocytosing, pathogens and segments of pathogens. Macrophages engulf pathogens with their large plasma membrane. They can deal with large numbers of pathogenic cells and thus are considered very efficient when attacking and inactivating bacteria, invading cells, or microbes. Neutrophils use cellular granules made up of toxic substances that kill or inhibit growth of bacteria. Neutrophils are the most abundant type of phagocyte, representing more than half of the total circulating leukocytes (Kolaczkowska and Kubes 2013). Neutrophils are usually the first cells to arrive at the site of an infection. Like macrophages and neutrophils, DCs are also capable of phagocytosis. These cells are typically in an immature state when found in peripheral blood. In tissues that work as barriers – such as skin, nose, and lung – DCs serve as an intermediate between the innate and adaptive immune systems (Iwasaki and Medzhitov 2015).

Given the importance of macrophages and other phagocytosing cells in regulating immune processes in health and disease, it should not be surprising that research has been undertaken to study nonhuman primate innate immune cells and to determine what differences exist when compared to those of humans. Much of this research has been motivated by the need to develop primate models for studying HIV/SIV infection, disease progression, and treatment. Based on studies of mice and humans, we know that macrophages differ in longevity, in their role in immune response, and in their tissue distribution (Gordon and Plüddemann 2017), but greater insights based on studies of nonhuman primates are likely to improve our understanding of human macrophage activity.

Due to the close genetic similarities between humans and nonhuman primates, many antibodies for humans cross-react with nonhuman primate cells, and flow cytometry can be used to identify and compare macrophages in humans and other primate species. Using comparative flow cytometry, researchers have noted variation across tissue types (Sopper et al. 1997) in rhesus macaques and variation when healthy and SIV-infected macaques are compared (Ortiz, Dinapoli, and Brenchley 2015). One tissue type of particular interest to virologists, alveolar macrophages (AMs), has been a major focus of research due to their role as targets of HIV/SIV infection (Jambo et al. 2014). Comparisons of humans with crab-eating macaques show that AMs have similar activity, but human AMs are larger in diameter and more than twice the volume of macaque AMs. This discovery may explain why there are species-specific differences in the size range of foreign particles that can be phagocytized and cleared by AMs (Krombach et al. 1997).

Comparative studies of DCs in humans and nonhuman primates are complex due to issues related to antibody cross-reactivity and subpopulations of the different DCs. The study of DCs in African green monkeys (Mortara et al. 2006) and rhesus macaques has been going on for more than a decade. In all primate species studied to date, DCs are a heterogeneous population involved in antigen presentation (Ju, Clark, and Hart 2010). Significant differences in phenotypic expression of receptor molecules on DCs is observed when humans and nonhuman primates are compared (Autissier et al. 2010).

Natural killer cells

Natural killer (NK) cells are the primary effector cell of the innate immune system (Carville, Evans, and Reeves 2013). They are large, granular, bone marrow–derived lymphocytes (Mandal and Viswanathan 2015). NK cells are well known for their cytotoxic potential used to destroy infected cells and their ability to release large quantities of cytokines, such as interferon, to stimulate maturation of more DCs and macrophages (Vivier et al. 2008). NK cells are controlled by inactivationg receptors, which help them discriminate between self and nonself. Healthy cells expressing major histocompatibility complex (MHC) class I molecules, inhibit NK cells (Parham et al. 2010). Cells that have been infected by virus downregulate MHC class I molecules, allowing NK cells to remove them (Mandal and Viswanathan 2015). Lack of MHC class I expression in tumor cells allows NK cells to target those as well. Despite the fact that the function of NK cells is fairly well characterized in humans, and some research has explored the role of human NK (hNK) cells in diseases other than cancer, an understanding of their role in health and disease is generally lacking (Mandal and Viswanathan 2015). Parham and Moffett (2013) argue that the human NK cell system is complicated by competing selection pressures that relate to their dual immunological and reproductive functions. NK cells have been found to promote angiogenesis in the placenta during pregnancy (Kopcow and Karumanchi 2007), which is counterintuitive given that normal function involves placental expression of some nonself MHC molecules. To date, the interrelationship between immunity and pregnancy in humans remains puzzling.

Relatively little research has been done to characterize nonhuman primate NK cells, despite the popularity of using nonhuman primates as models for disease. In recent years, there have been calls for the study of NK cells in nonhuman primates, especially where specific tissue can be obtained to examine more than just NK cells derived from peripheral blood, as is the case for most studies of hNK cells (see Hong et al. 2013). Using peripheral blood, lymph node tissue, and vaginal and rectal biopsies from rhesus macaques, Reeves et al. (2010) found that rhesus monkeys produce classical NK cells with the same markers found in humans (e.g. CD56 and CD16). They also found greater NK cell phenotypic heterogeneity than has been reported in humans. Similar results have been reported for sooty mangabeys (Pereira, Johnson, and Ansari 2008). Research on New World monkeys (such as common marmosets, cotton-topped tamarins, and squirrel monkeys) suggests that some functionality of specific NK cell subpopulations may be conserved in New and Old World primates and humans. However, some NK cell subsets shared by Old World and New World monkeys are not present in humans (Carville, Evans, and Reeves 2013).

Compared to NK cells, more research has been done on killer-cell immunoglobulin-like receptors (KIRs), which are expressed on the plasma membranes of NK cells. KIRs serve an important role in balancing effective defense and self-tolerance through NK inhibition. Like many immune genes, KIR molecules are highly polymorphic, meaning that their gene sequences differ greatly between individuals. KIRs are also polygenic, so that it is extremely rare for two unrelated individuals to possess the same KIR genotype. When KIR genes from different organisms are compared, it appears that KIRs evolve in a species-specific manner that may reflect unique pathogen pressure for the species under study (Guethlein et al. 2007). For many years, it was thought that KIRs did not even exist in species other than humans. However, recent research has shown that KIR-like sequences exist in many other mammals, including mice and rats (Hoelsbrekken et al. 2003).

Earlier studies of KIRs in nonhuman primates suggested that the receptors were limited to humans and apes, specifically common chimpanzees. It was found that some KIRs from humans could functionally recognize chimpanzee MHC class I molecules and that chimpanzee KIRs could do the same in the presence of human MHC-A and -B (class I) molecules (Khakoo et al. 2000). These results were compatible with the knowledge that humans and chimpanzees share similar MHC class I genes.

In humans, KIR haplotypes are characterized by the number of KIR genes on a single haplotype (between 6 and 12 loci) and by the number of genes inhibiting or activating KIRs. Studies of nonhuman primate haplotypes have helped to characterize KIR lineages and ancestral haplotypes throughout the primate order. One study of pygmy chimpanzees (*Pan paniscus*) identified nine KIR complementary DNA (cDNA) sequences, representing seven different KIR genes. When compared with humans and common chimpanzees, the *P. paniscus* sequences belonged to one of three KIR lineages previously identified, and they were more similar to common chimpanzee KIRs than to human KIRs (Rajalingam et al. 2001).

Over more than a decade, comparative studies have focused on studying KIR haplotypes to understand KIR gene and haplotype evolution in primates. Rajalingam et al. (2001) found that pygmy chimpanzees have a "short" KIR haplotype consisting of just three loci. These loci may be orthologous, or perhaps paralogous, to the three genes conserved in human KIR haplotypes. Future studies of nonhuman primate KIR haplotypes have supported the argument that the "short" haplotype represents the basic KIR haplotype framework found in all primates (Sambrook et al. 2006; Guethlein et al. 2007).

Based on cDNA sequences and *KIR* haplotypes, Sambrook et al. (2005) demonstrated that the orangutan KIR haplotype also resembled human and chimpanzee *KIR* haplotypes. Further study of orangutans found that the typical primate KIR haplotype included three major lineages of KIR genes, and the framework itself could serve as a template for duplication and expansion of KIR genes in primates (Guethlein et al. 2007). Studies of rhesus macaques and common marmosets show that KIR molecules evolve rapidly but within the limitations of MHC class I polymorphism (Hershberger et al. 2001). For example, common marmosets have limited MHC diversity, and, as a consequence, their KIR molecules reflect their unique pattern of MHC class I evolution. Humans, apes, and Old World monkeys have a wider range of MHC class I genes and polymorphisms, and therefore they are expected to have a more diverse set of KIR receptors (Carville, Evans, and Reeves 2013).

Toll-like receptors

Toll-like receptors (TLRs) are a set of ten proteins of the innate immune system encoded by evolutionarily conserved genes. Mice have TLRs 1–9, which are functionally equivalent to those of primates. This set of genes initiates the first response of the host to invading pathogens (Quach et al. 2013). Recognition occurs through nonspecific germline-encoded pattern recognition receptors (PRRs). The PRRs recognize broad categories of pathogen components called pathogen-associated molecular patterns (PAMPs).

TLR molecules can be broken up into two sets. TLRs 1, 2, 4, 5, 6, and 10 are located on cell membranes. They are responsible for recognizing bacterial, fungal, and parasitic cellular invaders through PAMPs such as lipopolysaccharide, peptidoglycan, and flagellin. The second set of TLRs (3, 7, 8, and 9) reside in the endosomal compartment membrane. These TLRs recognize single- and double-stranded RNA, and CpG DNA (Wlasiuk and Nachman 2010).

All the TLR proteins have transmembrane domains containing leucine-rich repeats that are responsible for the recognition of the PAMPs (Kawai and Akira 2010). Their intercompartmental domain is called the TIR domain, which recruits adaptor molecules upon activation. Upon activation, TLR2 heterodimerizes with either TLR1 or TLR6. The other TLR proteins 3, 4, 5, 7, 8, and 9 homodimerize. All of the TLR complexes except TLR3 recruit adaptor molecule MyD88; TLR 2, 4, and 6 recruit an additional adaptor molecule, TIRAP; and TLR3 recruits TRIF. While each pathway contains its own nuances, ultimately all of them will activate NF-κB, which will initiate the transcription of pro-inflammatory cytokines (Kawai and Akira 2010; Kumar, Kawai, and Akira 2011). Additionally, based on the PAMP each TLR recognizes, bacterial or viral, each will be expressed in a particular type of dendritic precursor cell. TLRs 1, 2, 4, 5, and 8 are primarily expressed in monocytes that secrete antibacterial compounds, such as TNF-alpha and IL6 upon activation. TLR7 and TLR9 are expressed in plasmacytoid pre-dendritic cells that secrete antiviral compounds such as interferons, which activate NK cells to destroy virus-infected cells (Kadowaki et al. 2001).

Since the TLR signaling pathways were discovered and characterized, they have been intensively studied (Kawai and Akira 2010). TLRs are the first line of defense against pathogens, making everything from their DNA sequence to variable expression of the pathway a strong candidate for pathogen susceptibility and immune system variability. Many studies have explored polymorphism linkage in humans to complex diseases. Polymorphisms in seven different TLRs have been correlated with increased risk for various types of cancer. Other polymorphisms in various TLR7s have been linked to increased viral and bacterial disease susceptibility from tuberculosis to HIV-1 (Trejo-de la O, Hernández-Sancén, and Maldonado-Bernal 2014).

TLR pathway expression differences and the effect on pathogenesis of complex diseases and autoimmune disease have been studied. Increased expression of TLR7 and TLR9 have been found to increase the likelihood of systemic lupus erythematosus. Low-level activation of the viral and bacterial TLRs by commensal gut organisms affects the ability of the TLRs to fight invading pathogens (Wilks and Golovkina 2012).

Nonhuman primate TLR studies are typically undertaken to better comprehend the evolution of the human innate immune response. Functional studies have compared TLR activation in humans and nonhuman primates. Overall, TLR responses are conserved in humans, chimpanzees, and rhesus macaques, but the regulation of the response due to viral immune challenge is lineage specific. This may reflect the necessity for a quickly adapting host response to rapid viral mutation (Barreiro et al. 2010).

Beyond gaining a general understanding of TLR immune response in primates, most other studies focus on identifying positive and purifying selection in primate TLRs in comparison to humans. There is evidence for primate lineage-specific positive selection. Overall, the endosomal TLRs are under stronger purifying selection than the cell surface TLRs in humans. However, when comparing selective landscapes in humans, gorillas, and chimpanzees, all three exhibit evidence for positive selection in TLR1, TLR6, and TLR10. The great ape TLRs are under stronger selective constraints than those in humans, however. Additionally, the endosomal TLRs are not exclusively under purifying selection pressure in great apes, as has been reported for humans (Quach et al. 2013).

The Adaptive Immune System

The adaptive immune system differs from the innate immune system in that adaptive immunity reflects tailored responses to specific pathogens. Adaptive immune responses are also known as acquired immune responses. The adaptive immune system relies on just two types of cells: B cells and T cells. B (McHeyzer-Williams et al. 2012) and T cells (Sprent and Surh 2002) have the ability to create memory cells that defend against future attacks by the same pathogen. This allows the B and T cells to produce stronger and faster adaptive immune responses against previously encountered pathogens before symptoms of infection arise. Both B and T cells are lymphocytes derived from stem cells in the bone marrow, but once they depart the bone marrow, they are significantly different in many ways. They differ in their structure, the genes that encode them, and the type of molecules they can bind.

B cells

B cells are also known as *B lymphocytes.* They represent only about 20% of the lymphocytes circulating in our blood. They have a short lifespan, and the antibodies on their surface bind viral or bacterial antigens immediately. When they initially move out of the bone marrow, they are immature (known as naïve) and have millions of surface antigen-specific receptors (Mauri and Bosma 2012). The B cell receptor is composed of an immunoglobulin (i.e. IgA, IgD, or IgE) and a B cell-specific transmembrane molecule. All immunoglobulins have the same general structure, but the tip of the immunoglobulin is extremely variable. When an antigen that fits the special receptor is encountered, a B cell will be stimulated to undergo a maturation process that leads to cell division. The product is another B cell that is like the naïve cell and also a plasma B cell that secretes antibodies that can bind to the antigen again directly (Mauri and Bosma 2012). Once the B cell binds the foreign antigen, it digests it to form fragments that are displayed on the cell surface of an MHC class II molecule.

T cells

T cells make up around 80% of the lymphocytes circulating in blood. They have a longer lifespan than B cells and possess a characteristic T-cell receptor (TCR) on the cell surface. The T cells undergo a process of rearrangement to increase their ability to bind diverse foreign antigens (Smith-Garvin, Koretzky, and Jordan 2009). TCRs are part of the immunoglobulin superfamily, and thus diversity is established in a way similar to antibodies and B-cell antigen receptors. TCRs are formed by recombination of three critical segments (**V**ariable, **D**iversity, and **J**oining), and this allows for the creation of many hundreds of unique receptors to antigens (Alt et al. 1992). Importantly, however, TCRs can only recognize antigens bound to certain receptor molecules, the MHC molecules. Therefore, mature T cells should recognize only foreign antigens combined with self-MHC molecules in order to mount an appropriate immune response.

Major histocompatibility complex

Major histocompatibility complex (MHC) molecules serve as receptors that bind fragments of foreign antigens, or peptides, in a peptide-binding region (PBR) (Bjorkman et al. 1987a; Brown et al. 1993). Infected cells signal their condition by displaying a small peptide from the infective agent in the PBR at the cell surface. The infected cell is then recognized as different from other, non-infected cells in the host. Killing the infected cell and activating appropriate immune response cells for antibody production prevent spread of the pathogen. If MHC molecules fail to bind or present some peptides derived from foreign proteins, then an effective immune response cannot be achieved and further infection is likely.

MHC genes fall into two basic classes. Class I genes encode proteins that are expressed on the surface of nearly all nucleated cells in the body. Most class I genes encode highly polymorphic molecules that bind peptides in the antigen recognition site and present them to cytotoxic T cells, triggering destruction of the target cell (Klein 1986). A subset of class I genes, the nonclassical MHC genes, exhibit limited polymorphism, plus more restricted tissue distribution and expression than the classical MHC molecules (Figueroa et al. 1994).

The second basic class of MHC genes are known as class II and they encode proteins expressed on the surface of B cells. MHC class II genes appear to be very ancient and orthologous in all primates and even most vertebrates. Both alpha (α) and beta (β) genes combine to form a single molecule expressed on the surface of immunocompetent cells. α and β genes vary between species and can even vary between individuals of the same species. As in the case of class I genes, knowledge about the exon and intron structure and patterns of nucleotide substitution of class II genes has helped to clarify the function of these loci.

The MHC has been most intensively studied in humans. In humans, it is also called human leukocyte antigens (HLAs). Within this gene-rich complex, classical HLA genes – HLA-A, HLA-B, and HLA-C – are the highly polymorphic molecules that bind peptides in the antigen recognition site and present them to cytotoxic T cells. Nonclassical HLA genes are known as HLA-E, HLA-F, and HLA-G. Human class II loci include HLA-DPα and HLA-DPβ, HLA-DQα and HLA-DQβ, and HLA-DRα and HLA-DRβ. In humans, different numbers of α and β genes vary between individuals. Only one HLA-DRα exists in humans, but four different HLA-DRβ genes (DRβ1, DRβ3, DRβ4, and DRβ5) can pair with the α to form a complete DR molecule. Other class II loci in humans include two functional DPα genes, one DPβ gene, and one each of the DQα and DQβ genes.

All primates possess MHC class I and II genes, and research suggests that most genes are orthologous, specifically homologous by descent from a common ancestral locus (Klein 1987). Given their common evolutionary origins, it should not be surprising that there is a great deal of structural similarity among MHC genes. Studies of the exon and intron structure of MHC molecules show that all regions of the gene do not exhibit the same patterns of nucleotide diversity, but these patterns of nucleotide substitution reveal important insights about the function of, and evolutionary pressures on, these genes. It should be noted, however, that shared sequence motifs may also exist as a result of convergence, rather than inheritance from a common ancestor (Klein 1987). In these cases, seemingly orthologous genes might not share common ancestry.

The high frequency of gene duplication events within the MHC has resulted in the presence of many paralogous genes in primates. Paralogous genes also share a common ancestor, but they are confined to members of a single species and are typically clustered together in a genome. These physical arrangements provide abundant opportunities for gene conversion and interlocus recombination.

Importantly, gene conversion, which is often referred to as ***concerted evolution***, homogenizes paralogous sequences and makes it especially difficult to study one locus at a time in closely related primate species.

Detailed surveys of the MHC have been undertaken in many primate species over the decades (de Groot, Bontrop, and Doxiadis 2017), and sophisticated laboratory methods have yielded comprehensive full-length complementary DNA (cDNA) sequences for some species (Hans, Bergl, and Vigilant 2017). Comparative studies of the differences in the number and variety of MHC loci among nonhuman primates have been undertaken to evaluate evolutionary relationships with HLA genes. Orthologs of each HLA class I locus have been identified in chimpanzees (Wroblewski et al. 2015) and gorillas (Adams, Cooper, and Parham 2001), but differences from humans have been noted. For example, gorillas possess an A-like ortholog and also an additional, presumably functional, A-like locus characterized by an unusual haplotype structure (Lawlor et al. 1991). Research on MHC class I gene intron variation suggests that chimpanzees have experienced a severe repertoire reduction, perhaps as a consequence of a "selective sweep" caused by SIV-causing pathogens (Doxiadis et al. 2016). This may be just one factor explaining differences between human and great ape class I A genes.

The most polymorphic MHC class I locus in humans is HLA-B, and the B loci of common and pygmy chimpanzees, as well as gorillas, are polymorphic like those of humans (de Groot et al. 2012). The human class I gene HLA-C is closely related to HLA-A and HLA-B, yet it demonstrates a lower degree of polymorphism and cell surface expression than either A or B (Colonna et al. 1993). Genes orthologous to HLA-C have been identified in the common and pygmy chimpanzees, and in gorilla; multiple copies have been described for orangutans (Adams et al. 2000; Adams, Thomson, and Parham 1999). While the type of selective forces maintaining primate MHC-C is not immediately apparent, the simplest explanation would be that this locus confers a selective advantage against specific pathogens due to its different peptide-binding affinities.

The nonclassical class I genes, HLA-E, -F, and -G, differ from the classical genes in that they are nearly monomorphic and expressed on a more restricted set of cell types. Two nonclassical class I genes, MHC-E and -F, are highly conserved among all primate taxa (Boyson et al. 1995; Knapp, Cadavid, and Watkins 1998), suggesting that they are under different selective pressures than classical class I genes (Adams, Cooper, and Parham 2001).

Old World monkeys, such as rhesus macaques (*Macaca* spp.) and baboons (*Papio cynocephalus*), possess classical and nonclassical MHC class I genes. Studies by Otting et al. (2007) show that macaque MHC-A loci are extremely old. Macaques and baboons have several expressed A-like loci (Boyson et al. 1996; Sidebottom, Kennedy, and William 2001). Surveys of rhesus macaque (*Macaca mulatta*) Mamu MHC indicate the presence of at least three B loci, with other haplotypes containing different numbers of B loci (Boyson et al. 1996). Research has shown that all Mamu-B alleles are not expressed at equivalent levels (Otting et al. 2005). Nonclassical MHC-F (Otting and Bontrop 1993), MHC-G (Boyson et al. 1996), and MHC-E (Knapp, Cadavid, and Watkins 1998) loci have close homologs in *M. mulatta*.

Among New World monkeys, there is generally a limited number and variety of MHC class I loci (Evans et al. 1999). For the most part, functional loci resemble orthologs of HLA-G (Cadavid, Mejia, and Watkins 1999) and HLA-B (Lugo and Cadavid 2015). Limited variability increases the likelihood of finding more similar DNA sequences within and between species. There is one case of an identically functional MHC class I molecule reported for two sympatric callitrichid primate species (Evans et al. 1998). Processed MHC pseudogenes are also frequent in New World monkeys (Cadavid, Hughes, and Watkins 1996), and gene duplication events (Kono et al. 2014) confer plasticity to species, such as callitrichids, with limited MHC class I region variability (Vogel et al. 1999). Other, non-callitrichid, New World monkeys exhibit traits characteristic of Old World monkey classical class I loci, such as high polymorphism and positive selection pressure within the PBR (Cadavid and Watkins 1997; Pascalis et al. 2003). Currently, the relationship between New and Old World MHC class I genes remains unclear, and direct comparisons may be premature.

Compared to haplorrhine primates, a relatively small number of strepsirrhine MHC class I gene sequences have been published thus far. Nucleotide sequence analysis revealed multiple duplications of loci and a divergent pattern of substitutions when compared to anthropoid primates (Flugge et al. 2002). In Malagasy lemurs, classical MHC class I genes appear to have a recent origin, and the most conserved nonclassical loci (MHC-E and -F) may be absent (Go et al. 2003). In the future, more research on strepsirrhine MHC should provide clearer insight into nonhuman primate MHC class I evolution.

MHC class II loci exhibit a high degree of conservation among the primates, and this may explain why so many studies of the MHC in primates have focused on class II loci such as MHC-DRβ (Loisel et al. 2006). Since most of the polymorphism of MHC-DRβ genes is concentrated in the peptide-binding regions of exon 2, MHC-DRβ alleles are identified according to the nucleotide sequence differences found in exon 2, with certain common sequence motifs assigned to specific “lineages” that reflect their common peptide-binding capabilities.

Functional and orthologous DRβ1, DRβ3, DRβ4, and DRβ5 loci have been identified in humans, chimpanzees, gorillas, and rhesus macaques. The presence of particular allelic lineages in anthropoid primates (Slierendregt et al. 1994) suggests inheritance of some DRβ alleles from a common ancestor. Old World monkey MHC-DRβ sequences that do not readily cluster with known HLA-DRβ sequences indicate uniqueness from known human class II sequences. Studies of mandrills and baboons (Huchard et al. 2006) support the conclusions that novel MHC lineages arise frequently in Old World monkeys and that transspecies evolution can be observed in distantly related species. Some New World monkey MHC-DRβ alleles, however, have changed dramatically since divergence, and the evolutionary history of platyrrhine MHC-DRβ genes still requires further study.

Based upon a relatively limited number of investigations, it has proven difficult to draw consensus conclusions about strepsirrhine MHC-DRβ genes. Figueroa et al. (1994) described MHC-DRβ sequences that were similar to modern human MHC-DRβ1 lineages, which suggests this locus was established more than 85 million years ago. A later study of four extant families of the lemurs suggested that MHC-DRβ variation was generated after the divergence of the lemurs from other primates (Go et al. 2002). Research focusing on cDNA sequences from MHC class II loci found that there was no clustering of anthropoid and prosimian sequences (Flugge et al. 2002). One of the earliest studies of MHC-DRβ in gray mouse lemurs found high MHC variability (Schad, Sommer, and Ganzhorn 2004). Large-scale genotyping of MHC DRβ and DQβ genes in gray mouse lemurs revealed the presence of single loci in tight equilibrium with relatively high polymorphism (Huchard et al. 2012). A study of three specialized wild Madame Berthe's mouse lemur populations, sympatric with the gray mouse lemur, found substantial polymorphism and also amino acid motifs very similar to those of the gray mouse lemur (Pechouskova et al. 2015). A later study (Grogan et al. 2016) genotyped 300 wild and captive ring-tailed lemurs using next-generation sequencing. This study identified a large number of previously undescribed MHC-DRβ sequences and found evidence for multiple MHC-DRβ loci in the ring-tailed lemur.

Based on decades of research, high polymorphism appears to be a hallmark of the MHC in most primates. There is substantial evidence that some of this genetic variation is the result of positive selection and the role of MHC molecules in presenting foreign peptides to T cells (Hughes and Nei 1988, 1989; but see Hughes 2007). Detailed studies of DNA sequences from MHC molecules have revealed that nucleotide substitution patterns for the PBR codons accumulate more nonsynonymous than synonymous substitutions, but vice versa for non-peptide-binding site codons in the same genes. New alleles in nonhuman primate species also seem to arise as a consequence of positive Darwinian selection, and many publications have demonstrated that balancing selection is an important influence on MHC genes (Aguilar et al. 2004; Prugnolle et al. 2005).

MHC genes are found in all primates studied to date, and there are many similarities between humans and nonhuman primates with respect to other features. Specifically, there is physical linkage of the class I and class II regions of the MHC, and groups of genes found on a single haplotype, or parental chromosome, are inherited together. Gene duplication is thought to be a significant force in the evolution of the MHC in mammals (Kulski et al. 2002). However, due to frequent gene duplication and inactivation events in primates, not all species share the same number or variety of genes on a haplotype (Flugge et al. 2002; Kono et al. 2014). Nei and coworkers (Nei, Gu, and Sitnikova 1997; Piontkivska and Nei 2003) have demonstrated that primate MHC genes follow a "birth-and-death" model of evolution in which new genes are created by repeated gene duplication, some duplicated genes are maintained in the genome, and others are deleted or become nonfunctional by deleterious mutations. This provides individuals with a large repertoire of peptide-binding abilities, but it makes the identification of alleles from single loci especially difficult (Huchard et al. 2006).

Gene duplication has probably also acted on the MHC on a smaller scale with the frequent duplication of single and, more rarely, multiple loci, by a process of tandem duplication. Since class II molecules are encoded by both an α and β chain gene, the creation of new class II proteins requires the tandem duplication of both the α and B chain in concert. The necessity for tandem duplication of both genes imposes a constraint on the occurrence of duplication and may explain the general stability of the class II region and its apparent conservation across primate species. MHC class I genes are not restrained by the need for tandem duplication, and this may explain why the MHC class I is less stable across the primate order.

Interallelic recombination, or gene conversion, can also enhance MHC polymorphism (Ohta 1998), and larger scale exchange of exons between different MHC genes (i.e. interlocus recombination) can promote even further MHC diversification (Hughes 1995). Two important consequences of interlocus recombination are the loss of locus-specific sequences (i.e. motifs) and, concomitantly, the failure to accurately determine phylogenetic relationships between different sequences from the same species. Gene conversion has been documented in humans (Parham et al. 1989) and may explain a great deal of the evolutionary history of MHC genes in primates. It is believed that many of the modern features in primate classical class I genes were generated by gene conversion (Parham, Adams, and Arnett 1995; Adams, Cooper, and Parham 2001) and intralocus recombination.

To appreciate the functional necessity of MHC genes, and gain a better understanding of the evolutionary relationships between MHC genes within and between species, we must obtain detailed knowledge about haplotype conformation and diversity, locus identity, and expression level of functional genes. It is especially important to determine patterns of expression and peptide-binding properties. The same information is required for other primate immune genes such as TLRs, KIRs, and even immunoglobulins.

Although individual MHC sequences have been identified in many different primates, in-depth studies of large genomic regions of the MHC are known for only a few species, such as humans, chimpanzees, and rhesus macaques. These studies have been motivated by a desire to clarify how disease resistance is conferred by the MHC. Unfortunately, however, because of the great variability found in the MHC, it is not always possible to extrapolate from chimpanzees or macaques to other primates. Based upon published studies describing MHC organization and haplotype composition, it is clear that many different strategies have evolved to maximize the ability of the MHC to protect individuals against disease. These evolutionary events make it difficult to determine if MHC loci are orthologous or paralogous, even between related primate species. They also complicate the assignment of alleles to loci within a species, making analysis of heterozygosity and calculation of genetic distance challenging.

Summary and Conclusion

The categories of innate and adaptive immunity may seem distinct, and they have been (for the most part) presented as separate processes in this chapter, but recent research demonstrates complex interactions between different immune molecules and highlights the fact that some molecules serve as intermediates in immune response. Innate and adaptive immune responses are not mutually exclusive. For example, interactions between innate and adaptive immunity can be mediated by dendritic cells, which are now known to be multifunctional. Dendritic cells detect pathogens, and they activate both an innate immune response and T cells, which signal B cells to initiate antigen-specific antibodies. Thus, the immune system represents a very complex interacting network that involves communication between the innate and adaptive immune systems through cell–cell interactions. Research in the future should account for the complex interactions between various arms of the immune system in nonhuman primates.

As exemplified in this chapter, comparative studies of human and nonhuman primate immune genes are increasing in number. Early on, genes such as MHC were the primary focus for many laboratories, and research primarily concentrated on genotyping methodology and DNA sequence collection. More recently, however, publications focusing on TLRs, KIRs, and NK cells have become more common, and functional studies are also increasing. Nevertheless, there is much to learn, and there are many questions that could be answered through the study of nonhuman primates.

To date, no research has yet explored the odd, counterintuitive role of NK cells in pregnancy or even the (better understood) role of NK cells in fighting tumor cells in nonhuman primates. TLR studies have been mostly limited to sequence-based analysis. Human and nonhuman primate comparative studies of the TLR functional pathways are lacking, and those that do exist are mostly undertaken to explain differential HIV/SIV pathogenesis in humans versus chimpanzees. So far, studies of the TLR functional pathways and genetic sequences are essentially limited to the great apes and rhesus macaques. Similarly, very few studies have examined any New World monkey species besides marmosets. Considering that the TLR pathways are evolutionarily conserved and very important to the survival of all living organisms, more comparative human and nonhuman primate research on the innate immune system, and how it works, seems essential. This research could illuminate the ways in which the innate immune system evolved to respond to a wide variety of pathogens and why there are individual differences in disease susceptibility. Given the limited divergence exhibited between primate species, a complete overview of immunological processes in nonhuman primate species could also offer critical information regarding mechanisms of disease progression and remission in human patients.

So far, research on nonhuman primates has improved our understanding of the diversity, evolution, and function of certain immune genes in humans. Due to their (i) high degree of genetic similarity to humans, (ii) natural susceptibility to a wide range of pathogens that are closely related to those that infect humans, and (iii) susceptibility to human pathogens, nonhuman primates can be valuable models for research on pathogenesis and immunity.

Notably, however, our close evolutionary relatives do respond differently to many types of infections. AIDS, malaria, and cancer kill millions of people each year, but nonhuman primates seem to be naturally resistant to these diseases. When studying species-specific immune responses, some researchers suggest that there are lineage-specific immune responses that explain known interspecies differences in susceptibility to infectious diseases. These lineage-specific responses may be due to unique alleles favored in certain habitats, or they may help to minimize pathogen exposure. Few studies have evaluated interspecies differences in immune response, and this is a critical task ahead. Future studies in primates could also address another neglected area of research, namely, the comparative immune response in single individuals to different infectious agents.

Nearly 60% of emerging diseases come from animals. One example is the well-known AIDS pandemic, which likely originated through hunting and butchering of wild African primates (Sharp and Hahn 2011). By identifying nonhuman primates that share diseases with us, we can learn about gene variants that provide resistance to diseases. For example, recent research has revealed that we share more diseases with baboons, macaques, and lemurs than expected. However, we share fewer diseases than expected with orangutans (Cooper et al. 2012). These findings challenge our assumptions that humans are most at risk of contracting diseases from nonhuman primates, and it also demonstrates the importance of learning more about disease, immune genes, and immune function in our nonhuman primate relatives.

CHAPTER 21

Genetics, Evolutionary Medicine, and the Evolution of Human Pathogens

Amanda VanSteelandt and Anne C. Stone

School of Human Evolution and Social Change, Arizona State University, Tempe, AZ, USA

Introduction

With mutually selective influence on growth, reproduction, and survival, pathogens and their host organisms are engaged in an evolutionary race to evade or attack the other. The history of human–pathogen interactions is written into the genomes of both humans and their pathogens and to some extent in our historical and archeological record. Thus, the evolutionary history of human disease has long been of interest to bioarchaeologists and anthropological geneticists as well as human biologists. In this chapter, we outline several key concepts in the study of human–pathogen co-evolution, describe genetic evidence used in these investigations, and discuss some interesting case studies and current research directions.

A Companion to Anthropological Genetics, First Edition. Edited by Dennis H. O'Rourke.

Evolutionary Medicine

Investigations in evolutionary medicine examine how evolutionary forces have shaped patterns of health and disease, including how humans and their pathogens have co-evolved through history. This approach shifts the focus to ultimate, rather than proximate, explanations of health and disease. An evolutionary perspective allows us to examine questions like: How have humans and their pathogens shaped each other's evolutionary trajectories? Why are some people more or less susceptible to infectious diseases? How and where did infectious diseases originate? And what tools can we use to combat infectious diseases? Several considerations for investigations of diseases have been outlined by scholars of evolutionary medicine (e.g. Williams and Nesse 1991; Stearns and Ebert 2001; Gluckman et al. 2011):

1. Evolutionary processes may come with trade-offs. A gain in function in one area may result in loss of function in other areas. For example, while an antibiotic-resistant phenotype will grant bacteria greater fitness in an environment with high antibiotic use, this phenotype often has a fitness cost, making the organism less competitive in antibiotic-free environments (Andersson and Levin 1999; Andersson and Hughes 2010).
2. Evolutionary processes are limited by biological constraints from phylogenetic history. Put another way: adaptations are limited by a population's existing genetic architecture. For example, pathogens can be adapted to specific host species' physiologies, and their ability to jump to another species may be limited by that species' similarities to or differences from their current host(s). Host shifts do take place, but they become considerably less likely with greater phylogenetic distance between hosts and decreased frequency of interactions of those species (Morse et al. 2012).
3. Phenotypes are shaped by gene–environment interactions. This is particularly remarkable in the case of the human immune system, where exposure to certain kinds of pathogens, especially early in life, conditions the immune system for future challenges (McDade 2003). Populations that are more exposed to macroparasites will develop immune systems biased toward the Th2 immune response system compared to the Th1 system in anticipation of greater challenges from macroparasites (Romagnani 1996, 1997; McDade 2003). The Th2 bias better prepares the immune system to fight these macroparasite infections, but it may create vulnerability to microparasite infections.
4. Pathogens evolve rapidly. The matchup between humans and their pathogens is an interesting dynamic, because pathogens – viruses, bacteria, fungi, nematodes, and ectoparasites – have much shorter generation times, often have higher mutation rates, and typically have vaster numbers of offspring. Human evolution occurs over a much longer time span; thus, humans are at a disadvantage when compared to the rapid genetic adaptation that their pathogens are capable of.

5. Humans can in part make up for their slow pace of genetic adaptation through phenotypic and cultural adaptations that allow a more rapid response to threats of infection. Phenotypic plasticity allows the immune system to "learn" about new pathogen challengers and mount a more efficient response to future attacks. Constructing more sanitary environments, creating social barriers to disease transmission, and caregiving for the sick are cultural adaptations to preventing and treating infections.

When viewed from an evolutionary perspective, host–pathogen interactions have some surprising dimensions that are often missed by conventional medical approaches. For example, many of the symptoms of infection that are targets of medical intervention – like fever, runny nose, or diarrhea – are host defenses that aim to destroy or expel pathogens from the body. Interfering with the body's natural defense systems is perhaps not the best approach to deal with all infections (Williams and Nesse 1991). In addition, many autoimmune diseases and allergic responses may be a legacy of past selection history on the immune system by pathogens that are less common today in places with modern sanitation, as postulated by the hygiene hypothesis (Strachan 1989; Versini et al. 2015).

Although evolutionary medicine as well as this chapter primarily focus on selection as the major evolutionary force affecting hosts and pathogens, it is important to consider the impact of the other forces of evolution as well (namely, mutation, genetic drift, and migration/gene flow). Mutation is the ultimate source of variation, producing new variants in the host or pathogen that can then be acted upon by selection. Genetic drift has been a significant force in human populations because of our low population size over most of our evolutionary history, and random events are also likely to be important in many populations of pathogens. Finally, migration or gene flow brings new combinations of variants together (via mechanisms such as recombination or horizontal gene transfer) in genomes, and, in the broader sense, host and/or pathogen migration can change the environment in which these interact.

Co-Evolutionary Theory

Host–pathogen co-evolution involves a process of reciprocal, adaptive changes in gene frequencies in both species (Woolhouse et al. 2002). The process depends on multiple aspects of the interaction: the combination of host and pathogen genotypes, the number of genes and alleles involved and how they interact with each other, fitness costs of alleles, and the mode of reproduction and generation time of the pathogen. Theoretical models of co-evolution predict that many outcomes are possible, including successive fixation of favorable alleles and polymorphisms that are stable or dynamic (which may or may not lead to allele fixation) and that are affected by balancing or fluctuating selection on alleles.

Following Woolhouse et al. (2002), two major patterns for co-evolution are described as the *arms race* (accumulated improvements in both host and pathogen populations) and *Red Queen dynamics* (a term first coined by Van Valen [1973] to describe "running as fast as you can to stay in the same place"), both involving reciprocal changes in host and pathogen (Figure 21.1). In the arms race pattern, selection is directional, and a series of selective sweeps are observed in host and pathogen alleles that arise through mutation or migration. Transient polymorphisms in the host and/or pathogen populations can be found during these alternating drives to fixation. In the Red Queen dynamics pattern, selection is fluctuating, and dynamic polymorphisms in the host and pathogen populations are persistent. Alleles do not become fixed in this pattern except through genetic drift.

Co-evolutionary theory typically focuses on a long-standing interaction between a host and pathogen that is not mutualistic or symbiotic. From the pathogen perspective, primary concerns are invading the host, evading the host's defenses, and transferring to a new host. This latter concern is tied to the

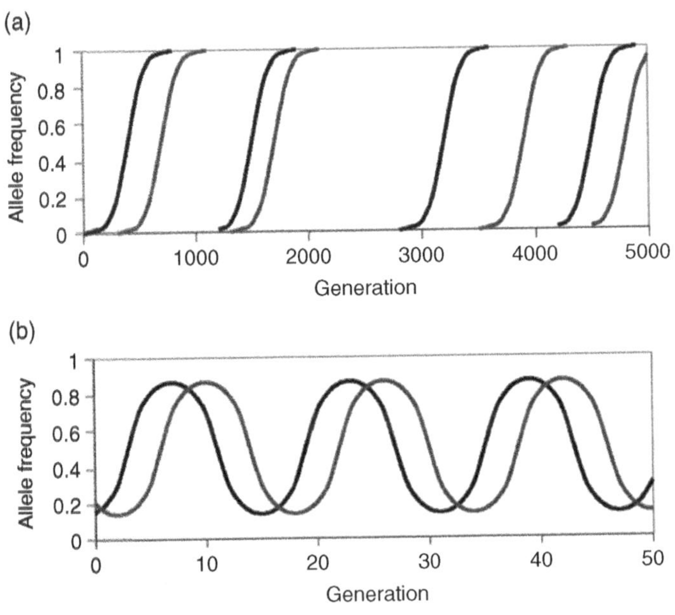

Figure 21.1 Allele frequency changes driven by co-evolution. (a) A series of selective sweeps by host (black line) and pathogen (grey line) alleles derived by mutation, which are then subject to positive selection. Genetic change accumulates in both populations. At any given time (represented in generations), there may be polymorphisms in one, both, or neither population. (b) Dynamic polymorphisms in both host and pathogen acting on existing genetic variation. At all stages, both populations are polymorphic. There can be local adaptation, but it can shift through time (and alleles that were once adaptive may become maladaptive). *Source*: Woolhouse et al. (2002).

pathogen's virulence factors – the traits that allow a pathogen to cause damage or disease in the host (and may or may not promote the spread of the pathogen). Virulence is the result of interactions between the host and pathogen that are complex. One prominent idea about virulence is that if a pathogen kills its host too rapidly, then it limits its opportunities to transfer to a new host; thus, over time, more virulent organisms are likely to die out more rapidly than those that take a more symbiotic approach to living within their hosts (May and Anderson 1983). This "conventional wisdom," however, is problematic from a population genetic perspective because it requires group selection to work over the long term, since natural selection at the level of the individual pathogen often favors virulence (Levin and Svanborg Eden 1990; Frank 1996; Levin 1996).

To replace the conventional wisdom about the evolution of virulence, several other hypotheses have been proposed. The *trade-off hypothesis* or model assumes that pathogen virulence and transmission are coupled traits (Anderson and May 1982; Ewald 1983) such that the level of transmission over the course of infection is tied to levels of virulence (i.e. increased virulence = increased transmission in a short time). Here, to reach an optimum, virulence adjusts to opportunities and limitations for successful transmission of the pathogen. A classic example of this dynamic has been the introduction of the myxoma virus to control rabbit populations in Australia (Fenner and Ratliffe 1966; Anderson and May 1982; Kerr et al. 2015), although some of the debates and complexities regarding the trade-off model are discussed by Alizon et al. (2009). The *coincidental evolution hypothesis* posits that virulence is the result (coincidentally) of some other phenotype of the pathogen that evolved for some other reason, such as to evade the immune defense of a different host (Levin and Svanborg Eden 1990). For example, *Clostridia botulinum* produces a toxin that causes botulism in humans, although it is a soil bacterium. According to the *short-sighted evolution hypothesis*, virulence "evolves within the microenvironment of individual hosts, without regard to the ultimate 'survival' (or transmission) of the pathogen in the population of hosts" (Levin and Bull 1994). In essence, the pathogens that survive within the host do so because of local advantages over other pathogen variants within the host, and this may lead ultimately to death of both host and pathogen. Pathogens that have been proposed to evolve in this manner include bacteria involved in meningitis and viruses such as poliovirus and HIV.

Demonstrations of host–pathogen co-evolution require evidence of (i) variable traits in host and pathogen populations, (ii) genetic heritability of these traits, (iii) reciprocal effects of these traits on the fitness (survival and reproduction) of the host and pathogen populations, and (iv) ongoing changes in both host and pathogen genotype frequencies. In vertebrate hosts, such as humans, clear evidence of host–pathogen evolution can be difficult to obtain given the long generation time; thus, the focus of investigation has been demonstrating patterns of variation in genotypes or phenotypes that are consistent with ongoing or past co-evolution with pathogens (Woolhouse et al. 2002). Exceptions include studies of recent host jumps (such as with myxoma virus and rabbits and with HIV and humans). Recent advances in ancient DNA methods offer the possibility of observing this process through time in humans and pathogens.

PATHOGEN EVOLUTION

Pathogens have a variety of strategies for avoiding detection and/or destruction by their host's immune system, including hiding in host cells, protective capsules, compounds that misdirect the immune response, binding antibodies, and surface markers (antigens) that change rapidly or mimic their host; not surprisingly, these are commonly subject to selection. Whether a microbe is a pathogen (and its level of virulence) often depends on circumstances such as host genetics or other aspects of the environment, including co-infection with another pathogen. This is illustrated by HIV, a well-studied lentivirus that "jumped" into humans fairly recently from other primates (Gao et al. 1992, 1999; Hahn et al. 2000). HIV's ability to infect a human and the length of infection (i.e. time to progression to AIDS and, ultimately, the death of the host) depend largely on host genetics (discussed further in Section 21.5), whether other pathogens co-infect the host, and host behavior (i.e. multiple sexual partners increases risk of infection). For example, some sexually transmitted diseases in the host may facilitate HIV transmission and infection (Johnson and Lewis 2008; Cohen et al. 2012), and the effects of HIV on the immune system can amplify the impact of other pathogens such as *Mycobacterium tuberculosis* or result in rare disease caused by opportunistic microbes such as *Pneumocystis carinii, Cytomegalovirus, Mycobacterium avium, and Karposi's sarcoma herpesvirus* (CDC 2015; see http://www.cdc.gov/hiv/basics/livingwithhiv/opportunisticinfections.html).

Because HIV is a retrovirus with a short generation time and a high mutation rate, it evolves rapidly. This is clear both from phylogenies of the virus, which include older samples sequenced using ancient DNA methods, and from studies of within-patient HIV variation (for a review, see Rambaut et al. 2004). The phylogenetic analyses of HIV have identified multiple jumps into humans and suggest that these jumps happened over the last 100–150 years (Zhu et al. 1998; Gilbert et al. 2007; Worobey et al. 2008; Wertheim and Worobey 2009). However, as these authors note, dating these events (and the antiquity of HIV and simian immunodeficiency virus [SIV] in general) is difficult because of selection, recombination, saturation (where the rapid rate of evolution causes repeat mutations at the same nucleotide), and rapid lineage turnover (where the rapid birth and death of lineages over time obscure the true divergence times of groups). The similar biology of humans and nonhuman primates likely allows pathogens such as SIV and HIV to "jump" relatively easily between species. Importantly, the greater diversity of these pathogens within nonhuman primates, while reflecting their longer evolutionary history in those species, also suggests that there may be specific mutations needed for successful cross-species transmission or that these other strains or pathogens have not yet had the opportunity to jump (Pedersen and Davies 2009; Calvignac-Spencer et al. 2012; Sharp, Rayner, and Hahn 2013).

Researchers have also tried to assess why SIV and HIV differ so greatly in virulence. While SIV is thought to have a relatively minor impact on the primates infected with it (but see Keele et al. [2009] for a study on the morbidity and mortality of wild chimps infected with SIV), HIV, ultimately, is almost uniformly fatal for humans. Is there a difference in the pathogen that accounts for this? Studies of within-patient HIV variation over the course of the disease have illustrated some of the selective pressures over time on the virus, such as the evolution of drug resistance and the shift from using only the CCR5 receptor to also using the CXCR4 receptor as a means to enter cells. In particular, positive selection in the HIV Env gene enables the virus to escape and avoid immune responses (e.g. Holmes et al. 1992; Richman et al. 2003). Interestingly, the selective pressures on the virus within a patient do not seem to be as important in transmission to another host because of the typical bottleneck that occurs at transmission (i.e. only a few randomly selected virions are transmitted to the next host) and perhaps because many of the variants that evolve later in the infection of one person may not survive the early stages of infection in another (Zhu et al. 1993; Delwart et al. 2002; Frange et al. 2013). There is also evidence that the HIV circulating within a population has been shaped by the human leukocyte antigen (HLA) alleles common in that population; for example, Moore et al. (2002) found evidence for selection in the reverse transcriptase gene enabling HIV to be better at escaping the cytotoxic (or killer) T cells key for the host immune response.

Another virus that has been the subject of recent attention is the Ebola virus (EBOV), which has been an intermittent zoonosis and, until recently, typically caused a relatively small and localized outbreak that was contained and/or burned itself out. However, in 2014, an outbreak in West Africa, which began in Guinea in late 2013, spread rapidly (Baize et al. 2014). As of March 31, 2016, the total number of cases (suspected, probable, and confirmed) was 28,652 with 11,325 deaths (https://www.cdc.gov/vhf/ebola/history/chronology.html). This rapid spread was fostered by traditional funeral practices, poor health infrastructure, limited drug treatment options (and no vaccines), and the time it took to mount a large response. This outbreak provided the opportunity to investigate how the pathogen evolved over time. From analyses of Ebola virus genomes from this as well as previous outbreaks, researchers found evidence that the virus originated in Central Africa, the site of the previous outbreaks, with a subsequent spread to West Africa within the last decade (Baize et al. 2014; Gire et al. 2014). In addition, these analyses show that the epidemic was initiated from a single zoonotic event from the reservoir host, which has long been suspected to be fruit bats.

After the spillover event into humans, the virus accumulated nonsynonymous mutations (i.e. mutations that cause a change in the amino acid sequence of a protein). Gire et al. (2014) find that this is consistent with incomplete purifying selection models but caution that these mutations could lead to viral adaptation to the new host (i.e. positive selection of some changes). Subsequent analyses of these data by Olabode et al. (2015) suggest that the nonsynonymous changes are consistent with the null hypothesis of neutral evolution since the changes do not have known important structural and functional effects (or do not represent changes that are unique to the 2014 outbreak), and thus, these changes do not appear to lead to functional adaptation of the virus to the new human hosts. Additional Ebola genome sequences, primarily from Sierra Leone, show significant diversity allowing the estimation of substitution rates and patterns of transmission within the country as well as to cases outside Sierra Leone (Park et al. 2015; Tong et al. 2015). Moreover, Park et al. found that nonsynonymous mutations, which are likely to be deleterious, are less likely to persist over the course of the epidemic, supporting the action of purifying selection. In an analysis of selection pressures in Ebola virus genes, Park et al. also conclude that although formal tests of selection are not significant, changes in the mucin-like domain of one gene, glycoprotein (GP), are suggestive of selection against the host humoral immune response since the changes are located in the B-cell epitope-binding region.

The longer term evolution of pathogens has also become the focus of ancient DNA research. Such analyses enable us to assess the timing of pathogen jumps into humans as well as investigate how pathogens have changed through time (i.e. illuminating the patterns of mutation, drift, and selection – and in some cases gene flow – reflected in their genomes). Pathogens including *M. tuberculosis*, HIV, *Yersinia pestis*, and *Treponema pallidum* were studied starting in the 1990s (e.g. Salo et al. 1994; Drancourt et al. 1998; Kolman et al. 1999; Reid et al. 2000, Basler et al. 2001; Haas et al. 2002), although most studies were limited to confirming the presence of the pathogen in archeological bone or other tissues. Many of these early studies were controversial because of the methodological challenges (for a recent review of ancient pathogen genomics, see Harkins and Stone 2015). The first pathogen genome sequences were the studies of the 1918 flu virus and early HIV biopsy samples (Tumpey et al. 2005; Worobey et al. 2008). Recently, technical advances have made it easier to study complete ancient genomes from samples with little remaining endogenous DNA (e.g. Green et al. 2009; Meyer et al. 2012, 2014; Dabney et al. 2013a). These protocols have been used to capture complete pathogen genomes that comprise a very small portion of the endogenous DNA in a sample and thus shed light on questions about the causative agents of disease and epidemics, the evolutionary history of pathogens, and co-infection with multiple strains or species. Such studies have provided insight into the evolution of *Y. pestis*, the cause of the Black Death and the Justinian Plague (Bos et al. 2011; Schuenemann et al. 2011; Wagner et al. 2014); the origins of leprosy (Schuenemann et al. 2013); the spread of *M. tuberculosis* complex strains around the world (Bos et al. 2014); and the relationship of the second-pandemic cholera strains to the classical biotype (Devault et al. 2014).

An example of the information that can be gleaned from ancient DNA about the evolution of pathogens is the research on the mycobacterial diseases tuberculosis (TB) and leprosy. Our knowledge of the temporal evolutionary dynamics for these pathogens in humans is remarkably limited. Recent research using ancient DNA has revealed that both *Mycobacterium leprae* and *M. tuberculosis* likely "jumped" from zoonotic sources and became human pathogens within the last 6000 years (Schuenemann et al. 2013; Bos et al. 2014). In addition, this research, along with that focused on extant mycobacteria, shows that zoonotic strains in primates and other animals continue to be sources of disease in humans.

For tuberculosis, Bos et al. (2014) provide clear evidence that a member of the *M. tuberculosis* complex (MTBC) – specifically, *Mycobacterium pinnipedii*, which was brought to South America by seals – caused disease in the pre-contact New World. These ancient samples provide a means of calibrating the molecular clock for the MTBC and indicate a most recent common ancestor for the *M. tuberculosis* complex less than 6000 years ago, which supports a dispersal of the disease during the Holocene (rather than deeper in the past as was expected). These results also focus attention on the role of zoonotic "jumps" by pathogens in ancient disease. While the pre–European contact cases of tuberculosis in the Americas may be the result of a single "spillover" event after which the strains adapted to humans, fostering the subsequent spread, it is also possible that there were multiple "jumps" from seals or sea lions to humans. This could be discerned based on strain diversity in phylogenetic analyses, although additional *M. pinnipedii* genome data from known geographic locations are needed. The initial comparison of the ancient Peruvian strains also suggests that the pathogen began adapting to humans, since the genomes of these strains differ from those of seal strains by five nonsynonymous changes but no synonymous changes (i.e. changes that do not affect protein structure). Further analyses of a larger number of ancient pre-contact TB genomes are needed to assess whether human re-adaptation occurred through time or whether *M. pinnipedi* was already sufficiently adapted to mammals in general and, thus, spread with patterns of genetic diversity more consistent with genetic drift. More virulent strains, specifically those from Europe, replaced local strains after contact, although it is possible that there were earlier introductions of new strains via trade or population expansion (e.g. into North America from Siberia via the Paleo-Eskimo expansion beginning ~3000 BCE).

Like *M. tuberculosis*, *M. leprae* likely "jumped" from zoonotic sources and became a human pathogen relatively recently. Genomic analyses of *M. leprae* reveal a remarkably reduced genome, with roughly 1300 pseudogenes (Cole et al. 2001). It is estimated that this loss of gene functionality occurred approximately 12–20 million years ago, long before the emergence of anatomically modern humans (Gomez-Valero et al. 2007). A major obstacle to research on leprosy has been the inability of the pathogen to replicate outside of its animal host or in laboratory media. Using modern and ancient sequence data, Schuenemann et al. (2013) calibrated substitution rates to infer that the all-human leprosy strains diverged at most ~5000 years ago, consistent with paleopathological evidence but much more recent than previous genetic estimates (Monot et al. [2005, 2009], but see Pinhasi, Foley, and Donoghue 2006). Their ancient data, along with those of Economou et al. (2013) and Mendum et al. (2014), showed that ancient European strains belonged to types 2 and 3. In addition, Schuenemann et al. (2013) were able to sequence additional modern strains from diverse geographic locations with the same enrichment methods used for the ancient samples (circumventing the challenge of culturing these strains to obtain sufficient DNA for conventional sequencing). This revealed that the most basal *M. leprae* strains (type 0) currently known are found in New Caledonia.

Because of the small number of cases of leprosy, and the limited genome data available, it is difficult to discern signals of selection in this pathogen. For *M. tuberculosis*, more information is currently available from analyses of modern strains. Analyses of both the MTBC and *M. leprae* have suggested that these organisms are primarily clonal (Cole et al. 2001; Smith et al. 2003; Supply et al. 2003; Achtman 2008). Lateral or horizontal gene transfer may affect the phylogeny and diversity of bacteria (e.g. Dykhuizen and Green 1991; Daubin, Moran, and Ochman 2003; Gogarten and Townsend 2005); genome comparisons among bacteria suggest that less than 5% of the *M. tuberculosis* genome is foreign (Ochman, Lawrence, and Groisman 2000; Pepperell et al. 2013), although others argue that recombination may have been important, particularly in the MTBC ancestor (Nakamura et al. 2004; Gutierrez et al. 2005; Veyrier et al. 2009; Namouchi et al. 2012). Initial analyses of diversity patterns in the MTBC suggested that genetic drift plays a major role, but recent research has indicated that selection (in particular purifying selection, which removes deleterious variants) has been important (Hershberg et al. 2008; Pepperell et al. 2013). In addition, positive selection for antibiotic resistance has occurred on multiple lineages of *M. tuberculosis* (e.g. Merker et al. 2013). Other signals of adaptation have also been detected in large samples of *M. tuberculosis* genomes. For example, Pepperell et al. (2013) in a study of global strains found that loci that clustered in the category of *defense mechanisms* showed evidence of diversifying selection or selective sweeps. In a study focused on the so-called Beijing family of strains, Merker et al. (2015) identify the mce (mammalian cell entry) and vapBC (virulence-associated protein) as two gene families that show signs of positive selection. Genes in these families are associated with virulence, growth, and modulation of the host immune response. These genes were found in addition to genes that are involved in drug resistance. In sum, despite the wealth of paleopathological, molecular, archeological, and modern genome data available today, many questions concerning the origin, exchange, and evolution of pathogenic mycobacteria remain unanswered.

Human Genetic Response to Pathogens

While pathogens have a variety of strategies for avoiding detection and/or destruction by their host's immune system, hosts have also evolved strategies to defend against, evade, and destroy pathogens generally as well as specific pathogens. It is not surprising, then, that many of the genes found to be under selection in the human genome are related to pathogen defense and the immune system (e.g. Barreiro and Quintana-Murci 2010; Fumagalli et al. 2011). For example, population allele frequencies for major histocompatibility complex (MHC) class I loci (Prugnolle et al. 2005), blood group antigen genes (Fumagalli et al. 2009a), and interleukin genes and their receptors (Fumagalli et al. 2009b) correlate with local pathogen diversity. The selective pressures of host–parasite interactions have been invoked to explain several phenomena that have important implications for human genetics. Hamilton (1980) suggested that the emergence of sexual reproduction was due to the advantage conferred by recombination in organisms pressured by pathogens. Similarly, the maintenance of genetic diversity in blood group polymorphisms (Haldane 1949) and protein polymorphisms (Clarke 1979) has been attributed to selective pressures of infectious diseases. Hamilton and Zuk (1982) suggested that female mate choice for fully expressed male display characteristics is a method of selecting mates who are healthier, and therefore have more resources to put into display rather than fighting off infection.

One specific (and classic) example of selection (both positive and balancing) in human hosts because of a pathogen is malaria. Malaria is the disease for which we arguably have the most information about the impact on the human host over the long term, and it was the driving force behind the first case of selection (at the HBB locus) identified in humans (Allison 1954b). Malaria is caused by the parasites in the genus *Plasmodium*, and four species cause malaria in humans (*Plasmodium falciparum*, *Plasmodium vivax*, *Plasmodium ovale*, and *Plasmodium malariae*). Of these, *P. falciparum* is typically the most severe and most common. Genes that show signals of selection in humans in response to malaria include red blood cell proteins, such as the α- and β- globin genes (HBA1, HBA2, and HBB); glycophorin A, B, and C (GYPA, GYPB, and GYPC), which determine the MN, Ss, and Gerbich blood types, respectively; glucose-6-phosphate dehydrogenase (G6PD); the Duffy chemokine receptor (DARC); as well as MHC genes (e.g. Allison 1954a; Flint et al. 1986; Hamblin and Di Rienzo 2000; Weatherall et al. 2002; Verrelli et al. 2006; Wilder, Hewett, and Gansner 2009; King et al. 2011). Much has been written about these variants (for a review, see Lopez et al. 2011) and the ways in which they hinder the parasite. For example, the Duffy chemokine receptor is required by *P. vivax* to enter red blood cells; thus, individuals who are homozygous for the Duffy null variant are resistant to this parasite, since it cannot complete a major portion of its life cycle in the host (Hamblin and Di Rienzo 2000). In addition, King et al. (2011) also found differences in the ability of *P. vivax* to bind with the other alleles (Fy^a and Fy^b) at this locus, with the parasite having decreased binding ability with Fy^a compared with Fy^b. Mutations associated with malaria resistance in HBA1, HBA2, HBB, and G6PD cause anemias, which shorten the lifespan of red blood cells and thus reduce the ability of the parasite to reproduce. Genes involved in immunity and host defense can also tell us how apes (including us) have differentially adapted to disease. Many primates are affected by malaria, but although humans show evidence for recent positive selection at the malarial resistance G6PD gene, chimpanzee DNA sequence data do not show this pattern, and comparisons with other primates show a long history of functional constraint (Verrelli et al. 2006). On the other hand, both humans and orangutans show signals of selection at the α-globin loci for malarial resistance (Steiper et al. 2005). These examples show that the same genes may or may not be associated with resistance to similar diseases among apes.

Although HIV jumped into humans quite recently and the virus was only identified in the 1980s, gene variants that confer resistance to infection and/or progression of the disease have already been identified (for a review, see McLaren and Carrington 2015). This illustrates how standing variation at immune and other loci can be important for fighting a new pathogen. For example, CCR5 is the primary co-receptor that HIV uses to enter T cells and macrophages, but some individuals, primarily of Northern European ancestry, have alleles of CCR5 that have a deletion that inactivates this receptor (the CCR5-Δ32 variant), such that homozygous individuals are resistant to HIV infection and heterozygotes progress to AIDS more slowly that those with wild-type alleles (Dean et al. 1996; Smith et al. 1997). However, one trade-off to having the deletion allele may be an increased risk of infection by other viruses, such as West Nile virus and tickborne encephalitis virus, which are both flaviviruses (e.g. Kindberg et al. 2008; Lim et al. 2008). CCR5 (along with MHC class I genes and the MIC gene) also shows evidence of recent positive selection in chimpanzees that could be the result of SIV (de Groot et al. 2005; Wooding et al. 2005).

Zhao et al. (2012) hypothesized that because of the evidence for the presence of SIV in chimpanzees, specifically the central subspecies *Pan troglodytes troglodytes*, for at least 20 000 years, SIV likely has crossed over to humans multiple times over that period. Thus, groups such as the Biaka, who live in the same part of Central Africa as central chimpanzees, may have evolved resistance to the virus. They examined genome-wide single nucleotide polymorphism (SNP) data from Biaka for signals of selection and compared them with SNP data from the related Mbuti as well as more distantly related Bantu groups, who also live in Central Africa but outside the range of infected central chimpanzees. Zhao et al. found that alleles at genes such as APOBEC3G, TRIM5, and CUL5, which have been shown to have protective effects against HIV1, are at higher frequency in Biaka compared to Mbuti. In addition, signals of selection in Biaka were highest for CUL5, a gene encoding the cullen 5 protein used by HIV to form a complex that suppresses host antiviral activity against HIV. Finally, other studies have found that

Biaka also have certain alleles of CCR5 as well as higher copy numbers of the gene CCL3L1 that encodes the normal ligand for CCR5, such that CCL3L1 competes with HIV to bind CCR5, both of which act to slow progression to AIDS in HIV-infected individuals (Gonzalez et al. 1999, 2005; Dolan et al. 2007).

Health Transitions and (Re-)Emerging Infectious Diseases

Many of the infectious diseases that afflict humans have changed over the course of human history as a result of changing human activities and environments. In particular, subsistence has shifted from hunting and gathering to agriculture, and urban settlements, industrialization, and globalization have increased population density and interactions. For most of human history, humans lived in small hunter-gatherer communities, and are thought to have been affected by pathogenic organisms with traits that favored their maintenance in small, mobile populations (Dobson and Carper 1996; Barrett et al. 1998). The hunter-gatherer lifestyle is expected to include infectious diseases resulting from exposure to animal reservoirs, as well as slow or chronic diseases that would allow the persistence of the pathogen in the population over the long term. In addition, hunter-gatherers likely would be affected by infectious diseases that cause incomplete immunity (allowing recovered individuals to become susceptible again), such as malaria (Wolfe, Dunavan, and Diamond 2007). Such diseases have been described as *heirloom diseases* since they may have plagued humans for much of our evolutionary history, whereas those appearing after the advent of agriculture have been termed *souvenir diseases* (Sprent 1962).

The advent of agriculture and establishment of more permanent human settlements around 11 000 years ago as well as the associated growth in population size likely favored the new souvenir pathogens. According to the domestic-origins hypothesis, the domestication of animals may have increased exposure to zoonotic infections and allowed these infections to jump the species barrier and adapt to human physiology (Cameron 1956; Weiss 2001; Diamond 2002). Examples of these pathogens are proposed to include those causing smallpox and measles. However, Pearce-Duvet (2006) suggests that agriculture had a more complex role in the emergence of new diseases. Specifically, she proposes that it changed the transmission ecology of existing human pathogens, increased the success of pathogen vectors, and provided a means for wildlife diseases to infect humans.

As in the past, emerging and re-emerging infectious diseases – those that have newly appeared in a population, or preexisting infections that are increasing in incidence or geographic range – today can be mostly attributed to human activities (Morse 1995). The majority of emerging infectious disease events are precipitated by contacts between human and animal populations that create opportunities for host-jumps, globalization and the ease of epidemic spread, and the misuse of medical tools that promote the evolution and resistance of pathogens

(Morse 1995; Jones et al. 2008). Humans, along with our domesticates, continue to encroach upon wild animal habitats, with opportunities for cross-species transmission (or "spillover") expanding as exposure increases (Weiss and McMichael 2004; Butler 2012; Loeffler et al. 2014). Many pathogens, such as *M. bovis*, can infect multiple hosts, and the spread of such a pathogen depends upon the route of transmission, abundance of each host species, and interactions among the different hosts (Nugent 2011). Once a pathogen has newly appeared in a human population, globalization and the ease of travel allow outbreaks to easily cross city, state, and national borders on airplanes, trains, and other vehicles. Historically, outbreaks of Ebola virus disease were quite small and limited to remote, rural villages in Africa. The large outbreak of Ebola virus disease in West Africa, which began in 2014, was largely a consequence of easily accessible transportation and migration across borders and to large urban centers where the virus was able to spread to larger populations (Pigott et al. 2014; Gatherer 2015).

The greatest concern about re-emerging diseases has been in regard to antibiotic resistance in pathogens that many considered "defeated." Since the introduction of antibiotic use in the treatment of pathogenic disease, pathogens have rapidly evolved traits for antibiotic resistance through artificial selection from incomplete treatment regimens or the widespread presence of antibiotics in the environment (Andersson and Hughes 2010). There is also evidence of preexisting resistance resulting from proposed "chemical warfare" with other microbes and fungi over hundreds of millions of years, although researchers are now also finding that the antibiotics produced in nature (and the resistance to them) are often mechanisms of communication among microbial community members affecting cellular activities, including metabolism and biofilm formation (Linares et al. 2006; Yim, Wang, and Davies 2007). The ability of many microbes to engage in horizontal gene transfer, for example through the exchange of plasmids within and across bacterial species, has allowed a more rapid adoption of antibiotic resistance traits than spontaneous mutation alone (Andersson and Hughes 2010). Antibiotic resistance is often associated with reduced bacterial fitness, usually expressed in reduced bacterial growth rate (Andersson and Hughes 2010). Experimental studies and theoretical modeling suggest that in an environment of reduced antibiotic use, the susceptible bacteria would be able to outcompete resistant strains and replace them over time (Levin et al. 1997; Levin 2002). However, compensatory evolution and co-selection for resistance to more than one antibiotic (due to genetic linkage of resistance genes, or when resistance to one drug confers resistance to other drugs of similar structure as well) can make resistance less costly and promote its persistence (Andersson and Hughes 2010). Thus, whether or not resistance is reversible depends on factors both within our control (the rate and pattern of antibiotic use) and out of our control (the fitness cost of bacterial resistance, and the rate and degree to which natural selection compensates for these costs) (Andersson and Levin 1999).

Conclusions and Future Directions

The goal of evolutionary medicine is to employ insights and methods from evolutionary biology to medical and epidemiological research to advance our understanding of human health (Williams and Nesse 1991; Stearns et al. 2010; Stearns 2012). The recent revolution in sequencing technology, as well as the decrease in cost for obtaining genome data from population samples, is furthering the application of evolutionary theory to understanding the underlying patterns of human and pathogen variation, including the impact of differing environments on that variation. In humans, this is leading to both a better sense of the ultimate causes of susceptibility and resistance to disease (as well as some trade-offs resulting from these) and practical applications in the realm of "personalized medicine." For pathogens, these data and the application of evolutionary theory are leading to a better understanding of the constraints on, patterns in, and pace of their evolution.

Important avenues of further research highlighted by Lively et al. (2014) include such questions as: what are the effects of host genetic diversity on the spread of pathogens, how is host–pathogen diversity maintained, what are the effects of external factors on virulence and the risk of infection, is our understanding of single host–parasite pairings generalizable to multihost–multipathogen networks that are common in nature, and what is the role of host microbiota in shaping disease ecology and evolution? With humans as hosts, an added environmental component is the cultural environment, which influences networks of interaction, diet, and other factors impacting risk of disease. This leads to questions, such as: how does cultural change affect risk of infection by different pathogens, how do vaccination and antibiotic use (or other uses of technology) affect pathogen diversity, and how do differing cultural responses to new diseases impact the probability that a new pathogen will adapt to humans? Research at the intersection of ecology, evolutionary biology, medicine, and anthropology will be key to addressing these questions.

References

1000 Genomes Project Consortium. 2010. A map of human genome variation from population-scale sequencing. *Nature* 467(7319):1061–1073. doi:10.1038/nature09991

1000 Genomes Project Consortium. 2012. An integrated map of genetic variation from 1,092 human genomes. *Nature* 491(7422):56–65. doi:10.1038/nature11632

23andMe. 2014. Oprah Winfrey's surprising DNA test. http://blogs.ancestry.com/cm/2014/05/27/the-surprising-facts-oprah-winfrey-learned-about-her-dna

Abdel-Malek, Z.A., A.L. Kadekaro, and V.B. Swope. 2010. Stepping up melanocytes to the challenge of UV exposure. *Pigment Cell & Melanoma Research* 23(2):171–186. doi:10.1111/j.1755-148X.2010.00679.x

Abecasis, G.R., D. Altshuler, A. Auton, et al. 2010. A map of human genome variation from population-scale sequencing. *Nature* 467(7319):1061–1073.

Abi-Rached, L., M.J. Jobin, S. Kulkarni, et al. 2011. The shaping of modern human immune systems by multiregional admixture with archaic humans. *Science* 334(6052):89–94.

Ach, R.A., H. Wang, and B. Curry. 2008. Measuring microRNAs: comparisons of microarray and quantitative PCR measurements, and of different total RNA prep methods. *BMC Biotechnology* 8:69.

Achilli, A., U.A. Perego, C.M. Bravi, et al. 2008. The phylogeny of the four pan-American MtDNA haplogroups: implications for evolutionary and disease studies. *PLoS One* 3:e1764.

Achilli, A., U.A. Perego, H. Lancioni, et al. 2013. Reconciling migration models to the Americas with the variation of North American native mitogenomes. *Proceedings of the National Academy of Sciences of the USA* 110:14308–14313.

Achtman, M. 2008. Evolution, population structure, and phylogeography of genetically monomorphic bacterial pathogens. *Annual Review of Microbiology* 62:53–70. doi:10.1146/annurev.micro.62.081307.162832

A Companion to Anthropological Genetics, First Edition. Edited by Dennis H. O'Rourke.

Ackermann, R.R. 2010. Phenotypic traits of primate hybrids: recognizing admixture in the fossil record. *Evolutionary Anthropology* 19:258–270.

Acuña-Alonzo, V., T. Flores-Dorantes, J.K. Kruit, et al. 2010. A functional *ABCA1* gene variant is associated with low HDL-cholesterol levels and shows evidence of positive selection in Native Americans. *Human Molecular Genetics* 19:2877–2885.

Adam, A. 1969. A further query on color blindness and natural selection. *Social Biology* 16:197–202.

Adam, A. 1973. Colorblindness and gene flow in Alaskans. *American Journal of Human Genetics* 25:564–566.

Adam, A. 1986. Polymorphisms of red-green vision among populations of the tropics. In: D.F. Roberts, and G.F. De Stefano (eds.), *Genetic Variation and Its Maintenance with Special Reference to Tropical Populations.* Cambridge University Press, New York.

Adams, E.J., S. Cooper, and P. Parham. 2001. A novel, nonclassical MHC class I molecule specific to the common chimpanzee. *Journal of Immunology* 167:3858–3869. doi:10.4049/jimmunol.167.7.3858

Adams, E.J., S. Cooper, G. Thomson, et al. 2000. Common chimpanzees have greater diversity than humans at two of the three highly polymorphic MHC class I genes. *Immunogenetics* 51(6):410–411. doi:10.1007/s002510050639

Adams, W.H., and L.J. Strang. 1975. Hemoglobin levels in persons of Tibetan ancestry living at high altitude. *Proceedings of the Society for Experimental Biology and Medicine* 149(4):1036–1039.

Adams, E.J., G. Thomson, and P. Parham. 1999. Evidence for an HLA-C-like locus in the orangutan *Pongo pygmaeus. Immunogenetics* 49(10):865–871. doi:10.1007/s002510050566

Adcock, G.J., E.S. Dennis, S. Easteal, et al. 2001. Mitochondrial DNA sequences in ancient Australians: implications for modern human origins. *Proceedings of the National Academy of Sciences of the USA* 98(2):537–542.

Adhikari, K., G. Reales, A.J.P. Smith, et al. 2015. A genome-wide association study identifies multiple loci for variation in human ear morphology. *Nature Communications* 6:7500.

Aeschbacher, S., M.A. Beaumont, and A. Futschik. 2012. A novel approach for choosing summary statistics in approximate Bayesian computation. *Genetics* 192:1027–1047. doi:10.1534/genetics.112.143164/-/DC1

Aguilar, A., G. Roemer, S. Debenham, et al. 2004. High MHC diversity maintained by balancing selection in an otherwise genetically monomorphic mammal. *Proceedings of the National Academy of Sciences* 101(10): 3490–3494. doi:10.1073/pnas.0306582101

Aikaike, H. 1974. A new look at the statistical model identification. *IEEE Transactions on Automatic Control* AC-19(6):716–723.

Akdis, M., A. Aab, C. Altunbulakli, et al. 2016. Interleukins (from IL-1 to IL-38), interferons, transforming growth factor B, and TNF-α: receptors, functions, and roles in diseases. *Journal of Allergy and Clinical Immunology* 138(4):984–1010. doi:10.1016/j.jaci.2016.06.033.

Akey, J.M. 2009. Constructing genomic maps of positive selection in humans: where do we go from here? *Genome Research* 19:711–722.

Akey, J.M., M.A. Eberle, M.J. Rieder, et al. 2004. Population history and natural selection shape patterns of genetic variation in 132 genes. *PLoS Biology* 2(10):e286.

Akey, J.M., H. Wang, M. Xiong, et al. 2001. Interaction between the melanocortin-1 receptor and P genes contributes to inter-individual variation in skin pigmentation phenotypes in a Tibetan population. *Human Genetics* 108(6):516–520.

Alaluf, S., D. Atkins, K. Barrett, et al. 2002a. Ethnic variation in melanin content and composition in photoexposed and photoprotected human skin. *Pigment Cell Research* 15(2):112–118.

Alaluf, S., D. Atkins, K. Barrett, et al. 2002b. The impact of epidermal melanin on objective measurements of human skin colour. *Pigment Cell Research* 15(2):119–126.

Alizon, S., A. Hurford, N. Mideo, et al. 2009. Virulence evolution and the trade-off hypothesis: history, current state of affairs and the future. *Journal of Evolutionary Biology* 22(2):245–259. doi:10.1111/j.1420-9101.2008.01658.x

Allen, J.A. 1877. The influence of physical conditions in the genesis of species. *Radical Review* 1:108–140.

Allendorf, F.W., P.A. Hohenlohe, and G. Luikart. 2010. Genomics and the future of conservation genetics. *Nature Reviews Genetics* 11:697–709.

Allentoft, M.E., M. Collins, D. Harker, et al. 2012. The half-life of DNA in bone: measuring decay kinetics in 158 dated fossils. *Proceedings of the Royal Society of London B: Biological Sciences* 279(1748):4724–4733.

Allentoft M.E., M. Sikora, K.-G. Sjögren, et al. 2015. Population genomics of bronze age Eurasia. *Nature* 522:167–172.

Allison, A.C. 1954a. The distribution of the sickle-cell trait in East Africa and elsewhere, and its apparent relationship to the incidence of subtertian malaria. *Transactions of the Royal Society of Tropical Medicine and Hygiene* 48(4):312–318.

Allison, A.C. 1954b. Protection afforded by sickle-cell trait against subtertian malarial infection. *The BMJ* 4857:290–294.

Almasy, L., and J. Blangero. 2004. Exploring positional candidate genes: linkage conditional on measured genotype. *Behavior Genetics* 34(2):173–177.

Alonso, S., N. Izagirre, I. Smith-Zubiaga, et al. 2008. Complex signatures of selection for the melanogenic loci *TYR*, *TYRP1* and *DCT* in humans. *BMC Evolutionary Biology* 8(1):74.

Alt, F.W., E.M. Oltz, F. Young, et al. 1992. VDJ recombination. *Immunology Today* 13(8):306–314. doi:10.1016/0167-5699(92)90043-7

Altmann, J., S.C. Alberts, S.A. Haines, et al. 1996. Behavior predicts genes structure in a wild primate group. *Proceedings of the National Academy of Sciences of the USA* 93:5797–5801.

Alves, I., A. Srámková Hanulová, M. Foll, and L. Excoffier. 2012. Genomic data reveal a complex making of humans. *PLoS Genetics* 8:e1002837.

American Society of Human Genetics. 2008. Ancestry testing statement. https://www.ashg.org/pdf/ASHGAncestryTestingStatement_FINAL.pdf

Amorim, C.E.G., J.T. Daub, F.M. Salzano, et al. 2015. Detection of convergent genome-wide signals of adaptation to tropical forests in humans. *PLoS One* 10(4):e0121557.

Amorim, C.E.G., K. Nunes, D. Meyer. 2017. Genetic signature of natural selection in first Americans. *Proceedings of the National Academy of Sciences of the USA* 114: 2195–2199.

Anderson, A., A.T. Bankier, B.G. Barrell, et al. 1981. Sequence and organization of the human mitochondrial genome. *Nature* 290:457–470.

Anderson, R.M., and R.M. May. 1982. Co-evolution of hosts and parasites. *Proceedings of the British Society for Parasitology* 85:411–426.

Andersson, D.I., and D. Hughes. 2010. Antibiotic resistance and its cost: is it possible to reverse resistance? *Nature Reviews Microbiology* 8(4):260–271. doi:10.1038/nrmicro2319

Andersson, D.I., and B.R. Levin. 1999. The biological cost of antibiotic resistance. *Current Opinion in Microbiology* 2(5):489–493.

Andrés, A.M. 2011. Balancing selection in the human genome. eLS. 15 March https://doi.org/10.1002/9780470015902.a0022863.

Andrés, A.M., M.J. Hubisz, A. Indap, et al. 2009. Targets of balancing selection in the human genome. *Molecular Biology and Evolution* 26(12):2755–2764. doi:10.1093/molbev/msp190

Anestis, S.F., T.H. Webster, J.S. Kamilar, et al. 2014. *AVPR1A* variation in chimpanzees (*Pan troglodytes*): population differences and association with behavioral style. *International Journal of Primatology* 35:305–324.

Ang, K.C., M.S. Ngu, K.P. Reid, et al. 2012. Skin color variation in orang Asli tribes of peninsular Malaysia. *PLoS One* 7(8):e42752. doi:10.1371/journal.pone.0042752

Angers, M., M. Uldry, D. Kong, et al. 2008. Mfsd2a encodes a novel major facilitator superfamily domain-containing protein highly induced in brown adipose tissue during fasting and adaptive thermogenesis. *Biochemical Journal* 416(3):347–355.

Ankel-Simons, F., and D.T. Rasmussen. 2008. Diurnality, nocturnality, and the evolution of primate visual systems. *American Journal of Physical Anthropology* 137:100–117.

Anthony, D.W. 2010. *The Horse, the Wheel, and Language: How Bronze-Age Riders from the Eurasian Steppes Shaped the Modern World.* Princeton University Press, Princeton, NJ.

Antón, S.C., R.S. Malhi, and A. Fuentes. 2018. The AAPA IDEAS program: influence on a science. Poster, American Association of Physical Anthropologists (AAPA) annual meeting, February.

Aoki, K. 2002. Sexual selection as a cause of human skin colour variation: Darwin's hypothesis revisited. *Annals of Human Biology* 29(6):589–608. doi:10.1080/03014460210000019144

Aran, D., G. Toperoff, M. Rosenberg, et al. 2011. Replication timing-related and gene body-specific methylation of active human genes. *Human Molecular Genetics* 20(4):670–680.

Armelagos, G.J., and K.N. Harper. 2005a. Genomics at the origin of agriculture, part two. *Evolutionary Anthropology* 14(3):109–121.

Armelagos, G.J., and K.N. Harper. 2005b. Genomics at the origins of agriculture, part one. *Evolutionary Anthropology* 14(2):68–77.

Armitage, S.J., S.A. Jasim, A.E. Marks, et al. 2011. The southern route "out of Africa": evidence for an early expansion of modern humans into Arabia. *Science* 331(6016):453–456.

Armitage, S.J., S.A. Jasim, A.E. Marks, et al. 2011. The southern route "out of Africa": evidence for an early expansion of modern humans into Arabia. *Science* 331(6016):453.

Árnason, Ú. 2016. The out of Africa hypothesis and the ancestry of recent humans: cherchez la femme (et l'homme). *Gene* 585:9–12.

Arsenault, P.R., F. Pei, R. Lee, et al. 2013. A knock-in mouse model of human *PHD2* gene-associated erythrocytosis establishes a haploinsufficiency mechanism. *Journal of Biological Chemistry* 288(47):33571–33584.

Aswegan, A.L., K.D. Josephson, R. Mowbray, et al. 1997. Autosomal dominant hypohidrotic ectodermal dysplasia in a large family. *American Journal of Medical Genetics* 72(4):462–467.

Athreya, S., and M.M. Glantz. 2003. Impact of character correlation and variable groupings on modern human population tree resolution. *American Journal of Physical Anthropology* 122:134–146.

Atkinson, Q.D. 2011. Phonemic diversity supports a serial founder effect model of language expansion from Africa. *Science* 332:346–349.

Atkinson, Q.D., R.D. Gray, and A.J. Drummond. 2008. mtDNA variation predicts population size in humans and reveals a major Southern Asian chapter in human prehistory. *Molecular Biology and Evolution* 25(2):468–474. doi:10.1093/molbev/msm277

Aubret, F., and R. Shine. 2009. Genetic assimilation and the postcolonization erosion of phenotypic plasticity in island tiger snakes. *Current Biology* 19(22):1932–1936.

Augusto, D.G., B.Z. Piovezan, L.T. Tsuneto, et al. 2013. KIR gene content in Amerindians indicates influence of demographic factors. *PLoS One* 8:e56755.

Autissier, P., C. Soulas, T.H. Burdo, et al. 2010. Immunophenotying of lymphocyte, monocyte and dendritic cell subsets in normal rhesus macaques by 12-color flow cytometry: clarification on DC heterogeneity. *Journal of Immunological Methods* 360:119–128. doi:10.1016/j.jim.2010.06.017.Immunophenotyping

Auton A., A. Fledel-Alon, S. Pfeifer, et al. 2012. A fine-scale chimpanzee genetic map from population sequencing. *Science* 336:193–198.

Avise, J.C. 1994. *Molecular Markers, Natural History and Evolution.* Chapman & Hall, New York. doi:10.1007/978-1-4615-2381-9

Awadalla, P., J. Gauthier, R.A. Myers, et al. 2010. Direct measure of the de novo mutation rate in autism and schizophrenia cohorts. *American Journal of Human Genetics* 87(3):316–324. doi:10.1016/j.ajhg.2010.07.019

Baab, K.L., J.M.G. Perry, F.J. Rohlf, et al. 2014. Phylogenetic, ecological, and allometric correlates of cranial shape in Malagasy lemuriforms. *Evolution* 68:1450–1468.

Bachmanov, A.A., N.P. Bosak, C. Lin, et al. 2014. Genetics of taste receptors. *Current Pharmaceutical Design* 20:2669–2683.

Bae, C.J. 2010. The late Middle Pleistocene hominin fossil record of eastern Asia: synthesis and review. *American Journal of Physical Anthropology* 143(Suppl 51):75–93.

Bailey, J.M. 1998. Can behavior genetics contribute to evolutionary behavioral science? In: C. Crawford, and D.L. Krebs (eds.), *Handbook of Evolutionary Psychology: Ideas, Issues, and Applications.* Psychology Press, New York, pp. 211–233.

Bailey, J.N., S.E. Breidenthal, M.J. Jorgensen, et al. 2007. The association of *DRD4* and novelty seeking is found in a nonhuman primate model. *Psychiatric Genetics* 17:23–27.

Baize, S., D. Pannetier, L. Oestereich, et al. 2014. Emergence of Zaire ebola virus disease in Guinea. *New England Journal of Medicine* 371(15):1418–1425. doi:10.1056/NE.J.M.oa1404505

Baker, P., and M. Little (eds.). 1976. *Man in the Andes: A Multidisciplinary Study of High-Altitude Quechua.* Dowden, Hutchinson, and Ross, Inc., Stroudsbourg, PA.

Ballestar, E., and A.P. Wolffe. 2001. Methyl-CpG-binding proteins. Targeting specific gene repression. *European Journal of Biochemistry* 268(1):1–6.

Ballinger, S.W., T.G. Schurr, A. Torroni, et al. 1992. Southeast Asian mitochondrial DNA analysis reveals genetic continuity of ancient mongoloid migrations. *Genetics* 130:139–152.

Bamshad, M., and S.P. Wooding. 2003. Signatures of natural selection in the human genome. *Nature Reviews Genetics* 4:99–111.

Bannister, A.J., and T. Kouzarides. 2011. Regulation of chromatin by histone modifications. *Cell Research* 21(3):381–395.

Barault, L., and R.C. Rancourt. 2012. Laboratory methods in epigenetic epidemiology. In: K.B. Michels (ed.), *Epigenetic Epidemiology*. Springer Verlag, Dordrecht, pp. 37–56.

Bardill, J., A.C. Bader, N.A. Garrison, et al. 2018. Advancing the ethics of paleogenomics. *Science* 360(6387):384–385.

Barker, D.J. 1988. Childhood causes of adult diseases. *Archives of Disease in Childhood* 63(7):867–869.

Barker, J. 2004. The human genome diversity project: "peoples," "populations," and the cultural politics of identification. *Cultural Studies* 18(4):578–613.

Barker, G., H. Barton, M. Bird, et al. 2007. The 'human revolution' in lowland tropical Southeast Asia: the antiquity and behavior of anatomically modern humans at Niah Cave (Sarawak, Borneo). *Journal of Human Evolution* 52(3):243–261.

Barker, E.D., H. Larsson, E. Viding, et al. 2009. Common genetic but specific environmental influences for aggressive and deceitful behaviors in preadolescent males. *Journal of Psychopathology and Behavioral Assessment* 31:299–308.

Barnes, C.P., S. Filippi, M.P.H. Stumpf, et al. 2012. Considerate approaches to constructing summary statistics for ABC model selection. *Statistics and Computing* 22(6):1181–1197. doi:10.1007/s11222-012-9335-7

Barnett, J.H., and J.W. Smoller. 2009. The genetics of bipolar disorder. *Neuroscience* 164(1):331–343.

Barnicot, N., and J. Woodburn. 1975. Colour-blindness and sensitivity to PTC in Hadza. *Annals of Human Biology* 2:61–68.

Barreiro, L.B., J.C. Marioni, R. Blekhman, et al. 2010. Functional comparison of innate immune signaling pathways in primates. *PLoS Genetics* 6(12):e1001249. doi:10.1371/journal.pgen.1001249

Barreiro, L.B., and L. Quintana-Murci. 2010. From evolutionary genetics to human immunology: how selection shapes host defence genes. *Nature Reviews Genetics* 11(1):17–30. doi:nrg2698 pii: 10.1038/nrg2698

Barrett, R., C.W. Kuzawa, T. McDade, et al. 1998. Emerging and re-emerging infectious diseases: the third epidemiologic transition. *Annual Review of Anthropology* 27:247–271.

Bartolomei, M.S. 2009. Genomic imprinting: employing and avoiding epigenetic processes. *Genes & Development* 23(18):2124–2133.

Basler, C.F., A.H. Reid, J.K. Dybing, et al. 2001. Sequence of the 1918 pandemic influenza virus nonstructural gene (NS) segment and characterization of recombinant viruses bearing the 1918 NS genes. *Proceedings of the National Academy of Sciences of the USA* 98(5):2746–2751.

Basu Mallick, C., F.M. Iliescu, M. Mols, et al. 2013. The light skin allele of *SLC24A5* in south Asians and Europeans shares identity by descent. *PLoS Genetics* 9(11):e1003912. doi:10.1371/journal.pgen.1003912

Batini, C., P. Hallast, D. Zadik, et al. 2015. Large-scale recent expansion of European patrilineages shown by population resequencing. *Nature Communications* 6:7152.

Beall, C.M., L.A. Almasy, J. Blangero, et al. 1999. Percent of oxygen saturation of arterial hemoglobin among Bolivian Aymara at 3,900-4,000 m. *American Journal of Physical Anthropology* 108(1):41–51.

Beall, C.M., J. Blangero, S. Williams-Blangero, et al. 1994. Major gene for percent of oxygen saturation of arterial hemoglobin in Tibetan highlanders. *American Journal of Physical Anthropology* 95(3):271–276.

Beall, C.M., G.M. Brittenham, F. Macuaga, et al. 1990. Variation in hemoglobin concentration among samples of high-altitude natives in the Andes and the Himalayas. *American Journal of Human Biology* 2(6):639–651.

Beall, C.M., G.M. Brittenham, K.P. Strohl, et al. 1998. Hemoglobin concentration of high-altitude Tibetans and Bolivian Aymara. *American Journal of Physical Anthropology* 106(3):385–400.

Beall, C.M., G.L. Cavalleri, L. Deng, et al. 2010. Natural selection on *EPAS1* (HIF2alpha) associated with low hemoglobin concentration in Tibetan highlanders. *Proceedings of the National Academy of Sciences of the USA* 107(25):11459–11464.

Beall, C.M., M.J. Decker, G.M. Brittenham, et al. 2002. An Ethiopian pattern of human adaptation to high-altitude hypoxia. *Proceedings of the National Academy of Sciences of the USA* 99(26):17215–17218.

Beall, C.M., and M.C. Goldstein. 1987. Hemoglobin concentration of pastoral nomads permanently resident at 4,850-5,450 meters in Tibet. *American Journal of Physical Anthropology* 73(4):433–438.

Beall, C.M., and A.B. Reichsman. 1984. Hemoglobin levels in a Himalayan high altitude population. *American Journal of Physical Anthropology* 63(3):301–306.

Beall, C.M., K.P. Strohl, J. Blangero, et al. 1997a. Ventilation and hypoxic ventilatory response of Tibetan and Aymara high altitude natives. *American Journal of Physical Anthropology* 104(4):427–447.

Beall, C.M., K.P. Strohl, J. Blangero, et al. 1997b. Quantitative genetic analysis of arterial oxygen saturation in Tibetan highlanders. *Human Biology* 69(5):597–604.

Beals, K.L. 1972. Head form and climatic stress. *American Journal of Physical Anthropology* 37(1):85–92.

Bearden, C.E., V.I. Reus, and N.B. Freimer. 2004. Why genetic investigation of psychiatric disorders is so difficult. *Current Opinion in Genetics & Development* 14:280–286.

Beaumont, M., W. Zhang, and D.J. Balding. 2002. Approximate Bayesian computation in population genetics. *Genetics* 162:2025–2035.

Beauval, C., B. Maureille, F. Lacrampe-Cuyaubere, et al. 2005. A late Neandertal femur from Les Rochers-de-Villeneuve, France. *Proceedings of the National Academy of Sciences of the USA* 102(20):7085–7090.

Beckwith, C. 2009. *Empires of the Silk Road: A History of Central Eurasia from the Bronze Age to the Present.* Princeton University Press, Princeton, NJ.

Behar, D.M., R. Villems, H. Soodyall, et al. 2008. The dawn of human matrilineal diversity. *American Journal of Human Genetics* 82(5):1130–1140. doi:10.1016/j.ajhg.2008.04.002

Behrens, M., H.C. Gunn, P.C.M. Ramos, et al. 2013. Genetic, functional, and phenotypic diversity in *TAS2R38*-mediated bitter taste perception. *Chemical Senses* 38:475–484.

Behrens, M., and W. Meyerhof. 2011. Gustatory and extragustatory functions of mammalian taste receptors. *Physiology & Behavior* 105:4–13.

Beinin, J. 2010. Knowing the "Other": Arabs, Islam, and the West. In: H.R. Markus, and P.M.L. Moya (eds.), *Doing Race: 21 Essays for the 21st Century.* W. W. Norton & Company, New York.

Beleza, S., N.A. Johnson, S.I. Candille, et al. 2013a. Genetic architecture of skin and eye color in an African-European admixed population. *PLoS Genetics* 9(3):e1003372. doi:10.1371/journal.pgen.1003372

Beleza, S., A.M. Santos, B. McEvoy, et al. 2013b. The timing of pigmentation lightening in Europeans. *Molecular Biology and Evolution* 30(1):24–35. doi:10.1093/molbev/mss207

Bennett, A., S.R. Sain, E. Vargas, et al. 2008. Evidence that parent-of-origin affects birth-weight reductions at high altitude. *American Journal of Human Biology* 20(5):592–597.

Benoist, J. 1964. Saint-Barthelemy: physical anthropology of an isolate. *American Journal of Physical Anthropology* 22:473–487.

Benton, M.J., and P.C. Donoghue. 2007. Paleontological evidence to date the tree of life. *Molecular Biology and Evolution* 24(1):26–53. doi:10.1093/molbev/msl150

Berg, J.J., and G. Coop. 2014. A population genetic signal of polygenic adaptation. *PLoS Genetics* 10(8):e1004412.

Bergmann, C. 1847. *Ueber die Verhaltnisse der Warmeokonomie der thiere zu ihrer grosse.* Gottinger Studien 3:595–708.

Bernal, V., S.I. Perez, and P.N. Gonzalez. 2006. Variation and causal factors of craniofacial robusticity in Patagonian hunter-gatherers from the late Holocene. *American Journal of Human Biology* 18:748–765.

Bersaglieri, T., P.C. Sabeti, N. Patterson, et al. 2004. Genetic signatures of strong recent positive selection at the lactase gene. *American Journal of Human Genetics* 74(6):1111–1120.

Betti, L., F. Balloux, W. Amos, et al. 2009. Distance from Africa, not climate, explains within-population phenotypic diversity in humans. *Proceedings of the Royal Society, B* 276:809–814.

Betti, L., F. Balloux, T. Hanihara, et al. 2010. The relative role of drift and selection in shaping the human skull. *American Journal of Physical Anthropology* 141:76–82.

Betti, L., N. von Cramon-Taubadel, A. Manica, et al. 2013. Global geometric morphometric analyses of the human pelvis reveal substantial neutral population history effects, even across sexes. *PLoS ONE* 8(2):e55909.

Betti, L., N. von Cramon-Taubadel, A. Manica, et al. 2014. The interaction of neutral evolutionary processes with climatically-driven changes in the 3D shape of the human os coxae. *Journal of Human Evolution* 73:64–74.

Bettinger, R.L., and M.A. Baumhoff. 1982. The Numic spread: Great Basin cultures in competition. *American Antiquity* 47:485–503. doi:10.2307/280231

Beunen, G., M. Thomis, H.H. Maes, et al. 2000. Genetic variance of adolescent growth in stature. *Annals of Human Biology* 27(2):173–186.

Bianchi, N.O., C.I. Catanesi, G. Bailliet, et al. 1998. Characterization of ancestral and derived Y-chromosome haplotypes of New World native populations. *American Journal of Human Genetics* 63:1862–1871.

Bigham, A., M. Bauchet, D. Pinto, et al. 2010. Identifying signatures of natural selection in Tibetan and Andean populations using dense genome scan data. *PLoS Genetics* 6:e1001116.

Bigham, A.W., C.G. Julian, M.J. Wilson, et al. 2014. Maternal *PRKAA1* and *EDNRA* genotypes are associated with birth weight, and *PRKAA1* with uterine artery diameter and metabolic homeostasis at high altitude. *Physiological Genomics* 46(18):687–697.

Bigham, A.W., M. Kiyamu, F. Leon-Velarde, et al. 2008. Angiotensin-converting enzyme genotype and arterial oxygen saturation at high altitude in Peruvian Quechua. *High Altitude Medicine & Biology* 9(2):167–178.

Bigham, A.W., and F.S. Lee. 2014. Human high-altitude adaptation: forward genetics meets the HIF pathway. *Genes & Development* 28(20):2189–2204.

Bigham, A.W., X. Mao, R. Mei, et al. 2009. Identifying positive selection candidate loci for high-altitude adaptation in Andean populations. *Human Genomics* 4(2): 79–90.

Bigham, A.W., M.J. Wilson, C.G. Julian, et al. 2013. Andean and Tibetan patterns of adaptation to high altitude. *American Journal of Human Biology* 25(2):190–197.

Bird, A.P. 1995. Gene number, noise reduction and biological complexity. *Trends in Genetics* 11(3):94–100.

Bird, A. 2002. DNA methylation patterns and epigenetic memory. *Genes & Development* 16(1):6–21.

Birdsell, J.B. 1958. On population structure in generalized hunting and gathering populations. *Evolution* 12:189–205. doi:10.2307/2406029

Birdsell, J.B. 1988. Some predictions for the Pleistocene based on equilibrium systems among recent hunter-gatherers. In: R.B. Lee, and I. DeVore (eds.), *Man the Hunter.* Aldine de Gruyter, New York, pp. 229–240.

Bishop, T., N.P. Talbot, P.J. Turner, et al. 2013. Carotid body hyperplasia and enhanced ventilatory responses to hypoxia in mice with heterozygous deficiency of *PHD2*. *Journal of Physiology* 591(Pt. 14):3565–3577.

Bisso-Machado, R., M.C. Bortolini, and F.M. Salzano. 2012. Uniparental genetic markers in South Amerindians. *Genetics and Molecular Biology* 35:365–387.

Bisso-Machado, R., V. Ramallo, F.M. Salzano, et al. 2016. La dispersión del *Homo sapiens* y el poblamiento temprano de América. In: L. Madrigal, and R. González-José (eds.), *Introducción a la Antropología Biológica*. Asociación Latinoamericana de Antropología Biológica, Santiago de Chile, Chile. pp. 467–494.

Bittles, A.H. 2004. Genetic aspects of inbreeding and incest. In: A.P. Wolf, and W.H. Durham (eds.), *Inbreeding, Incest, and the Incest Taboo.* Stanford University Press, Palo Alto, CA, pp. 38–60.

Bittles, A.H., and M.L. Black. 2010. Consanguinity, human evolution, and complex diseases. *Proceedings of the National Academy of Sciences* 107:1779–1786.

Bjarnason, A., A.T. Chamberlain, and C.A. Lockwood. 2011. A methodological investigation of hominoid craniodental morphology and phylogenetics. *Journal of Human Evolution* 60:47–57.

Bjorkman, P.J., M.A. Saper, B. Samraoui, et al. 1987a. The foreign antigen binding site and T cell recognition regions of class I histocompatibility antigens. *Nature* 329(October). doi:10.1038/329512a0

Bjorkman, P.J., M.A. Saper, B. Samraoui, et al. 1987b. Structure of the human class I histocompatibility antigen, HLA-A2. *Nature* 329(October). doi:10.1038/329506a0

Black, T.J. 2011. Extraction of mitochondrial DNA from prehistoric dental calculus. MS thesis, Central Washington University, Ellensburg, WA.

Blekhman, R., J.C. Marioni, P. Zumbo, et al. 2010. Sex-specific and lineage-specific alternative splicing in primates. *Genome Research* 20:180–189.

Blekhman, R., A. Oshlack, A.E. Chabot, et al. 2008. Gene regulation in primates evolves under tissue-specific selection pressures. *PLoS Genetics* 4:e1000271.

Blomquist, G.E. 2009. Environmental and genetic causes of maturational differences among rhesus macaque matrilines. *Behavioral Ecology and Sociobiology* 63:1345–1352.

Bloss, C.S., N.J. Schork, and E.J. Topol. 2011. Effect of direct-to-consumer genomewide profiling to assess disease risk. *New England Journal of Medicine* 364(6):524–534. doi:10.1056/NE.J.M.oa1011893

Blum, H.F. 1961. Does the melanin pigment of human skin have adaptive value? An essay in human ecology and the evolution of race. *Quarterly Review of Biology* 36:50–63.

Blum, M.G.B., M.A. Nunes, D. Prangle, et al. 2013. A comparative review of dimension reduction methods in approximate Bayesian computation. *Statistical Science* 28(2):189–208. doi:10.1214/12-sts406

Boas, F. 1912. Changes in bodily form of descendants of immigrants. *American Anthropologist* 14:530–562.

Boas, F. 1932. Studies in growth. *Human Biology* 4:307–350.

Bock, R.D., S.H.C. du Toit, and D. Thissen. 1994. *AUXAL: Auxological Analysis of Longitudinal Measurements of Human Stature.* Scientific Software International, Chicago.

Bock, R., S. du Toit, and D. Thissen. 2003. *Auxal: Auxological Analysis of Longitudinal Measurements of Human Stature, Version 3.* Scientific Software International, Chicago.

Bodmer, W.F., and L.L. Cavalli-Sforza. 1968. A migration matrix model for the study of random genetic drift. *Genetics* 59:565–592.

Bodmer, W.F., and L.L. Cavalli-Sforza. 1974. The analysis of genetic variation using migration matrices. In: J.F. Crow, and C. Denniston (eds.), *Genetic Distance.* Plenum Press, New York, pp. 45–61.

Bodmer, W.F., and L.L. Cavalli-Sforza. 1976. *Genetics, Evolution, and Man.* W.H. Freeman, San Francisco.

Boëda, E., I. Clemente-Conte, M. Fontugne, et al. 2014. A new Late Pleistocene archaeological sequence in South America: the Vale da Pedra Furada (Piauí, Brazil). *Antiquity* 88:927–955.

Boerwinkle, E., R. Chakraborty, and C.F. Sing. 1986. The use of measured genotype information in the analysis of quantitative phenotypes in man. I. Models and analytical methods. *Annals of Human Genetics*50(Pt 2):181–194.

Bogaard, A. 2004. *Neolithic Farming in Central Europe.* Routledge, New York.

Boissy, R.E., H. Zhao, W.S. Oetting, et al. 1996. Mutation in and lack of expression of tyrosinase-related protein-1 (*TRP-1*) in melanocytes from an individual with brown oculocutaneous albinism: a new subtype of albinism classified as *OCA3. American Journal of Human Genetics* 58(6):1145–1156.

Boks, M.P., E.M. Derks, D.J. Weisenberger, et al. 2009. The relationship of DNA methylation with age, gender and genotype in twins and healthy controls. *PLoS One* 4(8):e6767.

Bollongino, R., O. Nehlich, M.P. Richards, et al. 2013. 2000 years of parallel societies in stone age Central Europe. *Science* 342:479–481.

Bolnick D.A., H.M. Bonine, J. Mata-Míguez, et al. 2012. Nondestructive sampling of human skeletal remains yields ancient nuclear and mitochondrial DNA. *American Journal of Physical Anthropology* 147:293–300

Bolnick, D.A., D. Fullwiley, T. Duster, et al. 2007. The science and business of genetic ancestry testing. *Science* 318(5849):399–400.

Bolnick, D.A., J.A. Raff, L.C. Springs, et al. 2016. Native American genomics and population histories. *Annual Review of Anthropology* 45:319–340.

Bolnick, D.A., B.A. Shook, L. Campbell, et al. 2004. Problematic use of Greenberg's linguistic classification of the Americas in studies of Native American genetic variation. *American Journal of Human Genetics* 75:519–523.

Bolnick, D.A., and D.G. Smith. 2007. Migration and social structure among the Hopewell: evidence from ancient DNA. *American Antiquity* 72:627–644.

Bonatto, S.L., A.J. Redd, F.M. Salzano, et al. 1996. Lack of ancient Polynesian-Amerindian contact. *American Journal of Human Genetics* 59:253–258.

Bonatto, S.L., and F.M. Salzano. 1997a. A single and early migration for the peopling of the Americas supported by mitochondrial DNA sequence data. *Proceedings of the National Academy of Sciences of the USA* 94:1866–1871.

Bonatto, S.L., and F.M. Salzano. 1997b. Diversity and age of the four major mtDNA haplogroups, and their implications for the peopling of the New World. *American Journal of Human Genetics* 61:1413–1423.

Booij, L., D. Wang, M.L. Lévesque, et al. 2013. Looking beyond the DNA sequence: the relevance of DNA methylation processes for the stress-diathesis model of depression. *Philosophical Transactions of the Royal Society B: Biological Sciences* 368(1615):20120251.

Bortolini, M.C., R. González-José, S.L. Bonatto, et al. 2014. Reconciling pre-Columbian settlement hypotheses requires integrative, multidisciplinary, and model-bound approaches. *Proceedings of the National Academy of Sciences of the USA* 111:E213–E214.

Bortolini, M.C., F.M. Salzano, M.G. Thomas, et al. 2003. Y-chromosome for differing ancient demographic histories in the Americas. *American Journal of Human Genetics* 73:524–539.

Bos, K.I., K.M. Harkins, A. Herbig, et al. 2014. Pre-Columbian mycobacterial genomes reveal seals as a source of New World human tuberculosis. *Nature* 514:494–497.

Bos, K.I., V.J. Schuenemann, G.B. Golding, et al. 2011. A draft genome of *Yersinia pestis* from victims of the black death. *Nature* 478(7370):506–510. doi:10.1038/nature10549

Boss, O., S. Samec, A. Paoloni-Giacobino, et al. 1997. Uncoupling protein-3: a new member of the mitochondrial carrier family with tissue-specific expression. *FEBS Letters* 408(1):39–42.

Bosten, J.M., J.D. Robinson, G. Jordan, et al. 2005. Multidimensional scaling reveals a color dimension unique to 'color-deficient' observers. *Current Biology* 15:R950–R952.

Bouckaert, R., J. Heled, D. Kühnert, et al. 2014. BEAST2: a software platform for Bayesian evolutionary analysis. *PLoS Computational Biology* 10(4):e1003537. doi:10.1371/journal.pcbi.1003537

Bouckaert, R., P. Lemey, M. Dunn, et al. 2012. Mapping the origins and expansion of the Indo-European language family. *Science* 337(6097):957–960. doi:10.1126/science.1219669

Bouwman, A.S., S.L. Kennedy, R. Muller, et al. 2012. Genotype of a historic strain of *Mycobacterium tuberculosis*. *Proceedings of the National Academy of Sciences of the USA* 109:18511–18516.

Box, N.F., J.R. Wyeth, L.E. O'Gorman, et al. 1997. Characterization of melanocyte stimulating hormone receptor variant alleles in twins with red hair. *Human Molecular Genetics* 6(11):1891–1897.

Boyson, J.E., S.N. McAdam, A. Gallimore, et al. 1995. The MHC E locus in macaques is polymorphic and is conserved between macaques and humans. *Immunogenetics* 41:59–60. doi:10.1007/BF00182314

Boyson, J.E., C. Shufflebotham, L.F. Cadavid, et al. 1996. The MHC class I genes of the rhesus monkey. Different evolutionary histories of MHC class I and II genes in primates. *Journal of Immunology (Baltimore, Md.)* 156(12):4656–4665. http://www.ncbi.nlm.nih.gov/pubmed/8648109

Brace, C.L., A.R. Nelson, N. Seguchi, et al. 2001. Old World sources of the first New World human inhabitants: a comparative craniofacial view. *Proceedings of the National Academy of Sciences of the USA* 98:10017–10022.

Bradley, B.J. 2008. Reconstructing phylogenies and phenotypes: a molecular view of human evolution. *Journal of Anatomy* 212(4):337–353. doi:10.1111/j.1469-7580.2007.00840.xbrom

Bradley, B.J., and R.R. Lawler. 2011. Linking genotypes, phenotypes, and fitness in wild primate populations. *Evolutionary Anthropology* 20:104–119.

Bradley, B., and D. Stanford. 2004. The North Atlantic ice-edge corridor: a possible Palaeolithic route to the New World. *World Archaeology* 36:459–478.

Braendle, C., and T. Flatt. 2006. A role for genetic accommodation in evolution? *BioEssays* 28(9):868–873.

Braje, T.J., T.D. Dillehay, J.M. Erlandson, et al. 2017. Finding the first Americans. *Science* 358:592–594.

Bramanti, B., M.G. Thomas, W. Haak, et al. 2009. Genetic discontinuity between local hunter-gatherers and Central Europe's first farmers. *Science* 326(5949):137–140.

Branda, R.F., and J.W. Eaton. 1978. Skin color and nutrient photolysis: an evolutionary hypothesis. *Science* 201(4356):625–626.

Brandt, G., W. Haak, C.J. Adler, et al. 2013. Ancient DNA reveals key stages in the formation of central European mitochondrial genetic diversity. *Science* 342(6155):257–261.

Bräuer, G. 1981. New evidence on the transitional period between Neanderthal and modern man. *Journal of Human Evolution* 10(6):467–474.

Brent, L.J.N., S.R. Heilbronner, J.E. Horvath, et al. 2013. Genetic origins of social networks in rhesus macaques. *Scientific Reports* 3:1042.

Brent, L.J.N., and A.D. Melin. 2014. The genetic basis of primate behavior: genetics and genomics in field-based primatology. *International Journal of Primatology* 35:1–10.

Briggs, A.W., J.M. Good, R.E. Green, et al. 2009. Targeted retrieval and analysis of five Neandertal mtDNA genomes. *Science* 325(5938):318–321. doi:10.1126/science.1174462

Briggs, A.W., U. Stenzel, P.L. Johnson, et al. 2007. Patterns of damage in genomic DNA sequences from a Neandertal. *Proceedings of the National Academy of Sciences of the USA* 104:14616–14621.

Briggs A.W., U. Stenzel, M. Meyer, et al. 2010. Removal of deaminated cytosines and detection of in vivo methylation in ancient DNA. *Nucleic Acids Research* 38(6):e87

Brigham-Grette, J., L.M. Gualtieri, O.Y. Glushkova, et al. 2003. Chlorine-36 and 14C chronology support a limited last glacial maximum across Central Chukotka, Northeastern Siberia, and no Beringian ice sheet. *Qualitative Research* 59:386–398.

Broberg, K., S. Ahmed, K. Engström, et al. 2014. Arsenic exposure in early pregnancy alters genome-wide DNA methylation in cord blood, particularly in boys. *Journal of Developmental Origins of Health and Disease* 5(4):288–298.

Broman, K.W., J.C. Murray, V.C. Sheffield, et al. 1998. Comprehensive human genetic maps: individual and sex-specific variation in recombination. *American Journal of Human Genetics* 63(3):861–869. doi:10.1086/302011

Bromham, L. 2011. The genome as a life-history character: why rate of molecular evolution varies between mammal species. *Philosophical Transactions of the*

Royal Society B: Biological Sciences 366(1577):2503–2513. doi:10.1098/Rstb.2011.0014

Bromham, L., and D. Penny. 2003. The modern molecular clock. *Nature Reviews Genetics* 4:216–224.

Brotherton, P., W. Haak, J. Templeton, et al. 2013. Neolithic mitochondrial haplogroup H genomes and the genetic origins of Europeans. *Nature Communications* 4:1764. doi:10.1038/ncomms2656

Brown, P. 1999. The first modern East Asians? Another look at upper cave 101, Liujiang and Minatogawa 1. In: K. Omoto (ed.), *Interdisciplinary Perspectives on the Origins of the Japanese*. International Research Center for Japanese Studies, Kyoto, pp. 105–130.

Brown, M.D., S.H. Hosseini, A. Torroni, et al. 1998. mtDNA haplogroup X: an ancient link between Europe/Western Asia and North America? *American Journal of Human Genetics* 63:1852–61.

Brown, J.H., T.S. Jardetzky, J.C. Gorga, et al. 1993. Three-dimensional structure of the human class II histocompatibility antigen HLA-DR1. *Nature* 364(July). doi:10.1038/364033a0.

Brown, P., T. Sutikna, M.J. Morwood, et al. 2004. A new small-bodied hominin from the Late Pleistocene of Flores, Indonesia. *Nature* 431(7012):1055–1061.

Brown, R.P., and Z. Yang. 2010. Bayesian dating of shallow phylogenies with a relaxed clock. *Systematic Biology* 59(2):119–131. doi:10.1093/sysbio/syp082

Brues, A.M. 1940. Sibling resemblances as evidence for the genetic determination of traits of the eye, skin and hair in man. PhD thesis, Harvard University.

Brues, A.M. 1977. *People and Races*. Macmillan Press, New York.

Brutsaert, T.D., M. Araoz, R. Soria, et al. 2000. Higher arterial oxygen saturation during submaximal exercise in Bolivian Aymara compared to European sojourners and Europeans born and raised at high altitude. *American Journal of Physical Anthropology* 113(2):169–181.

Brutsaert, T.D., E.J. Parra, M.D. Shriver, et al. 2005. Ancestry explains the blunted ventilatory response to sustained hypoxia and lower exercise ventilation of Quechua altitude natives. *American Journal of Physiology: Regulatory, Integrative and Comparative Physiology* 289(1):R225–R234.

Bryc, L., A. Auton, M.R. Nelson, et al. 2010. Genome-wide patterns of population structure and admixture in west Africans and African Americans. *Proceedings of the National Academy of Sciences* 107:786–791.

Bryk, J., E. Hardouin, I. Pugach, et al. 2008. Positive selection in East Asians for an *EDAR* allele that enhances NF-kappaB activation. *PLoS One* 3(5):e2209–e2209.

Buck, L., and R. Axel. 1991. A novel multigene family may encode odorant receptors: a molecular basis for odor recognition. *Cell* 65:175–187.

Bufe, B., P.A.S. Breslin, C. Kuhn, et al. 2005. The molecular basis of individual differences in phenylthiocarbamide and propylthiouracil bitterness perception. *Current Biology* 15:322–327.

Bulmer, M.G. 1980. *The Mathematical Theory of Quantitative Genetics*. Oxford University Press, Oxford.

Bunce, J.A. 2015. Incorporating ecology and social system into formal hypotheses to guide field studies of color vision in primates. *American Journal of Primatology* 77:516–526.

Burbano, H.A., R.E. Green, T. Maricic, et al. 2012. Analysis of human accelerated DNA regions using archaic Hominin genomes. *PLoS One* 7(3):e32877.

Burggren, W.W., and D. Crews. 2014. Epigenetics in comparative biology: why we should pay attention. *Integrative and Comparative Biology* 54(1):7–20.

Burgner, D., S.E. Jamieson, and J.M. Blackwell. 2006. Genetic susceptibility to infectious diseases: big is beautiful, but will bigger be even better? *The Lancet Infectious Diseases* 6(10):653–663. doi:10.1016/S1473-3099(06)70601-6

Busby, G.B., F. Brisighelli, P. Sanchez-Diz, et al. 2012. The peopling of Europe and the cautionary tale of Y chromosome lineage R-M269. *Proceedings Biological Sciences/The Royal Society* 279(1730):884–892.

Buss, D. 1985. Human mate selection: opposites are sometimes said to attract, but in fact we are likely to marry someone who is similar to us in almost every variable. *American Scientist* 73(1):47–51.

Bustamante, C., and F. De La Vega. 2011. Genomics for the world. *Nature* 475: 163–165.

Butler, C.D. 2012. Infectious disease emergence and global change: thinking systemically in a shrinking world. *Infectious Diseases of Poverty* 1(1):5. doi:10.1186/2049-9957-1-5

Cabana, G.S., K. Hunley, and F. Kaestle. 2008. Population continuity or replacement? A novel computer simulation approach and its application to the numic expansion (Western Great Basin, USA). *American Journal of Physical Anthropology* 135:438–447. doi:10.1002/ajpa.20764

Cadavid, L.F., A.L. Hughes, and D.I. Watkins. 1996. MHC class I-processed pseudogenes in new world primates provide evidence for rapid turnover of MHC class I genes. *Journal of Immunology* 157:2403–2609. http://www.jimmunol.org/content/157/6/2403

Cadavid, L.F., B.E. Mejia, and D.I. Watkins. 1999. MHC class I genes in a new world primate, the cotton-top Tamarin (*Saguinus oedipus*), have evolved by an active process of loci turnover. *Immunogenetics* 49:196–205. doi:10.1007/s002510050480

Cadavid, L.F., and D.I. Watkins. 1997. The duplicative nature of the MHC class I genes an evolutionary perspective. *European Journal of Immunogenetics* 24:313–322. doi:10.1111/j.1365-2370.1997.tb00026.x

Cadien, J.D., E.F. Harris, W.P. Jones, et al. 1976. Biological lineages, skeletal populations, and microevolution. *Yearbook of Physical Anthropology* 18:194–201.

Cagliani, R., M. Fumagalli, M. Biasin, et al. 2010. Long-term balancing selection maintains trans-specific polymorphisms in the human *TRIM5* gene. *Human Genetics* 128:577–588.

Cagliani, R., M. Fumagalli, S. Riva, et al. 2008. The signature of long-standing balancing selection at the human defensin beta-1 promoter. *Genome Biology* 9(9):R143.1–R143.11.

Cahill, J.A., R.E. Green, T.L. Fulton, et al. 2013. Genomic evidence for island population conversion resolves conflicting theories of polar bear evolution. *PLoS Genetics* 9:e1003345.

Caine, N.G., and N.I. Mundy. 2000. Demonstration of a foraging advantage for trichromatic marmosets (*Callithrix geoffroyi*) dependent on food colour. *Proceedings of the Royal Society B: Biological Sciences* 267:439–444.

Caine, N.G., D. Osorio, and N.I. Mundy. 2010. A foraging advantage for dichromatic marmosets (*Callithrix geoffroyi*) at low light intensity. *Biology Letters* 6:36–38.

Calabrese, P. 2007. A population genetics model with recombination hotspots that are heterogeneous across the population. *Proceedings of the National Academy of Sciences of the USA* 104(11):4748–4752. doi:10.1073/pnas.0610195104

Calcagno, J.M., and A. Fuentes. 2012. What makes us human? Answers from evolutionary anthropology. *Evolutionary Anthropology* 21:182–194.

Calvignac-Spencer, S., S.A. Leendertz, T.R. Gillespie, et al. 2012. Wild great apes as sentinels and sources of infectious disease. *Clinical Microbiology and Infection* 18(6):521–527. doi:10.1111/j.1469-0691.2012.03816.x

Cameron, T.W.M. 1956. *Parasites and Parasitism*. Wiley, New York.

Campbell, M.C., A. Ranciaro, A. Froment, et al. 2012. Evolution of functionally diverse alleles associated with PTC bitter taste sensitivity in Africa. *Molecular Biology and Evolution* 29:1141–1153.

Campbell, M.C., A. Ranciaro, D. Zinshteyn, et al. 2014. Origin and differential selection of allelic variation at *TAS2R16* associated with salicin bitter taste sensitivity in Africa. *Molecular Biology and Evolution* 31:288–302.

de Campos Telles, M.P., and Diniz-Filho JAF. 2005. Multiple Mantel tests and isolation-by-distance, taking into account long-term historical divergence. *Genetics and Molecular Research* 4(4):742–748.

Candille, S.I., D.M. Absher, S. Beleza, et al. 2012. Genome-wide association studies of quantitatively measured skin, hair, and eye pigmentation in four European populations. *PLoS One* 7(10):e48294. doi:10.1371/journal.pone.0048294

Cann, R.L. 1994. mtDNA and Native Americans: a southern perspective. *American Journal of Human Genetics* 55:7–11.

Cann, H.M., C. de Toma, L. Cazes, et al. 2002. A human genome diversity cell line panel. *Science* 296:261–262.

Cann, R.L., M. Stoneking, and A.C. Wilson. 1987. Mitochondrial DNA and human evolution. *Nature* 325:31–36.

Cannon, B., and J. Nedergaard. 2011. Nonshivering thermogenesis and its adequate measurement in metabolic studies. *Journal of Experimental Biology* 214(Pt 2):242–253.

Caramelli, D., C. Lalueza-Fox, S. Condemi, et al. 2006. A highly divergent DNA sequence in a Neandertal individual from Italy. *Current Biology* 16.

Carbone, L., R.A. Harris, S. Gnerre, et al. 2014. Gibbon genome and the fast karyotype evolution of small apes. *Nature* 513:195–201.

Cardini, A., and S. Elton. 2008. Does the skull carry a phylogenetic signal? Evolution and modularity in the guenons. *Biological Journal of the Linnean Society* 93:813–834.

Cardona, A., L. Pagani, T. Antao, et al. 2014. Genome-wide analysis of cold adaptation in indigenous Siberian populations. *PLoS One* 9(5):e98076.

Carroll, S.B. 2008. Evo-devo and an expanding evolutionary synthesis: a genetic theory of morphological evolution. *Cell* 134(1):25–36.

Carson, E.A. 2006. Maximum likelihood estimation of human craniometric heritabilities. *American Journal of Physical Anthropology* 131:169–180.

Carvajal-Rodríguez, A. 2010. Simulation of genes and genomes forward in time. *Current Genomics* 2010(11):58–61.

Carville, A., T.I. Evans, and R.K. Reeves. 2013. Characterization of circulating natural killer cells in Neotropical primates. *PLoS One* 8(11). doi:10.1371/journal.pone.0078793

Castillo-Davis, C.I., F.A. Kondrashov, D.L. Hartl, et al. 2004. The functional genomic distribution of protein divergence in two animal phyla: coevolution, genomic conflict, and constraint. *Genome Research* 14(5):802–811. doi:10.1101/gr.2195604

Caswell, J.L., S. Mallick, D.J. Richter, et al. 2008. Analysis of chimpanzee history based on genome sequence alignments. *PLoS Genetics* 4:e1000057.

Cavalli, G., and R. Paro. 1998. The Drosophila Fab-7 chromosomal element conveys epigenetic inheritance during mitosis and meiosis. *Cell* 93(4):505–518.

Cavalli-Sforza, L.L. 1997. Genes, peoples, and languages. *Proceedings of the National Academy of Sciences of the USA* 94(15):7719–7724.

Cavalli-Sforza, L.L. 2005. The human genome diversity project: past, present and future. *Nature Reviews Genetics* 6(4):333–340.

Cavalli-Sforza, L.L., and F. Cavalli-Sforza. 1995. *The Great Human Diasporas: The History of Diversity and Evolution.* Basic Books, New York.

Cavalli-Sforza, L.L., P. Menozzi, and A. Piazza. 1993. Demic expansions and human evolution. *Science* 259:639–646.

Cavalli-Sforza, L.L., P. Menozzi, and A. Piazza. 1994. *The History and Geography of Human Genes.* Princeton University Press, Princeton, NJ.

Centers for Disease Control and Prevention. 2015. HIV/AIDS: Opportunistic infections. http://www.cdc.gov/hiv/basics/livingwithhiv/opportunisticinfections.html

Cepon, T.J., J.J. Snodgrass, W.R. Leonard, et al. 2011. Circumpolar adaptation, social change, and the development of autoimmune thyroid disorders among the Yakut (Sakha) of Siberia. *American Journal of Human Biology* 23(5):703–709.

Cerezo, M., A. Achilli, A. Olivieri, et al. 2012. Reconstructing ancient mitochondrial DNA links between Africa and Europe. *Genome Research* 22(5):821–826.

Chakraborty, R. 1986. Gene admixture in human populations: models and predictions. *Yearbook of Physical Anthropology* 29:1–43.

Chakraborty, R. 1990. Quantitative traits in relation to population structure: why and how are they used and what do they imply? *Human Biology* 62:147–162.

Chandrashekar, J., M.A. Hoon, N.J.P. Ryba, et al. 2006. The receptors and cells for mammalian taste. *Nature* 444:288–294.

Chang, S.H., S. Jobling, K. Brennan, et al. 2009. Enhanced Edar signalling has pleiotropic effects on craniofacial and cutaneous glands. *PLoS One* 4(10):e7591.

Chaplin, G. 2004. Geographic distribution of environmental factors influencing human skin coloration. *American Journal of Physical Anthropology* 125(3):292–302. doi:10.1002/ajpa.10263

Charlesworth, B. 1998. Adaptive evolution: the struggle for dominance. *Current Biology* 8(14):R502–R504.

Charlesworth, D. 2006. Balancing selection and its effects on sequences in nearby genome regions. *PLoS Genetics* 2(4):e64.

Charrier, C., K. Joshi, J. Coutinho-Budd, et al. 2012. Inhibition of *SRGAP2* function by its human-specific paralogs induces neoteny during spine maturation. *Cell* 149:923–935.

Chatters, J.C., D.J. Kennett, Y. Asmerom, et al. 2014. Late Pleistocene human skeleton and mtDNA link Paleoamericans and modern native Americans. *Science* 344(6185):750–754. doi:10.1126/science.1252619

Chaubey, G., M. Metspalu, Y. Choi, et al. 2011. Population genetic structure in Indian Austroasiatic speakers: the role of landscape barriers and sex-specific admixture. *Molecular Biology and Evolution* 28(2):1013–1024.

Chen, Q.-Y., S. Alarcon, A. Tharp, et al. 2009. Perceptual variation in umami taste and polymorphisms in *TAS1R* taste receptor genes. *American Journal of Clinical Nutrition* 90:770S–779S.

Chesler, E.J., L. Lu, S. Shou, et al. 2005. Complex trait analysis of gene expression uncovers polygenic and pleiotropic networks that modulate nervous system function. *Nature Genetics* 37:233–242.

Cheverud, J.M. 1982. Phenotypic, genetic, and environmental morphological integration in the cranium. *Evolution* 36(3):499–516.

Cheverud, J.M. 1988. A comparison of genetic and phenotypic correlations. *Evolution* 42:958–968.

Cheverud, J.M. 1995. Morphological integration in the saddle-back Tamarin (*Saguinus fuscicollis*). *The American Naturalist* 145(1):63–89.

Chikhi, L., R.A. Nichols, G. Barbujani, et al. 2002. Y genetic data support the Neolithic demic diffusion model. *Proceedings of the National Academy of Sciences of the USA* 99(17):11008–11013.

Chimpanzee Sequencing and Analysis Consortium. 2005. Initial sequence of the chimpanzee genome and comparison with the human genome. *Nature* 437:69–87.

Chiodi, H. 1957. Respiratory adaptations to chronic high altitude hypoxia. *Journal of Applied Physiology* 10(1):81–87.

Chu, Y., and D.R. Corey. 2012. RNA sequencing: platform selection, experimental design, and data interpretation. *Nucleic Acid Therapeutics* 22(4):271–274.

Church, G.M. 2005. The personal genome project. *Molecular Systems Biology* 1(1). doi:10.1038/msb4100040

Cirulli, E.T., and D.B. Goldstein. 2010. Uncovering the roles of rare variants in common disease through whole-genome sequencing. *Nature Reviews Genetics* 11(6):415–425.

Clark, G. 2007. Genetically capitalist? The Malthusian Era, institutions and the formation of modern preferences. Working paper, March. http://www.econ.ucdavis.edu/faculty/gclark/research.html

Clark, G. 2009. *A Farewell to Alms: A Brief Economic History of the World.* Princeton University Press, Princeton, NJ

Clark, A.A., C.D. Dotson, A.E.T. Elson, et al. 2015. *TAS2R* bitter taste receptors regulate thyroid function. *FASEB Journal* 29:164–172.

Clark, P.U., A.S. Dyke, J.D. Shakun, et al. 2009. The last glacial maximum. *Science* 325:710–714.

Clarke, B.C. 1979. The evolution of genetic diversity. *Proceedings of the Royal Society B: Biological Sciences* 205(1161):453–474.

Claw, K., and N. Garrison. 2015. Bringing indigenous researchers to the forefront of genomics. SACNAS News Magazine. January 15. http://sacnas.org/about/stories/sacnas-news/winter-2015-indigenous-genomics

Clement, K., J. Ruiz, A.M. Cassard-Doulcier, et al. 1996. Additive effect of A-->G (-3826) variant of the uncoupling protein gene and the Trp64Arg mutation of the beta 3-adrenergic receptor gene on weight gain in morbid obesity. *International Journal of Obesity and Related Metabolic Disorders* 20(12):1062–1066.

Clemente, F.J., A. Cardona, C.E. Inchley, et al. 2014. A selective sweep on a deleterious mutation in *CPT1A* in Arctic populations. *American Journal of Human Genetics* 95(5):584–589.

Cloud, J. 2010. Why your DNA isn't your destiny. Time, Vol. 175.

Cluzeau, C., S. Hadj-Rabia, E. Bal, et al. 2012. The EDAR370A allele attenuates the severity of hypohidrotic ectodermal dysplasia caused by EDA gene mutation. *British Journal of Dermatology* 166(3):678–681.

Cochran, G., J. Hardy, and H. Harpendingnry. 2006. Natural history of Askenazi intelligence. *Journal of Biosocial Science* 38(5):659–693.

Cochran, G., and H. Harpending. 2009. *The 10,000 Year Explosion: How Civilization Accelerated Human Evolution.* Basic Books, New York.

Cochran, G., J. Hawks, and H. Harpending. 2011. Overdominance and rapid adaptation. July. http://harpending.humanevo.utah.edu

Cohen, M.S., K.E. Muessig, M.K. Smith, et al. 2012. Antiviral agents and HIV prevention: controversies, conflicts, and consensus. *AIDS* 26(13):1585–1598. doi:10.1097/QAD.0b13e3283543e83

Cole, S.T., K. Eiglmeier, J. Parkhill, et al. 2001. Massive gene decay in the leprosy bacillus. *Nature* 409(6823):1007–1111. doi:10.1038/35059006

Collard, M., and B. Wood. 2001. Homoplasy and the early hominid masticatory system: inferences from analyses of extant hominoids and papionins. *Journal of Human Evolution* 41:167–194.

Collard, M., and B. Wood. 2007. Hominin homoiology: an assessment of the impact of phenotypic plasticity on phylogenetic analyses of humans and their fossil relatives. *Journal of Human Evolution* 52:573–584.

Collas, P. 2010. The current state of chromatin immunoprecipitation. *Molecular Biotechnology* 45(1):87–100.

Collins, L.J., B. Schonfeld, and X.S. Chen. 2011. The epigenetics of non-coding RNA. In: T.O. Tollefsbol (ed.), *Handbook of Epigenetics: The New Molecular and Medical Genetics*. Academic Press, Amsterdam, pp. 49–61.

Colonna, M., G. Borsellino, M. Falcot, et al. 1993. HLA-C is the inhibitory ligand that determines dominant resistance to lysis by NK1- and NK2-specific natural killer cells. *Proceedings of the National Academy of Sciences* 90(December):12000–12004.

Cone, R.D., D. Lu, S. Koppula, et al. 1996. The melanocortin receptors: agonists, antagonists, and the hormonal control of pigmentation. *Recent Progress in Hormone Research* 51:287–317; discussion 318.

Connallon, T., and A.G. Clark. 2014. Balancing selection in species with separate sexes: insights from Fisher's geometric model. *Genetics* 197(3):991–1006.

Conrad, D.F., J.E.M. Keebler, M.A. DePristo, et al. 2011. Variation in genome-wide mutation rates within and between human families. *Nature Genetics* 43(7):712–714. doi:10.1038/Ng.862

Cooper, A., K.W. Alt, W. Haak, et al. 2015. Eight thousand years of natural selection in Europe. bioRxiv. http://biorxiv.org/content/biorxiv/early/2015/03/13/016477.full

Cooper, N., R. Griffin, M. Franz, et al. 2012. Phylogenetic host specificity and understanding parasite sharing in primates. *Ecology Letters* 15:1370–1377. doi:10.1111/j.1461-0248.2012.01858.x

Cooper, A., and H.N. Poinar. 2000. Ancient DNA: do it right or not at all. *Science* 289(11):1139.

Cope, A. 2015. Hong Kong is using DNA analysis to publicly shame litterbugs. WeAreChange.Org. http://wearechange.org/hong-kong-is-using-dna-analysis-to-publicly-shame-litterbugs

Cordain, L., J. Brand Miller, S.B. Eaton, et al. 2000. Plant-animal subsistence ratios and macronutrient energy estimations in worldwide hunter-gatherer diets. *American Journal of Clinical Nutrition* 71(3):682–692.

Cordain, L., S.B. Eaton, J. Brand Miller, et al. 2002. The paradoxical nature of hunter-gatherer diets: meat-based, yet non-atherogenic. *European Journal of Clinical Nutrition* 56(1):S42.

Cortessis, V.K., D.C. Thomas, A.J. Levine, et al. 2012. Environmental epigenetics: prospects for studying epigenetic mediation of exposure-response relationships. *Human Genetics* 131(10):1565–1589.

Cosgrove, M.S., J.D. Boeke, and C. Wolberger. 2004. Regulated nucleosome mobility and the histone code. *Nature Structural & Molecular Biology* 11(11):1037–1043.

Costa, F.F. 2008. Non-coding RNAs, epigenetics and complexity. *Gene* 410(1):9–17.

Courtois, G., A. Smahi, and A. Israel. 2001. NEMO/IKKγ: linking NF-κB to human disease. *Trends in Molecular Medicine* 7(10):427–430.

Couzin-Frankel, J. 2010. Researchers to return blood samples to the Yanomamo. *Science* 328:1218.

Cox, M.P. 2008. Accuracy of molecular dating with the rho statistic: deviations from coalescent expectations under a range of demographic models. *Human Biology* 80:335–357.

Cox, M.P., B.R. Holland, M.C. Wilkins, et al. 2013. Reconstructing past changes in locus-specific recombination rates. *BMC Genetics* 14:11. doi:10.1186/1471-2156-14-11

Cox, M.P., T.M. Karafet, J.S. Lansing, et al. 2010. Autosomal and X-linked single nucleotide polymorphisms reveal a steep Asian-Melanesian ancestry cline in eastern Indonesia and a sex bias in admixture rates. *Proceedings of the Royal Society B* 277:1589–1596. doi:10.1098/rspb.2009.2041

Cramer, J.D., T. Gaetano, J.P. Gray, et al. 2013. Variation in scrotal color among widely distributed vervet monkey populations (*Chlorocebus aethiops pygerythrus* and *Chlorocebus aethiops sabaeus*). *American Journal of Primatology* 75:752–762.

Crawford, M.H. 1966. Hemoglobin polymorphism in *Macaca nemestrina*. *Science* 154:398–399.

Crawford, M.H. (ed.). 1976. *The Tlaxcaltecans: Prehistory, Demography, Morphology and Genetics.* University of Kansas Anthropology Series No. 7. University of Kansas Press, Lawrence.

Crawford, M.H. 1980a. Genetic epidemiology in anthropology. *Journal of Medical Anthropology* 43(3):415–422.

Crawford, M.H. 1980b. The breakdown of reproductive isolation in an alpine genetic isolate: Acceglio, Italy. In: A. Ericksson, H. Forsius, H.R. Nevanlinna, and P.L. Workman (eds.), *Population Structure and Genetic Diseases.* Academic Press, New York, pp. 57–71.

Crawford, M.H. 1983. The anthropological genetics of the black Caribs (Garifuna) of central America and the Caribbean. *Yearbook of Physical Anthropology* 25:155–186.

Crawford, M.H. (ed.) 1984. *Current Developments in Anthropological Genetics Vol. III. Black Caribs: A Case Study of Biocultural Adaptation.* Plenum Press, New York.

Crawford, M.H. 1987. Origin and maintenance of genetic variation in Black Carib populations of St. Vincent and Central America. In: D.F. Roberts, and G. DiStefano (eds.), *Genetic Variation and its Maintenance.* Cambridge University Press, Cambridge, pp. 157–180.

Crawford, M.H. 1998. *The Origins of Native Americans. Evidence from Anthropological Genetics.* Cambridge University Press, Cambridge.

Crawford, M.H. 2000a. Anthropological genetics in the 21st century: introduction. *Human Biology* 72(1):3–13.

Crawford, M.H. (ed.). 2000b. *Different Seasons: Biological Aging among the Mennonites of the Midwestern United States.* Publications in Anthropology 21. University of Kansas, Lawrence.

Crawford, M.H. 2007a. Foundations of anthropological genetics. In: M.H. Crawford (ed.), *Anthropological Genetics: Theory, Methods and Applications.* Cambridge University Press, Cambridge, pp. 1–16. doi:10.1017/CBO9781139167222.002

Crawford, M.H. 2007b. The importance of field research in anthropological genetics: methods, experiences and results. In: M.H. Crawford (ed.), *Anthropological Genetics.* Cambridge University Press, Cambridge, pp. 79–111.

Crawford, M.H., and K.G. Beaty. 2013. DNA fingerprints in anthropological genetics: past, present and future.(in honor of sir Alec Jeffreys) *Investigative Genetics* 4(1):23–33.

Crawford, M.H., A.G. Comuzzie, W.R. Leonard, and R.I. Sukernik. 1994. Molecular genetics, protein variation and the population structure of the Evenki. In: C.L. Markert, J.G. Scandalios, H.A. Lim, and O. Serov (eds.), *Isozymes: Organization and Roles in Evolution, Genetics and Physiology.* World Scientific, River Edge, NJ, pp. 227–241.

Crawford, M.H., D. Dykes, J. Mielke, et al. 1981. Population structure of circumpolar Alaskan and Siberian indigenous communities. *American Journal of Physical Anthropology* 55:167–186.

Crawford, M.H., and G. Gmelch. 1975. Demography, ethnohistory, and genetics of the Irish tinkers. *Social Biology* 21:321–331.

Crawford, M.H., and G.W. Lasker (eds.). 1989. Foundations of anthropological genetics. *Special Issue of Human Biology* 61(5–6):615–958.

Crawford, M.H., and J.H. Mielke (eds.). 1982. *Current Developments in Anthropological Genetics. Vol. II. Ecology and Population Structure.* Plenum Press, New York.

Crawford, M.H., and D.H. O'Rourke. 1978. Inbreeding, lymphoma, genetics and morphology of the *Papio hamadryas* colony of Sukhumi. *Journal of Medical Primatology* 7(6):355–360.

Crawford, M.H., and L. Rogers. 1982. Population genetic models in the study of aging and longevity in a Mennonite community. *Social Science and Medicine* 16:149–153.

Crawford, M.H., D. West, and D.H. O'Rourke. 2010. Special issue on the origins of the populations of the Aleutian Islands. *Human Biology* 82(5–6):481–774.

Crawford, M.H., and P.L. Workman (eds.). 1973. *Methods and Theories in Anthropological Genetics.* University of New Mexico Press, Albuquerque.

Crispo, E. 2007. The Baldwin effect and genetic assimilation: revisiting two mechanisms of evolutionary change mediated by phenotypic plasticity. *Evolution* 61(11):2469–2479.

Crow, J. 2008. Mid-century controversies in population genetics. *Annual Review of Genetics* 42:1–16.

Crow, J. 2010. On epistasis: why it is unimportant in polygenic directional selection. *Philosophical Transactions of the Royal Society, B: Biological Sciences* 365:1241–1244.

Crow, J., and J. Felsenstein. 1968. The effect of assortative mating on the genetic composition of the population. *Eugenics Quarterly* 15(1):85–97.

Crow, J.F., and M. Kimura. 1979. Efficiency of truncation selection. *Proceedings of the National Academy of Sciences* 76(1):696–399.

Crow, J.F., and A.P. Mange. 1965. Measurement of inbreeding from the frequency of marriages between persons of the same surname. *Eugenics Quarterly* 12:199–203.

Csilléry, K., M.G. Blum, O.E. Gaggiotti, et al. 2010. Approximate Bayesian computation (ABC) in practice. *Trends in Ecology & Evolution* 25(7):410–418. doi:10.1016/j.tree.2010.04.001

Cui, Y., J. Lindo, C.E. Hughes, et al. 2013. Ancient DNA analysis of mid-Holocene individuals from the northwest coast of North America reveals different evolutionary paths for mitogenomes. *PLoS One* 8(7):e66948.

Cypess, A.M., S. Lehman, G. Williams, et al. 2009. Identification and importance of brown adipose tissue in adult humans. *New England Journal of Medicine* 360(15):1509–1517.

Cysouw, M., D. Dediu, and S. Moran. 2012. Comment on "Phonemic diversity supports a serial founder effect model of language expansion from Africa". *Science* 335:657.

Czerwinski, S.A., M. Lee, A.C. Choh, et al. 2007. Genetic factors in physical growth and development and their relationship to subsequent health outcomes. *American Journal of Human Biology* 19(5):684–691.

Czerwinski, S.A., B. Towne, S.S. Guo, et al. 2000. Genetic and environmental influences on lipid and lipoprotein levels in pre-and post-pubertal children. *American Journal of Human Biology* 12(2):289.

Dabney, J., M. Knapp, I. Glocke, et al. 2013a. Complete mitochondrial genome sequence of a middle Pleistocene cave bear reconstructed from ultrashort DNA fragments. *Proceedings of the National Academy of Sciences of the USA* 110(39):15758–15763. doi:10.1073/pnas.1314445110 PMID: 24019490

Dabney, J., M. Meyer, and S. Pääbo. 2013b. Ancient DNA damage. *Cold Spring Harbor Perspectives in Biology* 5:a012567.

Daborn, P.J., J.L. Yen, M.R. Bogwitz, et al. 2002. A single P450 allele associated with insecticide resistance in Drosophila. *Science* 297(5590):2253–2256.

Dalén, L., L. Orlando, B. Shapiro, et al. 2012. Partial genetic turnover in Neandertals: continuity in the east and population replacement in the west. *Molecular Biology and Evolution* 29(8):1893–1897.

Damon, A., S. Damon, R.B. Reed, et al. 1969. Age at menarche of mothers and daughters with a note on accuracy of recall. *Human Biology* 41:160–175.

Danchin, E., A. Charmantier, F.A. Champagne, et al. 2011. Beyond DNA: integrating inclusive inheritance into an extended theory of evolution. *Nature Reviews Genetics* 12(7):475–486.

Dannemann, M., A.M. Andres, and J. Kelso. 2016. Introgression of Neandertal- and Denisovan-like haplotypes contribute to adaptive variation in human toll-like receptors. *American Journal of Human Genetics* 98:22–33.

Darwin, C. 1871. *The Descent of Man, and Selection in Relation to Sex.* John Murray, London.

Darwin, C., and A.R. Wallace. 1858. On the tendency of species to form varieties; and on the perpetuation of varieties and species by natural means of selection. *Proceedings of the Linnean Society of London* 3:45.

Daubin, V., N.A. Moran, and H. Ochman. 2003. Phylogenetics and the cohesion of bacterial genomes. *Science* 301(5634):829–832.

David, L.A., C.F. Maurice, R.N. Carmody, et al. 2013. Diet rapidly and reproducibly alters the human gut microbiome. *Nature* 505:559–563.

Davies, G., A. Tenesa, A. Payton, et al. 2011. Genome-wide association studies establish that human intelligence is highly heritable and polygenic. *Molecular Psychiatry* 16(10):996–1005.

Davis, R.S., and R.A. Knecht. 2010. Continuity and change in the Eastern Aleutian archaeological sequence. *Human Biology* 82:507–524.

Day, M.H. 1969. Omo human skeletal remains. *Nature* 222:1135–1138.

De Azevedo, S., A. Nocera, C. Paschetta, et al. 2011. Evaluating microevolutionary models for the early settlement of the New World: the importance of recurrent gene flow with Asia. *American Journal of Physical Anthropology* 146:116–129.

De Azevedo, S., M. Quinto-Sánchez, C. Paschetta, et al. 2015. The first human settlement of the New World: a closer look at craniofacial variation and evolution of early and late Holocene Native American groups. *Quaternary International* 431B:152–167.

De Saint Pierre, M., C.M. Bravi, J.M. Motti, et al. 2012. An alternative model for the early peopling of Southern South America revealed by analyses of three mitochondrial DNA haplogroups. *PLoS One* 7:e43486.

de Groot, N.G. de, R.E. Bontrop, and G.G.M. Doxiadis. 2017. Major histocompatibility complex (MHC). In: A. Fuentes (ed.), *The International Encyclopedia of Primatology.* Wiley Online Library, Hoboken, NJ.

de Groot, N.G., C.A. Garcia, E.J. Verschoor, et al. 2005. Reduced MIC gene repertoire variation in west African chimpanzees as compared to humans. *Molecular Biology and Evolution* 22(6):1375–1385.

de Groot, N.G. de, N. Otting, J. Robinson, et al. 2012. Nomenclature report on the major histocompatibility complex genes and alleles of great ape, old and new world monkey species. *Immunogenetics* 64(8):615–631.

Dean, M., M. Carrington, C. Winkler, et al. 1996. Genetic restriction of HIV-1 infection and progression to AIDS by a deletion allele of the *CKR5* structural gene. Hemophilia Growth and Development Study, Multicenter AIDS Cohort Study, Multicenter Hemophilia Cohort Study, San Francisco City Cohort, ALIVE Study.PG - 1856-62. *Science* 273(5283):1856–1862.

Deeb, S.S., and A.G. Motulsky. 1996. Molecular genetics of human color vision. *Behavior Genetics* 26:195–207.

DeGiorgio, M., M. Jakobsson, and N.A. Rosenberg. 2009. Explaining worldwide patterns of human genetic variation using a coalescent-based serial founder model of migration outward from Africa. *Proceedings of the National Academy of Sciences* 106:16057–16062.

Degnan, J.H., and N.A. Rosenberg. 2009. Gene tree discordance, phylogenetic inference and the multispecies coalescent. *Trends in Ecology & Evolution* 24:332–340.

Dekker, J., K. Rippe, M. Dekker, et al. 2002. Capturing chromosome conformation. *Science* 295(5558):1306–1311.

Deloria, V. 1988. *Custer Died for Your Sins: An Indian Manifesto.* University of Oklahoma Press, Norman, OK.

Delwart, E., M. Magierowska, M. Royz, et al. 2002. Homogeneous quasispecies in 16 out of 17 individuals during very early HIV-1 primary infection. *AIDS* 16(2):189–195.

Demerath, E.W., A.C. Choh, S.A. Czerwinski, et al. 2007. Genetic and environmental influences on infant weight and weight change: the Fels Longitudinal Study. *American Journal of Human Biology* 19(5):692–702.

Demerath, E., D. Reed, A. Choh, et al. 2009. Rapid postnatal weight gain and visceral adiposity in adulthood: the Fels longitudinal study. *Obesity* 17(11):2060–2066.

Demerath, E.W., B. Towne, S.A. Czerwinski, et al. 2000. Covariate effect of birth weight in a genetic analysis of fasting insulin and glucose concentrations in adulthood. *Diabetes* 49(S1):A183.

Demeter, F., L.L. Shackelford, A.M. Bacon, et al. 2012. Anatomically modern human in Southeast Asia (Laos) by 46 ka. *Proceedings of the National Academy of Sciences of the USA* 109(36):14375–14380.

Dennis, M.Y., X. Nuttle, P.H. Sudmant, et al. 2012. Evolution of human-specific neural *SRGAP2* genes by incomplete segmental duplication. *Cell* 149:912–922.

Denver, D.R., K. Morris, J.T. Streelman, 2005. The transcriptional consequences of mutation and natural selection in Caenorhabditis elegans. *Nature Genetics* 37(5):544–548.

Der Sarkissian, C., M.E. Allentoft, M.C. Ávila-Arcos, et al. 2014. Ancient genomics. *Philosophical Transactions of the Royal Society of London B: Biological Sciences* 370(1660).

Derenko, M., B. Malyarchuk, T. Grzybowski, et al. 2010. Origin and post-glacial dispersal of mitochondrial DNA haplogroups C and D in Northern Asia. *PLoS One* 5(12):e15214.

Deshpande, O., S. Batzoglou, M. Feldman, et al. 2009. A serial founder effect model for human settlement out of Africa. *Proceedings of the Royal Society B: Biological Sciences* 276(1655):291–300. doi:10.1098/rspb.2008.0750

Destro-Bisol, G., M.A. Jobling, J. Rocha, et al. 2010. Molecular anthropology in the genomic era. *Journal of Anthropological Sciences* 88:93–112.

Devault, A.M., G.B. Golding, N. Waglechner, et al. 2014. Second-pandemic strain of vibrio cholerae from the Philadelphia cholera outbreak of 1849. *New England Journal of Medicine* 370(4):334–340. doi:10.1056/NE.J.M.oa1308663

Di Fiore, A., A. Link, C.A. Schmitt. et al. 2009. Dispersal patterns in sympatric woolly and spider monkeys: integrating molecular and observational data. *Behaviour* 146:437–470.

Diamond, J. 1988. Express train to Polynesia. *Nature* 336:307–308.

Diamond, J. 2002. Evolution, consequences and future of plant and animal domestication. *Nature* 418(6898):700–707. doi:10.1038/nature01019

Dias, A.G., K.M. Eny, M. Cockburn, et al. 2015. Variation in the *TAS1R2* gene, sweet taste perception and intake of sugars. *Journal of Nutrigenetics and Nutrigenomics* 8:81–90.

Dickerson, R.E. 1971. The structures of cytochrome *c* and the rates of molecular evolution. *Journal of Molecular Evolution* 1(1):26–45. doi:10.1007/BF01659392

Dickins, T.E., and Q. Rahman. 2012. The extended evolutionary synthesis and the role of soft inheritance in evolution. *Proceedings Biological Sciences* 279(1740):2913–2921.

DiFiore, T., R.R. Lawler, and P. Gagneaux. 2011 Molecular primatology. In: C.J. Campbell, A. Fuentes, K.C. MacKinnon, M. Panger, and S.K. Bearder (eds.), *Primates in Perspective*. 2nd ed. Oxford University Press, Oxford, pp. 390–416.

DiGangi, E.A., and R.L. Jantz. 2004. Dermatoglyphic ridge counts compared to short tandem repeats as measures of population distance. *American Journal of Physical Anthropology Supplement* 38:87–88.

Dillehay, T.D. 2009. Probing deeper into first American studies. *Proceedings of the National Academy of Sciences of the USA* 106:971–978.

Dillehay, T.D., C. Ocampo, J. Saavedra, et al. 2015. New archaeological evidence for an early human presence at Monte Verde, Chile. *PLoS One* 10:e0141923.

Dillehay, T.D., C. Ramírez, M. Pino, et al. 2008. Monte Verde: seaweed, food, medicine, and the peopling of South America. *Science* 320:784–786.

Ding, Q., Y. Hu, S. Xu, et al. 2014. Neanderthal introgression at chromosome 3p21.31 was under positive natural selection in east asians. *Molecular Biology and Evolution* 31(3):683–695.

Diniz-Filho JAF, T.N. Soares, J.S. Lima, et al. 2013. Mantel test in population genetics. *Genetics and Molecular Biology* 36(4):475–485.

Dobson, A.P., and E.R. Carper. 1996. Infectious diseases and human population history. *Bioscience* 46(2):115–126.

Dodge, D.R. 2012. A molecular approach to Neanderthal extinction. *Quaternary International* 259:22–32.

Dolan, M.J., H. Kulkarni, J.F. Camargo, et al. 2007. *CCL3L1* and *CCR5* influence cell-mediated immunity and affect HIV-AIDS pathogenesis via viral entry-independent mechanisms. *Nature Immunology* 8(12):1324–1336. doi:10.1038/ni1521

Dominy, N.J., and P.W. Lucas. 2001. Ecological importance of trichromatic vision to primates. *Nature* 410:363–366.

Dominy, N.J., and P.W. Lucas. 2004. Significance of color, calories, and climate to the visual ecology of catarrhines. *American Journal of Primatology* 62:189–207.

Dominy, N.J., P.W. Lucas, D. Osorio, et al. 2001. The sensory ecology of primate food perception. *Evolutionary Anthropology* 10:171–186.

Donnelly, P., and S. Tavare. 1995. Coalescents and genealogical structure under neutrality. *Annual Review of Genetics* 29:401–421. doi:10.1146/annurev.ge.29.120195.002153

Dow, M.M., and J.M. Cheverud. 1985. Comparison of distance matrices in studies of population structure and genetic microdifferentiation: quadratic assignment. *American Journal of Physical Anthropology* 68:367–373.

Doxiadis, G.G.M., N. Otting, N.G. De Groot, et al. 2016. Unprecedented polymorphism of Mhc-DRB region configurations in rhesus macaques. *Journal of Immunology.* 164(6):3193–3199. doi:10.4049/jimmunol.164.6.3193

Drancourt, M., G. Aboudharam, M. Signoli, et al. 1998. Detection of 400-year-old *Yersinia pestis* DNA in human dental pulp: an approach to the diagnosis of ancient septicemia. *Proceedings of the National Academy of Sciences of the USA* 95:12637–13640.

Drummond, A.J., S.Y. Ho, M.J. Phillips, et al. 2006. Relaxed phylogenetics and dating with confidence. *PLoS Biology* 4(5):e88. doi:10.1371/journal.pbio.0040088

Drummond, A.J., O.G. Pybus, A. Rambaut, et al. 2003. Measurably evolving populations. *Trends in Ecology & Evolution* 18(9):481–488. doi:10.1016/S0169-5347(03)00216-7

Drummond, A.J., and A. Rambaut. 2007. BEAST: Bayesian evolutionary analysis by sampling trees. *BMC Evolutionary Biology* 7:214. doi:10.1186/1471-2148-7-214

Drummond, A.J., A. Rambaut, B. Shapiro, et al. 2005. Bayesian coalescent inference of past population dynamics from molecular sequences. *Molecular Biology and Evolution* 22(5):1185–1192. doi:10.1093/molbev/msi103

Dryomov, S.V., A.M. Nazhmidenova, S.A. Shalaurova, et al. 2015. Mitochondrial genome diversity at the Bering Strait area highlights prehistoric human migrations from Siberia to Northern North America. *European Journal of Human Genetics* 23:1399–1404.

Duchêne, S., R. Lanfear, and S.Y. Ho. 2014. The impact of calibration and clock-model choice on molecular estimates of divergence times. *Molecular Phylogenetics and Evolution* 78:277–289. doi:10.1016/j.ympev.2014.05.032

Duffy, V.B., J.E. Hayes, A.C. Davidson, et al. 2010. Vegetable intake in college-aged adults is explained by oral sensory phenotypes and *TAS2R38* genotype. *Chemosensory Perception* 3:137–148.

Duggleby, C.R., P.A. Haseley, R.G. Rawlins, et al. 1986. An overview of blood group genetic studies on the Cayo Santiago rhesus monkeys. In: R.G. Rawlins, and M.J. Kessler (eds.), *The Cayo Santiago Macaques: History, Behavior, and Biology*. SUNY Press, Albany, NY, pp. 269–282.

Dulik, M.C., A.C. Owings, J.B. Gaieski, et al. 2012b. Y-chromosome analysis reveals genetic divergence and new founding native lineages in Athapaskan- and Eskimoan-speaking populations. *Proceedings of the National Academy of Sciences of the USA* 109:8471–8476.

Dulik, M.C., S.I. Zhadanov, L.P. Osipova, et al. 2012a. Mitochondrial DNA and Y chromosome variation provides evidence for a recent common ancestry between Native Americans and indigenous Altaians. *American Journal of Human Genetics* 90:229–246.

Dulloo, A.G., M.J. Stock, G. Solinas, et al. 2002. Leptin directly stimulates thermogenesis in skeletal muscle. *FEBS Letters* 515(1–3):109–113.

Durand, E.Y., N. Patterson, D. Reich, et al. 2011. Testing for ancient admixture between closely related populations. *Molecular Biology and Evolution* 28(8):2239–2252.

Dykhuizen, D.E., and L. Green. 1991. Recombination in *Escherichia coli* and the definition of biological species. *Journal of Bacteriology* 173(22):7257–7268.

Eaaswarkhanth, M., P. Pavlidis, and O. Gokcumen. 2014. Geographic distribution and adaptive significance of genomic structural variants: an anthropological genetics perspective. *Human Biology* 86:260–275.

Eastern Band of Cherokee Indians. 2010. Enrollment application. http://nc-cherokee.com/enrollment/files/2010/09/enrollment-application.pdf

Eberle, R., and J.K. Hilliard. 1989. Serological evidence for variation in the incidence of Herpesvirus infections in different species of apes. *Journal of Clinical Microbiology* 27(6):1357–1366.

Economou, C., A. Kjellstrom, K. Liden, et al. 2013. Ancient-DNA reveals and Asian type of *Mycobacterium leprae* in medieval Scandinavia. *Journal of Archaeological Science* 40:465–470.

Eden, A., F. Gaudet, A. Waghmare, et al. 2003. Chromosomal instability and tumors promoted by DNA hypomethylation. *Science* 300(5618):455.

Edwards, M., A. Bigham, J. Tan, et al. 2010. Association of the *OCA2* polymorphism His615Arg with melanin content in east Asian populations: further evidence of convergent evolution of skin pigmentation. *PLoS Genetics* 6(3):e1000867. doi:10.1371/journal.pgen.1000867

Edwards, M., A. Gozdzik, K. Ross, et al. 2012. Technical note: quantitative measures of iris color using high resolution photographs. *American Journal of Physical Anthropology* 147(1):141–149. doi:10.1002/ajpa.21637

Egger, G., G. Liang, A. Aparicio, et al. 2004. Epigenetics in human disease and prospects for epigenetic therapy. *Nature* 429(6990):457–463.

Eiberg, H., J. Troelsen, M. Nielsen, et al. 2008. Blue eye color in humans may be caused by a perfectly associated founder mutation in a regulatory element located within the *HERC2* gene inhibiting *OCA2* expression. *Human Genetics* 123(2):177–187. doi:10.1007/s00439-007-0460-x

Enatftah, N.S., T. Sahi, E. Savilahti, et al. 2002. Identification of a variant associated with adult-type hypolactasia. *Nature Genetics* 30(2):233–237.

Endicott, P., and S.Y.W. Ho. 2008. A Bayesian evaluation of human mitochondrial substitution rates. *American Journal of Human Genetics* 82(4):895–902. doi:10.1016/j.ajhg.2008.01.019

Endicott, P., S.Y.W. Ho, M. Metspalu, et al. 2009. Evaluating the mitochondrial timescale of human evolution. *Trends in Ecology & Evolution* 24(9):515–521. doi:10.1016/j.tree.2009.04.006

Endicott, P., S.Y. Ho, and C. Stringer. 2010. Using genetic evidence to evaluate four palaeoanthropological hypotheses for the timing of Neanderthal and modern human origins. *Journal of Human Evolution* 59(1):87–95.

Enerback, S., A. Jacobsson, E.M. Simpson, et al. 1997. Mice lacking mitochondrial uncoupling protein are cold-sensitive but not obese. *Nature* 387(6628):90–94.

Eny, K.M., T.M.S. Wolever, P.N. Corey, et al. 2010. Genetic variation in *TAS1R2* (Ile191Val) is associated with consumption of sugars in overweight and obese individuals in 2 distinct populations. *American Journal of Clinical Nutrition* 92:1501–1510.

Epstein, A.C., J.M. Gleadle, L.A. McNeill, et al. 2001. *C. elegans* EGL-9 and mammalian homologs define a family of dioxygenases that regulate HIF by prolyl hydroxylation. *Cell* 107(1):43–54.

Eriksson, A., L. Betti, A.D. Friend, et al. 2012. Late pleistocene climate change and the global expansion of anatomically modern humans, *Proceedings of the National Academy of Sciences USA* 109:16089–16094.

Eriksson, A., and A. Manica. 2012. Effect of ancient population structure on the degree of polymorphism shared between modern human populations and ancient hominins. *Proceedings of the National Academy of Sciences* 109(35):13956–13960.

Erkmann, J. 2011. Histone modification research methods. *Materials and Methods* 1(92).

Ersts, P.J. 2015. [Internet]. Geographic Distance Matrix Generator v. 1.2.3. American Museum of Natural History, Center for Biodiversity and Conservation.

Eshed, V., A. Gopher, and I. Hershkovitz. 2006. Tooth wear and dental pathology at the advent of agriculture: new evidence from the Levant. *American Journal of Physical Anthropology* 130(2):145–281.

Esteller, M. 2003. Cancer epigenetics: DNA methylation and chromatin alterations in human cancer. *Advances in Experimental Medicine and Biology* 532:39–49.

Esterbauer, H., H. Oberkofler, Y.M. Liu, et al. 1998. Uncoupling protein-1 mRNA expression in obese human subjects: the role of sequence variations at the uncoupling protein-1 gene locus. *Journal of Lipid Research* 39(4):834–844.

Evans, D.T., L.A. Knapp, P. Jing, et al. Watkins. 1999. Rapid and slow Progressors differ by a single MHC class I haplotype in a family of MHC-defined rhesus macaques infected with SIV. *Immunology Letters* 66:53–59.

Evans, D.T., M.S. Piekarczyk, L. Cadavid, et al. 1998. Two different primate species express an identical functional MHC class I allele. *Immunogenetics* 47:206–207.

Eveleth, P.B., and J.M. Tanner. 1990. *Worldwide Variation in Human Growth.* 2nd ed Cambridge University Press, Cambridge.

Evteev, A., A.L. Cardini, I. Morozova, et al. 2014. Extreme climate, rather than population history, explains mid-facial morphology of Northern Asians. *American Journal of Physical Anthropology* 153(3):449–462.

Ewald, P.W. 1983. Host-parasite relations, vectors, and the evolution of disease severity. *Annual Review of Ecology, Evolution, and Systematics* 14:465–485. 10.1146/annurev.es.14.110183.002341

Ewing, G., and J. Hermisson. 2010. MSMS: a coalescent simulation program including recombination, demographic structure and selection at a single locus. *Bioinformatics* 26(16):2064–2065. doi:10.1093/bioinformatics/btq322

Faerman, M., G.K. Bar-Gal, D. Filon, et al. 1998. Determining the sex of infanticide victims from the late roman era through ancient DNA analysis. *Journal of Archaeological Science* 25:861–865.

Faerman, M., D. Filon, G. Kahila, C.L. Greenblatt, P. Smith, and A. Oppenheim. 1995. Sex identification of archaeological human remains based on amplification of the X and Y amelogenin alleles. *Gene* 167:327–332.

Fagundes, N.J.R. 2016. The mitochondrial genome and the origins of Native Americans: structuring variability in South America's initial settlements, moving forward toward integrated research approaches. In: UNESCO (eds.), Human Origin Sites and the World Heritage Convention in the Americas. World Heritage Papers 42, vol. II. UNESCO, Mexico City, pp. 1–13.

Fagundes, N.J.R., R. Kanitz, R. Eckert, et al. 2008. Mitochondrial population genomics supports a single pre-Clovis origin with a coastal route for the peopling of the Americas. *American Journal of Human Genetics* 82(3):583–592. doi:10.1016/j.ajhg.2007.11.013

Fagundes, N.J.R., N. Ray, M. Beaumont, et al. 2007. Statistical evaluation of alternative models of human evolution. *Proceedings of the National Academy of Sciences of the USA* 104:17614–17619.

Fairbanks, L.A., W.P. Melega, M.J. Jorgensen, et al. 2001. Social impulsivity inversely associated with C.S.F 5-HIAA and fluoxetine exposure in vervet monkeys. *Neuropsychopharmacology* 24:370–378.

Fairbanks, L.A., B.M. Way, S.E. Breidenthal, et al. 2012. Maternal and offspring dopamine D4 receptor genotypes interact to influence juvenile impulsivity in vervet monkeys. *Psychological Science* 23:1099–1104.

Falconer, D.S., and T.F.C. MacKay. 1996. *Introduction to Quantitative Genetics.* 4th ed. Prentice Hall, Upper Saddle River, NJ.

Falconer, D.S., and T.F.C. Mackay. 1996. *Introduction to Quantitative Genetics.* 4th ed. Longman, Harlow, UK.

Falconner, D.S. 1961. *Introduction to Quantitative Genetics.* Ronald Press, New York.

Fariña, R.A., P.S. Tambusso, L. Varela, et al. 2013. Arroyo del Vizcaíno, Uruguay: a fossil-rich 30-ka-old megafaunal locality with cut-marked bones. *Proceedings Biological Sciences* 281:20132211.

Fariña, R.A., P.S. Tambusso, L. Varela, et al. 2014. Among others, cut-marks are archaeological evidence: reply to 'Archaeological evidences are still missing: a comment on Fariña et al. Arroyo del Vizcaíno Site, Uruguay' by Suárez et al. *Proceedings Biological Sciences* 281:20141637.

Fatemi, M., M.M. Pao, S. Jeong, et al. 2005. Footprinting of mammalian promoters: use of a CpG DNA methyltransferase revealing nucleosome positions at a single molecule level. *Nucleic Acids Research* 33(20):e176.

Fears, S.C., W.P. Melega, S.K. Service, et al. 2009. Identifying heritable brain phenotypes in an extended pedigree of vervet monkeys. *Journal of Neuroscience* 29: 2867–2875.

Fears, S.C., K. Scheibel, Z. Abaryan, et al. 2011. Anatomic brain asymmetry in vervet monkeys. *PLoS One* 6:e28243.

Feeney, E. 2011. The impact of bitter perception and genotypic variation of *TAS2R38* on food choice. *Nutrition Bulletin* 36:20–33.

Felsenstein, J. 1985. Phylogenies and the comparative method. *American Naturalist* 125(1):1. doi:10.108s6/284325

Felsenstein, J. 2004. *Inferring Phylogenies.* Sinauer Associates Inc., Sunderland, MA.

Fenech, M. 2012. Folate (vitamin B9) and vitamin B12 and their function in the maintenance of nuclear and mitochondrial genome integrity. *Mutation Research* 733(1–2):21–33. doi:10.1016/j.mrfmmm.2011.11.003

Fenner, J.N. 2005. Cross-cultural estimation of the human generation interval for use in genetics-based population divergence studies. *American Journal of Physical Anthropology* 128(2):415–423. doi:10.1002/ajpa.20188

Fenner, F., and F.N. Ratliffe. 1966. *Myxomatosis.* Cambridge University Press, Cambridge.

Ferguson, W., S. Dvora, R.W. Fikes, et al. 2012. Long-term balancing selection at the antiviral gene *OAS1* in Central African chimpanzees. *Molecular Biology and Evolution* 29:1093–1103.

Fernandez, L., R. Milne, J. Bravo, et al. 2007. M.C.1R: three novel variants identified in a malignant melanoma association study in the Spanish population. *Carcinogenesis* 28(8):1659–1664. doi:10.1093/carcin/bgm084

Ferreira da Silva, M.J., T. Minhos, and R.M. Sa. 2012. Using genetics as a tool in primate conservation. Nature Education. http://www.nature.com/scitable/knowledge/library/using-genetics-as-a-tool-in-primate-59120330.

Figueroa, F., C. O'hUigin, H. Tichy, et al. 1994. The origin of the primate Mhc-DRB genes and allelic lineages as deduced from the study of prosimians. *Journal of Immunology* 152:4455–4465.

Fijarczyk, A., and W. Babik. 2015. Detecting balancing selection in genomes: limits and prospects. *Molecular Ecology* 24:3529–3545.

Fischer, A., Y. Gilad, O. Man, et al. 2005. Evolution of bitter taste receptors in humans and apes. *Molecular Biology and Evolution* 22:432–436.

Fisher, R.A. 1918.The correlation between relatives on the supposition of Mendelian inheritance. *Transactions of the Royal Society of Edinburgh* 52:399–433.

Fisher, R.A., E.B. Ford, and J. Huxley. 1939. Taste-testing the anthropoid apes. *Nature* 144:750.

Fix, A.G. 1977. The Demography of the Semai Senoi. Anthropological Papers No. 62. University of Michigan Museum of Anthropology, Ann Arbor.

Fix, A.G. 1999. *Migration and Colonization in Human Microevolution.* Cambridge University Press, Cambridge.

Fix, A.G. 2002. Colonization models and initial genetic diversity in the Americas. *Human Biology* 74:1–10.

Fix, A.G. 2004. Kin-structured migration: causes and consequences. *American Journal of Human Biology* 16:387–394. doi:10.1002/ajhb.20044

Fix, A.G. 2005. Rapid deployment of the five founding Amerind mtDNA haplogroups via coastal and riverine colonization. *American Journal of Physical Anthropology* 128:430–436.

Flanagan, N., E. Healy, A. Ray, et al. 2000. Pleiotropic effects of the melanocortin 1 receptor (*MC1R*) gene on human pigmentation. *Human Molecular Genetics* 9(17):2531–2537.

Flegontov, P., P. Changmai, A. Zidkova, et al. 2016. Genomic study of the Ket: a Paleo-Eskimo-related ethnic group with significant ancient North Eurasian ancestry. *Scientific Reports* 6:20768.

Fleury, C., M. Neverova, S. Collins, et al. 1997. Uncoupling protein-2: a novel gene linked to obesity and hyperinsulinemia. *Nature Genetics* 15(3):269–272.

Flint, J., A.V. Hill, D.K. Bowden, et al. 1986. High frequencies of alpha-thalassaemia are the result of natural selection by malaria. *Nature* 321(6072):744–750.

Flugge, P,, E. Zimmermann, A.L. Hughes, et al. 2002. Characterization and phylogenetic relationship of prosimian MHC class I genes. *Journal of Molecular Evolution* 55: 768–769. doi:10.1007/s00239-002-2372-7

Foerster, A.M., H.Q. Dinh, L. Sedman, et al. 2011. Genetic rearrangements can modify chromatin features at epialleles. *PLoS Genetics* 7(10):e1002331.

Foley, R., and C. Gamble. 2009. The ecology of social transitions in human evolution. *Philosophical Transactions of the Royal Society, B: Biological Sciences* 364(1533): 3267–3279.

Foll, M., O.E. Gaggiotti, J.T. Daub, et al. 2014. Widespread signals of convergent adaptation to high altitude in Asia and America. *American Journal of Human Genetics* 95(4):394–407.

Ford, E. 1964. *Ecological Genetics.* Metheun, London.

Forster, P. 2004. Ice ages and the mitochondrial DNA chronology of human dispersals: a review. *Philosophical Transactions of the Royal Society, B: Biological Sciences* 359(1442):255–264. doi:10.1098/rstb.2003.1394

Forster, P., R. Harding, A. Torroni, and H.-J. Bandelt. 1996. Origin and evolution of Native American mtDNA variation: a reappraisal. *American Journal of Human Genetics* 59(4):935–945.

Foster, P.L. 2007. Stress-induced mutagenesis in bacteria. *Critical Reviews in Biochemistry and Molecular Biology* 42(5):373–397.

Foster, D.O., and M.L. Frydman. 1979. Tissue distribution of cold-induced thermogenesis in conscious warm-acclimated cold-acclimated rats re-evaluated from changes in tissue blood-flow – dominant role of brown adipose-tissue in the replacement of shivering by nonshivering thermogenesis. *Canadian Journal of Physiology and Pharmacology* 57(3):257–270.

Fowler, C.S. 1992. In the Shadow of Fox Peak: An Ethnography of the Cattail-Eater Northern Paiute People of Stillwater Marsh. Cultural Resource Series No. 5. Fallon: U.S. Department of the Interior, Fish and Wildlife Service Region 1, Stillwater National Wildlife Refuge. US Government Printing Office, Washington, DC.

Franciscus, R.G., and J.C. Long. 1991. Variation in human nasal height and breadth. *American Journal of Physical Anthropology* 85(4):419–427.

Frange, P., L. Meyer, M. Jung, et al. 2013. Sexually-transmitted/founder HIV-1 cannot be directly predicted from plasma or PBMC-derived viral quasispecies in the transmitting partner. *PLoS One* 8(7):e69144. doi:10.1371/journal.pone.0069144

Frank, S.A. 1996. Models of parasite virulence. *Quarterly Review of Biology* 71(1):37–78.

Fraser, H.B., L.L. Lam, S.M. Neumann, et al. 2012. Population-specificity of human DNA methylation. *Genome Biology* 13(2):R8.

Frayer, D., M. Wolpoff, A. Thorne, et al. 1993. Theories of modern human origins: the paleontological test. *American Anthropologist* 95:14–50.

Freimer, N., and C. Sabatti. 2003. The human phenome project. *Nature Genetics* 34:15–21.

Fried, M.H. 1968. On the concepts of “tribe” and “tribal society”. In: J. Helm (ed.), *Essays on the Problem of Tribe: Proceedings of the 1967 Annual Spring Meeting of the American Ethnological Society*. University of Washington Press, Seattle, pp. 3–20. doi:10.1111/j.2164-0947.1966.tb02369.x

Fried, M.H. 1975. The Notion of Tribe. *Cummings Publishing Company*, Menlo Park, CA.

Friedlaender, J.S., F.R. Friedlaender, F.A. Reed, et al. 2008. The genetic structure of Pacific Islanders. *PLoS Genetics* 4(1):e19. doi:10.1371/journal.pgen.0040019

Friedlander, J. 1971. The population structure of south-Central Bougainville. *American Journal of Physical Anthropology* 35:13–26.

Friedlander, J. 2004. Commentary: changing standards of informed consent: raising the bar. In: T.R. Turner (ed.), *Biological Anthropology and Ethics: From Repatriation to Genetic Identity*. SUNY Press, Albany, NY.

Friso, S., and S.W. Choi. 2005. Gene-nutrient interactions in one-carbon metabolism. *Current Drug Metabolism* 6(1):37–46.

Frost, P. 2014. The puzzle of European hair, eye, and skin color. *Advances in Anthropology* 4:78–88.

Fu, Q., H. Li, P. Moorjani, et al. 2014. Genome sequence of a 45,000-year-old modern human from Western Siberia. *Nature* 514(7523):445–449.

Fu, Q., M. Meyer, X. Gao, et al. 2013a. DNA analysis of an early modern human from Tianyuan Cave, China. *Proceedings of the National Academy of Sciences of the USA* 110(6):2223–2227.

Fu, Q., A. Mittnik, P.L.F. Johnson, et al. 2013b. A revised timescale for human evolution based on ancient mitochondrial genomes. *Current Biology* 23(7):553–559. doi:10.1016/j.cub.2013.02.044

Fu, Q., C. Posth, M. Hajdinjak, M. Petr, S. Mallick, and D. Fernandes. 2016a. The genetic history of ice age Europe. *Nature* 534.

Fu, Q., C. Posth, M. Hajdinjak, et al. 2016b. The genetic history of ice age Europe. *Nature* 534 (7606). doi:10.1038/nature17993.

Fujimoto, A., R. Kimura, J. Ohashi, et al. 2008b. A scan for genetic determinants of human hair morphology: *EDAR* is associated with Asian hair thickness. *Human Molecular Genetics* 17(6):835–843.

Fujimoto, A., J. Ohashi, N. Nishida, et al. 2008a. A replication study confirmed the *EDAR* gene to be a major contributor to population differentiation regarding head hair thickness in Asia. *Human Genetics* 124(2):179–185.

Fukuda, K., K. Ichiyanagi, Y. Yamada, et al. 2013. Regional DNA methylation differences between humans and chimpanzees are associated with genetic changes, transcriptional divergence and disease genes. *Journal of Human Genetics* 58(7):446–454.

Fullerton, S.M., A. Bartoszewicz, G. Ybazeta, et al. 2002. Geographic and haplotype structure of candidate type 2 diabetes-susceptibility variants at the calpain-10 locus. *American Journal of Human Genetics* 70(5):1096–1106.

Fumagalli, M., R. Cagliani, U. Pozzoli, et al. 2009a. Widespread balancing selection and pathogen-driven selection at blood group antigen genes. *Genome Research* 19(2):199–212. doi:10.1101/gr.082768.108

Fumagalli, M., I. Moltke, N. Grarup, et al. 2015. Greenlandic Inuit show genetic signatures of diet and climate adaptation. *Science* 349:1343–1347.

Fumagalli, M., U. Pozzoli, R. Cagliani, et al. 2009b. Parasites represent a major selective force for interleukin genes and shape the genetic predisposition to autoimmune conditions. *Journal of Experimental Medicine* 206(6):1395–1408. doi:10.1084/jem.20082779

Fumagalli, M., M. Sironi, U. Pozzoli, et al. 2011. Signatures of environmental genetic adaptation pinpoint pathogens as the main selective pressure through human evolution. *PLoS Genetics* 7(11):e1002355. doi:10.1371/journal.pgen.1002355

Furusho, T. 1968. On the manifestations of the genotypes responsible for stature. *Human Biology* 40:437–455.

Gaetano T.J., J. Danzy, M.S. Mtshali, et al. 2014. Mapping correlates of parasitism in wild South African vervet monkeys (*Chlorocebus aethiops*). *South African Journal of Wildlife Research* 44:56–70.

Gallagher, J. 2015. UK approves three-person babies. BBC News, February 24. http://www.bbc.com/news/health-31594856.

Galloway, V.A., W.R. Leonard, and E. Ivakine. 2000. Basal metabolic adaptation of the Evenki reindeer herders of Central Siberia. *American Journal of Human Biology* 12(1):75–87.

Gamba, C., K. Hanghøj, C. Gaunitz, et al. 2015. Comparing the performance of three ancient DNA extraction methods for high-throughput sequencing. *Molecular Ecology Resources* 16:459–469.

Gamba, C., E.R. Jones, M.D. Teasdale, et al. 2014. Genome flux and stasis in a five millennium transect of European prehistory. *Nature Communications* 5:1–9. doi:10.1038/ncomms6257

Gao, F., E. Bailes, D.L. Robertson, et al. 1999. Origin of HIV-1 in the chimpanzee pan troglodytes troglodytes. *Nature* 397(6718):436–441.

Gao, F., L. Yue, A.T. White, et al. 1992. Human infection by genetically diverse SIVSM-related HIV-2 in West Africa. *Nature* 358(6386):495–499. doi:10.1038/358495a0

García-Martín, P., A. Hernández-Martín, and A. Torrelo. 2013. Ectodermal dysplasias: a clinical and molecular review. *Actas Dermo-Sifiliográficas (English ed.)* 104(6): 451–470.

Garn, S., S. Bailey, and P. Cole. 1976. Similarities between parents and their adopted children. *American Journal of Physical Anthropology* 45:539–543.

Garn, S., A.A. Dahlberg, A. Lewis, et al. 1966. Genetic independence of Carabelli's trait from tooth size or crown morphology. *Archives of Oral Biology* 11:745–747.

Garn, S., A. Lewis, and R. Kerewsky. 1963. Third molar agenesis and size reduction of the remaining teeth. *Nature* 200:488–489.

Garn, S., A. Lewis, and D. Polacheck. 1960. Sibling similarities in dental development. *Journal of Dental Research* 39:170–175.

Garn, S., A. Lewis, and A. Walenga. 1968. The genetic basis of the crown-size profile pattern. *Journal of Dental Research* 47:1190.

Garn, S.M., and C.G. Rohmann. 1966. Interaction of nutrition and genetics in the timing of growth and development. *Pediatric Clinics of North America* 13:353–379.

Gatherer, D. 2015. The unprecedented scale of the west African Ebola virus disease outbreak is due to environmental and sociological factors, not special attributes of the currently circulating strain of the virus. *Evidence-Based Medicine* 20(1):28. doi:10.1136/ebmed-2014-110127

Ge, R.L., Q.H. Chen, and L.G. He. 1994. [Characteristics of hypoxic ventilatory response in Tibetan living at moderate and high altitudes]. *Chinese Journal of Tuberculosis and Respiratory Diseases* 17(6):364–366, 384.

Geng, F., and W.P. Tansey. 2008. Polyubiquitylation of histone H_2B. *Molecular Biology of the Cell* 19(9):3616–3624.

Ghirotto, S., F. Tassi, E. Fumagalli, et al. 2013. Origins and evolution of the Etruscans' mtDNA. *PLoS One* 8(2). doi:10.1371/journal.pone.0055519.g001

Gibbons, A. 2002. In search of the first hominids. *Science* 295:1214–1219.

Gibbons, A. 2011. A new view of the birth of *Homo sapiens. Science* 331(6016):392.

Gibbs, J.R., M.P. van der Brug, D.G. Hernandez, et al. 2010. Abundant quantitative trait loci exist for DNA methylation and gene expression in human brain. *PLoS Genetics* 6(5):e1000952.

Gignoux, C.R., B.M. Henn, and J.L. Mountain. 2011. Rapid, global demographic expansions after the origins of agriculture. *Proceedings of the National Academy of Sciences of the USA* 108(15):6044–6049. doi:10.1073/pnas.0914274108

Gilbert, C.C., S.R. Frost, and D.S. Strait. 2009. Allometry, sexual dimorphism, and phylogeny: a cladistic analysis of extant African papionins using craniodental data. *Journal of Human Evolution* 57:298–320.

Gilbert, M.T., D.L. Jenkins, A. Götherstrom, et al. 2008b. DNA from pre-Clovis human coprolites in Oregon, North America. *Science* 320:786–789.

Gilbert, M.T., T. Kivisild, B. Grønnow, et al. 2008a. Paleo-Eskimo mtDNA genome reveals matrilineal discontinuity in Greenland. *Science* 320:1787–1789.

Gilbert, M.T., A. Rambaut, G. Wlasiuk, et al. 2007. The emergence of HIV/AIDS in the Americas and beyond. *Proceedings of the National Academy of Sciences of the USA* 104(47):18566–18570.

Gilbert, M.T., E. Willerslev, A.J. Hansen, et al. 2003. Distribution patterns of postmortem damage in human mitochondrial DNA. *American Journal of Human Genetics* 72(1):32–47.

Giles, E., E. Ogan, R.J. Walsh, et al. 1966. Microevolution in new Guinea: the role of genetic drift. *Annals of New York Academy of Science* 134:655–665.

Gill, P., P.L. Ivanov, C. Kimpton, et al. 1994. Identification of the remains of the Romanov family by DNA analysis. *Nature Genetics* 6:130–135.

Gillespie, J.H. 1991. *The Causes of Molecular Evolution.* Oxford University Press, Oxford.

Gingerich, P.D. 1983. Rates of evolution: effects of time and temporal scaling. *Science* 222(4620):159–161.

Gire, S.K., A. Goba, K.G. Andersen, et al. 2014. Genomic surveillance elucidates Ebola virus origin and transmission during the 2014 outbreak. *Science* 345(6202):1369–1372. doi:10.1126/science.1259657

Gluckman, P.D., F.M. Low, T. Buklijas, et al. 2011. How evolutionary principles improve the understanding of human health and disease. *Evolutionary Applications* 4(2):249–263. doi:10.1111/j.1752-4571.2010.00164.x

Go, Y., Y. Satta, Y. Kawamoto, et al. 2002. Mhc-DRB genes evolution in lemurs. *Immunogenetics* 54:403–417. doi:10.1007/s00251-002-0480-6

Go, Y., Y. Satta, Y. Kawamoto, et al. 2003. Frequent segmental sequence exchanges and rapid gene duplication characterize the MHC class I genes in lemurs. *Immunogenetics* 55:450–461. doi:10.1007/s00251-003-0613-6.

Go, Y., Y. Satta, O. Takenaka, et al. 2005. Lineage-specific loss of function of bitter taste receptor genes in humans and nonhuman primates. *Genetics* 170:313–326.

Goebel, T., M.R. Waters, and D.H. O'Rourke. 2008. The late Pleistocene dispersal of modern humans in the Americas. *Science* 319 (5869):1497–1502.

Goebel, T., M.R. Waters, H.L. Smith, et al. 2013. Serpentine Hot Springs, Alaska: results of excavations and implications for the age and significance of northern fluted points. *Journal of Archaeological Science* 40:4222–4233.

Gogarten, J.P., and J.P. Townsend. 2005. Horizontal gene transfer, genome innovation and evolution. *Nature Reviews Microbiology* 3(9):679–687.

Gokcumen, O., P.L. Babb, R.C. Iskow, et al. 2011. Refinement of primate copy number variation hotspots identifies candidate genomic regions evolving under positive selection. *Genome Biology* 12:R52.

Gokçümen, Ö., T. Gultekin, Y.D. Alakoc, et al. 2011. Biological ancestries, kinship connections, and projected identities in four central Anatolian settlements: insights from culturally contextualized genetic anthropology. *American Anthropologist* 113:116–131. doi:10.1111/j.1548-1433.2010.01310.x

Gokcumen, O., V. Tischler, J. Tica, et al. 2013a. Primate genome architecture influences structural variation mechanisms and functional consequences. *Proceedings of the National Academy of Sciences of the USA* 110:15764–15769.

Gokcumen, O., Q. Zhu, L.C.F. Mulder, et al. 2013b. Balancing selection on a regulatory region exhibiting ancient variation that predates human–neandertal divergence. *PLoS Genetics* 9(4):e1003404.

Gokhman, D., E. Lavi, K. Prüfer, et al. 2014. Reconstructing the DNA methylation maps of the Neandertal and the Denisovan. *Science* 344(6183):523–527. doi:10.1126/science.1250368

Gokhman, D., E. Meshorer, and L. Carmel. 2016. Epigenetics: it's getting old. Past meets future in paleoepigenetics. *Trends in Ecology and Evolution* 31(4):290–300.

Goldberg, A., and N.A. Rosenberg. 2015. Beyond 2/3 and 1/3: the complex signatures of sex-biased admixture on the X chromosome. *Genetics* 201(1):263–279. doi:10.1534/genetics.115.178509

Goldberg, A., P. Verdu, and N.A. Rosenberg. 2014. Autosomal admixture levels are informative about sex bias in admixed populations. *Genetics* 198(3):1209–1229. doi:10.1534/genetics.114.166793

Goll, M.G., and T.H. Bestor. 2005. Eukaryotic cytosine methyltransferases. *Annual Review of Biochemistry* 74:481–514.

Gomez-Valero, L., E.P. Rocha, A. Latorre, et al. 2007. Reconstructing the ancestor of *Mycobacterium leprae*: the dynamics of gene loss and genome reduction. *Genome Research* 17(8):1178–1185.

Gonçalves, V.F., J. Stenderup, C. Rodrigues-Carvalho, et al. 2013. Identification of Polynesian mtDNA haplogroups in remains of Botocudo Amerindians from Brazil. *Proceedings of the National Academy of Sciences of the USA* 110:6465–6469.

Gonder, M.K., H.M. Mortensen, F.A. Reed, A. de Sousa, et al. 2007. Whole-mtDNA genome sequence analysis of ancient African lineages. *Molecular Biology and Evolution* 24(3):757–768. doi:10.1093/molbev/msl209

Gonzalez, E., M. Bamshad, N. Sato, et al. 1999. Race-specific HIV-1 disease-modifying effects associated with CCR5 haplotypes. *Proceedings of the National Academy of Sciences of the USA* 96(21):12004–12009.

Gonzalez, E., H. Kulkarni, H. Bolivar, et al. 2005. The influence of CCL3L1 gene-containing segmental duplications on HIV-1/AIDS susceptibility. *Science* 307(5714): 1434–1440.

González-José, R., M.C. Bortolini, F.R. Santos, et al. 2008. The peopling of America: craniofacial shape variation on a continental scale and its interpretation from an interdisciplinary view. *American Journal of Physical Anthropology* 137:175–187.

González-José, R., S.L. Dahinten, M.A. Luis, et al. 2001. Craniometric variation and the settlement of the Americas: testing hypotheses by means of R-matrix and matrix correlation analyses. *American Journal of Physical Anthropology* 116:154–165.

González-José, R., A. González-Martín, M. Hernández, et al. 2003. Craniometric evidence for Palaeoamerican survival in Baja California. *Nature* 425:62–65.

González-José, R., W. Neves, M.M. Lahr, et al. 2005. Late Pleistocene/Holocene craniofacial morphology in Mesoamerican Paleoindians: implications for the peopling of the New World. *American Journal of Physical Anthropology* 128:772–780.

González-José, R., S. Van der Molen, E. Gonzalez-Jose, et al. 2004. Patterns of phenotypic covariation and correlation in modern humans as viewed from morphological integration. *American Journal of Physical Anthropology* 123:69–77.

Goodman, M. 1961. The role of immunochemical differences in the phyletic development of human behavior. *Human Biology* 33:131–162.

Goossens, B., L. Chikhi, M. Ancrenaz, et al. 2006. Genetic signature of anthropogenic population collapse in orang-utans. *PLoS Biology* 4:e25.

Gordon, S., and A. Plüddemann. 2017. Tissue macrophages: heterogeneity and functions. *BMC Biology* 15(1):53. doi:10.1186/s12915-017-0392-4

Gould, S.J. 1977. *Ontogeny and Phylogeny*. Belknap Press of Harvard University Press, Cambridge, MA.

Gould, S.J. 2000. The spice of life: an interview with Stephen Jay Gould. *Leader to Leader* 15(Winter):14–19.

Graf, J., R. Hodgson, and A. van Daal. 2005. Single nucleotide polymorphisms in the *MATP* gene are associated with normal human pigmentation variation. *Human Mutation* 25(3):278–284. doi:10.1002/humu.20143

Gravel, S., B. Henn, and R. Gutenkunst. 2011. Demographic history and rare alleles sharing among human populations. *Proceedings of the National Academy of Sciences of the USA* 108(29):11983–11988.

Gravel, S., F. Zakharia, A. Moreno-Estrada, et al. 2013. Reconstructing Native American migrations from whole-genome and whole-exome data. *PLoS Genetics* 9:e1004023.

Green, S.J. 2017. "A wrong had finally been righted": tribes bury remains of ancient ancestor known as Kennewick Man. Seattle Times, February 19.

Green, R.E., A.W. Briggs, J. Krause, et al. 2009. The Neandertal genome and ancient DNA authenticity. *EMBO Journal* 28(17):2494–2502.

Green, R.E., J. Krause, A.W. Briggs, et al. 2010. A draft sequence of the Neandertal genome. *Science* 328(5979):710–722. doi:10.1126/science.1188021

Green, R.E., J. Krause, S.E. Ptak, et al. 2006. Analysis of one million base pairs of Neanderthal DNA. *Nature* 444(7117):330–336.

Green, R.E., A.S. Malaspinas, J. Krause, et al. 2008. A complete neandertal mitochondrial genome sequence determined by high-throughput sequencing. *Cell* 134(3):416–426.

Greenberg, C.R., L.A. Dilling, G.R. Thompson, et al. 2009. The paradox of the carnitine palmitoyltransferase type Ia P479L variant in Canadian Aboriginal populations. *Molecular Genetics and Metabolism* 96(4):201–207.

Greenberg, J.H., C.G. Turner II, and S.L. Zegura. 1986. The settlement of the Americas: a comparison of the linguistic, dental and genetic evidence. *Current Anthropology* 27:477–495.

Greer, E.L., T.J. Maures, D. Ucar, et al. 2011. Transgenerational epigenetic inheritance of longevity in *Caenorhabditis elegans*. *Nature* 479(7373):365–371.

Griffiths, R.C., and S. Tavare. 1996. Monte Carlo inference methods in population genetics. *Mathematical and Computer Modelling* 23(8–9):141–158. doi:10.1016/0895-7177(96)00046-5

Grogan, K.E., G.J. McGinnis, M.L. Sauther, et al. 2016. Next-generation genotyping of hypervariable loci in many individuals of a non-model species: technical and theoretical implications. *BMC Genomics* 17(1):204.

Grossman, S.R., I. Shylakhter, E.K. Karlsson, et al. 2010. A composite of multiple signals distinguishes causal variants in regions of positive selection. *Science* 327(5967):883–886.

Grün, R., and C.B. Stringer. 1991. Electron spin resonance dating and the evolution of modern humans. *Archaeometry* 33(2):153–199. doi:10.1111/j.1475-4754.1991.tb00696.x

Guenther, C.A., B. Tasic, et al. 2014. A molecular basis for classic blond hair color in Europeans. *Nature Genetics* 46(7):748–752. doi:10.1038/ng.2991

Guethlein, L.A., A.M. Older Aguilar, L. Abi-Rached, et al. 2007. Evolution of killer cell Ig-like receptor (KIR) genes: definition of an orangutan KIR haplotype reveals expansion of lineage III KIR associated with the emergence of MHC-C. *The Journal of Immunology* 179(1):491–504.

Guevara, E.E., and M.E. Steiper. 2013. Molecular phylogenetic analysis of the Papionina using concatenation and species tree methods. *Journal of Human Evolution* 66:18–28.

Guglielmino-Matessi, C.R., P. Gluckman, and L.L. Cavalli-Sforza. 1979. Climate and the evolution of skull metrics in man. *American Journal of Physical Anthropology* 50:549–564.

Guillot, G., and F. Rousset. 2013. Dismantling the Mantel tests. *Methods in Ecology and Evolution* 4:336–344.

Gutenkunst, R.N., R.D. Hernandez, S.H. Williamson, et al. 2009. Inferring the joint demographic history of multiple populations from multidimensional SNP frequency data. *PLoS Genetics* 5(10):e1000695. doi:10.1371/journal.pgen.1000695

Gutierrez, M.C., S. Brisse, R. Brosch, et al. 2005. Ancient origin and gene mosaicism of the progenitor of *Mycobacterium tuberculosis*. *PLoS Pathogens* 1(1):e5.

Guttman, M., I. Amit, M. Garber, et al. 2009. Chromatin signature reveals over a thousand highly conserved large non-coding RNAs in mammals. *Nature* 458:223–227.

Haak, W., O. Balanovsky, J.J. Sanchez, et al. 2010. Ancient DNA from European early Neolithic farmers reveals their Near Eastern affinities. *PLoS Biology* 8(11):e1000536. doi:10.1371/journal.pbio.1000536

Haak, W., P. Forster, B. Bramanti, et al. 2005. Ancient DNA from the first European farmers in 7500-year-old Neolithic sites. *Science* 310(5750):1016–1018.

Haak, W., I. Lazaridis, N. Patterson, et al. 2015. Massive migration from the steppe was a source for Indo-European languages in Europe. *Nature* 522(7555):207–211

Haas, C.J., A. Zink, U. Szeimies, et al. 2002. Molecular evidence of *Mycobacterium leprae* in historic bone samples from South Germany. In: C.A. Roberts, M.E. Lewis, and K. Manchester (eds.), *The Past and Present of Leprosy. Archaeological, Historical, Palaeopathological and Clinical Approaches*. Archaeopress, Oxford, pp. 287–292.

Haber, M., D. Gauguier, S. Youhanna, et al. 2013. Genome-wide diversity in the levant reveals recent structuring by culture. *PLoS Genetics* 9:e1003316.

Hagelberg, E., M. Hofreiter, and C. Keyser. 2015 Ancient DNA: the first three decades. *Philosophical Transactions of the Royal Society B: Biological Sciences* 370:20130371. http://dx.doi.org/10.1098/rstb.2013.0371

Hahn, B.H., G.M. Shaw, K.M. De Cock, et al. 2000. AIDS as a zoonosis: scientific and public health implications. *Science* 287(5453):607–614.

Hailer, F., V.E. Kutschera, B.M. Hallström, et al. 2012. Nuclear genomic sequences reveal that polar bears are an old and distinct bear lineage. *Science* 336:344–347.

Haldane, J.B.S. 1922. Sex ratio and unisexual sterility in hybrid animals. *Journal of Genetics* 12(2):101–109.

Haldane, J.B.S. 1927. A mathematical theory of natural and artificial selection, part V: selection and mutation. *Mathematical Proceedings of the Cambridge Philosophical Society* 23(7):838–844.

Haldane, J.B.S. 1949. Disease and evolution. *La Ricerca Scientifica* 19(Suppl):68–76.

Halligan, J.J., M.R. Waters, A. Perrotti, et al. 2016. Pre-Clovis occupation 14,550 years ago at the Page-Ladson site, Florida, and the peopling of the Americas. *Science Advances* 2:e1600375.

Hamblin, M.T., and A. Di Rienzo. 2000. Detection of the signature of natural selection in humans: evidence from the Duffy blood group locus. *American Journal of Human Genetics* 66:1669–1679.

Hamilton, W.D. 1980. Sex versus non-sex versus parasite. *Oikos* 35:282–290.

Hamilton, G. 2005. Bayesian estimation of recent migration rates after a spatial expansion. *Genetics* 170(1):409–417. doi:10.1534/genetics.104.034199

Hamilton, W.D., and M. Zuk. 1982. Heritable true fitness and bright birds: a role for parasites? *Science* 218(4570):384–387.

Hammer, M.F., F.L. Mendez, M.P. Cox, et al. 2008. Sex-biased evolutionary forces shape genomic patterns of human diversity. *PLoS Genetics* 4(9):e1000202. doi:10.1371/journal.pgen.1000202

Hammer, M.F., A.B. Spurdle, T. Karafet, et al. 1997. The geographic distribution of human Y chromosome variation. *Genetics* 145:787–805.

Hancock, A.M., V.J. Clark, Y. Qian, et al. 2011a. Population genetic analysis of the uncoupling proteins supports a role for *UCP3* in human cold resistance. *Molecular Biology and Evolution* 28(1):601–614.

Hancock, A.M., D.B. Witonsky, G. Alkorta-Aranburu, et al. 2011b. Adaptations to climate-mediated selective pressures in humans. *PLoS Genetics* 7(4):e1001375.

Hancock, A.M., D.B. Witonsky, E. Ehler, et al. 2010. Human adaptations to diet, subsistence, and ecoregion are due to subtle shifts in allele frequency. *Proceedings of the National Academy of Sciences* 107(Suppl. 2):8924–8930.

Hancock, A.M., D.B. Witonsky, A.S. Gordon, et al. 2008. Adaptations to climate in candidate genes for common metabolic disorders. *PLoS Genetics* 4(2):e32.

Handley, L.J.L., A. Manica, J. Goudet, et al. 2007. Going the distance: human population genetics in a clinal world. *Trends in Genetics* 23:432–439.

Hans, J.B., R.A. Bergl, and L. Vigilant. 2017. Gorilla MHC class I gene and sequence variation in a comparative context. *Immunogenetics* 69(5):303–323. doi:10.1007/s00251-017-0974-x

Harding, R.M., E. Healy, A.J. Ray, et al. 2000. Evidence for variable selective pressures at *MC1R*. *American Journal of Human Genetics* 66(4):1351–1361. doi:10.1086/302863

Harkins, K.M., and A.C. Stone. 2015. Ancient pathogen genomics: insights into timing and adaptation. *Journal of Human Evolution* 79:137–149.

Harmon, A. 2010. Havasupai case highlights risks in DNA research. New York Times, April 21, pp. 1–2.

Harmon, L.J., and R.E. Glor. 2010. Poor statistical performance of the Mantel test in phylogenetic comparative analyses. *Evolution* 64:2173–2178.

Harpending, H. 1974. Genetic structure of small populations. *Annual Review of Anthropology* 3:229–243.

Harpending, H. 2013. Ongoing evolution in humans. In: CARTA (prod.), Culture-Gene Interactions in Human Origins. https://www.youtube.com/watch?v=IkoV3l4C_Bo&feature=youtu.be

Harpending, H., and G. Cochran. 2002. In our genes. *Proceedings of the National Academy of Sciences* 99(1):10–12.

Harpending, H., and G. Cochran. 2006. Genetic diversity and genetic burden in humans. *Infection, Genetics and Evolution* 6:154–162.

Harpending, H., and T. Jenkins. 1973. Genetic distance among southern African populations. In: M.H. Crawford, and P.L. Workman (eds.), *Methods and Theories of Anthropological Genetics*. University of New Mexico Press, Albuquerque, pp. 177–199.

Harpending, H., and T. Jenkins. 1973. Genetic distance among southern African populations. In: M.H. Crawford, and P.L. Workman (eds.), *Methods and Theories of Anthropological Genetics*. University of Mexico Press, Albuquerque, pp. 177–199.

Harpending, H., and A. Rogers. 2000. Genetic perspectives on human origins and differentiation. *Annual Review of Genomics and Human Genetics* 1:361–385.

Harris, K., and R. Nielsen. 2016. The genetic cost of Neanderthal Introgression. *Genetics* 203(2):881.

Harris, K., and R. Nielsen. 2016. The genetic cost of Neanderthal Introgression. *Genetics* 203:881–891.

Harritt, R.K. 1998. Paleo-Eskimo beginnings in North America: a new discovery at Kuzitria Lake, Alaska. *Etudes Inuit* 22:61–81.

Hart, S. 2003. Investigative report concerning the medical genetics project at Havasupai.

Hart, J.S., H.B. Sabean, J.A. Hildes, et al. 1962. Thermal and metabolic responses of coastal Eskimos during a cold night. *Journal of Applied Physiology* 17:953–960.

Hartl, D.L., and A.G. Clark. 2006. *Principles of Population Genetics*. Sinauer Associates, Sunderland, MA.

Harvati, K., and T.D. Weaver. 2006. Human cranial anatomy and the differential preservation of population history and climate signatures. *The Anatomical Record Part A* 288A:1225–1233.

Harvey, P.H., and M.D. Pagel. 1991. *The Comparative Method in Evolutionary Biology*. Oxford University Press, Oxford.

Hasegawa, M., and S. Horai. 1991. Time of the deepest root for polymorphism in human mitochondrial DNA. *Journal of Molecular Evolution* 32(1):37–42. doi:10.1007/BF02099927

Hassan, N.A.-M., K.A. Brown, J. Eyers, et al. 2014. Ancient DNA study of the remains of putative infanticide victims from the Yewden roman villa site at Hambleden, England. *Journal of Archaeological Science* 43:192–197.

Hauser, M.T., W. Aufsatz, C. Jonak, et al. 2011. Transgenerational epigenetic inheritance in plants. *Biochimica et Biophysica Acta* 1809(8):459–468.

Hawks, J., E.T. Wang, G.M. Cochran, et al. 2007. Recent acceleration of human adaptive evolution. *Proceedings of the National Academy of Sciences* 104(52):20753–20758.

Hayakawa, T., T. Sugawara, Y. Go, et al. 2012. Eco-geographical diversification of bitter taste receptor genes (*TAS2Rs*) among subspecies of chimpanzees (*Pan troglodytes*). *PLoS One* 7:e43277.

Hayakawa, T., N. Suzuki-Hashido, A. Matsui, et al. 2014. Frequent expansions of the bitter taste receptor gene repertoire during evolution of mammals in the Euarchontoglires clade. *Molecular Biology and Evolution* 31:2018–2031.

Hayashi, T., A.G. Motulsky, and S.S. Deeb. 1999. Position of a 'green-red' hybrid gene in the visual pigment array determines colour-vision phenotype. *Nature Genetics* 22:90–93.

Hayes, J.E., E.L. Feeney, and A.L. Allen. 2013. Do polymorphisms in chemosensory genes matter for human ingestive behavior? *Food Quality and Preference* 30:202–216.

Hayes, J.E., E.L. Feeney, A.A. Nolden, et al. 2015. Quinine bitterness and grapefruit liking associate with allelic variants in *TAS2R31*. *Chemical Senses* 40:437–443.

Headon, D.J., and P.A. Overbeek. 1999. Involvement of a novel Tnf receptor homologue in hair follicle induction. *Nature Genetics* 22(4):370–374.

Heard, E., and R.A. Martienssen. 2014. Transgenerational epigenetic inheritance: myths and mechanisms. *Cell* 157(1):95–109.

Heath, T.A., M.T. Holder, and J.P. Huelsenbeck. 2012. A dirichlet process prior for estimating lineage-specific substitution rates. *Molecular Biology and Evolution* 29(3):939–955. doi:10.1093/molbev/msr255

Heath, T.A., and B.R. Moore. 2014. Bayesian inference of species divergence times. In: M.-H. Chen, L. Kuo, and P.O. Lewis (eds.), *Bayesian Phylogenetics*. Chapman and Hall, London, pp. 277–318.

Hedrick, P.W. 2011. *Genetics of Populations*. Jones and Bartlett Publishers, Sudbury, MA.

Hedrick, P.W. 2011. Population genetics of malaria resistance in humans. *Heredity* 107(4):283–304.

Hedrick, P.W. 2012. What is the evidence for heterozygote advantage selection? *Trends in Ecology and Evolution* 27(12):698–704.

Heijmans, B.T., E.W. Tobi, A.D. Stein, et al. 2008. Persistent epigenetic differences associated with prenatal exposure to famine in humans. *Proceedings of the National Academy of Sciences of the USA* 105(44):17046–17049.

Heinbecker, P. 1928. Studies on the metabolism of Eskimos. *Journal of Biological Chemistry* 80(2):461–475.

Heintzman, P.D., D. Froese, J.W. Ives, et al. 2016. Bison phylogeography constrains dispersal and viability of the ice free corridor in Western Canada. *Proceedings of the National Academy of Sciences of the USA* 113:8057–8063.

Heintzman, P.D., A.E.R. Soares, D. Chang, et al. 2015. Paleogenomics. *Reviews in Cell Biology and Molecular Medicine* 1(3):243–267.

Hellenthal, G., G.B.J. Busby, G. Band, et al. 2014. A genetic atlas of human admixture history. *Science* 343(6172):747–751. doi:10.1126/science.1243518

Henn, B.M., L.L. Cavalli-Sforza, and M.W. Feldman. 2012. The great human expansion. *Proceedings of the National Academy of Sciences, USA* 109(44):17758–17764.

Henn, B.M., C.R. Gignoux, M.W. Feldman, and J.L. Mountain. 2009. Characterizing the time dependency of human mitochondrial DNA mutation rate estimates. *Molecular Biology and Evolution* 26(1):217–230. doi:10.1093/molbev/msn244ho

Hennig, W. 1966. *Phylogenetic Systematics.* University of Illinois Press, Urbana.

Herceg, Z., and R. Murr. 2011. Mechanisms of histone modifications. In: T.O. Tollefsbol (ed.), *Handbook of Epigenetics: The New Molecular and Medical Genetics.* Academic Press, Amsterdam, pp. 25–45.

Hermisson, J., and P.S. Pennings. 2005. Soft sweeps: molecular population genetics of adaptation from standing genetic variation. *Genetics* 169(4):2335–2352.

Hernandez, R.D. 2008. A flexible forward simulator for populations subject to selection and demography. *Bioinformatics* 24(23):2786–2787. doi:10.1093/bioinformatics/btn522

Hernando-Herraez, I., J. Prado-Martinez, P. Garg, et al. 2013. Dynamics of DNA methylation in recent human and great ape evolution. *PLoS Genetics* 9(9):e1003763.

Herrnstein, R.J., and C. Murray. 1994. *The Bell Curve: Intelligence and Class Structure in American Life.* Free Press, New York.

Hershberg, R., M. Lipatov, P.M. Small, et al. 2008. High functional diversity in *Mycobacterium tuberculosis* driven by genetic drift and human demography. *PLoS Biology* 6(12):e311.

Hershberger, K.L., R. Shyam, A. Miura, et al. 2001. Diversity of the killer cell Ig-like receptors of rhesus monkeys. *Journal of Immunology (Baltimore, Md.)* 166(7):4380–4390. doi:10.4049/jimmunol.166.7.4380

Hertzman, C., and T. Boyce. 2010. How experience gets under the skin to create gradients in developmental health. *Annual Review of Public Health* 31:329–347ff.

Hewitt, D. 1957. Some familial correlations in height, weight and skeletal maturity. *Annals of Human Genetics* 22:26–35.

Hewitt, G. 2000. The genetic legacy of the quaternary ice ages. *Nature* 405(6789): 907–913.

Hewlett, B., and B. Hewlett. 2007. *Ebola, Culture and Politics: The Anthropology of an Emerging Disease.* Cengage Learning, Boston.

Hey, J. 2005. On the number of new world founders: a population genetic portrait of the peopling of the Americas. *PLoS Biology* 3(6):0965–0975. doi:10.1371/journal.pbio.0030193.g001

Hey, J. 2010. Isolation with migration models for more than two populations. *Molecular Biology and Evolution* 27(4):905–920. doi:10.1093/molbev/msp296

Hey, J., and R. Nielsen. 2007. Integration within the Felsenstein equation for improved Markov chain Monte Carlo methods in population genetics. *Proceedings of the National Academy of Sciences of the USA* 104(8):2785–2790. doi:10.1073/pnas.0611164104

Heyer, E., R. Chaix, S. Pavard, et al. 2012. Sex-specific demographic behaviours that shape human genomic variation. *Molecular Ecology* 21(3):597–612. doi:10.1111/j.1365-294X.2011.05406.x

Heyer, E., E. Zietkiewicz, A. Rochowski, et al. 2001 Phylogenetic and familial estimates of mitochondrial substitution rates: study of control region mutations in deep-rooting pedigrees. *American Journal of Human Genetics* 69(5):1113–1126. doi:10.1086/324024

Heyn, H., S. Moran, I. Hernando-Herraez, et al. 2013. DNA methylation contributes to natural human variation. *Genome Research* 23(9):1363–1372.

Hickerson, M.J., G. Dolman, and C. Moritz. 2006. Comparative phylogeographic summary statistics for testing simultaneous vicariance. *Molecular Ecology* 15(1):209–223. doi:10.1111/j.1365-294X.2005.02718.x

Hider, J.L., R.M. Gittelman, T. Shah, et al. 2013. Exploring signatures of positive selection in pigmentation candidate genes in populations of east Asian ancestry. *BMC Evolutionary Biology* 13:150. doi:10.1186/1471-2148-13-150

Higham, J.P., C.S. Barr, C.L. Hoffman, et al. 2011. Mu-opioid receptor (*OPRM1*) variation, oxytocin levels and maternal attachment in free-ranging rhesus macaques *Macaca mulatta*. *Behavioral Neuroscience* 125:131–136.

Hill, W.G. 2005. A century of corn selection. *Science* 307(5710):683–684.

Hill, W.G., Goddard, M.E., and Visscher, P.M. 2008. Data and theory point to mainly additive genetic variance for complex traits. *PLoS Genetics* 4(2):e1000008.

Hill, K., and A.M. Hurtado. 1996. *Ache Life History: The Ecology and Demography of a Foraging People*. Aldine De Gruyter, New York.

Hipsley, C.A., and J. Muller. 2014. Beyond fossil calibrations: realities of molecular clock practices in evolutionary biology. *Frontiers in Genetics* 5:138. doi:10.3389/fgene.2014.00138

Hirsila, M., P. Koivunen, V. Gunzler, et al. 2003. Characterization of the human prolyl 4-hydroxylases that modify the hypoxia-inducible factor. *Journal of Biological Chemistry* 278(33):30772–30780.

Hiura, H., H. Okae, H. Chiba, et al. 2014. Imprinting methylation errors in ART. *Reproductive Medicine and Biology* 13(4):193–202.

Hiwatashi, T., A. Mikami, T. Katsumura, et al. 2011. Gene conversion and purifying selection shape nucleotide variation in gibbon L/M opsin genes. *BMC Evolutionary Biology* 11:312.

Hiwatashi, T., Y. Okabe, T. Tsutsui, et al. 2010. An explicit signature of balancing selection for color-vision variation in new world monkeys. *Molecular Biology and Evolution* 27:453–464.

Hlusko, L.J., J. Carlson, G. Chaplin, et al. 2018. Environmental selection during the last ice age on the mother-to-infant transmission of vitamin D and fatty acids through breast milk. *Proceedings of the National Academy of Sciences of the USA* 115(19):E4426–E4432.

Hlusko, L.J., N. Do, and M.C. Mahaney. 2007. Genetic correlations between mandibular molar cusp areas in baboons. *American Journal of Physical Anthropology* 132:445–454.

Hlusko, L., G. Suwa, R.T. Kono, et al. 2004. Genetics and the evolution of primate enamel thickness: a baboon model. *American Journal of Physical Anthropology* 124:223–233.

Ho, S.Y.W. 2014. The changing face of the molecular evolutionary clock. *Trends in Ecology & Evolution* 29(9):496–503. doi:10.1016/j.tree.2014.07.004

Ho, S.Y.W., and S. Duchêne. 2014. Molecular-clock models for estimating evolutionary rates and timescales. *Molecular Ecology* 23(24):5947–5965. doi:10.1111/mec.12953

Ho, S.Y., S. Duchêne, and D. Duchêne. 2014a. Simulating and detecting autocorrelation of molecular evolutionary rates among lineages. *Molecular Ecology Resources* 15(4):688–696. doi:10.1111/1755-0998.12320

Ho, S.Y.W., and P. Endicott. 2008. The crucial role of calibration in molecular date estimates for the peopling of the Americas. *American Journal of Human Genetics* 83(1):142–146. doi:10.1016/j.ajhg.2008.06.014

Ho, S.Y.W., P. Endicott, and M. Molak. 2014b. Human evolution: molecular timescale. In: C. Smith (ed.), *Encyclopedia of Global Archaeology*. Springer, New York, pp. 3511–3514.

Ho, S.Y.W., R. Lanfear, L. Bromham, et al. 2011. Time-dependent rates of molecular evolution. *Molecular Ecology* 20(15):3087–3101. doi:10.1111/j.1365-294X.2011.05178.x

Ho, S.Y., and G. Larson. 2006. Molecular clocks: when times are a-changin'. *Trends in Genetics* 22(2):79–83. doi:10.1016/j.tig.2005.11.006

Hoban, S., G. Bertorelle, and O.E. Gaggiotti. 2011. Computer simulations: tools for population and evolutionary genetics. *Nature Reviews Genetics* 13(2):110–122. doi:10.1038/nrg3130

Hoelsbrekken, S.E., Ø. Nylenna, P.C. Saether, et al. 2003. Cutting edge: molecular cloning of a killer cell Ig-like receptor in the mouse and rat. *The Journal of Immunology* 170(5):2259–2263.

Hoffecker, J.F., and S.A. Elias. 2003. Environment and archeology in Beringia. *Evolutionary Anthropology* 12:34–49.

Hoffecker, J.F., and S.A. Elias. 2007. *Human Ecology of Beringia*. Columbia University Press, New York.

Hoffecker, J.F., S.A. Elias, D.H. O'Rourke, et al. 2016. Beringia and the global dispersal of modern humans. *Evolutionary Anthropology* 25:64–78.

Hofreiter, M. 2012. Nondestructive DNA extraction from museum specimens. In: B. Shapiro, and M. Hofreiter (eds.), *Ancient DNA: Methods and Protocols*. Vol. 840. Humana, New York, pp. 93–100.

Höglind, A., I. Areström, C. Ehrnfelt, et al. 2017. Systematic evaluation of monoclonal antibodies and immunoassays for the detection of interferon-γ and Interleukin-2 in old and new world non-human primates. *Journal of Immunological Methods* 441:39–48. doi:10.1016/j.jim.2016.11.011

Holick, M.F. 2003. Vitamin D: a millenium perspective. *Journal of Cellular Biochemistry* 88(2):296–307. doi:10.1002/jcb.10338

Holick, M.F. 2004a. Sunlight and vitamin D for bone health and prevention of autoimmune diseases, cancers, and cardiovascular disease. *American Journal of Clinical Nutrition* 80(6 Suppl):1678S–1688S.

Holick, M.F. 2004b. Vitamin D: importance in the prevention of cancers, type 1 diabetes, heart disease, and osteoporosis. *American Journal of Clinical Nutrition* 79(3):362–371.

Holmes, E.C., L.Q. Zhang, P. Simmonds, et al. 1992. Convergent and divergent sequence evolution in the surface envelope glycoprotein of human immunodeficiency virus type 1 within a single infected patient. *Proceedings of the National Academy of Sciences of the USA* 89(11):4835–4839.

Holsinger, K.E., and B.S. Weir. 2009. Genetics in geographically structured populations: defining, estimating and interpreting FST. *Nature Reviews Genetics* 10(9):639–650. doi:10.1038/nrg2611

Hong, H.S., P.A. Rajakumar, J.M. Billingsley, et al. 2013. No monkey business: why studying NK cells in non-human primates pays off. *Frontiers in Immunology* 4:32.

Hoon, M.A., E. Adler, J. Lindemeier, et al. 1999. Putative mammalian taste receptors: a class of taste-specific GPCRs with distinct topographic selectivity. *Cell* 96:541–551.

Hooshiar Kashani, B., Perego U.A., A. Olivieri, et al. 2012. Mitochondrial haplogroup C4c: a rare lineage entering America through the ice-free corridor? *American Journal of Physical Anthropology* 147:35–39.

Horai, S., K. Hayasaka, R. Kondo, et al. 1995. Recent African origin of modern humans revealed by complete sequences of hominoid mitochondrial DNAs. *Proceedings of the National Academy of Sciences of the USA* 92(2):532–536. doi:10.1073/pnas.92.2.532

Horai, S., S. Kondo, Y. Nakagawa-Hattori, et al. 1993. Peopling of the Americas, founded by four major lineages of mitochondrial DNA. *Molecular Biology and Evolution* 10:23–47.

Horvath, S. 2013. DNA methylation age of human tissues and cell types. *Genome Biology* 14(10):R115.

Howell, N., C.B. Smejkal, D.A. Mackey, et al. 2003. The pedigree rate of sequence divergence in the human mitochondrial genome: there is a difference between phylogenetic and pedigree rates. *American Journal of Human Genetics* 72(3):659–670. doi:10.1086/368264

Howells, W.W. 1973. *Cranial Variation in Man: A Study by Multivariate Analysis of Patterns of Difference among Recent Human Populations. Peabody Museum of Archaeology and Ethnology*, Harvard University, Cambridge, MA.

Howells, W.W. 1996. Howells' craniometric data on the internet. *American Journal of Physical Anthropology* 101:441–442.

Hubbe, M., T. Hanihara, and K. Harvati. 2009. Climate signatures in the morphological differentiation of worldwide modern human populations. *The Anatomical Record* 292:1720–1733.

Hubbe, M., W.A. Neves, and K. Harvati. 2010. Testing evolutionary and dispersion scenarios for the settlement of the new world. *PLoS ONE* 5(6):e11105.

Hubbe, M., A. Strauss, A. Hubbe, et al. 2015. Early south Americans cranial morphological variation and the origin of American biological diversity. *PLoS One* 10:e0138090.

Hublin, J.J., A. Ben-Ncer, S.E. Bailey, et al. 2017. New fossils from Jebel Irhoud, Morocco and the pan-African origin of *Homo sapiens*. *Nature* 546:289–292.

Huchard, E., C. Albrecht, S. Schliehe-Diecks, et al. 2012. Large-scale MHC class II genotyping of a wild lemur population by next generation sequencing, *Immunogenetics* 895–913. doi:10.1007/s00251-012-0649-6

Huchard, E., A. Courtiol, J.A. Benavides, et al. 2009. Can fertility signals lead to quality signals? Insights from the evolution of primate sexual swellings. *Proceedings Biological Sciences* 276:1889–1897.

Huchard, E., G. Cowlishaw, M. Raymond, et al. 2006. Molecular study of Mhc-DRB in wild Chacma baboons reveals high variability and evidence for trans- species inheritance. *Immunogenetics* 805–816. doi:10.1007/s00251-006-0156-8

Hudjashov, G., T. Kivisild, P.A. Underhill, et al. 2007. Revealing the prehistoric settlement of Australia by Y chromosome and mtDNA analysis. *Proceedings of the National Academy of Sciences of the USA* 104(21):8726–8730. doi:10.1073/pnas.0702928104

Hudson, R.R. 1990. Gene genealogies and the coalescent process. In: D. Futuyma, and J. Antonovics (eds.), *Oxford Surveys in Evolutionary Biology*. Oxford University Press, New York, pp. 1–44.

Hudson, R.R. 1990. Gene genealogies and the coalescent process. *Oxford Surveys in Evolutionary Biology* 7:1–44.

Huerta-Sanchez, E., M. DeGiorgio, L. Pagani, et al. 2013. Genetic signatures reveal high-altitude adaptation in a set of Ethiopian populations. *Molecular Biology and Evolution* 30(8):1877–1888.

Huerta-Sánchez, E., X. Jin, B.Z. Asan, et al. 2014. Altitude adaptation in Tibetans caused by introgression of Denisovan-like DNA. *Nature*. 512(7513):194–197. doi:10.1038/nature13408

Huff, C.D., H.C. Harpending, and A.R. Rogers. 2010. Detecting positive selection from genome scans of linkage disequilibrium. *BMC Genomics* 11(1):8.

Hughes, A.L. 1995. Origin and evolution HLA class I pseudogenes. *Molecular Biology and Evolution* 12(2):247–258. doi:10.1093/oxfordjournals.molbev.a040201

Hughes, A.L. 2007. Looking for Darwin in all the wrong places: the misguided quest for positive selection at the nucleotide sequence level. *Heredity* 99(4):364–373. doi:10.1038/sj.hdy.6801031

Hughes, A.L., and M. Nei. 1988. Pattern of nucleotide substitution at major histocompatibility complex class-I loci reveals overdominant selection. *Nature* 335(6186):167–170. doi:10.1038/335167a0

Hughes, A.L., and M. Nei. 1989. Nucleotide substitution at major histocompatibility complex class II loci: evidence for overdominant selection. *Proceedings of the National Academy of Sciences* 86(February):958–962.

Hughes, D.A., K. Tang, R. Strotmann, et al. 2008. Parallel selection on T.R.PV6 in human populations. *PLoS One* 3(2):e1686.

HUGO Pan-Asian SNP Consortium, M.A. Abdulla, I. Ahmed, et al. 2009. Mapping human genetic diversity in Asia. *Science* 326(5959):1541–1545.

Hulse, F.S. 1955. Blood types and mating patterns among northwest coast Indians. *Southwest Journal of Anthropology* 11:93–104.

Hulse, F.S. 1964. Exogamy and heterosis. *Yearbook of Physical Anthropology* 9:240–257.

Hünemeier, T., C.E. Amorim, S. Azevedo, et al. 2012. Evolutionary responses to a constructed niche: ancient Mesoamericans as a model of gene-culture coevolution. *PLoS One* 7:e38862.

Hunley, K.L., G.S. Cabana, and J.C. Long. 2016. The apportionment of human diversity revisited. *American Journal of Physical Anthropology* 160(4):561–569.

Hunley, K.L., M.E. Healy, and J.C. Long. 2009. The global pattern of gene identity variation reveals a history of long-range migrations, bottlenecks, and local mate exchange: implications for biological race. *American Journal of Physical Anthropology* 139:35–46.

Hunt, D.M., L.S. Carvalho, J.A. Cowing, et al. 2009. Evolution and spectral tuning of visual pigments in birds and mammals. *Philosophical Transactions of the Royal Society, B: Biological Sciences* 364:2941–2955.

Hurrell, R.F. 2003. Influence of vegetable protein sources on trace element and mineral bioavailability. *Journal of Nutrition* 133(9):2973S–2977S.

Hutchinson, J., and M.H. Crawford. 1981. Genetic determinants of blood pressure level among the black Caribs of St. Vincent. *Human Biology* 53(3):453–466.

Huxley, J. 1943. *Evolution: The Modern Synthesis*. George Allen & Unwin Ltd, London.

Ibbotson, R.E., D.M. Hunt, J.K. Bowmaker, et al. 1992. Sequence divergence and copy number of the middle- and long-wave photopigment genes in Old World monkeys. *Proceedings of the Royal Society of London B: Biological Sciences* 247:145–154.

Ihara, Y., and K. Aoki. 1999. Sexual selection by male choice in monogamous and polygynous humans. *Theoretical Population Biology* 55:77–93.

Ingman, M., H. Kaessmann, S. Pääbo, et al. 2000. Mitochondrial genome variation and the origin of modern humans. *Nature* 408:708–13.

International HapMap Consortium. 2003. The International HapMap Project. *Nature* 426(6968):789–796. doi:http://www.nature.com/nature/journal/v426/n6968/suppinfo/nature02168_S1.html

International Society of Genetic Genealogy. 2014. Y-DNA haplogroup tree 2014. Version 9.22. http://www.isogg.org/tree

Ishida, Y. 2009. Sewall Wright and Gustave Malécot on isolation by distance. *Philosophy of Science* 76:784–796. doi:10.1086/605802

Iskow, R.C., O. Gokcumen, A. Abyzov, et al. 2012. Regulatory element copy number differences shape primate expression profiles. *Proceedings of the National Academy of Sciences of the USA* 109:12656–12661.

Ivan, M., K. Kondo, H. Yang, et al. 2001. HIFalpha targeted for VHL-mediated destruction by proline hydroxylation: implications for O_2 sensing. *Science* 292(5516): 464–468.

Iwasaki, A., and R. Medzhitov. 2015. Control of adaptive immunity by the innate immune system. *Nature Immunology* 16(4):1–7. doi:10.1038/ni.3123.Control

Izagirre, N., I. Garcia, C. Junquera, et al. 2006. A scan for signatures of positive selection in candidate loci for skin pigmentation in humans. *Molecular Biology and Evolution* 23(9):1697–1706. doi:10.1093/molbev/msl030

Jaakkola, P., D.R. Mole, Y.M. Tian, et al. 2001. Targeting of HIF-alpha to the von Hippel-Lindau ubiquitylation complex by O_2-regulated prolyl hydroxylation. *Science* 292(5516):468–472.

Jablonka, E., and J.J. Lamb. 2005. *Evolution in Four Dimensions: Genetic, Epigenetic, Behavioral, and Symbolic Variation in the History of Life*. MIT Press, Cambridge, MA.

Jablonski, N.G. 2004. The evolution of human skin and skin color. *Annual Reviews in Anthropology* 33:585–623.

Jablonski, N.G., and G. Chaplin. 2000. The evolution of human skin coloration. *Journal of Human Evolution* 39(1):57–106. doi:10.1006/jhev.2000.0403

Jablonski, N.G., and G. Chaplin. 2010. Colloquium paper: human skin pigmentation as an adaptation to UV radiation. *Proceedings of the National Academy of Sciences of the USA* 107(Suppl 2):8962–8968. doi:10.1073/pnas.0914628107

Jablonski, N.G., and G. Chaplin. 2014. Skin cancer was not a potent selective force in the evolution of protective pigmentation in early hominins. *Proceedings Biological Sciences* 281(1789):20140517. doi:10.1098/rspb.2014.0517

Jabot, F., T. Faure, N. Dumoulin, et al. 2013. EasyABC: performing efficient approximate Bayesian computation sampling schemes using R. *Methods in Ecology and Evolution* 4(7):684–687. doi:10.1111/2041-210x.12050

Jacobs, G.H. 2008. Primate color vision: a comparative perspective. *Visual Neuroscience* 25:619–633.

Jacobs, G.H. 2013. Losses of functional opsin genes, short-wavelength cone photopigments, and color vision – a significant trend in the evolution of mammalian vision. *Visual Neuroscience* 30:39–53.

Jacobs, G.H, J.F. Deegan II, Y. Tan, et al. 2002. Opsin gene and photopigment polymorphism in a prosimian primate. *Vision Research* 42:11–18.

Jacobs, G.H., and G.A. Williams. 2001. The prevalence of defective color vision in Old World monkeys and apes. *Color Research & Application* 26:S123–S127.

Jacobs R.L., MacFie, T.S., Spriggs, A.N., Baden, A.L., Morelli, T.L., Irwin, M.T., Lawler, R.R., Pastorini, J., Mayor, M., Lei, R., Culligan, R., Hawkins, M.T.R., Kappeler, P.M., Wright, P.C., Louis, E.E., Mundy, N.I., and Bradley, B.J. 2017. Novel opsin gene variation in large-bodied, diurnal lemurs. *Biology Letters* 13:20170050.

Jacovas, V.C., D.L. Rovaris, O. Peréz, et al. 2015. Genetic variations in the TP53 Pathway in Native Americans strongly suggest adaptation to the high altitudes of the Andes. *PLoS One* 10:e0137823.

Jaeger, T.F., D. Pontillo, and P. Graff. 2012. Comment on "Phonemic diversity supports a serial founder effect model of language expansion from Africa". *Science* 335:1042.

Jakobsson, M., S.W. Scholz, P. Scheet, et al. 2008. Genotype, haplotype and copy-number variation in worldwide human populations. *Nature* 451(7181):998–1003.

Jambo, K.C., D.H. Banda, A.M. Kankwatira, et al. 2014. Small alveolar macrophages are infected preferentially by HIV and exhibit impaired phagocytic function. *Mucosal Immunology* 7(5):1116–1126. doi:10.1038/mi.2013.127

Jameson, K.A., S.M. Highnote, and L.M. Wasserman. 2001. Richer color experience in observers with multiple photopigment opsin genes. *Psychonomic Bulletin & Review* 8:244–261.

Jang, H.-J., Z. Kokrashvili, M.J. Theodorakis, et al. 2007. Gut-expressed gustducin and taste receptors regulate secretion of glucagon-like peptide-1. *Proceedings of the National Academy of Sciences* 104:15069–15074.

Jantz, R.L., and D.W. Owsley. 2001. Variation among early North American Crania. *American Journal of Physical Anthropology* 114:146–155.

Jasinska, A.J., C.A. Schmitt, S.K. Service, et al. 2013. Systems biology of the vervet monkey. *IL.A.R Journal* 54:122–143.

Jeffreys, A.J. 1987. Highly variable minisatellites and DNA fingerprints. *Biochemical Society Transactions* 15:309–317.

Jeffreys, A.J., M. Allen, E. Hagelberg, A. Sonnberg. 1992. Identification of the skeletal remains of Josef Mengele by DNA analyses. *Forensic Science International* 56:65–76.

Jeffreys, A.J., and R. Neumann. 2009. The rise and fall of a human recombination hot spot. *Nature Genetics* 41(5):625–629. doi:10.1038/ng.346

Jenkins, D.L., L.G. Davis, T.W. Stafford Jr., et al. 2012. Clovis age Western Stemmed projectile points and human coprolites at the Paisley Caves. *Science* 337:223–228.

Jensen, A.R. 1998. *The g Factor: The Science of Mental Ability*. Praeger.

Jensen, G.M., and L.G. Moore. 1997. The effect of high altitude and other risk factors on birthweight: independent or interactive effects? *American Journal of Public Health* 87(6):1003–1007.

Jenuwein, T., and C.D. Allis. 2001. Translating the histone code. *Science* 293(5532):1074–1080.

Jiang, P., J. Josue, X. Li, et al. 2012. Major taste loss in carnivorous mammals. *Proceedings of the National Academy of Sciences* 109:4956–4961.

Jobling, M., E. Hollox, M. Hurles, et al. 2014. *Human Evolutionary Genetics*. 2nd ed. Garland Science, New York.

Jobling, M.A., and C. Tyler-Smith. 2003. The human Y chromosome: an evolutionary marker comes of age. *Nature Reviews Genetics* 4:598–612.

Johnson, L.F., and D.A. Lewis. 2008. The effect of genital tract infections on HIV-1 shedding in the genital tract: a systematic review and meta-analysis. *Sexually Transmitted Diseases* 35(11):946–959. doi:10.1097/OLQ.0b13e3181812d15

Johnson, M.E., National Institute of Health Intramural Sequencing Center Comparative Sequencing Program, Z. Cheng, et al. 2006. Recurrent duplication-driven transposition of DNA during hominoid evolution. *Proceedings of the National Academy of Sciences of the USA* 103:17626–17631.

Jones, P.A. 2012. Functions of DNA methylation: islands, start sites, gene bodies and beyond. *Nature Reviews Genetics* 13(7):484–492.

Jones, F.C., M.G. Grabherr, Y.F. Chan, et al. 2012. The genomic basis of adaptive evolution in threespine sticklebacks. *Nature* 484:55–61.

Jones, K.E., N.G. Patel, M.A. Levy, et al. 2008. Global trends in emerging infectious diseases. *Nature* 451(7181):990–993.

Jónsson, H., A. Ginolhac, M. Schubert, et al. 2013. mapDamage2.0: fast approximate Bayesian estimates of ancient DNA damage parameters. *Bioinformatics* 29(13): 1682–1684.

Jordan, G., S.S. Deeb, J.M. Bosten, et al. 2010. The dimensionality of color vision in carriers of anomalous trichromacy. *Journal of Vision* 10:12.

Jordan, G., and J.D. Mollon. 1993. A study of women heterozygous for colour deficiencies. *Vision Research* 33:1495–1508.

Jorde, L.B. 1980. The genetic structure of subdivided human populations. In J.H. Mielke, and M.H. Crawford (eds.), *Current Developments in Anthropological Genetics, Vol. 1: Theory and Methods.* Plenum Press, New York, pp. 135–208.

Joyce, P., and P. Marjoram. 2008. Approximately sufficient statistics and Bayesian computation. *Statistical Applications in Genetics and Molecular Biology* 7(1):art. 26.

Ju, X., G. Clark, and D.N. Hart. 2010. Review of human DC subtypes. *Journal of Molecular Biology and Methods* 595:3–20.

Julian, C.G., M.J. Wilson, M. Lopez, et al. 2009. Augmented uterine artery blood flow and oxygen delivery protect Andeans from altitude-associated reductions in fetal growth. *American Journal of Physiology: Regulatory, Integrative and Comparative Physiology* 296(5):R1564–R1575.

Juric, I., S. Aeschbacher, and G. Coop. 2016. The strength of selection against Neanderthal Introgression. *PLoS Genetics* 12(11):e1006340.

Jurmain, R., L. Kilgore, W. Trevathan, et al. 2013. *Introduction to Physical Anthropology.* Belmont, CA: Cengage Learning.

Kadekaro, A.L., R. Kavanagh, H. Kanto, et al. 2005. Alpha-Melanocortin and endothelin-1 activate antiapoptotic pathways and reduce DNA damage in human melanocytes. *Cancer Research* 65(10):4292–4299. doi:10.1158/0008-5472.C.A.N-04-4535

Kadekaro, A.L., S. Leachman, R.J. Kavanagh, et al. 2010. Melanocortin 1 receptor genotype: an important determinant of the damage response of melanocytes to ultraviolet radiation. *FASEB Journal* 24(10):3850–3860. doi:10.1096/fj.10-158485

Kadowaki, N., S. Ho, S. Antonenko, et al. 2001. Subsets of human dendritic cell precursors express different toll-like receptors and respond to different microbial antigens. *The Journal of Experimental Medicine* 194(6):863–869. doi:10.1084/jem.194.6.863

Kaestle, P.A., and K.A. Horsburgh. 2002. Ancient DNA in anthropology: methods, applications, and ethics. *Yearbook of Physical Anthropology* 45:92–130.

Kaestle, F.A., and D.G. Smith. 2001. Ancient mitochondrial DNA evidence for prehistoric population movement: the Numic expansion. *American Journal of Physical Anthropology* 115:1–12. doi:10.1002/ajpa.1051

Kaikkonen, M.U., M.T. Lam, and C.K. Glass. 2011. Non-coding RNAs as regulators of gene expression and epigenetics. *Cardiovascular Research* 90(3):430–440.

Kaiser, J. 2014. The epigenetics heretic. *Science* 343(6169):361–363.

Kamberov, G.Y., S. Wang, J. Tan, et al. 2013. Modeling recent human evolution in mice by expression of a selected *EDAR* variant. *Cell* 152(4):691–702.

Karafet, T.M., L.P. Osipova, M.A. Gubina, et al. 2002. High levels of Y-chromosome differentiation among native Siberian populations and the genetic signature of a boreal hunter-gatherer way of life. *Human Biology* 74(6):761–789.

Karafet, T.M., S.L. Zegura, O. Posukh, et al. 1999. Ancestral Asian source(s) of New World Y-chromosome founder haplotypes. *American Journal of Human Genetics* 64:817–831.

Karlsson, E.K., J.B. Harris, S. Tabrizi, et al. 2013. Natural selection in a Bangladeshi population from the cholera-endemic Ganges River delta. *Science Translational Medicine* 5(192):192ra186.

Karmin, M., L. Saag, M. Vicente, et al. 2015. A recent bottleneck of Y chromosome diversity coincides with a global change in culture. *Genome Research* 25(4):459–466.

Katz, D., M.N. Grote, and T.D. Weaver. 2016. A mixed model for the relationship between climate and human cranial form. *American Journal of Physical Anthropology* 160(4):593–603.

Kaur, T., J. Singh, S. Tong, et al. 2008. Descriptive epidemiology of fatal respiratory outbreaks and detection of a human-related metapneumovirus in wild chimpanzees (*Pan troglodytes*) at Mahale Mountains National Park, Western Tanzania. *American Journal of Primatology* 70(8):755–765. doi:10.1002/ajp.20565

Kawai, T., and S. Akira. 2010. The role of pattern-recognition receptors in innate immunity: update on toll-like receptors. *Nature Immunology* 11(5):373–384. doi:10.1038/ni.1863

Kawamura, S., and N. Kubotera. 2004. Ancestral loss of short wave-sensitive cone visual pigment in lorisiform prosimians, contrasting with its strict conservation in other prosimians. *Journal of Molecular Evolution* 58:314–321.

Kayser, M., S. Brauer, R. Cordaux, et al. 2006. Melanesian and Asian origins of Polynesians: mtDNA and Y chromosome gradients across the Pacific. *Molecular Biology and Evolution* 23:2234–2244.

Kayser, M., S. Brauer, G. Weiss, et al. 2000. Melanesian origin of Polynesian Y chromosomes. *Current Biology* 10:1237–1246.

Kayser, M., and P de Kniff. 2011. Improving human forensics through advances in genetics, genomics, and molecular biology. *Nature Reviews Genetics* 12:179–192.

Kayser, M., O. Lao, K. Saar, et al. 2008. Genome-wide analysis indicates more Asian than Melanesian ancestry of Polynesians. *American Journal of Human Genetics* 82:194–198.

Kayser, M., F. Liu, A.C. Janssens, et al. 2008. Three genome-wide association studies and a linkage analysis identify HER.C.2 as a human iris color gene. *American Journal of Human Genetics* 82(2):411–423. doi:10.1016/j.ajhg.2007.10.003

Keele, B.F., J.H. Jones, K.A. Terio, et al. 2009. Increased mortality and AIDS-like immunopathology in wild chimpanzees infected with SIVcpz. *Nature* 460(7254):515–519.

Keinan, A., J.C. Mullikin, N. Patterson, et al. 2009. Accelerated genetic drift on chromosome X during the human dispersal out of Africa. *Nature Genetics* 41(1): 66–70.

Keller, M.C., D. Thiessen, and R.K. Young. 1996. Mate assortment in dating and married couples. *Personality and Individual Differences* 21(2):217–221.

Kenny, E.E., N.J. Timpson, M. Sikora, et al. 2012. Melanesian blond hair is caused by an amino acid change in *TYRP1*. *Science* 336(6081):554. doi:10.1126/science.1217849

Kern, P.A., B.S. Finlin, B. Zhu, et al. 2014. The effects of temperature and seasons on subcutaneous white adipose tissue in humans: evidence for thermogenic gene induction. *Journal of Clinical Endocrinology and Metabolism* 99(12):E2772–E2779.

Kerr, P.J., J. Liu, I. Cattadori, et al. 2015. Myxoma virus and the Leporipoxviruses: an evolutionary paradigm. *Viruses* 7(3):1020–6106. doi:10.3390/v7031020

Key, F.M., J.C. Teixeira, C. de Filippo, et al. 2014. Advantageous diversity maintained by balancing selection in humans. *Current Opinion in Genetics and Development* 29:45–51.

Khakoo, S.I., R. Rajalingam, B.P. Shum, et al. 2000. Rapid evolution of NK cell receptor systems demonstrated by comparison of chimpanzees and humans. *Immunity* 12(6):687–698.

Khan, Z., M.J. Ford, D.A. Cusanovich, et al. 2013. Primate transcript and protein expression levels evolve under compensatory selection pressures. *Science* 342:1100–1104.

Khil, P.P., and R.D. Camerini-Otero. 2009. Variation in patterns of human meiotic recombination. *Genome Dynamics* 5:117–127. doi:10.1159/000166623

Khrameeva, E.E., K. Bozek, L. He, et al. 2014. Neanderthal ancestry drives evolution of lipid catabolism in contemporary Europeans. *Nature Communications* 5.

Kile, M.L., E.A. Houseman, A.A. Baccarelli, et al. 2014. Effect of prenatal arsenic exposure on DNA methylation and leukocyte subpopulations in cord blood. *Epigenetics* 9(5):774–782.

Kim, U., and D. Drayna. 2004. Genetics of individual differences in bitter taste perception: lessons from the PTC gene. *Clinical Genetics* 67:275–280.

Kim, U., E. Jorgenson, H. Coon, et al. 2003. Positional cloning of the human quantitative trait locus underlying taste sensitivity to phenylthiocarbamide. *Science* 299:1221–1225.

Kim, B.Y., and K.E. Lohmueller. 2015. Selection and reduced population size cannot explain higher amounts of Neandertal ancestry in east Asian than in European human populations. *American Journal of Human Genetics* 96(3):454–461.

Kim, U.-K., S. Wooding, N. Riaz, et al. 2006. Variation in the human *TAS1R* taste receptor genes. *Chemical Senses* 31:599–611.

Kim, U., S. Wooding, D. Ricci, et al. 2005. Worldwide haplotype diversity and coding sequence variation at human bitter taste receptor loci. *Human Mutation* 26:199–204.

Kimura, M. 1968. Evolutionary rate at the molecular level. *Nature* 217:624–626. doi:10.1038/217624a0

Kimura, M. 1983. *The Neutral Theory of Evolution*. Cambridge University Press, Cambridge. doi:10.1017/cbo9780511623486

Kimura, M., and G.H. Weiss. 1964. The stepping stone model of population structure and the decrease of genetic correlation with distance. *Genetics* 49:561–576.

Kimura, R., T. Yamaguchi, M. Takeda, et al. 2009. A common variation in *EDAR* is a genetic determinant of shovel-shaped incisors. *American Journal of Human Genetics* 85(4):528–535.

Kindberg, E., A. Mickiene, C. Ax, et al. 2008. A deletion in the chemokine receptor 5 (*CCR5*) gene is associated with tickborne encephalitis. *Journal of Infectious Diseases* 197(2):266–269. doi:10.1086/524709

King, C.L., J.H. Adams, J. Xianli, et al. 2011. Fy(a)/Fy(b) antigen polymorphism in human erythrocyte Duffy antigen affects susceptibility to plasmodium vivax malaria. *Proceedings of the National Academy of Sciences of the USA* 108(50):20113–20118. doi:10.1073/pnas.1109621108

King, T.E., G.G. Fortes, P. Balaresque, et al. 2014. Identification of the remains of King Richard III. *Nature Communications* 5:5631.

King, M.C., and A.C. Wilson. 2014. Evolution at two levels in humans and chimpanzees. In: F.C. Ayala, and J.C. Avise (eds.), *Essential Readings in Evolutionary Biology*. Johns Hopkins University Press, Baltimore, pp. 301–311.

Kingman, J.F.C. 1982. On the genealogy of large populations. *Journal of Applied Probability* 19:27–43.

Kinnally, E.L., J.P. Capitanio, R. Leibel, et al. 2010. Epigenetic regulation of serotonin transporter expression and behavior in infant rhesus macaques. *Genes, Brain and Behavior* 9(6):575–582.

Kitchen, A., M.M. Miyamoto, and C.J. Mulligan. 2008. A three-stage colonization model for the peopling of the Americas. *PLoS One* 3:e1596.

Klein, J. 1986. *Natural History of the Major Histocompatibility Complex*. John Wiley, NewYork.

Klein, J. 1987. Origin of major histocompatibility complex polymorphism: the trans-species hypothesis. *Human Immunology* 19(3):155–162.

Kleinjan, D.A., and V. van Heyningen. 2005. Long-range control of gene expression: emerging mechanisms and disruption in disease. *American Journal of Human Genetics* 76(1):8–32.

Knapp, L.A., L.F. Cadavid, and D.I. Watkins. 1998. The MHC-E locus is the most well conserved of all known primate class I histocompatibility genes. *Journal of Immunology (Baltimore)* 160(1) 189–196.

Koboldt, D.C., K.M. Steinberg, D.E. Larson, et al. 2013. The next-generation sequencing revolution and its impact on genomics. *Cell* 155:27–38.

Kodama, K. 2009. *Os Índios no Império do Brasil: a etnografia do IHGB entre as décadas de 1840 e 1860*. Editora Fiocruz, Rio de Janeiro.

Koerner, B.I. 2005. Blood feud. Wired, September. http://archive.wired.com/wired/archive/13.09/seminoles.html

Kolaczkowska, E., and P. Kubes. 2013. Neutrophil recruitment and function in health and inflammation. *Nature Reviews. Immunology* 13(3):159–175. doi:10.1038/nri3399

Kollias, N., R.M. Sayre, L. Zeise, et al. 1991. Photoprotection by melanin. *Journal of Photochemistry and Photobiology B* 9(2):135–160.

Kolman, C.J., A. Centurion-Lara, S.A. Lukehart, et al. 1999. Identification of *Treponema pallidum* subspecies *pallidum* in a 200-year-old skeletal sample. *Journal of Infectious Diseases* 180:2060–2063.

Kolopenuk, J. 2014. Wiindigo incarnate: consuming "Native American DNA." GeneWatch. http://www.councilforresponsiblegenetics.org/GeneWatch/GeneWatch Page.aspx?pageId=531&archive=yes

Kondrashov, A.S. 2003. Direct estimates of human per nucleotide mutation rates at 20 loci causing Mendelian diseases. *Human Mutation* 21(1):12–27. doi:10.1002/humu.10147

Kong, A., M.L. Frigge, G. Masson, et al. 2012. Rate of *de novo* mutations and the importance of father's age to disease risk. *Nature* 488(7412):471–475. doi:10.1038/nature11396

Kong, A., D.F. Gudbjartsson, J. Sainz, et al. 2002. A high-resolution recombination map of the human genome. *Nature Genetics* 31(3):241–247. doi:10.1038/ng917

Konigsberg, L.W. 1990. Analysis of prehistoric biological variation under a model of isolation by geographic and temporal distance. *Human Biology* 62(1):49–70.

Konigsberg, L.W. 1997. Comments on matrix permutation tests in the evaluation of competing models for modern human origins. *Journal of Human Evolution* 32:479–488.

Kono, A., M. Brameier, C. Roos, et al. 2014. Genomic sequence analysis of the MHC class IG/F segment in common marmoset (*Callithrix jacchus*). *Journal of Immunology* 192(7):3239–3246.

Kopcow, H.D., and S.A. Karumanchi. 2007. Angiogenic factors and natural killer (NK) cells in the pathogenesis of preeclampsia. *Journal of Reproductive Immunology* 76(1):23–29.

Kopelman, N.M., L. Stone, O. Gascuel, et al. 2013. The behavior of admixed populations in neighbor-joining inference of population trees. *Pacific Symposium on Biocomputing* 2013:273–284.

Kouzarides, T. 2007. Chromatin modifications and their function. *Cell* 128(4):693–705.

Kowal, E. 2013. Orphan DNA: indigenous samples, ethical biovalue and postcolonial science. *Social Studies of Science* 43:577–597.

Kowalczyk-Quintas, C., S. Schuepbach-Mallepell, L. Willen, et al. 2014. Pharmacological stimulation of *EDAR* signaling in the adult enhances sebaceous gland size and function. *Journal of Investigative Dermatology* 135:359–368.

Krause, J., A.W. Briggs, M. Kircher, et al. 2010. A complete mtDNA genome of an early modern human from Kostenki, Russia. *Current Biology: CB* 20(3):231–236.

Krause, J., Q. Fu, J.M. Good, et al. 2010. The complete mitochondrial DNA genome of an unknown hominin from southern Siberia. *Nature* 464:894–897.

Krause, J., L. Orlando, D. Serre, et al. 2007. Neanderthals in Central Asia and Siberia. *Nature* 449:902–904.

Krauss, S., C.Y. Zhang, and B.B. Lowell. 2005. The mitochondrial uncoupling-protein homologues. *Nature Reviews Molecular Cell Biology* 6(3):248–261.

Kreitman, M. 2000. Methods to detect selection in populations with applications to the human. *Annual Review of Genomics and Human Genetics* 1:539–559.

Krings, M., C. Capelli, F. Tschentscher, et al. 2000. A view of Neandertal genetic diversity. *Nature Genetics* 26:144.

Krings, M., H. Geisert, R.W. Schmitz, et al. 1999. DNA sequence of the mitochondrial hypervariable region II from the Neandertal type specimen. *Proceedings of the National Academy of Sciences USA* 96:5581–5585.

Krings, M., A. Stone, R.W. Schmitz, et al. 1997. Neanderthal DNA sequences and the origin of modern humans. *Cell* 90:19–30.

Krombach, F., S. Munzing, A.-M. Allmeling, et al. 1997. Cell size of alveolar macrophages: an interspecies comparison. *Environmental Health Perspectives* 105 (September):1261–1263.

Kuhlwilm, M., I. Gronau, M.J. Hubisz, et al. 2016. Ancient gene flow from early modern humans into eastern Neanderthals. *Nature* 530(7591):429–433.

Kulski, J.K., T. Shiina, T. Anzai, et al. 2002. Comparative genomic analysis of the MHC: the evolution of class I duplication blocks, diversity and complexity from shark to man. *Immunological Reviews* 190(1):95–122.

Kumar, H., T. Kawai, and S. Akira. 2011. Pathogen recognition by the innate immune system. *International Reviews of Immunology* 30(1):16–34. doi:10.3109/08830185.2010.529976

Kurokawa, R., M.G. Rosenfeld, and C.K. Glass. 2009. Transcriptional regulation through noncoding RNAs and epigenetic modifications. *RNA Biology* 6(3):233–236.

Kutz, G. 2010. *Direct-to-Consumer Genetic Tests: Misleading Test Results are Further Complicated by Deceptive Marketing and Other Questionable Practices.* General Accountability Office, Washington, DC.

Kvam, E., and R.M. Tyrrell. 1997. Induction of oxidative DNA base damage in human skin cells by UV and near visible radiation. *Carcinogenesis* 18(12):2379–2384.

Kwiatkowski, D.P. 2005. How malaria has affected the human genome and what human genetics can teach us about malaria. *American Journal of Human Genetics* 77(2):171–192.

Laayouni, H., L. Montanucci, M. Sikora, et al. 2011. Similarity in recombination rate estimates highly correlates with genetic differentiation in humans. *PLoS One* 6(3):e17913. doi:10.1371/journal.pone.0017913

Lachance, J., and S.A. Tishkoff. 2013. SNP ascertainment bias in population genetic analyses: why it is important, and how to correct it. *BioEssays* 35:780–786.

Lachance, J., B. Vernot, C.C. Elbers, et al. 2012. Evolutionary history and adaptation from high-coverage whole-genome sequences of diverse African hunter-fatherers. *Cell* 150:457–469.

Lahaye, C., M. Hernandez, E. Boëda, et al. 2013. Human occupation in South America by 20,000 BC: the Toca da Tira Peia site, Piauí, Brazil. *Journal of Archaeological Science* 40:2840–2847.

Lahr, M.M. 1995. Patterns of modern human diversification: implications for Amerindian origins. *Yearbook of Physical Anthropology* 38:163–198.

Lahr, M.M. 1996. *The Evolution of Modern Human Diversity: A Study of Cranial Variation.* Cambridge University Press, Cambridge.

Lai, E.C. 2005. miRNAs: whys and wherefores of miRNA-mediated regulation. *Current Biology* 15(12):R458–R460.

Laland, K.N., J. Odling-Smee, and S. Myles. 2010. How culture shaped the human genome: bringing genetics and human sciences together. *Nature Reviews Genetics* 11(2):314–334.

Lalueza-Fox, C., E. Gigli, M. de la Rasilla, et al. 2008. Genetic characterization of the ABO blood group in Neandertals. *BMC Evolutionary Biology* 8(1):342.

Lalueza-Fox, C., E. Gigli, M. de la Rasilla, et al. 2009. Bitter taste perception in Neanderthals through the analysis of the *TAS2R38* gene. *Biology Letters* 5(6):809–811.

Lalueza-Fox, C., H. Rompler, D. Caramelli, et al. 2007. A Melanocortin 1 receptor allele suggests varying pigmentation among Neanderthals. *Science* 318(5855):1453–1455.

Lalueza-Fox, C., A. Rosas, A. Estalrrich, et al. 2011. Genetic evidence for patrilocal mating behavior among Neandertal groups. *Proceedings of the National Academy of Sciences* 108(1):250–253.

Lalueza-Fox, C., M.L. Sampietro, D. Caramelli, et al. 2005. Neandertal evolutionary genetics: mitochondrial DNA data from the Iberian peninsula. *Molecular Biology and Evolution* 22(4):1077–1081.

Lam, L.L., E. Emberly, H.B. Fraser, et al. 2012. Factors underlying variable DNA methylation in a human community cohort. *Proceedings of the National Academy of Sciences of the USA* 109(Suppl. 2):17253–17260.

Lamason, R.L., M.A. Mohideen, J.R. Mest, et al. 2005. *SLC24A5*, a putative cation exchanger, affects pigmentation in zebrafish and humans. *Science* 310(5755):1782–1786. doi:10.1126/science.1116238

Lamb, S.M. 1958. Linguistic prehistory in the Great Basin. *International Journal of American Linguistics* 24:95–100. doi:10.1086/464442

Lande, R. 1977. Statistical tests for natural selection on quantitative characters. *Evolution* 31(2):442–444.

Lande, R. 2009. Adaptation to an extraordinary environment by evolution of phenotypic plasticity and genetic assimilation. *Journal of Evolutionary Biology* 22(7):1435–1446.

Lang, P.O., N. Samaras, D. Samaras, et al. 2013. How important is vitamin D in preventing infections? *Osteoporosis International* 24(5):1537–1553. doi:10.1007/s00198-012-2204-6

Langergraber, K.E., K. Prüfer, C. Rowney, et al. 2012. Generation times in wild chimpanzees and gorillas suggest earlier divergence times in great ape and human evolution. *Proceedings of the National Academy of Sciences of the USA* 109(39):15716–15721. doi:10.1073/pnas.1211740109

Lao, O., J.M. de Gruijter, K. van Duijn, et al. 2007. Signatures of positive selection in genes associated with human skin pigmentation as revealed from analyses of single nucleotide polymorphisms. *Annals of Human Genetics* 71(Pt 3):354–369. doi:10.1111/j.1469-1809.2006.00341.x

Lappalainen, T., M. Sammeth, M.R. Friedländer, et al. 2013. Transcriptome and genome sequencing uncovers functional variation in humans. *Nature* 501:506–511.

Lari, M., E. Rizzi, L. Milani, et al. 2010. The microcephalin ancestral allele in a Neanderthal individual. *PLoS One* 5.

Laribee, R.N., S.M. Fuchs, and B.D. Strahl. 2007. H2B ubiquitylation in transcriptional control: a FACT-finding mission. *Genes & Development* 21(7):737–743.

Larsen, C.S. 1981. Skeletal and dental adaptations to the shift to agriculture on the Georgia coast. *Current Anthropology* 22(4):422–423.

Larsen, C.S. 1995. Biological changes in human populations with agriculture. *Annual Review of Anthropology* 24(1):185–213.

Larsen, C.S. 2002. Bioarchaeology: the lives and lifestyles of past people. *Journal of Archaeological Research* 10(2):119–166.

Larsen, C.S. 2003. Animal source foods and human health during evolution. *Journal of Nutrition* 133(11):3893S–3897S.

Larsen, C.S. 2006. The agricultural revolution as environmental catastrophe: implications for health and lifestyle in the Holocene. *Quaternary International* 150(1):12–20.

Larsen, C.S. 2015. *Bioarchaeology: Interpreting Behavior from the Human Skeleton*. 2nd ed. Cambridge University Press, Cambridge.

LaSalle, J.M. 2013. Epigenomic strategies at the interface of genetic and environmental risk factors for autism. *Journal of Human Genetics* 58(7):396–401.

Laska, M., and L.T. Hernandez Salazar. 2004. Gustatory responsiveness to monosodium glutamate and sodium chloride in four species of nonhuman primates. *Journal of Experimental Zoology. Part A, Comparative Experimental Biology* 301:898–905.

Laska, M., E.C. Sanchez, J.A.R. Rivera, et al. 1996. Gustatory thresholds for food-associated sugars in the spider monkey (*Ateles geoffroyi*). *American Journal of Primatology* 39:189–193.

Lasker, G.W. 1985. *Surnames and Genetic Structure*. Cambridge University Press, New York.

Lasker, G.W. 1989. Genetics in the journal Human Biology. *Human Biology* 61:615–627.

Lasker, G.W., and B. Kaplan. 1964. The coefficient of breeding isolation: population size, migration rates, and possibilities for random genetic drift in six human communities in northern Peru. *Human Biology* 36:327–338.

Laughlin, W.S. 1951. The Alaska gateway viewed from the Aleutian Islands. In: W.S. Laughlin (ed.), *The Physical Anthropology of the American Indian.* Viking Fund, New York.

Laughlin, W.S. 1963. Eskimos and Aleuts: their origin and evolution. *Science* 142: 633–635.

Laurie, C.C., S.D. Chasalow, J.R. LeDeaux, et al. 2004. The genetic architecture of response to long-term artificial selection for oil concentration in the maize kernel. *Genetics* 168(4):2141–2155.

Laurikkala, J., J. Pispa, H.-S. Jung, et al. 2002. Regulation of hair follicle development by the TNF signal ectodysplasin and its receptor Edar. *Development* 129(10):2541–2553.

Lawlor, D.A., E. Warren, P. Taylor, et al. 1991. Gorilla class I major histocompatibility complex alleles: comparison to human and chimpanzee class I. *Journal of Experimental Medicine* 174(6):1491–1509.

Lawson, D.J., G. Hellenthal, S. Myers, et al. 2012. Inference of population structure using dense haplotype data. *PLoS Genetics* 8(1):e1002453.

Lazaridis, I., N. Patterson, A. Mittnik, et al. 2014. Ancient human genomes suggest three ancestral populations for present-day Europeans. *Nature* 513(7518):409–413.

Lee, S.M. 2015. Your 23andMe DNA can be used in racist, discriminatory ways. BuzzFeed News. https://www.buzzfeednews.com/article/stephaniemlee/your-23andme-dna-can-be-used-in-racist-discriminatory-ways

Lee, S.S.J., D.A. Bolnick, T. Duster, P. Ossorio, and K. TallBear. 2009. The illusive gold standard in genetic ancestry testing. *Science* 325(5936):38.

Lee, J.J., and C.C. Chow. 2014. Conditions for the validity of SNP-based heritability estimation. *Human Genetics* 133(8):1011–1022.

Lee, R.J., and N.A. Cohen. 2013. The emerging role of the bitter taste receptor *T2R38* in upper respiratory infection and chronic rhinosinusitis. *American Journal of Rhinology & Allergy* 27:283–286.

Leffler, E.M., Z. Gao, S. Pfeifer, et al. 2013. Multiple instances of ancient balancing selection shared between humans and chimpanzees. *Science* 339:1578–1582.

Legendre, P., and M.-J. Fortin. 2010. Comparison of the Mantel test and alternative approaches for detecting complex multivariate relationships in the spatial analysis of genetic data. *Molecular Ecology Resources* 10:831–844.

Legendre, P., and L. Legendre. 1998. *Numerical ecology.* 2nd English ed. Elsevier, Amsterdam.

Leite, T.K., R.M. Fonseca, N.M. de Franca, et al. 2011. Genomic ancestry, self-reported color and quantitative measures of skin pigmentation in Brazilian admixed siblings. *PLoS One* 6(11):e27162. doi:10.1371/journal.pone.0027162

Lell, J.T., M.D. Brown, T.G. Schurr, et al. 1997. Y chromosome polymorphisms in Native American and Siberian populations: identification of native Y chromosome haplotypes. *Human Genetics* 100:536–543.

Lell, J.T., R.I. Sukernik, Y.B. Starikovskaya, et al. 2002. The dual origin and Siberian affinities of Native American Y chromosomes. *American Journal of Human Genetics* 70:192–206.

Leonard, W.R., M.H. Crawford, A.G. Comuzzie, et al. 1994. Correlates of low serum-lipid levels among the Evenki herders of Siberia. *American Journal of Human Biology* 6(3):329–338.

Leonard, W.R., V.A. Galloway, E. Ivakine, et al. 1999. Nutrition, thyroid function and basal metabolism of the Evenki of Central Siberia. *International Journal of Circumpolar Health* 58(4):281–295.

Leonard, W.R., J.J. Snodgrass, and M.V. Sorensen. 2005. Metabolic adaptation in indigenous Siberian populations. *Annual Review of Anthropology* 34:451–471.

Leonard, W.R., M.V. Sorensen, V.A. Galloway, et al. 2002. Climatic influences on basal metabolic rates among circumpolar populations. *American Journal of Human Biology* 14(5):609–620.

Lepage, T., D. Bryant, H. Philippe, et al. 2007. A general comparison of relaxed molecular clock models. *Molecular Biology and Evolution* 24(12):2669–2680. doi:10.1093/molbev/msm193

Lesch, K.P., J. Meyer, K. Glatz, et al. 1997. The 5-HT transporter gene-linked polymorphic region (5-HTTLPR) in evolutionary perspective: alternative biallelic variation in rhesus monkeys. *Journal of Neural Transmission* 104:1259–1266.

Leslie, P.W. 1985. Potential mates analysis and the study of human population structure. *Yearbook of Physical Anthropology* 28:53–78.

Leslie, S., B. Winney, G. Hellenthal, et al. 2015. The fine-scale genetic structure of the British population. *Nature* 519(7543):309–314.

Levin, B.R. 1996. The evolution and maintenance of virulence in microparasites. *Emerging Infectious Diseases* 2(2):93–102. doi:10.3201/eid0202.960203

Levin, B.R. 2002. Models for the spread of resistant pathogens. *Netherlands Journal of Medicine* 60(7 Suppl.):58–64; discussion 64-66.

Levin, B.R., and J.J. Bull. 1994. Short-sighted evolution and the virulence of pathogenic microorganisms. *Trends in Microbiology* 2(3):76–81.

Levin, B.R., M. Lipsitch, V. Perrot, et al. 1997. The population genetics of antibiotic resistance. *Clinical Infectious Diseases* 24(Suppl. 1):S9–S16.

Levin, B.R., and C. Svanborg Eden. 1990. Selection and evolution of virulence in bacteria: an ecumenical excursion and modest suggestion. *Parasitology* 100(Suppl.):S103–S115.

Lewontin, R. 1972. The apportionment of human diversity. *Evolutionary Biology* 6:381–398.

Lewontin, R.C. 1974. *The Genetic Basis of Evolutionary Change*. Vol. 560. Columbia University Press, New York.

Ley, R.E., M. Hamady, C. Lozupone, et al. 2008. Evolution of mammals and their gut microbes. *Science* 320:1647–1651.

Li, J.Z., D.M. Abscher, H. Tang, et al. 2008. Worldwide human relationships inferred from genome-wide patterns of variation. *Science* 319:1100–1104.

Li, H., and R. Durbin. 2011. Inference of human population history from individual whole-genome sequences. *Nature* 475(7357):493–496.

Li, M., L. Huang, K. Li, et al. 2016. Adaptive evolution of interleukin-3 (*IL3*), a gene associated with brain volume variation in general human populations. *Human Genetics* 135(4):377–392. doi:10.1007/s00439-016-1644-z

Li, J., H. Li, M. Jakobsson, et al. 2012. Joint analysis of demography and selection in population genetics: where do we stand and where could we go? *Molecular Ecology* 21(1):28–44. doi:10.1111/j.1365-294X.2011.05308.x

Li, B., L.A. Nolte, J.S. Ju, et al. 2000. Skeletal muscle respiratory uncoupling prevents diet-induced obesity and insulin resistance in mice. *Nature Medicine* 6(10):1115–1120.

Li, Z.-Y., X.-J. Wu, L.-P. Zhou, et al. 2017. Late Pleistocene archaic human crania from Xuchang, China. *Science* 355(6328):969.

Lieberman-Aiden, E., N.L. van Berkum, L. Williams, et al. 2009. Comprehensive mapping of long-range interactions reveals folding principles of the human genome. *Science* 326(5950):289–293.

Lim, L.P., N.C. Lau, P. Garrett-Engele, et al. 2005. Microarray analysis shows that some microRNAs downregulate large numbers of target mRNAs. *Nature* 433(7027):769–273.

Lim, J.K., C.Y. Louie, C. Glaser, et al. 2008. Genetic deficiency of chemokine receptor CCR5 is a strong risk factor for symptomatic West Nile virus infection: a meta-analysis of 4 cohorts in the US epidemic. *Journal of Infectious Diseases* 197(2):262–265. doi:10.1086/524691

Lin, Y.-L., P. Pavlidis, E. Karakoc, et al. 2015. The evolution and functional impact of human deletion variants shared with archaic hominin genomes. *Molecular Biology and Evolution*. doi:10.1093/molbev/msu405

Linares, J.F., I. Gustafsson, F. Baquero, et al. 2006. Antibiotics as intermicrobial signaling agents instead of weapons. *Proceedings of the National Academy of Sciences of the USA* 103(51):19484–19489. doi:10.1073/pnas.0608949103

Lindahl, T. 1993. Instability and decay of the primary structure of DNA. *Nature* 362:709–715. PMID:8469282

Lindo, J., A. Achilli, U.A. Perego, et al. 2017. Ancient individuals from the North American Northwest Coast reveal 10,000 years of regional genetic continuity. *Proceedings of the National Academy of Sciences USA* 114(16):4093–4098.

Lipson, M., P.-R. Loh, N. Patterson, et al. 2014. Reconstructing Austronesian population history in Island Southeast Asia. *Nature Communications* 5:4689. doi:10.1038/ncomms5689

Little, A.C., D.M. Burt, and D.I. Perrett. 2006. Assortative mating for perceived facial personality traits. *Personality and Individual Differences* 40(5):973–984.

Little, B.B., P.H. Buschang, M.E. Peña Reyes, et al. 2006. Craniofacial dimensions in children in rural Oaxaca, Southern Mexico: secular change, 1968-2000. *American Journal of Physical Anthropology* 131:127–136.

Liu, W., C.Z. Jin, Y.Q. Zhang, et al. 2010. Human remains from Zhirendong, South China, and modern human emergence in East Asia. *Proceedings of the National Academy of Sciences of the USA* 107(45):19201–19206.

Liu, S., E.D. Lorenzen, M. Fumagalli, et al. 2014. Population genomics reveal recent speciation and rapid evolutionary adaptation in polar bears. *Cell* 157:785–794.

Liu, W., M. Martinon-Torres, Y.J. Cai, et al. 2015. The earliest unequivocally modern humans in southern China. *Nature* 526(7575):696–699. doi:10.1038/nature15696

Liu, H., F. Prugnolle, A. Manica, et al. 2006. A geographically explicit genetic model of worldwide human-settlement history. *American Journal of Human Genetics* 79:230–237.

Liu, G., L. Walter, S. Tang, et al. 2014. Differentiated adaptive evolution, episodic relaxation of selective constraints, and pseudogenization of umami and sweet taste genes *TAS1Rs* in catarrhine primates. *Frontiers in Zoology* 11:79.

Liu, F., A. Wollstein, P.G. Hysi, et al. 2010. Digital quantification of human eye color highlights genetic association of three new loci. *PLoS Genetics* 6(5):e1000934. doi:10.1371/journal.pgen.1000934

Lively, C.M., J.C. de Roode, M.A. Duffy, et al. 2014. Interesting open questions in disease ecology and evolution. *The American Naturalist* 184(Suppl. 1):S1–S8. doi:10.1086/677032

Livingstone, F.B. 1958. Anthropological implications of sickle-cell gene distributions in West Africa. *American Anthropologist* 60:533–562.

Liyanage, V.R., J.S. Jarmasz, N. Murugeshan, et al. 2014. DNA modifications: function and applications in normal and disease states. *Biology (Basel)* 3(4):670–723.

Llamas, B., L. Fehren-Schmitz, G. Valverde. 2016. Ancient mitochondrial DNA provides high-resolution time scale of the peopling of the Americas. *Sci Adv* 2:e1501385.

Llamas, B., M.L. Holland, K. Chen, et al. 2012a. High-resolution analysis of cytosine methylation in ancient DNA. *PLoS One* 7(1):1–6.

Llamas, B., M.L. Holland, K. Chen, et al. 2012b. High-resolution analysis of cytosine methylation in ancient DNA. *PLoS One* 7(1):e30226.

Locke, D.P., L.W. Hillier, W.C. Warren, et al. 2011. Comparative and demographic analysis of orang-utan genomes. *Nature* 469:529–533.

Loeffler, S.H., G.W. de Lisle, M.A. Neill, et al. 2014. The seal tuberculosis agent, *Mycobacterium pinnipedii*, infects domestic cattle in New Zealand: epidemiologic factors and DNA strain typing. *Journal of Wildlife Diseases* 50(2):180–187. doi:10.7589/2013-09-237

Lohmueller, K.E., A.R. Indap, S. Schmidt, et al. 2008. Proportionally more deleterious genetic variation in European than in African populations. *Nature* 451:994–997.

Lohse, K., and L.A.F. Frantz. 2014. Neandertal admixture in Eurasia confirmed by maximum likelihood analysis of three genomes. *Genetics* 196:1241–1251.

Loisel, D.A., M.V. Rockman, G.A. Wray, et al. 2006. Ancient polymorphism and functional variation in the primate MHC-DQA1 5′ cis-regulatory region. *Proceedings of the National Academy of Sciences* 103(44):16331–16336.

Long, J.C., J. Li, and M.E. Healy. 2009. Human DNA sequences: more variation and less race. *American Journal of Physical Anthropology* 139:23–34.

Loomis, W.F. 1967. Skin-pigment regulation of vitamin-D biosynthesis in man. *Science* 157(3788):501–506.

Loos, R.J., T. Rankinen, Y. Chagnon, et al. 2006. Polymorphisms in the leptin and leptin receptor genes in relation to resting metabolic rate and respiratory quotient in the Quebec family study. *International Journal of Obesity* 30(1):183–190.

Lopez, M.A., H.T. Nguyen, M. Oberholzer, et al. 2011. Social parasites. *Current Opinion in Microbiology* 14:642–648.

Lopez-Valenzuela, M., O. Ramírez, A. Rosas, et al. 2012. An ancestral miR-1304 allele present in Neanderthals regulates genes involved in enamel formation and could explain dental differences with modern humans. *Molecular Biology and Evolution* 29(7):1797–1806.

Lordkipanidze, D., M.S. Ponce de Leon, A. Margvelashvili, et al. 2013. A complete skull from Dmanisi, Georgia, and the evolutionary biology of early *Homo*. *Science* 342(6156):326–331.

Lorenzo, F.R., C. Huff, M. Myllymaki, et al. 2014. A genetic mechanism for Tibetan high-altitude adaptation. *Nature Genetics* 46(9):951–956.

Lu, C., and E. Fuchs. 2014. Sweat gland progenitors in development, homeostasis, and wound repair. *Cold Spring Harbor Perspectives in Medicine* 4(2):a015222.

Lucock, M., I. Daskalakis, D. Briggs, et al. 2000. Altered folate metabolism and disposition in mothers affected by a spina bifida pregnancy: influence of 677c --> t methylenetetrahydrofolate reductase and 2756a --> g methionine synthase genotypes. *Molecular Genetics and Metabolism* 70(1):27–44. doi:10.1006/mgme.2000.2994

Luger, K., A.W. Mäder, R.K. Richmond, et al. 1997. Crystal structure of the nucleosome core particle at 2.8 A resolution. *Nature* 389(6648):251–260.

Lugo, J.S., and L.F. Cadavid. 2015. Patterns of MHC-G-like and MHC-B diversification in new world monkeys. *PLoS One* 10(6):1–20. doi:10.1371/journal.pone.0131343

Lukacs, J.R. 1992. Dental paleopathology and agricultural intensification in South Asia: new evidence from Bronze Age Harappa. *American Journal of Physical Anthropology* 87(2):133–150.

Luke, A., A. Adeyemo, H. Kramer, et al. 2004. Association between blood pressure and resting energy expenditure independent of body size. *Hypertension* 43(3): 555–560.

Lukic, S., and J. Hey. 2012. Demographic inference using spectral methods on SNP data, with an analysis of the human out-of-Africa expansion. *Genetics* 192:619–639.

Lumsden, C.J. 1980. *Genes, Mind, and Culture.* Harvard University Press, Cambridge, MA.

Lycett, S.J., and M. Collard. 2005. Do homoiologies impede phylogenetic analyses of the fossil hominids? An assessment based on extant papionin craniodental morphology. *Journal of Human Evolution* 49(5):618–642.

Lynch, M. 1989. Phylogenetic hypotheses under the assumption of neutral quantitative-genetic variation. *Evolution* 43(1):1–17.

Lynch, M. 2010. Rate, molecular spectrum, and consequences of human mutation. *Proceedings of the National Academy of Sciences of the USA* 107(3):961–968. doi:10.1073/pnas.0912629107

Lynch, K.E., and D.J. Kemp. 2014. Nature-via-nurture and unravelling causality in evolutionary genetics. *Trends in Ecology & Evolution* 29(1):2–4.

Lynch, M., and B. Walsh. 1998. *Genetic Analysis of Quantitative Traits.* Sinauer Associates, Sunderland, MA.

Ma, D., A.J. Jasinska, F. Feyertag, et al. 2014. Factors associated with siman immunodeficiency virus transmission in a natural African nonhuman primate host in the wild. *Journal of Virology* 88:5687–5705.

Ma, D., A. Jasinska, J. Kristoff, et al. 2013. SIVagm infection in wild African green monkeys from South Africa: epidemiology, natural history, and evolutionary considerations. *PLoS Pathogens* 9:e1003011.

Macaulay, V., C. Hill, A. Achilli, et al. 2005. Single, rapid coastal settlement of Asia revealed by analysis of complete mitochondrial genomes. *Science* 308(5724):1034–1036. doi:10.1126/science.1109792

MacDonald, D.H., and B.S. Hewlett. 1999. Reproductive interests and forager mobility. *Current Anthropology* 40:501–523. doi:10.2307/2991382

Macholán, M. 2006. A geometric morphometric analysis of the shape of the first upper molar in mice of the genus *Mus* (Muridae, Rodentia). *Journal of Zoology* 270:672–681.

Mack, G.S. 2007. MicroRNA gets down to business. *Nature Biotechnology* 25(6):631–638.

Madrigal, L., and W. Kelly. 2007. Human skin-color sexual dimorphism: a test of the sexual selection hypothesis. *American Journal of Physical Anthropology* 132(3):470–482. doi:10.1002/ajpa.20453

Madsen, D.B., and D. Rhode (eds.). 1994. *Across the West: Human Population Movement and the Expansion of the Numa.* University of Utah Press, Salt Lake City.

Maes, M., K. Mommen, D. Hendrickx, et al. 1997. Components of biological variation, including seasonality, in blood concentrations of TSH, TT3, FT4, PRL, cortisol and testosterone in healthy volunteers. *Clinical Endocrinology* 46(5):587–598.

Majumder, P.P., W.S. Laughlin, and R.E. Ferrell. 1988. Variation in the Aleuts of the Pribilof Islands and Eskimos of Kodiak Island. *American Journal of Physical Anthropology* 76:481–488.

Malaspinas, A.S., O. Lao, H. Schroeder, et al. 2014. Two ancient human genomes reveal Polynesian ancestry among the indigenous Botocudos of Brazil. *Current Biology* 24:R1035–1037.

Malaspinas A.-S., M.C. Westaway, C. Muller, et al. 2016. A genomic history of aboriginal Australia. *Nature* 538(7624):207–214.

Malécot, G. 1967. Identical loci and relationship. In: L.M. Le Cam, and J. Neyman (eds.), *Proceedings of the Fifth Berkeley Symposium on Mathematical Statistics and Probability,* vol. 4. University of California Press, Berkeley, pp. 317–332.

Malécot, G. 1973. Isolation by distance. In: N.E. Morton (ed.), *Genetic Structure of Populations*. University of Hawaii Press, Honolulu, pp. 72–75.

Malhi, R.S. 2009. Implications of the Genographic project for molecular anthropologists. *International Journal of Cultural Property* 16:193–194.

Malhi, R.S., and A.C. Bader. 2015. Engaging Native Americans in genomics research. *American Anthropologist* 117(4):743–744.

Mallick, S., H. Li, M. Lipson, et al. 2016. The Simons genome diversity project: 300 genomes from 142 diverse populations. *Nature* 538(7624):201–206.

Malyarchuk, B., M. Derenko, G. Denisova, et al. 2011. Ancient links between Siberians and Native Americans revealed by subtyping the Y chromosome haplogroup Q1a. *Journal of Human Genetics* 56:583–588.

Mandal, A., and C. Viswanathan. 2015. Natural killer cells: in health and disease. *Hematology/Oncology and Stem Cell Therapy* 8(2):47–55. doi:10.1016/j.hemonc.2014.11.006

Mandryk, C.A.S., H. Josenhans, D.W. Fedje, et al. 2001. Late quaternary paleoenvironments of Northwestern North America: implications for inland versus coastal migration routes. *Quaternary Science Reviews* 20:301–314.

Manica, A., W. Amos, F. Balloux, et al. 2007. The effect of ancient population bottlenecks on human phenotypic variation. *Nature* 448:346–349.

Mann, C.C. 2006. *1491: New Revelations of the Americas before Columbus*. Random House Digital, New York.

Mannucci, A., K.M. Sullivan, P.L. Ivanov, et al. 1994. Forensic application of a rapid and quantitative DNA sex test by amplification of the X-Y homologous gene amelogenin. *International Journal of Legal Medicine* 106:190–193.

Manolio, T.A., L.D. Brooks, and F.S. Collins. 2008. A HapMap harvest of insights into the genetics of common disease. *Journal of Clinical Investigation* 118(5):1590–1605. doi:10.1172/J.C.I34772

Manolio, TA, F.S. Collins, N.J. Cox, et al. 2009. Finding the missing heritability of complex diseases. *Nature* 461(7265): 747–753.

Manry, J., G. Laval, E. Patin, et al. 2011. Evolutionary genetic dissection of human interferons. *Journal of Experimental Medicine* 208(13):2747–2759.

Mantel, N.A. 1967. The detection of disease clustering and a generalized regression approach. *Cancer Research* 27:209–220.

Marangoni, A., D. Caramelli, and G. Manzi. 2014. *Homo sapiens* in the Americas: overview of the earliest human expansion in the New World. *Journal of Anthropological Sciences* 92:79–97.

Mare, R.D. 1991. Five decades of educational assortative mating. *American Sociological Review* 56(1):15–32.

Marjoram, P., and S. Tavaré. 2006. Modern computational approaches for analysing molecular genetic variation data. *Nature Reviews Genetics* 7(10):759–770. doi:10.1038/nrg1961.

Marques-Bonet, T., J.M. Kidd, M. Ventura, et al. 2009. A burst of segmental duplications in the genome of the African great ape ancestor. *Nature* 457:877–881.

Marroig, G., L.T. Shirai, A. Porto, et al. 2009. The evolution of modularity in the mammalian skull II: evolutionary consequences. *Evolutionary Biology* 36:136–148.

Martínez-Abadías, N., M. Esparza, T. Sjovold, et al. 2009. Heritability of human cranial dimensions: comparing the evolvability of different cranial regions. *Journal of Anatomy* 214:19–35.

Martinez-Cadenas, C., S. Lopez, G. Ribas, et al. 2013. Simultaneous purifying selection on the ancestral *MC1R* allele and positive selection on the melanoma-risk allele V60L

in south Europeans. *Molecular Biology and Evolution* 30(12):2654–2665. doi:10.1093/molbev/mst158
Martins, E.P., and T. Garland. 1991. Phylogenetic analyses of the correlated evolution of continuous characters: a simulation study. *Evolution* 45(3):534–557.
Martorell, R., F. Mendoza, and R. Caastillo. 1988. Poverty and stature in children. In: J.C. Waterlow (ed.), *Linear Growth Retardation in Less Developed Countries.* Vol. 14. Raven Press, New York, pp. 57–73.
Mata-Míguez, J., L. Overholtzer, E. Rodriguez-Alegria, et al. 2012. The genetic impact of Aztec imperialism: ancient mitochondrial DNA evidence from Xaltocan, Mexico. *American Journal of Physical Anthropology* 149(4):504–516. doi:10.1002/ajpa.22152
Mathieson, I., I. Lazaridis, N. Rohland, et al. 2015. Eight thousand years of natural selection in Europe. *bioRxiv* 016477. https://doi.org/10.1101/016477
Mathur, U., S.L. Datta, and B.B. Mathur. 1977. The effect of aminopterin-induced folic acid deficiency on spermatogenesis. *Fertility and Sterility* 28(12):1356–1360.
Matsumura, S., and P. Forster. 2008. Generation time and effective population size in Polar Eskimos. *Proceedings Biological Sciences* 275(1642):1501–1508. doi:10.1098/rspb.2007.1724
Matsushita, Y., H. Oota, B.J. Welker, et al. 2014. Color vision variation as evidenced by hybrid L/M opsin genes in wild populations of trichromatic Alouatta new world monkeys. *International Journal of Primatology* 35:71–87.
Maunakea, A.K., I. Chepelev, K. Cui, et al. 2013. Intragenic DNA methylation modulates alternative splicing by recruiting MeCP2 to promote exon recognition. *Cell Research* 23(11):1256–1269.
Mauri, C., and A. Bosma. 2012. Immune regulatory function of B cells. *Annual Review of Immunology* 30(1):221–241. doi:10.1146/annurev-immunol-020711-074934
Maxwell, P.H., M.S. Wiesener, G.W. Chang, et al. 1999. The tumour suppressor protein VHL targets hypoxia-inducible factors for oxygen-dependent proteolysis. *Nature* 399(6733):271–275.
May, R.M., and R.M. Anderson. 1983. Parasite-host co-evolution. In: D.J. Futuyma, and M. Slatkin (eds.), *Coevolution.* Sinauer Associates, Sunderland, MA, pp. 186–206.
Mayr, E. 1942. *Systematics and the Origin of Species.* Columbia University Press, New York.
Mayr, E. 1959. Darwin and the evolutionary theory in biology. In: B.J. Meggers (ed.), *Evolution and Anthropology: A Centennial Appraisal.* Theo. Gaus' Sons, Inc., Brooklyn, NY, pp. 1–10.
Mayr, E. 1963. *Animal Species and Evolution.* Belknap Press, Cambridge, MA.
Mays, S., and M. Cox. 2000. Sex determination in skeletal remains. In: M. Cox, and S. Mays (eds.), *Human Osteology in Archaeology and Forensic Science.* Cambridge University Press, Cambridge, pp. 117–130.
Mays, S., and M. Faerman. 2001. Sex identification in some putative infanticide victims from roman Britain using ancient DNA. *Journal of Archaeological Science* 28:555–559.
McComb, J., N. Blagitko, A. Comuzzie, et al. 1995. VNTR variation in Siberian indigenous populations. *Human Biology* 67(2):217–229.
McCullough, J.M. 1982. Secular trend for stature in adult male Yucatec Maya to 1968. *American Journal of Physical Anthropology* 58(2):221–225.

McCullough, J.M., K.M. Heath, and A.M. Smith. 2015. Hemochromatosis: niche construction and the genetic domino effect in the European Neolithic. *Human Biology* 87(1):39–58

McDade, T.W. 2003. Life history theory and the immune system: steps toward a human ecological immunology. *American Journal of Physical Anthropology* 37(Suppl.): 100–125. doi:10.1002/ajpa.10398

McDougall, I., F.H. Brown, and J.G. Fleagle. 2005. Stratigraphic placement and age of modern humans from Kibish, Ethiopia. *Nature* 433:733–736.

McEvoy, B.P., J.E. Powell, M.E. Goddard, et al. 2011. Human population dispersal "Out of Africa" estimated from linkage disequilibrium and allele frequencies of SNPs. *Genome Research* 21:821–829.

McHeyzer-Williams, M., S. Okitsu, N. Wang, et al. 2012. Molecular programming of B cell memory. *Nature Reviews Immunology* 12(1):24–34.

McInnes, R.R. 2011. 2010 Presidential Address: Culture: the silent language geneticists must learn-genetic research with indigenous populations. *American Journal of Human Genetics* 88:254–261.

McLaren, P.J., and M. Carrington. 2015. The impact of host genetic variation on infection with HIV-1. *Nature Immunology* 16(6):577–583. doi:10.1038/ni.3147

McLean, C.Y., P.L. Reno, A.A. Pollen, et al. 2011. Human-specific loss of regulatory DNA and the evolution of human-specific traits. *Nature* 471:216–219.

McManus, K.F., J.L. Kelley, S. Song, et al. 2015. Inference of gorilla demographic and selective history from whole-genome sequence data. *Molecular Biology and Evolution* 32:600–612.

McMichael, A.J. 2004. Environmental and social influences on emerging infectious diseases: past, present and future. *Philosophical Transactions B* 359(1447):1049–1058.

McNutt, M. 2015. Editorial: breakthrough to genome editing. *Science* 350:1445.

Medzhitov, R., and C.A. Janeway Jr. 2002. Decoding the patterns of self and nonself by the innate immune system. *Science* 296(April):298–300.

Meiklejohn, C., and J. Babb. 2011. Long bone length, stature and time in the European late Pleistocene and early Halocene. In: R. Pinhasi, and J.T. Stock (eds.), *Human Bioarchaeology of the Transition to Agriculture*. Wiley-Blackwell, Hoboken, NJ, pp. 153–175.

Meiri, M., A.M. Lister, M.J. Collins, et al. 2013. Faunal record identifies Bering isthmus conditions as constraint to end-Pleistocene migration to the New World. *Proceedings of the Royal Society B* 281:20132167.

Meissner, A. 2010. Epigenetic modifications in pluripotent and differentiated cells. *Nature Biotechnology* 28(10):1079–1088.

Melin, A.D., L.M. Fedigan, H.C. Young, et al. 2010. Can color vision variation explain sex differences in invertebrate foraging by capuchin monkeys? *Current Zoology* 56:300–312.

Melin, A.D., C. Hiramatsu, N.A. Parr, et al. 2014. The behavioral ecology of color vision: considering fruit conspicuity, detection distance and dietary importance. *International Journal of Primatology* 35:258–287.

Mellars, P., and J.C. French. 2011. Tenfold population increase in Western Europe at the Neandertal-to-modern human transition. *Science* 333(6042):623–627.

Mellars, P., K.C. Gori, M. Carr, et al. 2013. Genetic and archaeological perspectives on the initial modern human colonization of southern Asia. *Proceedings of the National Academy of Sciences of the USA* 110(26):10699–10704.

Meloni, M. 2014a. Remaking local biologies in an epigenetic time. Somatosphere. http://somatosphere.net/2014/08/remaking-local-biologies-in-an-epigenetic-time.html

Meloni, M. 2014b. The social brain meets the reactive genome: neuroscience, epigenetics and the new social biology. *Frontiers in Human Neuroscience* 8:309.

Meltzer, D.J. 1993. Pleistocene peopling of the Americas. *Evolutionary Anthropology* 1:157–169.

Meltzer, D.J. 2009. *First Peoples in a New World. Colonizing Ice Age America*. London. University of California Press.

Mendez Fernando L, Poznik GD, Castellano S, and Bustamante Carlos D. 2016. The divergence of Neandertal and modern human Y chromosomes. *The American Journal of Human Genetics* 98(4):728–734.

Mendez, F.L., J.C. Watkins, and M.F. Hammer. 2012a. A haplotype at *STAT2* introgressed from Neanderthals and serves as a candidate of positive selection in Papua New Guinea. *American Journal of Human Genetics* 91(2):P265–P274. https://doi.org/10.1016/j.ajhg.2012.06.015

Mendum, T.A., V.J. Schuenemann, S. Roffey, et al. 2014. *Mycobacterium leprae* genomes from a British medieval leprosy hospital: towards understanding an ancient epidemic. *BMC Genomics* 15(1):270. doi:10.1186/1471-2164-15-270

Menozzi, P., A. Piazza, and L.L. Cavalli-Sforza. 1978. Synthetic maps of human gene frequencies in Europe. *Science* 201:786–792.

Mercer, T.R., M.E. Dinger, and J.S. Mattick. 2009. Long non-coding RNAs: insights into functions. *Nature Reviews Genetics* 10(3):155–159.

Mercier, N., H. Valladas, O. Bar-Yosef, et al. 1993. Thermoluminescence date for the Mousterian burial site of Es-Skhul, Mt. Carmel. *Journal of Archaeological Science* 20(2):169–174. doi:10.1006/jasc.1993.1012

Meredith, P., and T. Sarna. 2006. The physical and chemical properties of eumelanin. *Pigment Cell Research* 19(6):572–594. doi:10.1111/j.1600-0749.2006.00345.x

Merker, M., C. Blin, S. Mona, et al. 2015. Evolutionary history and global spread of the *Mycobacterium tuberculosis* Beijing lineage. *Nature Genetics* 47(3):242–249. doi:10.1038/ng.3195

Merker, M., T.A. Kohl, A. Roetzer, et al. 2013. Whole genome sequencing reveals complex evolution patterns of multidrug-resistant *Mycobacterium tuberculosis* Beijing strains in patients. *PLoS One* 8(12):e82551. doi:10.1371/journal.pone.0082551

Merkestein, M., F.R. Cagampang, and D. Sellayah. 2014. Fetal programming of adipose tissue function: an evolutionary perspective. *Mammalian Genome* 25(9–10):413–423.

Merriwether, D.A. 1999. Freezer anthropology: new uses for old blood. *Philosophical Transactions of the Royal Society of London B* 354:121–129.

Merriwether, D.A., W.W. Hall, A. Vahlne, et al. 1996. mtDNA variation indicates Mongolia may have been the source for the founding population for the New World. *American Journal of Human Genetics* 59:204–212.

Merriwether, D.A., F. Rothhammer, and R.E. Ferrell. 1995. Distribution of the four founding lineage haplotypes in Native Americans suggests a single wave of migration for the new world. *American Journal of Physical Anthropology* 98:411–430.

Messer, P.W. 2013. SLiM: simulating evolution with selection and linkage. *Genetics* 194:1037–1039. doi:10.1534/genetics.113.152181/-/DC1

Metspalu, M., I.G. Romero, B. Yunusbayev, et al. 2011. Shared and unique components of human population structure and genome-wide signals of positive selection in South Asia. *American Journal of Human Genetics* 89(6):731–744.

Meyer, C.G., G.K. Amedofu, J.M. Brandner, et al. 2002. Selection for deafness? *Nature Medicine* 8(12):1332–1333.

Meyer, M., J.-L. Arsuaga, C. de Filippo, et al. 2016. Nuclear DNA sequences from the middle Pleistocene Sima de los Huesos hominins. *Nature* 531(7595):504–507.

Meyer, M.B., N.A. Benkusky, and J.W. Pike. 2015. Profiling histone modifications by chromatin immunoprecipitation coupled to deep sequencing in skeletal cells. *Methods in Molecular Biology* 1226:61–70.

Meyer, M., Q.M. Fu, A. Aximu-Petri, et al. 2014. A mitochondrial genome sequence of a hominin from Sima de los Huesos. *Nature* 505(7483):403–406. doi:10.1038/nature12788

Meyer, M., M. Kircher, M.-T. Gansauge, et al. 2012. A high-coverage genome sequence from an archaic Denisovan individual. *Science* 338(6104):222–226. doi:10.1126/science.1224344

Meyerhof, W. 2005. Elucidation of mammalian bitter taste. *Reviews of Physiology, Biochemistry and Pharmacology* 154:37–72.

Meyerhof, W., C. Batram, C. Kuhn, et al. 2010. The molecular receptive ranges of human *TAS2R* bitter taste receptors. *Chemical Senses* 35:157–170.

Meyer-Lindenberg, A., G. Domes, P. Kirsch, and M. Heinrichs. 2011. Oxytocin and vasopressin in the human brain: social neuropeptides for translational medicine. *Nature Reviews Neuroscience* 12:524–538.

Michels, K.B., A.M. Binder, S. Dedeurwaerder, et al. 2013. Recommendations for the design and analysis of epigenome-wide association studies. *Nature Methods* 10(10):949–955.

Michener, C.D., and R.R. Sokal. 1957. A quantitative approach to a problem of classification. *Evolution* 11(2):130–162.

Mielke, J.H., and M.H. Crawford (eds.). 1980. *Current Developments in Anthropological Genetics Vol. 1. Theory and Methods*. Plenum Press, New York.

Mill, J., and B.T. Heijmans. 2013. From promises to practical strategies in epigenetic epidemiology. *Nature Reviews Genetics* 14(8):585–594.

Miller, C.T., S. Beleza, A.A. Pollen, et al. 2007. Cis-regulatory changes in kit ligand expression and parallel evolution of pigmentation in sticklebacks and humans. *Cell* 131(6):1179–1189. doi:10.1016/j.cell.2007.10.055

Miller, W., S.C. Schuster, A.J. Welch, et al. 2012. Polar and brown bear genomes reveal ancient admixture and demographic footprints of past climate change. *Proceedings of the National Academy of Sciences of the USA* 109:E2382–90.

Mills, L. 2003. Mitochondrial DNA Analysis of the Ohio Hopewell of the Hopewell Mound Group. PhD dissertation. The Ohio State University, Columbus.

Minwalla, L., Y. Zhao, I.C. Le Poole, et al. 2001. Keratinocytes play a role in regulating distribution patterns of recipient melanosomes in vitro. *Journal of Investigative Dermatology* 117(2):341–347. doi:10.1046/j.0022-202x.2001.01411.x

Miró-Herrans, A.T., A. Al-Meeri, and C.J. Mulligan. 2014. Human migration patterns in Yemen and implications for reconstructing prehistoric population movements. *PLoS One* 9(4):e95712. doi:10.1371/journal.pone.0095712

Miró-Herrans, A.T., and C.J. Mulligan. 2013. Human demographic processes and genetic variation as revealed by mtDNA simulations. *Molecular Biology and Evolution* 30:244–252.

Misarti, N., B.P. Finney, J.W. Jordan, et al. 2012. Early retreat of the Alaska Peninsula glacier complex and the implications for coastal migrations of first Americans. *Quaternary Science Reviews* 48:1–6.

Misner, L.M., A.C. Halvorson, J.L. Dreier, et al. 2009. The correlation between skeletal weathering and DNA quality and quantity. *Journal of Forensic Sciences* 54:822–828. doi:10.1111/j.1556-4029.2009.01043.x PMID: 19368622

Moczek, A.P., S. Sultan, S. Foster, et al. 2011. The role of developmental plasticity in evolutionary innovation. *Proceedings of the Royal Society B* 278:2705–2713.

Monot, M., N. Honore, T. Garnier, et al. 2005. On the origin of leprosy. *Science* 308(5724):1040–1042.

Monot, M., N. Honore, T. Garnier, et al. 2009. Comparative genomic and phylogeographic analysis of *Mycobacterium leprae*. *Nature Genetics* 41:1282–1289.

Monreal, A.W., B.M. Ferguson, D.J. Headon, et al. 1999. Mutations in the human homologue of mouse *dl* cause autosomal recessive and dominant hypohidrotic ectodermal dysplasia. *Nature Genetics* 22(4):366–369.

Montagna, W., and K. Carlisle. 1991. The architecture of black and white facial skin. *Journal of the American Academy of Dermatology* 24(6 Pt 1):929–937.

Monteiro, L.R., and Dos Reis SF. 2005. Morphological evolution in the mandible of spiny rates, genus *Trinomy* (Rodentia: Echimyidae). *Journal of Zoological Systematics and Evolutionary Research* 43(4):332–338.

Moore, C.B., M. John, I.R. James, et al. 2002. Evidence of HIV-1 adaptation to HLA-restricted immune responses at a population level. *Science* 296(5572):1439–1443. doi:10.1126/science.1069660

Moore, L.G., M. Shriver, L. Bemis, et al. 2006. An evolutionary model for identifying genetic adaptation to high altitude. *Advances in Experimental Medicine and Biology* 588:101–118.

Moore, L.G., D. Young, R.E. McCullough, et al. 2001. Tibetan protection from intrauterine growth restriction (IUGR) and reproductive loss at high altitude. *American Journal of Human Biology* 13(5):635–644.

Moorjani, P., N. Patterson, J.N. Hirschhorn, et al. 2011. The history of African gene flow into Southern Europeans, Levantines, and Jews. *PLoS Genetics* 7(4):e1001373.

Morak, M., H.K. Schackert, N. Rahner, et al. 2008. Further evidence for heritability of an epimutation in one of 12 cases with *MLH1* promoter methylation in blood cells clinically displaying HNPCC. *European Journal of Human Genetics* 16(7):804–811.

Morgan, H.D., F. Santos, K. Green, et al. 2005. Epigenetic reprogramming in mammals. *Human Molecular Genetics* 14 1:R47–R58.

Morgan, H.D., H.G. Sutherland, D.I. Martin, et al. 1999. Epigenetic inheritance at the agouti locus in the mouse. *Nature Genetics* 23(3):314–318.

Moritz, G.L., and N.J. Dominy. 2010. Selective advantages of mono- and dichromatic vision among nocturnal primates. *Journal of Vision* 10:1–1.

Morse, S.S. 1995. Factors in the emergence of infectious diseases. *Emerging Infectious Diseases* 1(1):7–15. doi:10.3201/eid0101.950102

Morse, S.S., J.A. Mazet, M. Woolhouse, et al. 2012. Prediction and prevention of the next pandemic zoonosis. *Lancet* 380(9857):1956–1965. doi:10.1016/S0140-6736(12)61684-5

Mortara, L., J. Ploquin, A. Faye, et al. 2006. Phenotype and function of myeloid dendritic cells derived from African green monkey blood monocytes. *Journal of Immunological Methods* 308:138–155. doi:10.1016/j.jim.2005.10.005

Morton, N.E. 1969. Human population structure. *Annual Review of Genetics* 3:53–74.

Morton, N.E. (ed.). 1973a. *Genetic Structure of Populations.* University Press of Hawaii, Honolulu.

Morton, N.E. 1973b. Isolation by distance. In N.E. Morton (ed.), *Genetic Structure of Populations*. University Press of Hawaii, Honolulu, pp. 76–79.

Morton, N.E., J.F. Crow, and H.J. Muller. 1956. An estimate of the mutational damage in man from data on consanguineous marriages. *Proceedings of the National Academy of Sciences* 42:855–863.

Mou, C., H.A. Thomason, P.A. Willan, et al. 2008. Enhanced ectodysplasin-A receptor (*EDAR*) signaling alters multiple fiber characteristics to produce the East Asian hair form. *Human Mutation* 29(12):1405–1411.

Mueller, W.H. 1976. Parent-child correlations for stature and weight among school aged children: a review of 24 studies. *Human Biology* 48(2):379–397.

Mueller, W.H. 1998. Genetic and environmental influences on fetal growth. In: S.J. Ulijaszek, F.E. Johnston, and M.A. Preece (eds.), *The Cambridge Encyclopedia of Human Growth and Development*. Cambridge University Press, Cambridge, pp. 133–136.

Mulligan, C.J., N.C. D'Errico, J. Stees, et al. 2012. Methylation changes at *NR3C1* in newborns associate with maternal prenatal stress exposure and newborn birth weight. *Epigenetics* 7(8):853–857.

Mulligan, C.J., A. Kitchen, and M.M. Miyamoto. 2008. Updated three-stage model for the peopling of the Americas. *PLoS One* 3:e3199.

Mulligan, C.J., and E.J.E. Szathmáry. 2017. The peopling of the Americas and the origin of the Beringian occupation model. *American Journal of Physical Anthropology* 162:403–408.

Mullis, K., F. Faloona, S. Scharf, et al. 1986. Specific enzymatic amplification of DNA in vitro: the polymerase chain reaction. *Cold Spring Harbor Symposium on Quantitative Biology* 51:263.

Mundade, R., H.G. Ozer, H. Wei, et al. 2014. Role of ChIP-seq in the discovery of transcription factor binding sites, differential gene regulation mechanism, epigenetic marks and beyond. *Cell Cycle* 13(18):2847–2852.

Murray, F. 1934. Pigmentation, sunlight, and nutritional disease. *American Anthropologist* 36(3):438–445.

Murray, C. 2013. *Coming Apart: The State of White America, 1960–2010*. Three Rivers Press, New York.

Myles, S., M. Somel, K. Tang, et al. 2007. Identifying genes underlying skin pigmentation differences among human populations. *Human Genetics* 120(5):611–621.

Myres, N.M., S. Rootsi, A.A. Lin, et al. 2011. A major Y-chromosome haplogroup R1b Holocene era founder effect in Central and Western Europe. *European Journal of Human Genetics* 19(1):95–101.

Nagai, N., N. Sakane, A. Fujishita, et al. 2007. The -3826 A --> G variant of the uncoupling protein-1 gene diminishes thermogenesis during acute cold exposure in healthy children. *Obesity Research & Clinical Practice* 1(2):I–II.

Nagel, M., J. Dischinger, M. Turck, et al. 2012. Human-associated *Staphylococcus aureus* strains within great ape populations in Central Africa (Gabon). *Journal of Bacteriology* 19:1072–1077.

Nagylaki, T. 1982. Assortative mating for a quantitative character. *Journal of Mathematical Biology* 16(1):57–74.

Nagylaki, T. 1992. *Introduction to Theoretical Population Genetics*. Springer, New York.

Nakamura, Y., T. Itoh, H. Matsuda, et al. 2004. Biased biological functions of horizontally transferred genes in prokaryotic genomes. *Nature Genetics* 36(7):760–766.

Namouchi, A., X. Didelot, U. Schock, et al. 2012. After the bottleneck: genome-wide diversification of the mycobacterium tuberculosis complex by mutation, recombination, and natural selection. *Genome Research* 22(4):721–734. doi:10.1101/gr.129544.111

Nassir, R., R. Kosoy, C. Tian, et al. 2009. An ancestry informative marker set for determining continental origin: validation and extension using human genome diversity panels. *BMC Genetics* 10(1):39. doi:10.1186/1471-2156-10-39

Nater, A., M.P. Greminger, N. Arora, et al. 2015. Reconstructing the demographic history of orang-utans using approximate Bayesian computation. *Molecular Ecology* 24(2):310–327. doi:10.1111/mec.13027

Nathans, J., T.P. Piantanida, R.L. Eddy, et al. 1986a. Molecular genetics of inherited variation in human color vision. *Science* 232:203–210.

Nathans, J., D. Thomas, and D.S. Hogness. 1986b. Molecular genetics of human color vision: the genes encoding blue, green, and red pigments. *Science* 232:193–202.

Naugler, C. 2008. Hemochromatosis: a Neolithic adaptation to cereal grain diets. *Medical Hypotheses* 70(3):691–692.

Necsulea, A., M. Soumillon, M. Warnefors, et al. 2014. The evolution of lncRNA repertoires and expression patterns in tetrapods. *Nature* 505:635–640.

Neel, J. 1957. Human hemoglobin types: their epidemiologic implications. *New England Journal of Medicine* 256(4):161–171.

Neel, J., and W. Schull. 1954. *Human Heredity*. University of Chicago Press, Chicago.

Nei, M., G. Xun, and T. Sitnikova. 1997. Evolution by the birth-and-death process in multigene families of the vertebrate immune system. *Proceedings of the National Academy of Sciences* 94(15):7799–1806.

Neitz, J., and M. Neitz. 2011. The genetics of normal and defective color vision. *Vision Research* 51:633–651.

Nelson, G., M.A. Hoon, J. Chandrashekar, et al. 2001. Mammalian sweet taste receptors. *Cell* 106:381–390.

Nelson, C.J., H. Santos-Rosa, and T. Kouzarides. 2006. Proline isomerization of histone H3 regulates lysine methylation and gene expression. *Cell* 126(5):905–916.

Neumann, R., and A.J. Jeffreys. 2006. Polymorphism in the activity of human crossover hotspots independent of local DNA sequence variation. *Human Molecular Genetics* 15(9):1401–1411. doi:10.1093/hmg/ddl063

Neves, W.A., and M. Hubbe. 2005. Cranial morphology of early Americans from Lagoa Santa, Brazil: implications for the settlement of the New World. *Proceedings of the National Academy of Sciences of the USA* 102:18309–18314.

Neves, W.A., M. Hubbe, and G. Correal. 2007a. Human skeletal remains from Sabana de Bogotá, Colombia: a case of Paleoamerican morphology late survival in South America? *American Journal of Physical Anthropology* 133:1080–1098.

Neves, W.A., M. Hubbe, and L.B. Piló. 2007b. Early Holocene human skeletal remains from Sumidouro Cave, Lagoa Santa, Brazil: history of discoveries, geological and chronological context, and comparative cranial morphology. *Journal of Human Evolution* 52:16–30.

Neves, W.A., J.F. Powell, and E.G. Ozolins. 1999. Extra-continental morphological affinities of Palli Aike, Southern Chile. *Interciencia* 24:258–263.

Neves, W.A., A. Prous, R. Gonzáles-José, et al. 2003. Early Holocene human skeletal remains from Santana do Riacho, Brazil: implications for the settlement of the New World. *Journal of Human Evolution* 45:19–42.

Neves, W.A., and H.M. Pucciarelli. 1991. Morphological affinities of the first Americans: an exploratory analysis based on South American human remains. *Journal of Human Evolution* 21:261–273.

New, M., D. Lister, M. Hulme, et al. 2002. A high-resolution data set of surface climate over global land areas. *Climate Research* 21:1–25.

Newton, J.M., O. Cohen-Barak, N. Hagiwara, et al. 2001. Mutations in the human orthologue of the mouse underwhite gene (*uw*) underlie a new form of oculocutaneous albinism, *OCA4. American Journal of Human Genetics* 69(5):981–988. doi:10.1086/324340

Nicholls, R.D. 1993. Genomic imprinting and uniparental disomy in Angelman and Prader-Willi syndromes: a review. *American Journal of Medical Genetics* 46(1):16–25.

Nielsen, R., J.M. Akey, M. Jakobsson, et al. 2017. Tracing the peopling of the world through genomics. *Nature* 541(7637):302–310.

Nielsen, R., I. Hellmann, M. Hubisz, et al. 2007. Recent and ongoing selection in the human genome. *Nature Reviews Genetics* 8(11):857–868.

Non, A.L., and Z.M. Thayer. 2015. Epigenetics for anthropologists: an introduction to methods. *American Journal of Human Biology* 27:295–303.

Noonan, J.P., G. Coop, S. Kudaravalli, et al. 2006. Sequencing and analysis of Neanderthal genomic DNA. *Science* 314(5802):1113–1118.

Norton, H.L., E.A. Correa, G. Koki, et al. 2014. Distribution of an allele associated with blond hair color across Northern Island Melanesia. *American Journal of Physical Anthropology* 153(4):653–662. doi:10.1002/ajpa.22466

Norton, H.L., J.S. Friedlaender, D.A. Merriwether, et al. 2006. Skin and hair pigmentation variation in island Melanesia. *American Journal of Physical Anthropology* 130(2):254–268. doi:10.1002/ajpa.20343

Norton, H.L., R.A. Kittles, E. Parra, et al. 2007. Genetic evidence for the convergent evolution of light skin in Europeans and east Asians. *Molecular Biology and Evolution* 24(3):710–722. doi:10.1093/molbev/msl203

Novembre, J. 2015. Ancient DNA steps into the language debate. *Nature* 522:164–165.

Nugent, G. 2011. Maintenance, spillover and spillback transmission of bovine tuberculosis in multi-host wildlife complexes: a New Zealand case study. *Veterinary Microbiology* 151:34–42.

Nunes, M.A., and D.J. Balding. 2010. On optimal selection of summary statistics for approximate Bayesian computation. *Statistical Applications in Genetics and Molecular Biology* 9(1):art. 34.

Nunes, M.A., and D. Prangle. 2015. Abctools: an R package for tuning approximate Bayesian computation analyses. *The R Journal* 7(2):189–205.

Nunn, C.L. 2011. *The Comparative Method in Evolutionary Anthropology and Biology.* University of Chicago Press, Chicago.

O'Connell, J.F., and J. Allen. 2004. Dating the colonization of Sahul (Pleistocene Australia–New Guinea): a review of recent research. *Journal of Archaeological Science* 31:835–853. doi:10.1016/j.jas.2003.11.005

O'Roak, B.J., L. Vives, S. Girirajan, et al. 2012. Sporadic autism exomes reveal a highly interconnected protein network of *de novo* mutations. *Nature* 485(7397):246–250. doi:10.1038/nature10989

O'Rourke, D.H. 2003. Anthropological genetics in the genomic era: a look back and ahead. *American Anthropologist* 105(1):101–109.

O'Rourke, D.H., and J.A. Raff. 2010. The human genetic history of the Americas: the final frontier. *Current Biology* 20(4):R202–R207. doi:10.1016/j.cub.2009.11.051

O'Rourke, D.H. 2006. Blood groups, immunoglobulins, and genetic variation. In: W.C. Sturtevant, and D. Ubelaker (eds.), *Handbook of North American Indians, Vol.3: Environment, Origins, and Population.* Smithsonian Institution Press, Washington, DC, pp. 762–776.

O'Rourke, D.H. 2011. Contradictions and concordances in American colonization models. *Evolution Education and Outreach* 4:244–253.

Ochman, H., J.G. Lawrence, and E.A. Groisman. 2000. Lateral gene transfer and the nature of bacterial innovation. *Nature* 405(6784):299–304.

Oden, N.L. 1992. Spatial autocorrelation invalidates the Dow-Cheverud test. *American Journal of Physical Anthropology* 89:257–264.

Oden, N.L., and R.R. Sokal. 1992. An investigation of three-matrix permutation tests. *Journal of Classification* 9:275–290.

Oetting, W.S., and R.A. King. 1999. Molecular basis of albinism: mutations and polymorphisms of pigmentation genes associated with albinism. *Human Mutation* 13:99–115.

Off, M.K., A.E. Steindal, A.C. Porojnicu, et al. 2005. Ultraviolet photodegradation of folic acid. *Journal of Photochemistry and Photobiology B* 80(1):47–55. doi:10.1016/j.jphotobiol.2005.03.001

Ohgane, J., S. Yagi, and K. Shiota. 2008. Epigenetics: the DNA methylation profile of tissue-dependent and differentially methylated regions in cells. *Placenta* 29 Suppl A:S29–S35.

Ohta, T. 1992. The nearly-neutral theory of molecular evolution. *Annual Review of Ecology, Evolution, and Systematics* 23:263–286.

Ohta, T. 1998. On the pattern of polymorphisms at major histocompatibility complex loci. *Journal of Molecular Evolution* 46(6):633–638.

Olabode, A.S., X. Jiang, D.L. Robertson, et al. 2015. Ebolavirus is evolving but not changing: no evidence for functional change in EBOV from 1976 to the 2014 outbreak. *Virology* 482:202–207. doi:10.1016/j.virol.2015.03.029

Oleksyk, T.K., M.W. Smith, and S.J. O'Brien. 2010. Genome-wide scans for footprints of natural selection. *Philosophical Transactions of the Royal Society, B: Biological Sciences* 365(1537):185–205.

Onishi, A., S. Koike, M. Ida, et al. 1999. Vision: dichromatism in macaque monkeys. *Nature* 402:139–140.

Onishi, A., S. Koike, M. Ida-Hosonuma, et al. 2002. Variations in long- and middle-wavelength-sensitive opsin gene loci in crab-eating monkeys. *Vision Research* 42:281–292.

Oppenheimer, S., B. Bradley, and D. Stanford. 2014. Solutrean hypothesis: genetics, the mammoth in the room. *World Archaeology* 46:752–774.

Orlando, L., P. Darlu, M. Toussaint, D. Bonjean, et al. 2006. Revisiting Neandertal diversity with a 100,000 year old mtDNA sequence. *Current Biology* 16(11):R400–R402.

Orlando, L., A. Ginolhac, G. Zhang, D. Froese, et al. 2013. Recalibrating Equus evolution using the genome sequence of an early middle Pleistocene horse. *Nature* 499(7456):74–78.

Orlando, L., and E. Willerslev. 2014. An epigenetic widow into the past? *Science* 345:511–512.

Orlando, L., and E. Willerslev. 2014. An epigenetic window into the past? *Science* 345(6196):511–512.

Orr, H.A. 1999. The evolutionary genetics of adaptation: a simulation study. *Genetics Research* 74(3):207–214.

Orr, H.A. 2005. The genetic theory of adaptation: a brief history. *Nature Reviews Genetics* 6:119–127.

Orr, H.A. 2007. The population genetics of adaptation: the adaptation of DNA sequences. *Evolution* 56(7):1317–1330.

Ortiz, A.M., S.R. Dinapoli, and J.M. Brenchley. 2015. Macrophages are phenotypically and functionally diverse across tissues in simian immunodeficiency virus-infected and uninfected Asian macaques. *Journal of Virology* 89(11):5883–5894. doi:10.1128/JV.I.00005-15

Otting, N., and R.E. Bontrop. 1993. Characterization of the rhesus macaque (*Macaca mulatta*) equivalent of HLA-F. *Immunogenetics* 38:141–145.

Otting, N., M.C. Heijmans, R.C. Noort, et al. 2005. Unparalleled complexity of the MHC class I region in rhesus macaques. *Proceedings of the National Academy of Sciences* 102(5):1626–1631.

Otting, N., A.J.M. Vos-rouweler, C.M.C. Heijmans, et al. 2007. MHC class I a region diversity and polymorphism in macaque species. *Immunogenetics* 59:367–375. doi:10.1007/s00251-007-0201-2

Otto, S.P., and A.C. Gerstein. 2008. The evolution of haploidy and diploidy. *Current Biology* 18(24):R1121–R1124.

Otto, S.P., and M.C. Whitlock. 1997. The probability of fixation in populations of changing size. *Genetics* 146(2):723–733.

Ouellet, V., S.M. Labbe, D.P. Blondin, et al. 2012. Brown adipose tissue oxidative metabolism contributes to energy expenditure during acute cold exposure in humans. *Journal of Clinical Investigation* 122(2):545–552.

Ouellet-Morin, I., C.C. Wong, A. Danese, et al. 2013. Increased serotonin transporter gene (*SERT*) DNA methylation is associated with bullying victimization and blunted cortisol response to stress in childhood: a longitudinal study of discordant monozygotic twins. *Psychological Medicine* 43(9):1813–1823.

Ovchinnikov, I.V., A. Götherström, G.P. Romanova, et al. 2000. Molecular analysis of Neanderthal DNA from the northern Caucasus. *Nature* 404:490–493.

Overholtzer, L. 2011. Otomies at Xaltocan: a case of ethnic change. Presented at the 76th AnnualMeeting of the Society for American Archaeology in Sacramento, California, March 30–April 3, 2011.

Overholtzer, L. 2012. Empire and everyday material practices: a household archaeology of Aztec and Spanish imperialism at Xaltocan, Mexico. PhD dissertation, Northwestern University, Chicago, IL.

Owen, M.J., and M.C. O'Donovan. 2003. Schizophrenia and genetics. In: R. Plomin (ed.), *Behavioral Genetics in the Postgenomic Era*. American Psychological Association, Washington, DC, pp. 463–480.

Owsley, D., and R. Jantz. 2014. *Kennewick Man: The Scientific Investigation of an Ancient American Skeleton*. Texas A&M University Press, College Station.

Paabo, S., H. Poinar, D. Serre, et al. 2004. Genetic analyses from ancient DNA. *Annual Review of Genetics* 38:645–679.

Pagani, L., D.J. Lawson, E. Jagoda, et al. 2016. Genomic analyses inform on migration events during the peopling of Eurasia. *Nature* 538(7624):238–242.

Pai, A.A., J.T. Bell, J.C. Marioni, et al. 2011. A genome-wide study of DNA methylation patterns and gene expression levels in multiple human and chimpanzee tissues. *PLoS Genetics* 7:e1001316.

Paige, S. 2008. *The Difference: How the Power of Diversity Creates Better Groups, Firms, Schools, and Societies*. Princeton University Press, Princeton, NJ.

Paixão-Côrtes, V.R., D. Meyer, T.V. Pereira, et al. 2011. Genetic variation among major human geographic groups supports a peculiar evolutionary trend in *PAX9*. *PLoS One* 6:e15656.

Paradis, E. 2013. Molecular dating of phylogenies by likelihood methods: a comparison of models and a new information criterion. *Molecular Phylogenetics and Evolution* 67(2):436–444. doi:10.1016/j.ympev.2013.02.008

Pardoe, C. 1991. Isolation and evolution in Tasmania. *Current Anthropology* 32:1–21. doi:10.1086/203909

Parham, P., L. Abi-Rached, L. Matevosyan, et al. 2010. Primate-specific regulation of natural killer cells. *Journal of Medical Primatology* 39(4):194–212. doi:10.1111/j.1600-0684.2010.00432.x

Parham, P., E.J. Adams, and K.L. Arnett. 1995. The origins of HLA-A, B,C polymorphism. *Immunological Reviews* 143:141–180.

Parham, P., D.A. Lawlor, C.E. Lomen, et al. 1989. Diversity and diversification of HLA-A,B,C alleles. *Journal of Immunology* 142:3937–3950.

Parham, P., and A, Moffett. 2013. How did variable NK-cell receptors and MHC class I ligands influence immunity, reproduction and human evolution? *Nature Reviews Immunology* 13(2):133–134. doi:10.1038/nri3370.How

Park, D.J., G. Dudas, S. Wohl, et al. 2015. Ebola virus epidemiology, transmission, and evolution during seven months in Sierra Leone. *Cell* 161(7):1516–1526. doi:10.1016/j.cell.2015.06.007

Park, J.-H., T. Yamaguchi, C. Watanabe, et al. 2012. Effects of an Asian-specific nonsynonymous *EDAR* variant on multiple dental traits. *Journal of Human Genetics* 57(8):508–514.

Parra, E.J. 2007. Human pigmentation variation: evolution, genetic basis, and implications for public health. *American Journal of Physical Anthropology* 45(Suppl.):85–105. doi:10.1002/ajpa.20727

Parra, E.J., R.A. Kittles, G. Argyropoulos, et al. 2001. Ancestral proportions and admixture dynamics in geographically defined African Americans living in South Carolina. *American Journal of Physical Anthropology* 114:18–29.

Parra, E.J., A. Marcini, J. Akey, et al. 1998. Estimating African American admixture proportions by use of population-specific alleles. *American Journal of Human Genetics* 63:1839–1851.

Parry, C.M., A. Erkner, and J. le Coutre. 2004. Divergence of T2R chemosensory receptor families in humans, bonobos, and chimpanzees. *Proceedings of the National Academy of Sciences* 101:14830–14834.

Pascalis, H., J.-M. Heraud, R. Fendel, et al. 2003. Molecular characterization of major histocompatibility complex class 1 (MHC-I) from squirrel monkeys (*Saimiri sciureus*). *Immunogenetics* 55:633–639. doi:10.1007/s00251-003-0607-4

Patterson, N., P. Moorjani, Y. Luo, et al. 2012. Ancient admixture in human history. *Genetics* 192(3):1065–1093.

Patterson, N., D.J. Richter, S. Gnerre, et al. 2006. Genetic evidence for complex speciation of humans and chimpanzees. *Nature* 441:1103–1108. doi:10.1038/nature04789

Paul, A. 2002. Sexual selection and mate choice. *International Journal of Primatology* 23:877–904.

PBS. 2015. Roots of freedom. http://www.pbs.org/wnet/finding-your-roots/roots-freedom-full-episode/11903

Pearce-Duvet, J.M. 2006. The origin of human pathogens: evaluating the role of agriculture and domestic animals in the evolution of human disease. *Biological Reviews of the Cambridge Philosophical Society* 81(3):369–382.

Pearson, K., and A. Lee. 1903. On the laws of inheritance in man. I. Inheritance of physical characteristics. *Biometrika* 2:356–462.

Pearson, O.M., D.F. Royer, F.E. Grine, et al. 2008a. A description of the Omo 1 postcranial skeleton, including newly discovered fossils. *Journal of Human Evolution* 55:421–437.

Pearson, O.M., D.F. Royer, F.E. Grine, et al. 2008b. Brief report on new hominin fossils from elsewhere in the Kibish formation, Southwestern Ethiopia. *Journal of Human Evolution* 55:444–447.

Pechouskova, E., M. Dammhahn, M. Brameier, et al. 2015. MHC class II variation in a rare and ecological specialist mouse lemur reveals lower allelic richness and contrasting selection patterns compared to a generalist and widespread sympatric congener. *Immunogenetics* 67(4):229–245.

Pedersen, A.B., and T.J. Davies. 2009. Cross-species pathogen transmission and disease emergence in primates. *Ecohealth* 6(4):496–508. doi:10.1007/s10393-010-0284-3

Pedersen, M.W., S. Overballe-Petersen, L. Ermini, et al. 2014. Ancient and modern environmental DNA. Philosophical transactions of the Royal Society of London B. *Biological Sciences* 370(1660).

Pedersen, M.W., A. Ruter, C. Schweger, et al. 2016. Postglacial viability and colonization in North America's ice-free corridor. *Nature* 537:45–49.

Pedersen, J.S., E. Valen, A.M. Vargas Velazquez, et al. 2014. Genome-wide nucleosome map and cytosine methylation levels of an ancient human genome. *Genome Research* 24(3):454–466. doi:10.1101/gr.163592.113 PMID: 24299735

Pellegrini, M., and R. Ferrari. 2012. Epigenetic analysis: ChIP-chip and ChIP-seq. *Methods in Molecular Biology* 802:377–387.

Peltier, W.R. 2002. On eustatic sea level history: last glacial maximum to Holocene. *Quaternary Science Reviews* 24:377–396.

Pemberton, T.J., M. DeGiorgio, and N.A. Rosenberg. 2013. Population structure in a comprehensive genomic data set on human microsatellite variation. *G3 Genes, Genomes, Genetics* 3:891–907.

Pena, S.D., F.R. Santos, N.O. Bianchi, et al. 1995. A major founder Y-chromosome haplotype in Amerindians. *Nature Genetics* 11:15–16.

Pendleton, L.S.A., A.R. McLane, and D.H. Thomas. 1982. *Cultural Resource Overview, Carson City District, West Central Nevada.* Cultural Resource Series 5. Bureau of Land Management, Washington, DC.

Peng, Y., Z. Yang, H. Zhang, et al. 2011. Genetic variations in Tibetan populations and high-altitude adaptation at the Himalayas. *Molecular Biology and Evolution* 28(2):1075–1081.

Penrose, L. 1954. Some recent trends in human genetics. *Caryologia* 6(Suppl.): 521–530.

Pepperell, C.S., A.M. Casto, A. Kitchen, et al. 2013. The role of selection in shaping diversity of natural *M. tuberculosis* populations. *PLoS Pathogens* 9(8):e1003543. doi:10.1371/journal.ppat.1003543

Perego, U.A., A. Achilli, N. Angerhofer, et al. 2009. Distinctive Paleo-Indian migration routes from Beringia marked by two rare mtDNA haplogroups. *Current Biology* 19:1–8.

Perego, U.A., N. Angerhofer, M. Pala, et al. 2010. The initial peopling of the Americas: a growing number of founding mitochondrial genomes from Beringia. *Genome Research* 20:1174–1179.

Pereira, L.E., R.P. Johnson, and A.A. Ansari. 2008. Sooty mangabeys and rhesus macaques exhibit significant divergent natural killer cell responses during both acute and chronic phases of SIV infection. *Cellular Immunology* 254(1):10–19.

Perez, S.I., V. Bernal, and P.N. Gonzalez. 2007. Morphological differentiation of aboriginal human populations from Tierra del Fuego (Patagonia): implications for South American peopling. *American Journal of Physical Anthropology* 133:1067–1079.

Perez, S.I., V. Bernal, P.N. Gonzalez, et al. 2009. Discrepancy between cranial and DNA data of early Americans: implications for American peopling. *PLoS One* 4:e5746.

Perry, G.H., N.J. Dominy, K.G. Claw, et al. 2007. Diet and the evolution of human amylase gene copy number variation. *Nature Genetics* 39:1256–1260.

Perry, G.H., L. Kistler, M.A. Kelaita, et al. 2015. Insights into hominin phenotypic and dietary evolution from ancient DNA sequence data. *Journal of Human Evolution* 79:55–63.

Perry, G.H., P. Melsted, J.C. Marioni, et al. 2012. Comparative RNA sequencing reveals substantial genetic variation in endangered primates. *Genome Research* 22:602–610.

Peter, B.M., E. Huerta-Sanchez, and R. Nielsen. 2012. Distinguishing between selective sweeps from standing variation and from a *de novo* mutation. *PLoS Genetics* 8(10):e1003011. doi:10.1371/journal.pgen.1003011

Petousi, N., Q.P. Croft, G.L. Cavalleri, et al. 2014. Tibetans living at sea level have a hyporesponsive hypoxia-inducible factor system and blunted physiological responses to hypoxia. *Journal of Applied Physiology* 116(7):893–904.

Petraglia, M.D., P. Ditchfield, S. Jones, et al. 2012. The Toba volcanic super-eruption, environmental change, and hominin occupation history in India over the last 140,000 years. *Quaternary International* 258:119–134.

Petraglia, M., R. Korisettar, N. Boivin, et al. 2007. Middle Paleolithic assemblages from the Indian subcontinent before and after the Toba super-eruption. *Science* 317(5834):114–116. doi:10.1126/science.1141564

Pickrell, J.K., G. Coop, J. Novembre, et al. 2009. Signals of recent positive selection in a worldwide sample of human populations. *Genome Research* 19(5):826–837.

Pickrell, J.K., and J.K. Pritchard. 2012. Inference of population splits and mixtures from genome-wide allele frequency data. *PLoS Genetics* 8(11):e1002967.

Pickrell, J.K., and D. Reich. 2014. Toward a new history and geography of human genes informed by ancient DNA. *Trends in Genetics* 30(9):377–389.

Pielou, E.C. 1992. *After the Ice Age: The Return of Life to Glaciated North America.* University of Chicago Press, Chicago, IL.

Pier, G.B., M. Grout, T. Zaidi, et al. 1998. *Salmonella typhi* uses CFTR to enter intestinal epithelial cells. *Nature* 393(6680):79–82.

Pigliucci, M., C.J. Murranand, and C.D. Schlichting. 2006. Phenotypic plasticity and evolution by genetic assimilation. *Journal of Experimental Biology* 209:2362–2367.

Pigliucci, M., and C.J. Murren. 2003. Genetic assimilation and a possible evolutionary paradox: can macroevolution sometimes be so fast as to pass us by? *Evolution* 57:1455–1464.

Pigott, D.M., N. Golding, A. Mylne, et al. 2014. Mapping the zoonotic niche of Ebola virus disease in Africa. *eLife* 3:e04395. doi:10.7554/eLife.04395

Pinhasi, R., D. Fernandes, K. Sirak, et al. 2015 Optimal ancient DNA yields from the inner ear part of the human petrous bone. *PLoS One* 10(6):e0129102. doi:10.1371/journal.pone.0129102

Pinhasi, R., R. Foley, and H.D. Donoghue. 2006. Reconsidering the antiquity of leprosy. *Science* 312:846.

Pinhasi, R., and N. von Cramon-Taubadel. 2009. Craniometric data supports demic diffusion model for the spread of agriculture into Europe. *PLoS ONE* 4(8):e6747.

Pinho, C., and J. Hey. 2010. Divergence with gene flow: models and data. *Annual Review of Ecology, Evolution, and Systematics* 41:215–230. doi:10.1146/Annurev-Ecolsys-102209-144644

Piontkivska, H., and M. Nei. 2003. Birth-and-death evolution in primate MHC class I genes: divergence time estimates. *Molecular Biology and Evolution* 20(4):601–609. doi:10.1093/molbev/msg064

Pitulko, V.V., P.A. Nikolsky, E.Y. Girya, et al. 2004. The Yana RHS site: humans in the Arctic before the last glacial maximum. *Science* 303(5654):52–56.

Poinar, H., M. Kuch, K.D. Sobolik, et al. 2001. A molecular analysis of dietary diversity for three archaic Native Americans. *Proceedings National Academies of Sciences USA* 98:4317–4322.

Pollack, A. 2015. Building a face, and a case, on DNA. New York Times. http://www.nytimes.com/2015/02/24/science/building-face-and-a-case-on-dna.html

Polly, P.D. 2001. On morphological clocks and paleophylogeography: towards a timescale for *Sorex* hybrid zones. *Genetica* 112–113:339–357.

Pomery, C. 2009. The advantages of a dual DNA/documentary approach to reconstruct the family trees of a surname. *Journal of Genetic Genealogy* 5(2):86–95.

Porto, A., F.B. De Oliveira, L.T. Shirai, et al. 2009. The evolution of modularity in the mammalian skull I: morphological integration patterns and magnitudes. *Evolutionary Biology* 36:118–135.

Post, R.H. 1962. Population differences in red and green color vision deficiency: a review, and a query on selection relaxation. *Biodemography and Social Biology* 9:131–46.

Prabhakar, S., A. Visel, J.A. Akiyama, et al. 2008. Human-specific gain of function in a developmental enhancer. *Science* 321:1346–1350.

Prado-Martinez, J., I. Hernando-Herraez, B. Lorente-Galdos, et al. 2013a. The genome sequencing of an albino western lowland gorilla reveals inbreeding in the wild. *BMC Genomics* 14:363.

Prado-Martinez, J., P.H. Sudmant, J.M. Kidd, et al. 2013b. Great ape genetic diversity and population history. *Nature* 499:471–475.

Prangle, D., P. Fearnhead, M.P. Cox, et al. 2014. Semi-automatic selection of summary statistics for ABC model choice. *Statistical Applications in Genetics and Molecular Biology* 13(1):67–82. doi:10.1515/sagmb-2013-0012

Prendergast, J.G., E.V. Chambers, and C.A. Semple. 2014. Sequence-level mechanisms of human epigenome evolution. *Genome Biology and Evolution* 6(7):1758–1771.

Presgraves, D.C., and S.V. Yi. 2009. Doubts about complex speciation between humans and chimpanzees. *Trends in Ecology & Evolution* 24(10):533–540. doi:10.1016/j.tree.2009.04.007

Pritchard, J.K., J.K. Pickrell, and G. Coop. 2010. The genetics of human adaptation: hard sweeps, soft sweeps, and polygenic adaptation. *Current Biology* 20(4):R208–R215.

Pritchard, J.K., and M. Przeworski. 2001. Linkage disequilibrium in humans: models and data. *American Journal of Human Genetics* 69(1):1–14.

Pritchard, J.K., M.T. Seielstad, A. Perez-Lezaun, et al. 1999. Population growth of human Y chromosomes: a study of Y chromosome microsatellites. *Molecular Biology and Evolution* 16(12):1791–1798.

Provine, W. 2001. *The Origins of Theoretical Population Genetics.* University of Chicago Press, Chicago.

Prüfer, K., C. de Filippo, S. Grote, et al. 2017. A high-coverage Neandertal genome from Vindija cave in Croatia. *Science* 358(6363):655.

Prüfer, K., K. Munch, I. Hellmann, et al. 2012. The bonobo genome compared with the chimpanzee and human genomes. *Nature* 486:527–531. doi:10.1038/nature11128

Prüfer, K., F. Racimo, N. Patterson, et al. 2014. The complete genome sequence of a Neanderthal from the Altai Mountains. *Nature* 505(7481):43–49.

Prugnolle, F., A. Manica, and F. Balloux. 2005. Geography predicts neutral genetic diversity of human populations. *Current Biology* R129.

Prugnolle, F., A. Manica, M. Charpentier, et al. 2005. Pathogen-driven selection and worldwide HLA class I diversity. *Current Biology* 15(11):1022–1027. doi:10.1016/j.cub.2005.04.050

Ptak, C., and A. Petronis. 2008. Epigenetics and complex disease: from etiology to new therapeutics. *Annual Review of Pharmacology and Toxicology* 48:257–276.

Pucciarelli, H.M., W.A. Neves, R. González-José, et al. 2006. East-west cranial differentiation in pre-Columbian human populations of South America. *Homo* 57:133–150.

Pucciarelli, H., M.L. Sardi, J.C. Jimenez Lopez, et al. 2003. Early peopling and evolutionary diversification in America. *Quaternary International* 109–110:123–132.

Pugach, I., F. Delfin, E. Gunnarsdóttir, et al. 2013. Genome-wide data substantiate Holocene gene flow from India to Australia. *Proceedings of the National Academy of Sciences of the USA* 110(5):1803–1808.

Pugach, I., R. Matveyev, A. Wollstein, et al. 2011. Dating the age of admixture via wavelet transform analysis of genome-wide data. *Genome Biology* 12:R19. doi:10.1186/gb-2011-12-2-r19

Pugach, I., and M. Stoneking. 2015. Genome-wide insights into the genetic history of human populations. *Investigative Genetics* 6:6.

Pummila, M., I. Fliniaux, R. Jaatinen, et al. 2007. Ectodysplasin has a dual role in ectodermal organogenesis: inhibition of Bmp activity and induction of Shh expression. *Development* 134(1):117–125.

Pusey, A. 2004. Inbreeding avoidance in primates. In: A.P. Wolf, and D.H. Durham (eds.), *Inbreeding, Incest, and the Incest Taboo.* Stanford University Press, Palo Alto, C.A., pp. 61–75.

Quach, H., D. Wilson, G. Laval, et al. 2013. Different selective pressures shape the evolution of toll-like receptors in human and African great ape populations. *Human Molecular Genetics* 22(23):4829–4840. doi:10.1093/hmg/ddt335

R Development Core Team. 2008. R: A Language and Environment for Statistical Computing. R Foundation for Statistical Computing, Vienna, Austria.

Rabinowitch, I.M., and F.C. Smith. 1936. Metabolic studies of Eskimos in the Canadian Eastern Arctic. *Journal of Nutrition* 12(4):337–356.

Racimo, F., G. Renaud, and M. Slatkin. 2016. Joint estimation of contamination, error and demography for nuclear DNA from ancient humans. *PLoS Genetics* 12(4):e1005972.

Racimo F, S. Sankararaman, R. Nielsen, et al. 2015. Evidence for archaic adaptive introgression in humans. *Nature Reviews Genetics* 16:359.

Racimo, F., S. Sankararaman, R. Nielsen, et al. 2015. Evidence for archaic adaptive introgression in humans. *Nature Reviews Genetics* 16:359–371.

Radtke, K.M., M. Ruf, H.M. Gunter, et al. 2011. Transgenerational impact of intimate partner violence on methylation in the promoter of the glucocorticoid receptor. *Translational Psychiatry* 1:e21.

Raff, J.A. 2008. An ancient DNA perspective on the prehistory of the lower Illinois Valley. PhD dissertation. Indiana University, Bloomington.

Raff, J.A., and D.A. Bolnick. 2015. Does mitochondrial haplogroup X indicate ancient trans-Atlantic migration to the Americas? A critical re-evaluation. *PaleoAmerica* 1:297–304

Raff, J.A., D.A. Bolnick, J. Tackney, et al. 2011. Ancient DNA perspectives on American colonization and population history. *American Journal of Physical Anthropology* 146:503 514.

Raff, J.A., M. Rzhetskaya, J. Tackney, et al. 2015. Mitochondrial diversity of Iñupiat people from the Alaskan North Slope provides evidence for the origins of the Paleo- and Neo-Eskimo peoples. *American Journal of Physical Anthropology* 157:603–614.

Raghavan, M., M. DeGiorgio, A. Albrechtsen, et al. 2014a. The genetic prehistory of the New World Arctic. *Science* 345(6200):1255832.

Raghavan, M., P. Skoglund, K.E. Graf, et al. 2014b. Upper Palaeolithic Siberian genome reveals dual ancestry of Native Americans. *Nature* 505(7481):87–91.

Raghavan, M., M. Steinrucken, K. Harris, et al. 2015. Genomic evidence for the Pleistocene and recent population history of native Americans. *Science* 349(6250):aab3884. doi:10.1126/science.aab3884

Rajakumar, C., M.R. Ban, H. Cao, et al. 2009. Carnitine palmitoyltransferase IA polymorphism P479L is common in Greenland Inuit and is associated with elevated plasma apolipoprotein A-I. *Journal of Lipid Research* 50(6):1223–1228.

Rajalingam, R., M. Hong, E.J. Adams, et al. 2001. Short KIR haplotypes in pygmy chimpanzee (bonobo) resemble the conserved framework of diverse human KIR haplotypes. *Journal of Experimental Medicine* 193(1):135–146.

Ramachandran, S., O. Deshpande, C.C. Roseman, et al. 2005. Support from the relationship of genetic and geographic distance in human populations for a serial founder effect originating in Africa. *Proceedings of the National Academy of Sciences of the USA* 102(44):15942–15947.

Ramachandran, S., N.A. Rosenberg, L.A. Zhivotovsky, et al. 2004. Robustness of the inference of human population structure: a comparison of X-chromosomal and autosomal microsatellites. *Human Genomics* 1(2):87–97.

Rambaut, A., D. Posada, K.A. Crandall, et al. 2004. The causes and consequences of HIV evolution. *Nature Reviews Genetics* 5(1):52–61.

Rana, B.K., D. Hewett-Emmett, L. Jin, et al. 1999. High polymorphism at the human melanocortin 1 receptor locus. *Genetics* 151(4):1547–1557.

Rancourt, R.C., H.R. Harris, and K.B. Michels. 2012. Methylation levels at imprinting control regions are not altered with ovulation induction or in vitro fertilization in a birth cohort. *Human Reproduction* 27(7):2208–2216.

Rando, O.J. 2012. Daddy issues: paternal effects on phenotype. *Cell* 151:702–708.

Raoult, D., G. Aboudharam, E. Crube'zy, et al. 2000. Molecular identification by "suicide PCR" of *Yersinia pestis* as the agent of Medieval Black Death. *Proceedings of the National Academy of Sciences of the USA* 97:12800–12803.

Rasmussen, M., S.L. Anzick, M.R. Waters, et al. 2014. The genome of a late Pleistocene human from a Clovis burial site in western Montana. *Nature* 506(7487):225–229. doi:10.1038/nature13025

Rasmussen, M., X. Guo, Y. Wang, et al. 2011. An aboriginal Australian genome reveals separate human dispersals into Asia. *Science* 334(6052):94–98. doi:10.1126/science.1211177

Rasmussen, M., Y. Li, S. Lindgreen, et al. 2010. Ancient human genome sequence of an extinct Palaeo-Eskimo. *Nature* 463(7282):757–762. doi:10.1038/nature08835 PMID: 20148029

Rasmussen, M., M. Sikora, A. Albrechtsen, et al. 2015. The ancestry and affiliations of Kennewick man. *Nature* 523(7561):455–458. doi:10.1038/nature14625

Raspe, P.D. 1988. Models of human migration: an inter-island example. In: C.G. Nicholas Mascie-Taylor, and G.W. Lasker (eds.), *Biological Aspects of Human Migration*. Cambridge University Press, Cambridge, pp. 70–89. doi:10.1017/CBO9780511897801.004

Rasteiro, R., P.-A. Bouttier, V.C. Sousa, et al. 2012. Investigating sex-biased migration during the Neolithic transition in Europe, using an explicit spatial simulation framework. *Proceedings of the Royal Society B* 279:2409–2416.

Ray, N., M. Currat, M. Foll, et al. 2010. SPLATCHE2: a spatially explicit simulation framework for complex demography, genetic admixture and recombination. *Bioinformatics* 26(23):2993–2994. doi:10.1093/bioinformatics/btq579

Ray, N., D. Wegmann, N.J.R. Fagundes, et al. 2010. A statistical evaluation of models for the initial settlement of the American continent emphasizes the importance of gene flow with Asia. *Molecular Biology and Evolution* 27:337–345.

Reardon, J. 2005. *Race to the Finish: Identity and Governance in an Age of Genomics*. Information series. Princeton University Press, Princeton, NJ.

Reardon, J., and K. TallBear. 2012. Your DNA is our history: genomics, anthropology, and the construction of whiteness as property. *Current Anthropology* 53(S5):S233–S245. doi:10.1086/662629

Reed, T.E. 1969. Caucasian genes in American negroes. *Science* 165:762–768.

Rees, J.L. 2004. The genetics of sun sensitivity in humans. *American Journal of Human Genetics* 75(5):739–751. doi:10.1086/425285

Reeves, R.K., J. Gillis, F.E. Wong, et al. 2010. CD16- natural killer cells: enrichment in mucosal and secondary lymphoid tissues and altered function during chronic SIV infection. *Blood* 115(22):4439–4446.

Regalado, A. 2015. 23andMe's new formula: patient consent = $. MIT Technology Review, January 6. http://www.technologyreview.com/view/534006/23andmes-new-formula-patient-consent

Reich, D., R.E. Green, M. Kircher, et al. 2010. Genetic history of an archaic hominin group from Denisova Cave in Siberia. *Nature* 468(7327):1053–1060. doi:10.1038/nature09710

Reich, D., N. Patterson, D. Campbell, et al. 2012. Reconstructing Native American population history. *Nature* 488(7411):370–374. doi:10.1038/nature11258

Reich, D., N. Patterson, M. Kircher, et al. 2011a. Denisova admixture and the first modern human dispersals into Southeast Asia and Oceania. *American Journal of Human Genetics* 89(4):516–528. doi:10.1016/j.ajhg.2011.09.005

Reich, D., N. Patterson, M. Kircher, et al. 2011b. Denisova admixture and the first modern human dispersals into Southeast Asia and Oceania. *American Journal of Human Genetics* 89:1–13.

Reich, D., K. Thangaraj, N. Patterson, et al. 2009. Reconstructing Indian population history. *Nature* 461(7263):489–494.

Reid, A.H., T.G. Fanning, T.A. Janczewski, et al. 2000. Characterization of the 1918 Spanish influenza virus neuraminidase gene. *Proceedings of the National Academy of Sciences of the USA* 97(12):6785–6790.

Relethford, J.H. 1986. A gravity model of human population structure. *Human Biology* 58:801–815.

Relethford, J.H. 1990. Effect of population size on marital migration distance. *Human Biology* 63:95–98.

Relethford, J.H. 1991. Genetic drift and anthropometric variation in Ireland. *Human Biology* 63(2):155–165.

Relethford, J.H. 1994. Craniometric variation among modern human populations. *American Journal of Physical Anthropology* 95:53–62.
Relethford, J.H. 1996. Genetic drift can obscure population history: problem and solution. *Human Biology* 68:29–44.
Relethford, J.H. 1997. Hemispheric difference in human skin color. *American Journal of Physical Anthropology* 104(4):449–457. doi:10.1002/(SICI)1096-8644(199712)104:4<449::AID-AJPA2>3.0.CO;2-N
Relethford, J.H. 2000. Human skin color diversity is highest in sub-Saharan African populations. *Human Biology* 72(5):773–780.
Relethford, J. 2002. Apportionment of global human genetic diversity based on craniometrics and skin color. *American Journal of Physical Anthropology* 118:393–398.
Relethford, J. 2003. Anthropomorphic data and population history. In: D.A. Herring, and A.C. Swedlund (eds.), *Human Biologists in the Archives: Demography, Health, Nutrition and Genetics in Historical Populations.* Cambridge University Press, Cambridge, pp. 32–52.
Relethford, J.H. 2004a. Global patterns of isolation by distance based on genetic and morphological data. *Human Biology* 76:499–513.
Relethford, J.H. 2004b. Boas and beyond: migration and craniometric variation. *American Journal of Human Biology* 16:379–386.
Relethford, J. 2007. The use of quantitative traits in anthropological genetics studies of population structure and history. In: M.H. Crawford (ed.), *Anthropological Genetics: Theory, Methods and Applications.* Cambridge University Press. Cambridge, pp. 187–209.
Relethford, J.H. 2010. Population-specific deviations of global human craniometric variation from a neutral model. *American Journal of Physical Anthropology* 142:105–111.
Relethford, J.H. 2012. *Human Population Genetics.* Wiley-Blackwell, Hoboken, NJ.
Relethford, J.H. 2013. Understanding human cranial variation in light of modern human origins. In F.H. Smith, and J.C.M. Ahern (eds.), *The Origins of Modern Humans: Biology Reconsidered.* Wiley-Blackwell, Hoboken, NJ, pp. 321–338.
Relethford, J.H., and J. Blangero. 1990. Detection of differential gene flow from patterns of quantitative variation. *Human Biology* 62(1):5–25.
Relethford, J.H., and M.H. Crawford. 2013. Genetic drift and the population history of the Irish travellers. *American Journal of Physical Anthropology* 150:184–189.
Relethford, J., and H. Harpending. 1994. Craniometric variation, genetic theory, and modern human origins. *American Journal of Physical Anthropology* 95:249–270.
Reyes-Centeno, H., S. Ghirotto, F. Détroit, et al. 2014. Genomic and cranial phenotype data support multiple modern human dispersals from Africa and a southern route into Asia, *Proceedings of the National Academy of Sciences USA* 111(20):7248–7253.
Reynolds, E. 1943. Degree of kinship and pattern of ossification. *American Journal of Physical Anthropology* 1:405–416.
Reynolds, E., and G. Schoen. 1947. Growth patterns of identical triplets from 8 through 18 years. *Child Development* 18:130–151.
Rhesus Macaque Genome Sequencing and Analysis Consortium, R.A. Gibbs, J. Rogers, et al. 2007. Evolutionary and biomedical insights from the rhesus macaque genome. *Science* 316:222–234.
Riba-Hernández, P., K.E. Stoner, and P.W. Lucas. 2005. Sugar concentration of fruits and their detection via color in the Central American spider monkey (*Ateles geoffroyi*). *American Journal of Primatology* 67:411–23.

Richards, M., H. Corte-Real, P. Forster, et al. 1996. Paleolithic and neolithic lineages in the European mitochondrial gene pool. *American Journal of Human Genetics* 59(1):185–203.

Richards, M., V. Macaulay, E. Hickey, et al. 2000. Tracing European founder lineages in the Near Eastern mtDNA pool. *American Journal of Human Genetics* 67(5):1251–1276.

Richerson, P.J., R. Boyd, and J. Henrich. 2010. Gene-culture coevolution in the age of genomics. *Proceedings of the National Academy of Sciences* 107(Suppl. 2):8985–8992.

Richman, D.D., T. Wrin, S.J. Little, et al. 2003. Rapid evolution of the neutralizing antibody response to HIV type 1 infection. *Proceedings of the National Academy of Sciences of the USA* 100(7):4144–4149. doi:10.1073/pnas.0630530100

Richter, D., R. Grün, R. Joannes-Boyau, et al. 2017. The age of the hominin fossils from Jebel Irhoud, Morocco, and the origins of the Middle Stone Age. *Nature* 546:293–296.

Rieux, A., A. Eriksson, M. Li, et al. 2014. Improved calibration of the human mitochondrial clock using ancient genomes. *Molecular Biology and Evolution* 31(10):2780–2792. doi:10.1093/molbev/msu222

Rifkin, S.A., D. Houle, J. Kim, et al. 2005. A mutation accumulation assay reveals a broad capacity for rapid evolution of gene expression. *Nature* 438:220–223.

Roach, J.C., G. Glusman, A.F.A. Smit, et al. 2010. Analysis of genetic inheritance in a family quartet by whole-genome sequencing. *Science* 328(5978):636–639. doi:10.1126/science.1186802

Robbins, A.H. 1991. *Biological Perspectives on Human Pigmentation*. Cambridge University Press, Cambridge.

Robert, C.P., J.M. Cornuet, J.M. Marin, et al. 2011. Lack of confidence in approximate Bayesian computation model choice. *Proceedings of the National Academy of Sciences of the USA* 108(37):15112–15117. doi:10.1073/pnas.1102900108

Roberts, D.F. 1953. Body weight, race and climate. *American Journal of Physical Anthropology* 11(4):533–558.

Roberts, D.F. 1956. A demographic study of a Dinka village. *Human Biology* 28:323–349.

Roberts, D.F. 1968. Genetic effects of population size reduction. *Nature* 220:1084–1088.

Roberts, D.F. 1973. Anthropological genetics: problems and pitfalls. In: M.H. Crawford, and P.L. Workman (eds.), *Methods and Theories of Anthropological Genetics*. University of New Mexico Press, Albuquerque, p. 2.

Roberts, D.F. 1978. *Climate and Human Variability*. Cummings Pub. Co., Menlo Park, CA.

Roberts, D.F., and R.W. Hiorns. 1962. The dynamics of racial admixture. *American Journal of Human Genetics* 14:261–277.

Rockman, M.V., M.W. Hahn, N. Soranzo, et al. 2004. Positive selection on MMP3 regulation has shaped heart disease risk. *Current Biology* 14(17):1531–1539.

Rodahl, K. 1952. Basal metabolism of the Eskimo. *Journal of Nutrition* 48(3):359–368.

Rode, A., and R.J. Shephard. 1995. Modernization of lifestyle, body fat content and body fat distribution: a comparison of Igloolik Inuit and Volochanka Nganasan. *International Journal of Obesity and Related Metabolic Disorders* 19(10):709–716.

Rodríguez, R.L., J.D. Cramer, C.A. Schmitt, et al. 2015. The static allometry of sexual and non-sexual traits in vervet monkeys. *Biological Journal of the Linnean Society of London* 114:527–537.

Rodriguez-Paredes, M., and M. Esteller. 2011. Cancer epigenetics reaches mainstream oncology. *Nature Medicine* 17(3):330–339.

Roewer, L., M. Nothnagel, L. Gusmão, et al. 2013. Continent-wide decoupling of Y-chromosomal genetic variation from language and geography in Native South Americans. *PLoS Genetics* 9:e1003460.

Roff, D.A. 1995. The estimation of genetic correlations from phenotypic correlations: a test of Cheverud's conjecture. *Heredity* 74:481–490.

Rogaev, E.I., A.P. Grigorenko, Y.K. Moliaka, et al. 2009. Genomic identification in the historical case of Nicholas II royal family. *Proceedings of the National Academy of Sciences* 106(13):5258–5263.

Rogers, A.R. 1988. Three components of genetic drift in subdivided populations. *American Journal of Physical Anthropology* 77:435–449. doi:10.1002/ajpa.1330770405

Rogers, A.R., and H.C. Harpending. 1986. Migration and genetic drift in human populations. *Evolution* 40:1312–1327. doi:10.2307/2408956

Rogers, A., D. Iltis, and S. Wooding. 2004. Genetic variation at the *MC1R* locus and the time since loss of human body hair. *Current Anthropology* 45(1):105–108.

Rogers, A., and A. Prügel-Bennet. 2000. Evolving populations with overlapping generations. *Theoretical Population Biology* 57:121–129. doi:10.1006/tpbi.1999.1446

Rogers, J., and R.A. Gibbs. 2014. Comparative primate genomics: emerging patterns of genome content and dynamics. *Nature Reviews Genetics* 15:347–359.

Rohland, N., and M. Hofreiter. 2007. Comparison and optimization of ancient DNA extraction. *Biotechniques* 42:343–352. PMID: 17390541

Rohland, N., H. Siedel, and M. Hofreiter. 2010. A rapid column-based ancient DNA extraction method for increased sample throughput. *Molecular Ecology Resources* 10:677–683. doi:10.1111/j.1755-0998.2009.02824.x PMID: 21565072

Rohlf, F.J. 2000. Phylogenetic models and reticulations. *Journal of Classification* 17:185–189.

Romagnani, S. 1996. Th1 and Th2 in human diseases. *Clinical Immunology and Immunopathology* 80(3 Pt 1):225–235.

Romagnani, S. 1997. The Th1/Th2 paradigm. *Immunology Today* 18(6):263–266.

Romero, I.G., A. Manica, J. Goudet, et al. 2009. How accurate is the current picture of human genetic variation? *Heredity* 102:120–126.

Romero, I.G., I. Ruvinsky, and Y. Gilad. 2012. Comparative studies of gene expression and the evolution of gene regulation. *Nature Reviews Genetics* 13:505–516.

Roseman, C.C. 2004. Detecting interregionally diversifying natural selection on modern human cranial form by using matched molecular and morphometric data. *Proceedings of the National Academy of Sciences* 101(35):12824–12829.

Roseman, C.C. 2016. Random genetic drift, natural selection, and noise in human cranial evolution. *American Journal of Physical Anthropology* 160(4):582–592.

Roseman, C.C., and B.M. Auerbach. 2015. Ecogeography, genetics, and the evolution of human body form. *Journal of Human Evolution* 78:80–90.

Roseman, C.C., and T.D. Weaver. 2004. Multivariate apportionment of global human craniometric diversity. *American Journal of Physical Anthropology* 125(3):257–263.

Roseman, C.C., and T.D. Weaver. 2007. Molecules versus morphology? Not for the human cranium. *BioEssays* 29:1185–1188.

Roseman, C.C., K.E. Willmore, J. Rogers, et al. 2010. Genetic and environmental contributions to variation in baboon cranial morphology. *American Journal of Physical Anthropology* 143:1–12.

Rosenberg, N.A., J.K. Pritchard, J.L. Weber, et al. 2002. Genetic structure of human populations. *Science* 298(5602):2381–2385. doi:10.1126/science.1078311

Rosmond, R., Y.C. Chagnon, G. Holm, et al. 2000. Hypertension in obesity and the leptin receptor gene locus. *Journal of Clinical Endocrinology and Metabolism* 85(9):3126–3131.

Royal, C.D., J. Novembre, S.M. Fullerton, et al. 2010. Inferring genetic ancestry: opportunities, challenges, and implications. *American Journal of Human Genetics* 86(5):661–673. doi:10.1016/j.ajhg.2010.03.011

Rubin, P. 2004. Indian givers: the Havasupai trusted the white man to help with a diabetes epidemic. Instead, *ASU* tricked them into bleeding for academia. Phoenix New Times, 1–15.

Ruff, C.B. 1991. Climate and body shape in hominid evolution. *Journal of Human Evolution* 21(2):81–105.

Ruff, C.B. 1994. Morphological adaptation to climate in modern and fossil hominids. *Yearbook of Physical Anthropology* 37:65–107.

Ruiz-Linares, A., D. Ortíz-Barrientos, M. Figueroa, et al. 1999. Microsatellites provide evidence for Y chromosome diversity among the founders of the New World. *Proceedings of the National Academy of Sciences of the USA* 96:6312–6317.

Runcie, D.E., R.T. Wiedmann, E.A. Archie, et al. 2013. Social environment influences the relationship between genotype and gene expression in wild baboons. *Philosophical Transactions of the Royal Society B: Biological Sciences* 368:20120345.

Russo, V.E.A., R.A. Martienssen, and A.D. Riggs. 1996. *Epigenetic Mechanisms of Gene Regulation*. Cold Spring Harbor Laboratory Press, Plainview, NY.

Sabeti, P.C., P. Varilly, B. Fry, et al. 2007. Genome-wide detection and characterization of positive selection in human populations. *Nature* 449(7164):913–918. doi:10.1038/nature06250

Sabeti, P.C., E. Walsh, S.F. Schaffner, et al. 2005. The case for selection at CCR5-Delta32. *PLoS Biology* 3(11):e378.

Sage, E. 1993. Distribution and repair of photolesions in DNA: genetic consequences and the role of sequence context. *Photochemistry and Photobiology* 57(1):163–164.

Saglican, E., E. Ozkurt, H. Hu, et al. 2014. Heterochrony explains convergent testis evolution in primates. bioRxiv. http://biorxiv.org/content/early/2014/10/21/010553.abstract

Saillard, J., P. Forster, N. Lynnerup, et al. 2000. mtDNA variation among Greenland Eskimos: the edge of the Beringian expansion. *American Journal of Human Genetics* 67:718–726.

Saito, M., Y. Okamatsu-Ogura, M. Matsushita, et al. 2009. High incidence of metabolically active brown adipose tissue in healthy adult humans: effects of cold exposure and adiposity. *Diabetes* 58(7):1526–1531.

Saitou, N., and M. Nei. 1987. The neighbor-joining method: a new method for reconstructing phylogenetic trees. *Molecular Biology and Evolution* 4:406–425.

Salo, W.L., A.C. Aufderheide, J. Buikstra, et al. 1994. Identification of *Mycobaterium tuberculosis* DNA in a pre-Columbian Peruvian mummy. *Proceedings of the National Academy of Sciences of the USA* 91:2091–2094.

Salzano, F.M. 1972. Visual acuity and color blindness among Brazilian Cayapo Indians. *Human Heredity* 22:72–79.

Salzano, F.M. 2011. The prehistoric colonization of the Americas: evidence and models. *Evolution Education and Outreach* 4:199–204.

Salzano, F., and S.M. Callegari-Jacques. 1988. *South American Indians: A Case Study in Evolution*. Clarendon Press, Oxford.

Salzano, F.M., H. Gershowitz, H. Mohrenweiser, et al. 1986. Gene flow across tribal barriers and its effect among the Amazonian Içana River Indians. *American Journal of Physical Anthropology* 69:3–14.

Sambrook, J.G., A. Bashirova, H. Andersen, et al. 2006. Identification of the ancestral killer immunoglobulin-like receptor gene in primates. *BMC Genomics* 7(1):209.

Sambrook, J.G., A. Bashirova, S. Palmer, et al. 2005. Single haplotype analysis demonstrates rapid evolution of the killer immunoglobulin-like receptor (KIR) loci in primates. *Genome Research* 15(1):25–35. doi:10.1101/gr.2381205

Samonte, R.V., and E.E. Eichler. 2002. Segmental duplications and the evolution of the primate genome. *Nature Reviews Genetics* 3:65–72.

Sánchez-Quinto, F., L.R. Botigué, S. Civit, et al. 2012. North African populations carry the signature of admixture with Neandertals. *PLoS One* 7:e47765.

Sander, C.S., H. Chang, F. Hamm, et al. 2004. Role of oxidative stress and the antioxidant network in cutaneous carcinogenesis. *International Journal of Dermatology* 43(5):326–335. doi:10.1111/j.1365-4632.2004.02222.x

Sanders, S.J., M.T. Murtha, A.R. Gupta, et al. 2012. *De novo* mutations revealed by whole-exome sequencing are strongly associated with autism. *Nature* 485(7397):237–241. doi:10.1038/nature10945

Sanderson, J., H. Sudoyo, T.M. Karafet, et al. 2015. Reconstructing past admixture processes from local genomic ancestry using wavelet transformation. *Genetics* 200(2):469–481. doi:10.1534/genetics.115.176842

Sanger, F., S. Micklen, and A.R. Coulson. 1977. DNA sequencing and chain-terminating inhibitors. *Proceedings of the National Academy of Sciences USA* 74:5463–5467.

Sankararaman, S., S. Mallick, M. Dannemann, et al. 2014. The genomic landscape of Neanderthal ancestry in present-day humans. *Nature* 507(7492):354–357.

Sankararaman, S., S. Mallick, N. Patterson, et al. 2016. The combined landscape of Denisovan and Neanderthal ancestry in present-day humans. *Current Biology* 26(9):1241–1247.

Sankararaman, S., N. Patterson, H. Li, et al. 2012. The date of interbreeding between Neandertals and modern humans. *PLoS Genetics* 8(10):e1002947.

Santos, F.R., S.L. Bonatto, and M.C. Bortolini. 2007. Molecular evidence from contemporary indigenous populations to the peopling of America. In: C. Santos, and M. Lima (eds.), *Recent Advances in Molecular Biology and Evolution: Applications to Biological Anthropology*. Research Signpost, Kerala, pp. 1–13.

Santos, C., R. Montiel, B. Sierra, et al. 2005. Understanding differences between phylogenetic and pedigree-derived mtDNA mutation rate: a model using families from the Azores Islands (Portugal). *Molecular Biology and Evolution* 22(6):1490–1505. doi:10.1093/molbev/msi141

Santos, F.R., A. Pandya, C. Tyler-Smith, et al. 1999. The Central Siberian origin for Native American Y chromosomes. *American Journal of Human Genetics* 64:619–628.

Santos, F.R., L. Rodriguez-Delfin, S.D. Pena, et al. 1996. North and South Amerindians may have the same major founder Y chromosome haplotype. *American Journal of Human Genetics* 58:1369–1370.

Sarata, A.K. 2008. Genetic ancestry testing. In: *CRS Report for Congress*. Congressional Research Service, Washington, DC.

Sardi, M.L., P.S. Novellino, and H.M. Pucciarelli. 2006. Craniofacial morphology in the Argentine Center-West: consequences of the transition to food production. *American Journal of Physical Anthropology* 130:333–343.

Sarich, V.M., and A.C. Wilson. 1967. Immunological time scale for hominid evolution. *Science* 158(3805):1200–1203. doi:10.1126/science.158.3805.1200

Sarrafzadegan, N., and C. Gotay. 2015. CVD prevention in 2014: advances in the prevention of cardiovascular disease. *Nature Reviews Cardiology* 12(2):71–73.

Sawyer, S., J. Krause, K. Guschanski, et al. 2012. Temporal patterns of nucleotide misincorporations and DNA fragmentation in ancient DNA. *PLoS One* 7(3):e34131. doi:10.1371/journal.pone.0034131

Sawyer, S., G. Renaud, B. Viola, et al. 2015. Nuclear and mitochondrial DNA sequences from two Denisovan individuals. *Proceedings of the National Academy of Sciences* 112(51):15696–15700.

Sazzini, M., G. Schiavo, S. De Fanti, et al. 2014. Searching for signatures of cold adaptations in modern and archaic humans: hints from the brown adipose tissue genes. *Heredity (Edinburgh)* 113(3):259–267.

Scally, A., and R. Durbin. 2012. Revising the human mutation rate: implications for understanding human evolution. *Nature Reviews Genetics* 13(10):745–753. doi:10.1038/nrg3353

Scally, A., J.Y. Dutheil, L.W. Hillier, et al. 2012. Insights into hominid evolution from the gorilla genome sequence. *Nature* 483:169–175.

Schad, J., S. Sommer, and J.U. Ganzhorn. 2004. MHC variability of a small lemur in the littoral forest fragments of Southeastern Madagascar. *Conservation Genetics* 5(3):299–309.

Scheinfeldt, L.B., S. Soi, S. Thompson, et al. 2012. Genetic adaptation to high altitude in the Ethiopian highlands. *Genome Biology* 13(1):R1.

Schiffels, S., and R. Durbin. 2014. Inferring human population size and separation history from multiple genome sequences. *Nature Genetics* 46(8):919–925.

Schillaci, M.A. 2008. Human cranial diversity and evidence for an ancient lineage of modern humans. *Journal of Human Evolution* 54:814–826.

Schlossberg, D. (ed.). 2008. *Clinical Infectious Disease.* Cambridge University Press, Cambridge.

Schmitz, R.W., D. Serre, G. Bonani, et al. 2002. The Neandertal type site revisited: interdisciplinary investigations of skeletal remains from the Neander Valley, Germany. *Proceedings of the National Academy of Sciences* 99(20):13342–13347.

Schneider, E., N. El Hajj, S. Richter, et al. 2014. Widespread differences in cortex DNA methylation of the "language gene" *CNTNAP2* between humans and chimpanzees. *Epigenetics* 9(4):533–545.

Scholander, P.F., H.T. Hammel, J.S. Hart, et al. 1958. Cold adaptation in Australian aborigines. *Journal of Applied Physiology* 13(2):211–218.

Schreiber, V., F. Dantzer, J.C. Ame, et al. 2006. Poly(ADP-ribose): novel functions for an old molecule. *Nature Reviews Molecular Cell Biology* 7(7):517–528.

Schreider, E. 1950. Geographical distribution of the body-weight/body-surface ratio. *Nature* 165(4190):286.

Schreider, E. 1964. Ecological rules body-heat regulation + human-evolution. *Evolution* 18(1):1–9.

Schroeder, K.B., T.G. Schurr, J.C. Long, et al. 2007. A private allele ubiquitous in the Americas. *Biology Letters* 3:218–223.

Schuenemann, V.J., K. Bos, S. DeWitte, et al. 2011. Targeted enrichment of ancient pathogens yielding the pPCP1 plasmid of *Yersinia pestis* from victims of the Black Death. *Proceedings of the National Academy of Sciences of the USA* 108(38):E746–E752. doi:10.1073/pnas.1105107108

Schuenemann, V.J., P. Singh, T.A. Mendum, et al. 2013. Genome-wide comparison of medieval and modern *Mycobacterium leprae*. *Science* 341(6142):179–183. doi:10.1126/science.1238286

Schull, W.J.R.F. (ed.). 1990. *The Aymara: Strategies in Human Adaptation to a Rigorous Environment*. Kluwer Academic Publishers, Boston.

Schurr, T.G., and S.T. Sherry. 2004. Mitochondrial DNA and the peopling of the Americas: evolutionary and demographic evidence. *American Journal of Human Biology* 16:420–439.

Schurr, T.G., R.I. Sukernik, Y.B. Starikovskaya, et al. 1999. Mitochondrial DNA variation in Koryaks and Itel'men: population replacement in the Okhotsk Sea-Bering Sea region during the Neolithic. *American Journal of Physical Anthropology* 108:1–39.

Schwartz, J.H., and I. Tattersall. 2010. Fossil evidence for the origin of Homo sapiens. *American Journal of Physical Anthropology* 143:94–121.

Schwartz, J.H., I. Tattersall, and Z. Chi. 2014. Comment on "A complete skull from Dmanisi, Georgia, and the evolutionary biology of early Homo". *Science* 344(6182):360.

Scliar, M.O., M.H. Gouveia, A. Benazzo, et al. 2014. Bayesian inferences suggest that Amazon Yunga natives diverged from Andeans less than 5000 ybp: implications for South American prehistory. *BMC Evolutionary Biology* 14(1):174.

Scott. 2012. 23andMe's newest feature explores your ancestry. http://blog.23andme.com/ancestry/23andmes-newest-feature-explores-your-ancestry

Scott, S., and C.J. Duncan. 1998. *Human Demography and Disease*. Cambridge University Press, Cambridge.

Scwartz, C.R., and R.D. Mare. 2005. Trends in educational assortative marriage from 1940 to 2003. *Demography* 42(4):621–646.

Seguin-Orlando, A., T.S. Korneliussen, M. Sikora, et al. 2014. Paleogenomics. Genomic structure in Europeans dating back at least 36,200 years. *Science* 346(6213):1113–1118.

Ségurel, L., E.E. Thompson, T. Flutre, et al. 2012. The ABO blood group is a trans-species polymorphism in primates. *Proceedings of the National Academy of Sciences of the USA* 109:18493–18498.

Ségurel, L., M.J. Wyman, and M. Przeworski. 2014. Determinants of mutation rate variation in the human germline. *Annual Review of Genomics and Human Genetics* 15(1):47–70. doi:10.1146/annurev-genom-031714-125740

Seielstad, M., N. Yuldasheva, N. Singh, P. Underhill, et al. 2003. A novel Y- chromosome variant puts an upper limit on the timing of first entry into the Americas. *American Journal of Human Genetics* 73:700–705.

Sellis, D., B.J. Callahan, D.A. Petrov, et al. 2011. Heterozygote advantage as a natural consequence of adaptation in diploids. *Proceedings of the National Academy of Sciences* 108(51):20666–20671.

Seong, K.H., D. Li, H. Shimizu, et al. 2011. Inheritance of stress-induced, ATF-2-dependent epigenetic change. *Cell* 145(7):1049–1061.

Serre, D., A. Langaney, M. Chech, et al. 2004. No evidence of Neandertal mtDNA contribution to early modern humans. *Plos Biology* 2(3):313–317.

Shapiro, B., A.J. Drummond, A. Rambaut, et al. 2004. Rise and fall of the Beringian Steppe Bison. *Science* 306:1561–1565. doi:10.1126/science.1101074

Sharp, P., and B. Hahn. 2011. Origins of HIV and the AIDS pandemic. *Cold Springs Harbor Perspectives in Medicine* 1:1–22. doi:10.1101/cshperspect.a006841

Sharp, P.M., J.C. Rayner, and B.H. Hahn. 2013. Evolution. Great apes and zoonoses. *Science* 340(6130):284–286. doi:10.1126/science.1236958

Sharpe, L.T., A. Stockman, H. Jägle, et al. 1999. Opsin genes, cone photopigments, color vision, and color blindness. In: K.R. Gegenfurtner, and L.T. Sharpe (eds.), *Color Vision: From Genes to Perception*. Cambridge University Press, Cambridge, 3–52.

Shea, J.J. 2013. Transitions or turnovers? Climatically-forced extinctions of *Homo sapiens* and Neanderthals in the East Mediterranean Levant. *Quaternary Science Reviews* 27:2253–2270.

Sherry, S.T., and M.A. Batzer. 1997. Modelling human evolution – to tree or not to tree? *Genome Research* 7:947–949.

Sherwood, R.J., D.L. Duren, L.M. Havill, et al. 2008. A genomewide linkage scan for quantitative trait loci influencing the craniofacial complex in baboons (*Papio hamadryas* spp.). *Genetics* 180:619–628.

Sherwood, R.J., D.L. Duren, M.C. Mahaney, et al. 2011. A genome-wide linkage scan for quantitative trait loci influencing the craniofacial complex in humans (*Homo sapiens sapiens*). *The Anatomical Record* 294:664–675.

Shi, P., and J. Zhang. 2006. Contrasting modes of evolution between vertebrate sweet/umami receptor genes and bitter receptor genes. *Molecular Biology and Evolution* 23:292–300.

Shibata, Y., N.C. Sheffield, O. Fedrigo, et al. 2012. Extensive evolutionary changes in regulatory element activity during human origins are associated with altered gene expression and positive selection. *PLoS Genetics* 8:e1002789.

Shigemura, N., S. Shirosaki, K. Sanematsu, et al. 2009. Genetic and molecular basis of individual differences in human umami taste perception. *PLoS One* 4:e6717.

Shiio, Y., and R.N. Eisenman. 2003. Histone sumoylation is associated with transcriptional repression. *Proceedings of the National Academy of Sciences of the USA* 100(23):13225–13230.

Shimkin, D.B., and R.M. Reid. 1970. Socio-cultural persistence among Shoshoneans of the Carson River Basin, Nevada. In: E.H. Swanson, Jr. (ed.), *Languages and Cultures of Western North America. Essays in Honor of Sven S. Liljeblad*. Idaho State University Press, Pocatello, pp. 172–200.

Shrey, K., A. Suchit, M. Nishant, et al. 2009. RNA interference: emerging diagnostics and therapeutics tool. *Biochemical and Biophysical Research Communications* 386(2):273–277.

Shriver, M.D., and E.J. Parra. 2000. Comparison of narrow-band reflectance spectroscopy and tristimulus colorimetry for measurements of skin and hair color in persons of different biological ancestry. *American Journal of Physical Anthropology* 112(1):17–27. doi:10.1002/(SICI)1096-8644(200005)112:1<17::AID-AJPA3>3.0.CO;2-D

Shriver, M.D., E.J. Parra, S.S. Dios, et al. 2003. Skin pigmentation, biogeographical ancestry and admixture mapping. *Human Genetics* 112(4):387–399. doi:10.1007/s00439-002-0896-y

Sidebottom, D.A., R. Kennedy, and H. William. 2001. Class I MHC expression in the yellow baboon. *Journal of Immunology* 166:3983–3993. doi:10.4049/jimmunol.166.6.3983

Sigurðardóttir, S., A. Helgason, J.R. Gulcher, et al. 2000. The mutation rate in the human mtDNA control region. *American Journal of Human Genetics* 66(5):1599–1609. doi:10.1086/302902

Sikora, M., A. Seguin-Orlando, V.C. Sousa, et al. 2017. Ancient genomes show social and reproductive behavior of early upper Paleolithic foragers. *Science* 358(6363):659.

Silventoinen, K., J. Kaprio, E. Lahelma, et al. 2003. Assortative mating by body height and BMI: Finnish twins and their spouses. *American Journal of Human Biology* 15(5):620–627.

Simmen, B. 1994. Taste discrimination and diet differentiation among New World primates. In: D.J. Chivers, and P. Langer (eds.), *The Digestive System in Mammals: Food, Form, and Function*. Cambridge University Press, New York, pp. 150–165.

Simmen, B., and C.M. Hladik. 1998. Sweet and bitter taste discrimination in primates: scaling effects across species. *Folia Primatologica* 69:129–138.

Simmen, B., and A. Hladik. 1999. Taste discrimination in lemurs and other primates, and the relationships to distribution of plant allelochemicals in different habitats of Madagascar. In: B. Rakotosamimanana, H. Rasamimanana, J.U Ganzhorn, and S.M. Goodman (eds.), *New Directions in Lemur Studies*. Kluwer Academic/Plenum Publishers, New York, pp. 201–219.

Simonson, T.S., Y.Z. Yang, C.D. Huff, et al. 2010. Genetic evidence for high-altitude adaptation in Tibet. *Science* 329(5987):72–75.

Simonti C.N., B. Vernot, L. Bastarache, et al. 2016. The phenotypic legacy of admixture between modern humans and Neandertals. *Science* 351(6274):737.

Sjøvold, T. 1984. A report on the heritability of some cranial measurements and non-metric traits. In: G.N. Van Vark, and W.W. Howells (eds.), *Multivariate Statistical Methods in Physical Anthropology*. D. Reidel Publishing Co., Dordrecht, pp. 223–246.

Skoglund, P., and M. Jakobsson. 2011. Archaic human ancestry in East Asia. *Proceedings of the National Academy of Sciences of the USA* 108(45):18301–18306.

Skoglund, P., S. Mallick, M.C. Bortolini, et al. 2015. Genetic evidence for two founding populations of the Americas. *Nature* 525(7567):104–108. doi:10.1038/nature14895

Skoglund, P., H. Malmstrom, A. Omrak, et al. 2014. Genomic diversity and admixture differs for Stone-Age Scandinavian foragers and farmers. *Science* 344(6185):747–750.

Skoglund, P., H. Malmstrom, M. Raghavan, et al. 2012. Origins and genetic legacy of Neolithic farmers and hunter-gatherers in Europe. *Science* 336(6080):466–469.

Skoglund, P., B.H. Northoff, M.V. Shunkov, et al. 2014. Separating endogenous ancient DNA from modern day contamination in a Siberian Neandertal. *Proceedings of the National Academy of Sciences of the USA* 111(6):2229–2234.

Skoglund, P., J. Storå, A. Götherström, et al. 2013. Accurate sex identification fo ancient human remains using DNA shotgun sequencing. *Journal of Archaeological Science* 40(12):4477–4482.

Slatkin, M. 1995. A measure of population subdivision based on microsatellite allele frequencies. *Genetics* 139:457–462.

Slatkin, M., and G. Bertorelle. 2001. The use of intraallelic variability for testing neutrality and estimating population growth rate. *Genetics* 158(2):865–874.

Slatkin, M., and F. Racimo. 2016. Ancient DNA and human history. *Proceedings of the National Academy of Sciences of the USA* 113(23):6380–6387.

Slierendregt, B.L., N. Otting, N. Van Besouw, et al. 1994. Expansion and contraction of rhesus macaque DRB regions by duplication and deletion. *Journal of Immunology* 152:2298–2307.

Slon, V., C. Hopfe, C.L. Weiß, et al. 2017a. Neandertal and Denisovan DNA from Pleistocene sediments. *Science*.

Slon, V., B. Viola, G. Renaud, et al. 2017b. A fourth Denisovan individual. *Science Advances* 3(7):e170018.

Smith, H.F. 2009. Which cranial regions reflect molecular distances reliably in humans? Evidence from three-dimensional morphology. *American Journal of Human Biology* 21:36–47.

Smith, O., A.J. Clapham, P. Rose, et al. 2014. Genomic methylation patterns in archaeological barley show de-methylation as a time-dependent diagenetic process. *Scientific Reports* 4:5559.

Smith, N.H., J. Dale, J. Inwald, et al. 2003. The population structure of *Mycobacterium bovis* in great Britain: clonal expansion. *Proceedings of the National Academy of Sciences of the USA* 100(25):15271–15275.

Smith, M.W., M. Dean, M. Carrington, et al. 1997. Contrasting genetic influence of CCR2 and CCR5 variants on HIV-1 infection and disease progression. Hemophilia Growth and Development Study (HGDS), Multicenter AIDS Cohort Study (MACS), Multicenter Hemophilia Cohort Study (MHCS), San Francisco City Cohort (SFCC), ALIVE Study. *Science* 277(5328):959–965.

Smith, H.F., and F.E. Grine. 2008. Cladistic analysis of early *Homo* crania from Swartkrans and Sterkfontein, South Africa. *Journal of Human Evolution* 54:684–704.

Smith, J.M., and J. Haigh. 1974. The hitch-hiking effect of a favourable gene. *Genetics Research* 23(1):23–35. doi:10.1017/S0016672300014634

Smith, S.E., M.G. Hayes, G.S. Cabana, et al. 2009. Inferring population continuity versus replacement with aDNA: a cautionary tale from the Aleutian Islands. *Human Biology* 81:407–426.

Smith, F., I. Janković, and I. Karavanić. 2005. The assimilation model, modern human origins in Europe, and the extinction of Neanderthals. *Quaternary International* 137:7–19.

Smith, R.W.A., C. Monroe, and D.A. Bolnick. 2015. Detection of cytosine methylation in ancient DNA from five native American populations using bisulfite sequencing. *PLoS One* 10(5):e0125344. doi:10.1371/journal.pone.0125344

Smith, H.F., C.E. Terhune, and C.A. Lockwood. 2007. Genetic, geographic, and environmental correlates of human temporal bone variation. *American Journal of Physical Anthropology* 134:312–322.

Smith-Garvin, J.E., G.A. Koretzky, and M.S. Jordan. 2009. T cell activation. *Annual Review of Immunology* 27(1):591–619. doi:10.1146/annurev.immunol.021908.132706

Smouse, P.E. 2000. Reticulation inside the species boundary. *Journal of Classification* 17:165–173.

Smouse, P.E., and J.C. Long. 1992. Matrix correlation analysis in anthropology and genetics. *Yearbook of Physical Anthropology* 35:187–213.

Smouse, P.E., J.C. Long, and R.R. Sokal. 1986. Multiple regression and correlation extensions of the Mantel test of matrix correspondence. *Systematic Zoology* 35:627–632.

Sneath, P.H.A., and R.R. Sokal. 1973. *Numerical Taxonomy*. W.H. Freeman and Co., San Francisco.

Snodgrass, J.J., W.R. Leonard, M.V. Sorensen, et al. 2008. The influence of basal metabolic rate on blood pressure among indigenous Siberians. *American Journal of Physical Anthropology* 137(2):145–155.

Snodgrass, J.J., W.R. Leonard, L.A. Tarskaia, et al. 2005. Basal metabolic rate in the Yakut (Sakha) of Siberia. *American Journal of Human Biology* 17(2):155–172.

Snodgrass, J.J., W.R. Leonard, L.A. Tarskaia, et al. 2010. Impaired fasting glucose and the metabolic syndrome in an indigenous Siberian population. *International Journal of Circumpolar Health* 69(1):87–98.

Snodgrass, J.J., M.V. Sorensen, L.A. Tarskaia, et al. 2007. Adaptive dimensions of health research among indigenous Siberians. *American Journal of Human Biology* 19(2):165–180.

Soares, P., F. Alshamali, J.B. Pereira, et al. 2012. The expansion of mtDNA haplogroup L3 within and out of Africa. *Molecular Biology and Evolution* 29(3):915–927. doi:10.1093/molbev/msr245

Soares, P., L. Ermini, N. Thomson, et al. 2009. Correcting for purifying selection: an improved human mitochondrial molecular clock. *American Journal of Human Genetics* 84(6):740–759. doi:10.1016/j.ajhg.2009.05.001

Soejima, M., and Y. Koda. 2007. Population differences of two coding SNPs in pigmentation-related genes *SLC24A5* and *SLC45A2*. *International Journal of Legal Medicine* 121(1):36–39.

Soejima, M., H. Tachida, T. Ishida, et al. 2006. Evidence for recent positive selection at the human AIM1 locus in a European population. *Molecular Biology and Evolution* 23(1):179–188. doi:10.1093/molbev/msj018

Sokal, R.R., N.L. Oden, J. Walker, et al. 1997. Using distance matrices to choose between competing theories and an application to the origins of modern humans. *Journal of Human Evolution* 32:501–522.

Somel, M., H. Franz, Z. Yan, et al. 2009. Transcriptional neoteny in the human brain. *Proceedings of the National Academy of Sciences of the USA* 106:5743–5748.

Somel, M., X. Liu, L. Tang, et al. 2011. MicroRNA-driven developmental remodeling in the brain distinguishes humans from other primates. *PLoS Biology* 9:e1001214.

Song, D., L.S. Li, P.R. Arsenault, et al. 2014. Defective Tibetan PHD2 binding to p23 links high altitude adaption to altered oxygen sensing. *Journal of Biological Chemistry* 289(21):14656–14665.

Song, X., N. Mosby, J. Yang, et al. 2009. α-MSH activates immediate defense responses to UV-induced oxidative stress in human melanocytes. *Pigment Cell and Melanoma Research* 22(6):809–818.

Sontag, L.W. 1946. Biological and medical studies at the Samuel S. Fels Research Institute. *Child Development* 17(1–2):81–84.

Sontag, L., and G. Comstock. 1938. Striae in bones of a set of monozygotic triplets. *American Journal of Diseases of Children* 56:301–308.

Sontag, L., and V. Nelson. 1933. A study of identical triplets part I. Comparison of the physical and mental traits of a set of monozygotic dichorionic triplets. *Journal of Heredity* 24:473–480.

Sontag, L., and E. Reynolds. 1944. Ossification sequences in identical triplets. A longitudinal study of resemblences and differences in the ossification patterns of a set of monozygotic triplets. *Journal of Heredity* 35:57–64.

Sopper, S., C. Stahl-Hennig, M. Demuth, et al. 1997. Lymphocyte subsets and expression of differentiation markers in blood and lymphoid organs of rhesus monkeys. *Cytometry* 29(4):351–362.

Soranzo, N., B. Bufe, P.C. Sabeti, et al. 2005. Positive selection on a high-sensitivity allele of the human bitter-taste receptor *TAS2R16*. *Current Biology* 15:1257–1265.

Soria, R., C.G. Julian, E. Vargas, et al. 2013. Graduated effects of high-altitude hypoxia and highland ancestry on birth size. *Pediatric Research* 74(6):633–638.

Sousa, V., S. Peischl, and L. Excoffier. 2014. Impact of range expansions on current human genomic diversity. *Current Opinion in Genetics & Development* 29:22–30.

Spitze, K. 1993. Population structure in *Daphnia obtusa*: quantitative genetic and allozymic variation. *Genetics* 135:367–374.

Sprent, J.F.A. 1962. Parasitism, immunity and evolution. In: G.S. Leeper (ed.), *The Evolution of Living Organisms.* Melbourne University Press, Melbourne, pp. 149–165.

Sprent, J., and C.D. Surh. 2002. T cell memory. *Annual Review of Immunology* 20(1):551–579.

Spuhler, J.N. 1956 Estimation of mutation rates in man. *Clinical Orthopaedics* 8:34–43.

Spuhler, J.N. 1963 The scope for natural selection in man. In: W.J. Schull (ed.), *Genetic Selection in Man.* University of Michigan Press, Ann Arbor, pp. 1–111.

Spuhler, J. 1972. Genetic, linguistic and geographical distances in native North America. In: J. Weiner, and J. Huizinga (eds.), *The Assessment of Population Affinities in Man.* Oxford University Press, Oxford, pp. 73–95.

Spuhler, J.N., and P.J. Clark. 1961. Migration into the human breeding population of Ann Arbor, Michigan, 1900–1950. *Human Biology* 25:223–236.

Spuhler, J.N., and C. Kluckhohn. 1953. Inbreeding coefficients of the Ramah Navajo populations. *Human Biology* 25:295–317.

Stankiewicz, P., S.S. Park, K. Inoue, et al. 2001. The evolutionary chromosome translocation 4;19 in *Gorilla gorilla* is associated with microduplication of the chromosome fragment syntenic to sequences surrounding the human proximal CMT1A-REP. *Genome Research* 11:1205–1210.

Starikovskaya, E.B., R.I. Sukernik, O.A. Derbeneva, et al. 2005. Mitochondrial DNA diversity in indigenous populations of the southern extent of Siberia, and the origins of Native American haplogroups. *Annals of Human Genetics* 69:67–89.

Starikovskaya, Y.B., R.I. Sukernik, T.G. Schurr, et al. 1998. mtDNA diversity in Chukchi and Siberian Eskimos: implication for the genetic history of ancient Beringia and the peopling of the New World. *American Journal of Human Genetics* 63:1473–1491.

Stearns, S.C. 2012. Evolutionary medicine: its scope, interest and potential. *Proceedings Biological Sciences* 279(1746):4305–4521. doi:10.1098/rspb.2012.1326

Stearns, S.C., and D. Ebert. 2001. Evolution in health and disease: work in progress. *Quarterly Review of Biology* 76(4):417–432.

Stearns, S.C., R.M. Nesse, D.R. Govindaraju, et al. 2010. Evolution in health and medicine Sackler colloquium: evolutionary perspectives on health and medicine. *Proceedings of the National Academy of Sciences of the USA* 107(Suppl. 1):1691–1695. doi:10.1073/pnas.0914475107

Steckel, R.H., J.C. Rose, C.S. Larsen, et al. 2002. Skeletal health in the Western Hemisphere from 4000 BC to the present. *Evolutionary Anthropology* 11:142–155.

Steele, D.G., and J.F. Powell. 1993. Paleobiology of the first Americans. *Evolutionary Anthropology* 2:138–146.

Steinmann, G.G., F. Rosenkaimer, and G. Leitz. 2013. Clinical experiences with interferon-U and interferon-Y. *Cytokine-Induced Pathology: Inflammatory Cytokines, Receptors, and Diseases* 34:193.

Steiper, M.E., N.D. Wolfe, W.B. Karesh, et al. 2005. The population genetics of the alpha-2 globin locus of orangutans (*Pongo pygmaeus*). *Journal of Molecular Evolution* 60(3):400–408.

Steward, J.H. 1940. Native cultures of the intermontane (Great Basin) area. *Smithsonian Miscellaneous Collections* 100:445–502.

Stewart, O.C. 1939. The Northern Paiute Bands. *University of California Anthropological Records* 2:127–149.

Stojanowski, C.M. 2004. Population history of native groups in pre- and postcontact Spanish Florida: aggregation, gene flow, and genetic drift on the southeastern US.

Atlantic Coast. *American Journal of Physical Anthropology* 123:316–332. doi:10.1002/ajpa.10320

Stojanowski, C.M. 2005. Unhappy trails: forensic examination of ancient remains sheds new light on the emergence of Florida's Seminole Indians. *Natural History* 114(6):38–44.

Stokowski, R.P., P.V. Pant, T. Dadd, et al. 2007. A genomewide association study of skin pigmentation in a south Asian population. *American Journal of Human Genetics* 81(6):1119–1132. doi:10.1086/522235

Stone, A.C., G.R. Milner, S. Pääbo, et al. 1996. Sex determination of ancient human skeletons using DNA. *American Journal of Physical Anthropology* 99:231–238.

Stone, A.C., and M. Stoneking. 1998. mtDNA analysis of a prehistoric Oneota population: implications for the peopling of the New World. *American Journal of Human Genetics* 62:1153–1170.

Stoneking, M. 1994. In defense of "Eve" – a response to Templeton's critique. *American Anthropologist* 96:131–141.

Stoneking, M. 1994. Mitochondrial DNA and human evolution. *Journal of Bioenergetics and Biomembranes* 26:251–259.

Stoneking, M., S.T. Sherry, A.J. Redd, et al. 1992. New approaches to dating suggest a recent age for the human mtDNA ancestor. *Philosophical Transactions of the Royal Society B: Biological Sciences* 337(1280):167–175. doi:10.1098/rstb.1992.0094

Strachan, D.P. 1989. Hay fever, hygiene, and household size. *The BMJ* 299(6710): 1259–1260.

Strait, D.S., and F.E. Grine. 2004. Inferring hominoid and early hominid phylogeny using craniodental characters: the role of fossil taxa. *Journal of Human Evolution* 47:399–452.

Straus, L.G., D.J. Meltzer, and T. Goebel. 2005. Ice age Atlantis? Exploring the Solutrean-Clovis "Connection". *World Archaeology* 37:507–532.

Strauss, A., M. Hubbe, W.A. Neves, et al. 2015. The cranial morphology of the Botocudo Indians, Brazil. *American Journal of Physical Anthropology* 157:202–216.

Stringer, C. 2002. Modern human origins: progress and prospects. *Philosophical Transactions of the Royal Society of London. Series B, Biological Sciences* 357(1420):563–579.

Stringer, C. 2016. The origin and evolution of Homo sapiens. *Philosophical Transactions of the Royal Society B: Biological Sciences* 371:20150237.

Stringer, C.B., and P. Andrews. 1988. Genetic and fossil evidence for the origin of modern humans. *Science* 239:1263–1268.

Stringer, C.B., and I. Barnes. 2015. Deciphering the Denisovans. *Proceedings of the National Academy of Sciences* 112(51):15542–15543.

Stringer, C., and J. Galway-Witham. 2017. On the origin of our species. *Nature* 546: 212–214.

Stringer, C., and J. Galway-Witham. 2017. Palaeoanthropology: on the origin of our species. *Nature* 546(7657):212–214.

Stuart-Macadam, P. 1998. Iron deficiency anemia: exploring the difference. In: A.L. Grauer, and P. Stuart-Macadam (eds.), *Sex and Gender in Paleopathological Perspective*. Cambridge University Press, Cambridge, pp. 45–63.

Sturm, R.A., D.L. Duffy, Z.Z. Zhao, et al. 2008. A single SNP in an evolutionary conserved region within intron 86 of the *HERC2* gene determines human blue-brown eye color. *American Journal of Human Genetics* 82(2):424–431. doi:10.1016/j.ajhg.2007.11.005

Suárez, R., L.A. Borrero, K. Borrazzo, et al. 2014. Archaeological evidences are still missing: a comment on Fariña et al. Arroyo del Vizcaíno site, Uruguay. *Proceedings Biological Sciences* 281:20140449.

Subramanian, S. 2012. The abundance of deleterious polymorphisms in humans. *Genetics* 190:1579–1583.

Sulem, P., D.F. Gudbjartsson, S.N. Stacey, et al. 2007. Genetic determinants of hair, eye and skin pigmentation in Europeans. *Nature Genetics* 39(12):1443–1452. doi:10.1038/ng.2007.13

Sun, J.X., J.C. Mullikin, N. Patterson, et al. 2009. Microsatellites are molecular clocks that support accurate inferences about history. *Molecular Biology and Evolution* 26(5):1017–1027. doi:10.1093/Molbev/Msp025

Suomi, S.J. 2006. Risk, resilience, and gene × environment interactions in rhesus monkeys. *Annals of the New York Academy of Sciences* 1994:52–62.

Supply, P., R.M. Warren, A.L. Banuls, et al. 2003. Linkage disequilibrium between minisatellite loci supports clonal evolution of mycobacterium tuberculosis in a high tuberculosis incidence area. *Molecular Microbiology* 47(2):529–538.

Surridge, A.K., D. Osorio, and N.I. Mundy. 2003. Evolution and selection of trichromatic vision in primates. *Trends in Ecology & Evolution* 18:198–205.

Suzuki, M.M., and A. Bird. 2008. DNA methylation landscapes: provocative insights from epigenomics. *Nature Reviews Genetics* 9(6):465–476.

Suzuki, Y., and H.F. Nijhout. 2008. Genetic basis of adaptive evolution of a polyphenism by genetic accommodation. *Journal of Evolutionary Biology* 21(1):57–66.

Suzuki-Hashido, N., T. Hayakawa, A. Matsui, et al. 2015. Rapid expansion of phenylthiocarbamide non-tasters among Japanese macaques. *PLoS One* 10:e0132016.

Svensson, E., L. Olsen, C. Morck, et al. 2011. The adiponectin receptor homologs in *C. elegans* promote energy utilization and homeostasis. *PLoS One* 6(6):e21343.

Szabo, G., A.B. Gerald, M.A. Pathak, et al. 1969. Racial differences in the fate of melanosomes in human epidermis. *Nature* 222(5198):1081–1082.

Szathmáry, E.J.E. 1993. Genetics of aboriginal North Americans. *Evolutionary Anthropology* 1:202–220.

Szecsenyi-Nagy, A., G. Brandt, W. Haak, et al. 2015. Tracing the genetic origin of Europe's first farmers reveals insights into their social organization. *Proceedings Biological sciences/The Royal Society* 282(1805): pii: 20150339.

Tackney, J.C., B.A. Potter, J. Raff, et al. 2015. Two contemporaneous mitogenomes from terminal Pleistocene burials in eastern Beringia. *Proceedings of the National Academy of Sciences of the USA* 112:13833–13838.

Takahata, N., and Y. Satta. 1997. Evolution of the primate lineage leading to modern humans: phylogenetic and demographic inferences from DNA sequences. *Proceedings of the National Academy of Sciences of the USA* 94(9):4811–4815.

Takenaka, A., S. Nakamura, F. Mitsunaga, et al. 2012. Human-specific SNP in obesity genes, adrenergic receptor beta2 (*ADRB2*), Beta3 (*ADRB3*), and PPAR gamma2 (*PPARG*), during primate evolution. *PLoS One* 7(8):e43461.

TallBear, K. 2007. Narratives of race and indigeneity in the Genographic project. *The Journal of Law, Medicine and Ethics* 35(3):412–424. doi:10.1111/j.1748-720X.2007.00164.x

TallBear, K. 2013. Genomic articulations of indigeneity. *Social Studies of Science* 43:509–533.

TallBear, K. 2013. *Native American DNA: Tribal Belonging and the False Promise of Genetic Science*: University of Minnesota Press, Minneapolis.

Tamm, E., T. Kivisild, M. Reidla, et al. 2007. Beringian standstill and spread of Native American founders. *PLoS One* 2:e829.

Tamura, K., and M. Nei. 1993. Estimation of the number of nucleotide substitutions in the control region of mitochondrial DNA in humans and chimpanzees. *Molecular Biology and Evolution* 10(3):512–526.

Tan, J., Y. Yang, K. Tang, et al. 2013. The adaptive variant *EDARV370A* is associated with straight hair in East Asians. *Human Genetics* 132(10):1187–1191.

Tanner, J.M. 1962. *Growth at Adolescence*. Blackwell, Oxford.

Tanner, J., H. Goldstein, and R. Whitehouse. 1970. Standards for childrens height at ages 2-9 years allowing for height of the parents. *Archives of Disease in Childhood* 45:755–762.

Tarazona-Santos, E., and F.R. Santos. 2002. The peopling of the Americas: a second major migration? *American Journal of Human Genetics* 70:1377–1380.

Taubenberger, J.K., A.H. Reid, A.E. Krafft, et al. 1997. Initial genetic characterization of the 1918 "Spanish" influenza virus. *Science* 275:1793–1796.

Taubenberger, J.K., A.H. Reid, R.M. Lourens, et al. 2005. Characterization of the 1918 influenza virus polymerase genes. *Nature* 437:889–893.

Tavaré, S., D.J. Balding, R.C. Griffiths, et al. 1997. Inferring coalescence times from DNA sequence data. *Genetics* 145:505–518.

Tayles, N., Domett, K. and K. Nelsen. 2000. Agriculture and dental caries? The case of rice in prehistoric Southeast Asia. *World Archaeology* 32(1):68–83.

Taylor, K. 2011. Bitter fight to determine who is an American Indian turns to DNA testing – ICTMN.Com. Indian Country Today, October 13, pp. 1–6.

Teh, A.L., H. Pan, L. Chen, et al. 2014. The effect of genotype and *in utero* environment on interindividual variation in neonate DNA methylomes. *Genome Research* 24(7):1064–1074.

Teixeira de Filippo, J.C., C. Weihmann, A. Meneu, et al. 2014. Long-term balancing selection in *LAD1* maintains a missense trans-species polymorphism in humans, chimpanzees and bonobos. *bioRxiv*. doi:10.1101/006684

Temple, D.H., and C.S. Larsen. 2007. Dental caries prevalence as evidence for agriculture and subsistence variation during the Yayoi period in prehistoric Japan: biocultural interpretations of an economy in transition. *American Journal of Physical Anthropology* 134(4):501–512.

Templeton, A.R. 1993. The "eve" hypotheses: a genetic critique and reanalysis. *American Anthropologist* 95:51–72.

Templeton, A.R. 2005. Haplotype trees and modern human origins. *Yearbook of Physical Anthropology* 48 48:33–59.

Templeton, A.R. 2007. Genetics and recent human evolution. *Evolution* 61:1507–1519.

Templeton, A.R., and B. Read. 1994. Inbreeding: one word, several meanings, much confusion. In: V. Loeschcke, J. Tomiuk, and S.K. Jain (eds.), *Conservation Genetics*. Birkauser Verlag, Basel, pp. 91–105. doi:10.1002/zoo.1430030302

Tenesa, A., P. Navarro, B.J. Hayes, et al. 2007. Recent human effective population size estimated from linkage disequilibrium. *Genome Research* 17(4):520–526. doi:10.1101/gr.6023607

Terao, K., A. Mikami, A. Saito, et al. 2005. Identification of a protanomalous chimpanzee by molecular genetic and electroretinogram analyses. *Vision Research* 45:1225–1235.

Thayer, Z.M., and C.W. Kuzawa. 2011. Biological memories of past environments: epigenetic pathways to health disparities. *Epigenetics* 6(7):798–803.

Thayer, Z.M., and A. Non. 2015. Anthropology meets epigenetics: current and future directions. *American Anthropologist* 177(4):722–735.

Thomson, A., and L.H.D. Buxton. 1923. Man's nasal index in relation to certain climatic conditions. *Journal of the Royal Anthropological Institute of Great Britain and Ireland* 53:92–122.

Thong, H.Y., S.H. Jee, C.C. Sun, et al. 2003. The patterns of melanosome distribution in keratinocytes of human skin as one determining factor of skin colour. *British Journal of Dermatology* 149(3):498–505.

Tindale, N.B. 1953. Tribal and intertribal marriage among Australian Aborigines. *Human Biology* 25:169–190.

Tishkoff, S.A., and M.K. Gonder. 2007. Human origins within and out of Africa. In M.H. Crawford (ed.), *Anthropological Genetics: Theory, Methods and Applications.* Cambridge University Press, Cambridge, pp. 337–379.

Tishkoff, S.A., F.A. Reed, F.R. Friedlaender, et al. 2009. The genetic structure and history of Africans and African Americans. *Science* 324:1035–1044.

Tishkoff, S.A, F.A. Reed, A. Ranciaro, et al. 2007. Convergent adaptation of human lactase persistence in Africa and Europe. *Nature Genetics* 39(1):31–40.

Toda, Y., T. Nakagita, T. Hayakawa, et al. 2013. Two distinct determinants of ligand specificity in *T1R1/T1R3* (the umami taste receptor). *Journal of Biological Chemistry* 288:36863–36877.

Tollefsbol, T.O. 2011. Epigenetics: the new science of genetics. In: T.O. Tollefsbol (ed.), *Handbook of Epigenetics: The New Molecular and Medical Genetics.* Academic Press, Amsterdam, pp. 1–6.

Tong, Y.G., W.F. Shi, D. Liu, et al. 2015. Genetic diversity and evolutionary dynamics of Ebola virus in Sierra Leone. *Nature* 524(7563):93–96. doi:10.1038/nature14490

Torroni, A., T.G. Schurr, M.F. Cabell, et al. 1993a. Asian affinities and continental radiation of the four founding Native American mtDNAs. *American Journal of Human Genetics* 53:563–590.

Torroni, A., T.G. Schurr, C.-C. Yang, et al. 1992. Native American mitochodrial DNA analysis indicates that the Amerind and the Nadene populations were founded by two independent migrations. *Genetics* 130:153–162.

Torroni, A., R.I. Sukernik, T.G. Schurr, et al. 1993a. Mitochondrial DNA variation of aboriginal Siberians reveals distinct genetic affinities with native Americans. *American Journal Human Genetics* 53(3):563–590.

Torroni, A., R.I. Sukernik, T.G. Schurr, et al. 1993b. MtDNA variation of aboriginal Siberians reveals distinct genetic affinities with Native Americans. *American Journal of Human Genetics* 53:591–608.

Tosi, A.J., K.M. Detwiler, and T.R. Disotell. 2005. X-chromosomal window into the evolutionary history of the guenons (primates: Cercopithecini). *Molecular Phylogenetics and Evolution* 36:58–66.

Towne, B., S.A. Czerwinski, E.W. Demerath, et al. 2005. Heritability of age at menarche in girls from the Fels Longitudinal Study. *American Journal of Physical Anthropology* 128(1):210–219.

Towne, B., E.W. Demerath, and S.A. Czerwinski. 2012. The genetic epidemiology of growth and development. In: N. Cameron, and B. Bogin (eds.), *Human Growth and Development.* 2nd ed. Academic Press, London, pp. 173–223.

Towne, B., S.S. Guo, A.F. Roche, et al. 1993. Genetic analysis of patterns of growth in infant recumbent length. *Human Biology* 65(6):977–989.

Towne, B., J.S. Parks, M.R. Brown, et al. 2000. Effect of a luteinizing hormone b-subunit polymorphism on growth in stature. *Acta Medica Auxologica* 32(1):43–44.

Towne, B., K.D. Williams, J. Blangero, et al. 2008. Presentation, heritability, and genome-wide linkage analysis of the mid-childhood growth spurt in healthy children from the Fels Longitudinal Study. *Human Biology* 80:623–636.

Trefilov, A., J. Berard, M. Krawczak, et al. 2000. Natal dispersal in rhesus macaques is related to serotonin transporter gene promoter variation. *Behavior Genetics* 30:295–301.

Trejo-de la O, A., P. Hernández-Sancén, and C. Maldonado-Bernal. 2014. Relevance of single-nucleotide polymorphisms in human TLR genes to infectious and inflammatory diseases and cancer. *Genes and Immunity* 15(4):199–209. doi:10.1038/gene.2014.10

Treves, A., and P. Palmqvist. 2007. Reconstructing hominin interactions with mammalian carnivores (6.0–1.8 Ma). In: S. Gursky, and K.A.I. Nekaris (eds.), *Primate Anti-Predator Strategies*. Springer, New York, pp. 355–381.

Trinkaus, E. 1981. Neanderthal limb proportions and cold adaptation. In: C.B. Stringer (ed.), *Aspects of Human Evolution*. Taylor and Francis, London, pp. 187–224.

Tumpey, T.M., C.F. Basler, P.V. Aguilar, et al. 2005. Characterization of the reconstructed 1918 Spanish influenza pandemic virus. *Science* 310(5745):77–80.

Tung, J., S.C. Alberts, and G.A. Wray. 2010. Evolutionary genetics in wild primates: combining genetic approaches with field studies of natural populations. *Trends in Genetics* 26:353–362.

Tung, J., L.B. Barreiro, M.B. Burns, et al. 2015. Social networks predict gut microbiome composition in wild baboons. *eLife* 4. doi:10.7554/eLife.05224

Tung, J., M.J.E. Charpentier, S. Mukherjee, et al. 2012. Genetic effects on mating success and partner choice in a social mammal. *The American Naturalist* 180:113–129.

Tung, J., X. Zhou, S.C. Alberts, et al. 2015. The genetic architecture of gene expression levels in wild baboons. *Elife* 4. doi:10.7554/eLife.04729

Turkheimer, E. 2000. Three laws of behavior genetics and what they mean. *Current Directions in Psychological Science* 9(5):160–164.

Turkheimer, E. 2011. Still missing. *Research in Human Development* 8(3–4):227–241.

Udpa, N., R. Ronen, D. Zhou, et al. 2014. Whole genome sequencing of Ethiopian highlanders reveals conserved hypoxia tolerance genes. *Genome Biology* 15(2):R36.

Ulivi, S., M. Mezzavilla, and P. Gasparini. 2013. Genetics of eye colours in different rural populations on the Silk Road. *European Journal of Human Genetics* 21(11):1320–1323. doi:10.1038/ejhg.2013.41

Ullastres, A., M. Farré, L. Capilla, et al. 2014. Unraveling the effect of genomic structural changes in the rhesus macaque – implications for the adaptive role of inversions. *BMC Genomics* 15:530.

Underhill, P.A., L. Jin, R. Zemans, et al. 1996. A pre-Columbian Y chromosome-specific transition and its implications for human evolutionary history. *Proceedings of the National Academy of Sciences of the USA* 93:196–200.

Underhill, P.A., and T. Kivisild. 2007. Use of Y chromosome and mitochondrial DNA population structure in tracing human migrations. *Annual Review of Genetics* 41(1):539–564. doi:10.1146/annurev.genet.41.110306.130407

Underhill, P.A., G.D. Poznik, S. Rootsi, et al. 2015. The phylogenetic and geographic structure of Y-chromosome haplogroup R1a. *European Journal of Human Genetics* 23(1):124–131.

US Department of State Bureau of Consular Affairs. 2015. DNA relationship testing procedures. http://travel.state.gov/content/visas/english/immigrate/family/dna-test-procedures.html

Valenta, K., R.J. Burke, S.A. Styler, et al. 2013. Colour and odour drive fruit selection and seed dispersal by mouse lemurs. *Scientific Reports* 3:2424.

Valenzuela, R.K., M.S. Henderson, M.H. Walsh, et al. 2010. Predicting phenotype from genotype: normal pigmentation. *Journal of Forensic Sciences* 55(2):315–322. doi:10.1111/j.1556-4029.2009.01317.x

Valverde, P., E. Healy, I. Jackson, et al. 1995. Variants of the melanocyte-stimulating hormone receptor gene are associated with red hair and fair skin in humans. *Nature Genetics* 11(3):328–330. doi:10.1038/ng1195-328

Van Tuyl, R., and A. Pereltsvaig. 2012. Comment on "Phonemic diversity supports a serial founder effect model of language expansion from Africa". *Science* 335:657.

Van Valen, L. 1973. A new evolutionary law. *Evolutionary Theory* 1:1–30.

Vaquero, A., A. Loyola, and D. Reinberg. 2003. The constantly changing face of chromatin. *Science of Aging Knowledge Environment* 2003(14):RE4.

Varki, A., and T.K. Altheide. 2005. Comparing the human and chimpanzee genomes: searching for needles in a haystack. *Genome Research* 15:1746–1758.

Varki, A., C. Wills, D. Perlmutter, et al. 1998. Great ape phenome project? *Science* 282:239–240.

Vattathil, S., and J.M. Akey. 2015. Small amounts of archaic admixture provide big insights into human history. *Cell* 163(2):281–284.

van den Berghe, P.L., and P. Frost. 1986. Skin color preference, sexual dimorphism, and sexual selection: a case of gene culture co-evolution? *Ethnic and Racial Studies* 9:87–113.

van Dommelen, P., M.C. de Gunst, A.W. van der Vaart, et al. 2004. Genetic study of the height and weight process during infancy. *Twin Research and Human Genetics* 7(6):607–616.

Veeramah, K.R., and M.F. Hammer. 2014. The impact of whole-genome sequencing on the reconstruction of human population history. *Nature Reviews Genetics* 15(3):149–162. doi:10.1038/nrg3625

Veilleux, C.C, and D.A Bolnick. 2009. Opsin gene polymorphism predicts trichromacy in a cathemeral lemur. *American Journal of Primatology* 71:86–90.

Veilleux, C.C., R.L. Jacobs, M.E. Cummings, et al. 2013. Opsin genes and visual ecology in a nocturnal folivorous lemur. *International Journal of Primatology* 35:88–107.

Veilleux, C.C., and E.C. Kirk. 2014. Visual acuity in mammals: effects of eye size and ecology. *Brain, Behavior and Evolution* 83:43–53.

Veilleux, C.C., E.E. Louis, and D.A. Bolnick. 2013. Nocturnal light environments influence color vision and signatures of selection on the *OPN1SW* opsin gene in nocturnal lemurs. *Molecular Biology and Evolution* 30:1420–1437.

Venkataram, S., D. Sellis, and D.A. Petrov. 2015. Predictability of adaptive evolution under the successive fixation assumption. *bioRXiv*. http://biorxiv.org/content/early/2015/07/28/001016

Vernot, B., and J.M. Akey. 2014. Resurrecting surviving Neandertal lineages from modern human genomes. *Science* 343(6174):1017–1021.

Vernot, B., and J.M. Akey. 2015. Complex history of admixture between modern humans and Neandertals. *American Journal of Human Genetics* 96(3):448–453. doi:10.1016/j.ajhg.2015.01.006

Vernot, B., S. Tucci, J. Kelso, et al. 2016. Excavating Neandertal and Denisovan DNA from the genomes of Melanesian individuals. *Science* 17 March:aad9416. doi:10.1126/science.aad9416

Verrelli, B.C., C.M. Lewis, A.C. Stone, et al. 2008. Different selective pressures shape the molecular evolution of color vision in chimpanzee and human populations. *Molecular Biology and Evolution* 25:2735–2743.

Verrelli, B.C., and S.A. Tishkoff. 2004. Signatures of selection and gene conversion associated with human color vision variation. *American Journal of Human Genetics* 75:363–375.

Verrelli, B.C., S.A. Tishkoff, A.C. Stone, et al. 2006. Contrasting histories of G6PD molecular evolution and malarial resistance in humans and chimpanzees. *Molecular Biology and Evolution* 23(8):1592–1601.

Versini, M., P.Y. Jeandel, T. Bashi, et al. 2015. Unraveling the hygiene hypothesis of helminthes and autoimmunity: origins, pathophysiology, and clinical applications. *BMC Medicine* 13:81. doi:10.1186/s12916-015-0306-7

Veyrier, F., D. Pletzer, C. Turenne, et al. 2009. Phylogenetic detection of horizontal gene transfer during the step-wise genesis of *Mycobacterium tuberculosis. BMC Evolutionary Biology* 9:196. doi:10.1186/1471-2148-9-196

Vialou, D., M. Benabdelhadi, J. Feathers, et al. 2017. Peopling South America's Centre: the Late Pleistocene site of Santa Elina. *Antiquity* 91:865–884.

Vigilant, L., R. Pennington, H. Harpending, et al. 1989. Mitochondrial DNA sequences in single hairs from a southern African population. *Proceedings of the National Academy of Sciences* 86(23):9350–9354.

Vigilant, L., M. Stoneking, H. Harpending, et al. 1991. African populations and the evolution of human mitochondrial DNA. *Science* 253:1503–1507. doi:10.1126/science.1840702

Virtanen, K.A., M.E. Lidell, J. Orava, et al. 2009. Functional brown adipose tissue in healthy adults. *New England Journal of Medicine* 360(15):1518–1525.

Visscher, P.M., M.A. Brown, M.I. McCarthy, et al. 2012. Five years of GWAS discovery. *American Journal of Human Genetics* 90(1):7–24.

Visscher, P.M., W.G. Hill, and N.R. Wray. 2008. Heritability in the genomics era– concepts and misconceptions. *Nature Reviews Genetics* 9(4):255–266.

Vitzthum, V.J. 2003. A number no greater than the sum of its parts: the use and abuse of heritability. *Human Biology* 75:539–558.

Vivier, E., E. Tomasello, M. Baratin, et al. 2008. Functions of natural killer cells. *Nature Immunology* 9(5):503–510. doi:10.1038/ni1582

Vogel, T.U., D.T. Evans, J.A. Urvater, et al. 1999. Major histocompatibility complex class I genes in primates: co-evolution with pathogens. *Immunological Reviews* 167:327–337. doi:10.1111/j.1600-065X.1999.tb1402.x

Voight, B.F., S. Kudaravalli, X. Wen, et al. 2006a. A map of recent positive selection in the human genome. *PLoS Biology* 4(3):446.

Voight, B.F., S. Kudaravalli, X. Wen, et al. 2006b. A map of recent positive selection in the human genome. *PLoS Biology* 4(3):e72. doi:10.1371/journal.pbio.0040072

Volodko, N.V., E.B. Starikovskaya, I.O. Mazunin, et al. 2008. Mitochondrial genome diversity in Arctic Siberians, with particular reference to the evolutionary history of Beringia and Pleistocenic peopling of the Americas. *American Journal of Human Genetics* 82:1084–1100.

von Cramon-Taubadel, N. 2009a. Congruence of individual cranial bone morphology and neutral molecular affinity patterns in modern humans. *American Journal of Physical Anthropology* 140:205–215.

von Cramon-Taubadel, N. 2009b. Revisiting the homoiology hypothesis: the impact of phenotypic plasticity on the reconstruction of human population history from craniometric data. *Journal of Human Evolution* 57:179–190.

von Cramon-Taubadel, N. 2011. Global human mandibular variation reflects differences in agricultural and hunter-gatherer subsistence strategies. *Proceedings of the National Academy of Sciences, USA* 108:19546–19551.

von Cramon-Taubadel, N. 2014a. Evolutionary insights into global patterns of human cranial diversity: population history, climatic and dietary effects. *Journal of Anthropological Sciences* 92:43–77.

von Cramon-Taubadel, N. 2014b. The microevolution of modern human cranial variation: implications for hominin and primate evolution. *Annals of Human Biology* 41(4):323–335.

von Cramon-Taubadel, N., and S.J. Lycett. 2008. Brief communication: human cranial variation fits iterative founder effect model with African origin. *American Journal of Physical Anthropology* 136:108–113.

von Cramon-Taubadel, N., and S.J. Lycett. 2014. A comparison of catarrhine genetic distances with pelvic and cranial morphology: implications for determining hominin phylogeny. *Journal of Human Evolution* 77:179–186.

von Cramon-Taubadel, N., and R. Pinhasi. 2011. Craniometric data support a mosaic model of demic and cultural Neolithic diffusion to outlying regions of Europe. *Proceedings of the Royal Society, B* 278:2874–2880.

von Cramon-Taubadel, N., and H.F. Smith. 2012. The relative congruence of cranial and genetic estimates of hominoid taxon relationships: implications for the reconstruction of hominin phylogeny. *Journal of Human Evolution* 62:640–653.

von Cramon-Taubadel, N., J.T. Stock, and R. Pinhasi. 2013. Skull and limb morphology differentially track population history and environmental factors in the transition to agriculture in Europe. *Proceedings of the Royal Society B* 280:20131337.

von Cramon-Taubadel, N., and T.D. Weaver. 2009. Insights from a quantitative genetic approach to human morphological evolution. *Evolutionary Anthropology* 18:237–240.

Voutilainen, M., P.H. Lindfors, S. Lefebvre, et al. 2012. Ectodysplasin regulates hormone-independent mammary ductal morphogenesis via NF-κB. *Proceedings of the National Academy of Sciences* 109(15):5744–5749.

Waddington, C.H. 1942. Canalization of development and the inheritance of acquired characters. *Nature* 150(3811):563–565.

Waddington, C.H. 1953. Genetic assimilation of an acquired character. *Evolution* 7(2):118–126.

Waddington, C.H. 1956. Genetic assimilation of the bithorax phenotype. *Evolution* 10(1):1–13.

Waddington, C.H. 2011. The epigenotype. 1942. *International Journal of Epidemiology* 41(1):10–13.

Waddle, D.M. 1994. Matrix correlation tests support a single origin for modern humans. *Nature* 368:452–454.

Wagner, J.K., J.D. Cooper, R. Sterling, et al. 2012. Tilting at windmills no longer: a data-driven discussion of DTC DNA ancestry tests. *Genetics in Medicine* 14(6):586–93. doi:10.1038/gim.2011.77

Wagner, D.M., J. Klunk, M. Harbeck, et al. 2014. *Yersinia pestis* and the plague of Justinian 541-543 A.D.: a genomic analysis. *Lancet Infectious Diseases* 14(4):319–326. doi:10.1016/S1473-3099(13)70323-2

Waguespack, N.M. 2007. Why we're still arguing about the Pleistocene occupation of the Americas. *Evolutionary Anthropology* 16:63–74.

Wakamatsu, K., R. Kavanagh, A.L. Kadekaro, et al. 2006. Diversity of pigmentation in cultured human melanocytes is due to differences in the type as well as quantity of melanin. *Pigment Cell Research* 19(2):154–162. doi:10.1111/j.1600-0749.2006.00293.x

Wakeley, J. 2008a. *Coalescent Theory: An Introduction*. Roberts and Company Publishers, Greenwood Village, CO.

Wakeley, J. 2008b. Complex speciation of humans and chimpanzees. *Nature* 452(7184):E3–E4. doi:10.1038/nature06805

Wald, P. 2006. Blood and stories: how genomics is rewriting race, medicine and human history. *Patterns of Prejudice* 40(4–5):303–333. doi:10.1080/00313220601020064

Wall, J.D., M.P. Cox, F.L. Mendez, et al. 2008. A novel DNA sequence database for analyzing human demographic history. *Genome Research* 18:1354–1361. doi:10.1101/gr.075630.107

Wall, J.D., R. Jiang, C. Gignoux, et al. 2011. Genetic variation in Native Americans, inferred from Latino SNP and resequencing data. *Molecular Biology and Evolution* 28:2231–2237.

Wall, J.D., and S.K. Kim. 2007. Inconsistencies in Neanderthal genomic DNA sequences. *PLoS Genetics* 3.

Wall, J.D., M.A. Yang, F. Jay, et al. 2013. Higher levels of Neanderthal ancestry in east Asians than in Europeans. *Genetics* 194(1):199.

Walsh, G.P., E.V. Tan, E.C. dela Cruz, et al. 1996. The Philippine cynomolgus monkey (*Macaca fasicularis*) provides a new nonhuman primate model of tuberculosis that resembles human disease. *Nature Medicine* 2(4):430–436. doi:10.1038/nm0496-430

Wang, C.C., Q.L. Ding, H. Tao, et al. 2012. Comment on "Phonemic diversity supports a serial founder effect model of language expansion from Africa". *Science* 335:657.

Wang, Z., M. Gerstein, and M. Snyder. 2009. RNA-Seq: a revolutionary tool for transcriptomics. *Nature Reviews Genetics* 10(1):57–63.

Wang, S., C.M. Lewis, M. Jakobsson, et al. 2007. Genetic variation and population structure in Native Americans. *PLoS Genetics* 3:e185.

Wang, X., S.D. Thomas, and J. Zhang. 2004. Relaxation of selective constraint and loss of function in the evolution of human bitter taste receptor genes. *Human Molecular Genetics* 13:2671–2678.

Wang, B., Y.B. Zhang, F. Zhang, et al. 2011. On the origin of Tibetans and their genetic basis in adapting to high-altitude environments. *PLoS One* 6(2):e17002.

Ward, R.H. 1972. The genetic structure of a tribal population, the Yanomama Indians. V. Comparisons of a series of genetic networks. *Annals of Human Genetics* 36:21–43.

Ward, R.H., B.L. Frazier, K. Dew-Jager, et al. 1991. Extensive mitochondrial diversity within a single Amerindian tribe. *Proceedings of the National Academy of Sciences of the USA* 88:8720–8724.

Ward, R.H., and J.V. Neel. 1976. The genetic structure of a tribal population, the Yanomama Indians. XIV. Clines and their interpretation. *Genetics* 82(1):103–121.

Warinner, C., J.F.M. Rodrigues, R. Vyas, et al. 2014. Pathogens and host immunity in the ancient human oral cavity. *Nature Genetics* 46(4):336–344.

Warren, W.C., A.J. Jasinska, R. García-Pérez, et al. 2015. The genome of the vervet (*Chlorocebus aethiops sabaeus*). *Genome Research* 25(12):1921–1933. doi:10.1101/gr.192922.115

Washburn, S.L. 1951. The new physical anthropology. *Transactions of the New York Academy of Sciences Series II* 13:298–304.

Watanabe, T. 2009. Called back to Africa by DNA. Los Angeles Times, February 18. http://articles.latimes.com/2009/feb/18/local/me-africa18

Waterland, R.A., M. Travisano, and K.G. Tahiliani. 2007. Diet-induced hypermethylation at agouti viable yellow is not inherited transgenerationally through the female. *FASEB Journal* 21(12):3380–3385.

Waters, M.R., S.L. Forman, T.A. Jennings, et al. 2011. The Buttermilk Creek complex and the origins of Clovis at the Debra L. Friedkin site, Texas. *Science* 331:1599–1603.

Waters, M.R., and T.W. Stafford Jr. 2007. Redefining the age of Clovis: implications for the peopling of the Americas. *Science* 315:1122–1126.

Watson, J.T. 2008. Prehistoric dental disease and the dietary shift from cactus to cultigens in Northwest Mexico. *International Journal of Osteoarchaeology* 18(2):202–212.

Watson, D., E.C. Klohnen, A. Casillas, et al. 2004. Match makers and deal breakers: analyses of assortative mating in newlywed couples. *Journal of Personality* 72(5):1029–1068.

Weatherall, I.L., and B.D. Coombs. 1992. Skin color measurements in terms of CIELAB color space values. *Journal of Investigative Dermatology* 99(4):468–473.

Weatherall, D.J., L.H. Miller, D.I. Baruch, et al. 2002. Malaria and the red cell. *Hematology: American Society of Hematology: Education Program*:35–57.

Weaver, I.C., N. Cervoni, F.A. Champagne, et al. 2004. Epigenetic programming by maternal behavior. *Nature Neuroscience* 7(8):847–854.

Wegmann, D., C. Leuenberger, and L. Excoffier. 2009. Efficient approximate Bayesian computation coupled with Markov chain Monte Carlo without likelihood. *Genetics* 182(4):1207–1218. doi:10.1534/genetics.109.102509

Wegmann, D., C. Leuenberger, S. Neuenschwander, et al. 2010. ABCtoolbox: a versatile toolkit for approximate Bayesian computations. *BMC Bioinformatics* 11:116.

Weiss, K.M. 1973. Demographic Models for Anthropology. Memoirs of the Society for American Archaeology No. 27.

Weiss, R.A. 2001. The Leeuwenhoek lecture 2001. Animal origins of human infectious disease. *Philosophical Transactions of the Royal Society B: Biological Sciences* 356(1410):957–977. doi:10.1098/rstb.2001.0838

Weiss, R.A., and A.J. McMichael. 2004. Social and environmental risk factors in the emergence of infectious diseases. *Nature Medicine* 10(12 Suppl):S70–S76. doi:10.1038/nm1150

Weiss, G., and A. von Haeseler. 1998. Inference of population history using a likelihood approach. *Genetics* 149:1541 –1546.

Weitz, C.A., and R.M. Garruto. 2007. A comparative analysis of arterial oxygen saturation among Tibetans and Han born and raised at high altitude. *High Altitude Medicine & Biology* 8(1):13–26.

Weller, M., M. Tanieri, J.C. Pereira, et al. 2012. Consanguineous unions and the burden of disability: a population-based study in communities of northeastern Brazil. *American Journal of Human Biology* 24:835–840.

Wells, R.S., N. Yuldasheva, R. Ruzibakiev, et al. 2001. The Eurasian heartland: a continental perspective on Y-chromosome diversity. *Proceedings of the National Academy of Sciences of the USA* 98:10244–10249.

Welon, Z., and T. Bielicki. 1971. Further investigations of parent-child similarity in stature as assessed from longitudinal data. *Human Biology* 43:517–525.

Wentz, R.K., and N.T. de Grummond. 2009. Life on horseback: paleopathology of two Scythian skeletons from Alexandropol, Ukraine. *International Journal of Osteoarchaeology* 19(1):107–115.

Wertheim, J.O., and M. Worobey. 2009. Dating the age of the SIV lineages that gave rise to HIV-1 and HIV-2. *PLoS Computational Biology* 5(5):e1000377. doi:10.1371/journal.pcbi.1000377

West-Eberhard, M.J. 2003. *Developmental Plasticity and Evolution.* Oxford University Press, Oxford.

Wetterstrand, K.A. 2015. DNA sequencing costs: data from the NHGRI genome sequencing program (GSP). National Human Genome Research Institute. http://www.genome.gov/sequencingcosts

White House: Office of the Press Secretary. 2015. President Obama's Precision Medicine Initiative. January 20. https://www.whitehouse.gov/the-press-office/2015/01/30/fact-sheet-president-obama-s-precision-medicine-initiative

Widdig, A., M.J. Kessler, F.B. Bercovitch, et al. 2015. Genetic studies on the Cayo Santiago rhesus macaques: a review of 40 years of research. *American Journal of Primatology*. doi:10.1002/ajp.22424

Wilde, S., A. Timpson, K. Kirsanow, et al. 2014. Direct evidence for positive selection of skin, hair, and eye pigmentation in Europeans during the last 5,000 y. *Proceedings of the National Academy of Sciences* 111(13):4832–4837. doi:10.1073/pnas.1316513111

Wilder, J.A., E.K. Hewett, and M.E. Gansner. 2009. Molecular evolution of *GYPC*: evidence for recent structural innovation and positive selection in humans. *Molecular Biology and Evolution* 26(12):2679–1687. doi:10.1093/molbev/msp183

Wilks, J., and T. Golovkina. 2012. Influence of microbiota on viral infections. *PLoS Pathogens* 8(5):5–7. doi:10.1371/journal.ppat.1002681

Willems, T., M. Gymrek, G. Highnam, et al. 2014. The landscape of human STR variation. *Genome Research* 24(11):1894–1904.

Williams, A.L., S.B.R. Jacobs, H. Moreno-Macías, et al. 2013. Sequence variants in *SLC16A11* are a common risk factor for type 2 diabetes in Mexico. *Nature* 506(7486):97–101.

Williams, J.M., E.V. Lonsdorf, M.L. Wilson, et al. 2008. Causes of death in the Kasekela chimpanzees of Gombe National Park, Tanzania. *American Journal of Primatology* 70(April):766–777. doi:10.1002/ajp.20573

Williams, G.C., and R.M. Nesse. 1991. The dawn of Darwinian medicine. *Quarterly Review of Biology* 66(1):1–22.

Wilson, R. 1976. Concordance in physical growth for monozygotic and dizygotic twins. *Annals of Human Biology* 3:1–10.

Wilson, M.J., M. Lopez, M. Vargas, et al. 2007. Greater uterine artery blood flow during pregnancy in multigenerational (Andean) than shorter-term (European) high-altitude residents. *American Journal of Physiology: Regulatory, Integrative and Comparative Physiology* 293(3):R1313–R1324.

Wilson, J.N., K. Rockett, B. Keating, et al. 2006. A hallmark of balancing selection is present at the promoter region of interleukin 10. *Genes and Immunity* 7(8):680–683.

Winderickx, J., D.T. Lindsey, E. Sanocki, et al. 1992. Polymorphism in red photopigment underlies variation in colour matching. *Nature* 356:431–433.

Winkler, C.A., G.W. Nelson, and M.W. Smith. 2010. Admixture mapping comes of age. *Annual Review of Genomics and Human Genetics* 11(1):65–89. doi:10.1146/annurev-genom-082509-141523

Witt, K.E., K. Judd, A. Kitchen, et al. 2015. DNA analysis of ancient dogs of the Americas: identifying possible founding haplotypes and reconstructing population histories. *Journal of Human Evolution* 79:105–118.

Wittwer-Backofen, U., and N. Tomo. 2008. From health to civilization stress? In search for traces of a health transition during the early Neolithic in Europe. In: J.-P. Bocquet-Appel, and O. Bar-Yosef (eds.), *The Neolithic Demographic Transition and Its Consequences.* Springer, New York, pp. 501–538.

Wiwchar, D. 2004. Nuu-Chah-Nulth blood returns to West Coast. Ha-Shilth-Sa Newspaper, pp. 1–3.

Wlasiuk, G., and M.W. Nachman. 2010. Adaptation and constraint at toll-like receptors in primates. *Molecular Biology and Evolution* 27(9):2172–2186. doi:10.1093/molbev/msq104

Wobst, H.M. 1974. Boundary conditions for Paleolithic social systems: a simulation approach. *American Antiquity* 39:147–178. doi:10.2307/279579

Wobst, H.M. 1976. Locational relationships in Paleolithic society. *Journal of Human Evolution* 5(1):49–58. doi:10.1016/0047-2484(76)90099-3

Woelk, C.H., S.D.W. Frost, D.D. Richman, et al. 2007. Evolution of the interferon alpha gene family in Eutherian mammals. *Gene* 397(1–2):38–50. doi:10.1016/j.gene.2007.03.018

Woerner, A.E., M.P. Cox, and M.F. Hammer. 2007. Recombination-filtered genomic datasets by information maximization. *Bioinformatics* 23:1851–1853. doi:10.1093/bioinformatics/btm253

Wolfe, N.D., C.P. Dunavan, and J. Diamond. 2007. Origins of major human infectious diseases. *Nature*, 447(7142):279–283.

Wolpoff, M., X.Z. Wu, and A. Thorne. 1984. Modern *Homo sapiens* origins: a general theory of hominid evolution involving the fossil evidence from East Asia. In: F. Smith, and F. Spencer (eds.), *The Origins of Modern Humans: A World Survey of the Fossil Evidence.* Alan R Liss, New York, pp. 411–483.

Wood, E.T., D.A. Stover, M. Slatkin, et al. 2005. The beta-globin recombinational hotspot reduces the effects of strong selection around HbC, a recently arisen mutation providing resistance to malaria. *American Journal of Human Genetics* 77(4):637–642.

Wooding, S. 2005. Evolution: a study in bad taste? *Current Biology* 15:R805–R807.

Wooding, S. 2006. Phenylthiocarbamide: a 75-year adventure in genetics and natural selection. *Genetics* 172:2015–2023.

Wooding, S., B. Bufe, C. Grassi, et al. 2006. Independent evolution of bitter-taste sensitivity in humans and chimpanzees. *Nature* 440:930–934.

Wooding, S., H. Gunn, P. Ramos, et al. 2010. Genetics and bitter taste responses to goitrin, a plant toxin found in vegetables. *Chemical Senses* 35:685–692.

Wooding, S., U.-K. Kim, M.J. Bamshad, et al. 2004. Natural selection and molecular evolution in PTC, a bitter-taste receptor gene. *American Journal of Human Genetics* 74:637–646.

Wooding, S., A.C. Stone, D.M. Dunn, et al. 2005. Contrasting effects of natural selection on human and chimpanzee CC chemokine receptor 5. *American Journal of Human Genetics* 76(2):291–301.

Woolhouse, M.E., J.P. Webster, E. Domingo, et al. 2002. Biological and biomedical implications of the co-evolution of pathogens and their hosts. *Nature Genetics* 32(4):569–577. doi:10.1038/ng1202-569

Workman, P.L., H. Harpending, J.M. Lalouel, et al. 1973. Population studies of Southwestern Indian Tribes VI. Papago population structure: a comparison of genetic and migration analyses. In: N.E. Morton (ed.), *Genetic Structure of Populations*. University Press of Hawaii, Honolulu, pp. 166–194.

Worobey, M., M. Gemmel, D.E. Teuwen, et al. 2008. Direct evidence of extensive diversity of HIV-1 in Kinshasa by 1960. *Nature* 455(7213):661–664.

Wray, N.R., and P.M. Visscher. 2010. Narrowing the boundaries of the genetic architecture of schizophrenia. *Schizophrenia Bulletin* 36(1):14–23.

Wright, S. 1931. Evolution in Mendelian populations. *Genetics* 16:97–159. doi:10.1007/bf02459575

Wright, S. 1943. Isolation by Distance. *Genetics* 28:114–138.

Wright, S. 1951. The genetical structure of populations. *Annals of Eugenics* 15:323–354. doi:10.1111/j.1469-1809.1949.tb02451.x

Wright, S. 1969. *Evolution and the Genetics of Populations*. Vol. 2: The Theory of Gene Frequencies. University of Chicago Press, Chicago.

Wroblewski, E.E., P.J. Norman, L.A. Guethlein, et al. 2015. Signature patterns of MHC diversity in three Gombe communities of wild chimpanzees reflect fitness in reproduction and immune defense against SIVcpz. *PLoS Biology* 13(5):e1002144.

Wu, X., and S. Athreya. 2013. A description of the geological context, discrete traits, and linear morphometrics of the middle Pleistocene hominin from Dali, Shaanxi Province, China. *American Journal of Physical Anthropology* 150(1):141–157.

Wu, T., and B. Kayser. 2006. High altitude adaptation in Tibetans. *High Altitude Medicine & Biology* 7(3):193–208.

Xia, X. 2013. DAMBE5: a comprehensive software package for data analysis in molecular biology and evolution. *Molecular Biology and Evolution* 30(7):1720–1728. doi:10.1093/molbev/mst064

Xia, J., L. Han, and Z. Zhao. 2012. Investigating the relationship of DNA methylation with mutation rate and allele frequency in the human genome. *BMC Genomics* 13(Suppl 8):S7.

Xiang, K., Y.P. Ouzhuluobu, Z. Yang, et al. 2013. Identification of a Tibetan-specific mutation in the hypoxic gene *EGLN1* and its contribution to high-altitude adaptation. *Molecular Biology and Evolution* 30(8):1889–1898.

Xu, S., W. Huang, J. Qian, et al. 2008. Analysis of genomic admixture in Uyghur and its implication in mapping strategy. *American Journal of Human Genetics* 82(4):883–894.

Xu, S., S. Li, Y. Yang, et al. 2011. A genome-wide search for signals of high-altitude adaptation in Tibetans. *Molecular Biology and Evolution* 28(2):1003–1011.

Xu, S., I. Pugach, M. Stoneking, et al. 2012. Genetic dating indicates that the Asian-Papuan admixture through Eastern Indonesia corresponds to the Austronesian expansion. *Proceedings of the National Academy of Sciences of the USA* 109(12):4574–4579. doi:10.1073/pnas.1118892109

Yamaguchi, K., C. Watanabe, A. Kawaguchi, et al. 2012. Association of melanocortin 1 receptor gene (*MC1R*) polymorphisms with skin reflectance and freckles in Japanese. *Journal of Human Genetics* 57(11):700–708. doi:10.1038/jhg.2012.96

Yang, Z. 2014. *Molecular Evolution: A Statistical Approach*. Oxford University Press, Oxford.

Yang, J., B. Benyamin, B.P. McEvoy, et al. 2010. Common SNPs explain a large proportion of the heritability for human height. *Nature Genetics* 42(7):565–569.

Yang, N., H. Li, L.A. Criswell, et al. 2005. Examination of ancestry and ethnic affiliation using highly informative diallelic DNA markers: application to diverse and admixed populations and implications for clinical epidemiology and forensic medicine. *Human Genetics* 118(3):382–392.

Yang, M.A., A.-S. Malaspinas, E.Y. Durand, et al. 2012. Ancient structure in Africa unlikely to explain Neanderthal and non-African genetic similarity. *Molecular Biology and Evolution* 29(10):2987–2995.

Yang, D.Y., and K. Watt. 2005. Contamination controls when preparing archaeological remains for ancient DNA analysis. *Journal of Archaeological Science* 32:331–336.

Yasuda, N., and N.E. Morton. 1967. Studies on human population structure. In: J.F. Crow, and J.V. Neel (eds.), *Proceedings of the Third International Congress of Human Genetics.* Johns Hopkins University Press, Baltimore, pp. 249–265.

Yi, X., Y. Liang, E. Huerta-Sanchez, et al. 2010. Sequencing of 50 human exomes reveals adaptation to high altitude. *Science* 329(5987):75–78.

Yildirim, S., C.J. Yeoman, M. Sipos, et al. 2010. Characterization of the fecal microbiome from non-human wild primates reveals species specific microbial communities. *PLoS One* 5:e13963.

Yim, G., H.H. Wang, and J. Davies. 2007. Antibiotics as signalling molecules. *Philosophical Transactions of the Royal Society B: Biological Sciences* 362(1483):1195–1200. doi:10.1098/rstb.2007.2044

Yong, E. 2013. Americas' natives have European roots. *Nature* doi:10.1038/nature.2013.14213

Yu, N., F.C. Chen, S. Ota, et al. 2002. Larger genetic differences within Africans than between Africans and Eurasians. *Genetics* 161:269–274.

Zacharias, L., W. Rand, and R. Wurtman. 1976. A prospective study of sexual development and growth in American girls: the statistics of menarche. *Obstetrical and Gynecological Survey* 31:325–337.

Zamudio, S., T. Droma, K.Y. Norkyel, et al. 1993. Protection from intrauterine growth-retardation in Tibetans at high-altitude. *American Journal of Physical Anthropology* 91(2):215–224.

Zegura, S.L., T.M. Karafet, L.A. Zhivotovsky, et al. 2004. High-resolution SNPs and microsatellite haplotypes point to a single, recent entry of Native American Y chromosomes into the Americas. *Molecular Biology and Evolution* 21:164–175.

Zeng, J., G. Konopka, B.G. Hunt BG, et al. 2012. Divergent whole-genome methylation maps of human and chimpanzee brains reveal epigenetic basis of human regulatory evolution. *American Journal of Human Genetics* 91(3):455–465.

Zhadanov, S.I., M.C. Dulik, M. Markley, et al. 2010. Genetic heritage and native identity of the Seaconke Wampanoag tribe of Massachusetts. *American Journal of Physical Anthropology* 142(4):579–589. doi:10.1002/ajpa.21281

Zhang, Y. 2003. Transcriptional regulation by histone ubiquitination and deubiquitination. *Genes & Development* 17(22):2733–2740.

Zhang, Z., and B.F. Pugh. 2011. High-resolution genome-wide mapping of the primary structure of chromatin. *Cell* 144(2):175–176.

Zhang, Y., J. Shu, J. Si, et al. 2012. Repetitive elements and enforced transcriptional repression co-operate to enhance DNA methylation spreading into a promoter CpG-island. *Nucleic Acids Research* 40(15):7257–7268.

Zhao, K., Y. Ishida, T.K. Oleksyk, et al. 2012. Evidence for selection at HIV host susceptibility genes in a west central African human population. *BMC Evolutionary Biology* 12:237. doi:10.1186/1471-2148-12-237

Zhao, H., J.-R. Yang, H. Xu, et al. 2010. Pseudogenization of the umami taste receptor gene *Tas1r1* in the giant panda coincided with its dietary switch to bamboo. *Molecular Biology and Evolution* 27:2669–2673.

Zhou, X., C.E. Cain, M. Myrthil, et al. 2014. Epigenetic modifications are associated with inter-species gene expression variation in primates. *Genome Biology* 15:547.

Zhou, D., N. Udpa, R. Ronen, et al. 2013. Whole-genome sequencing uncovers the genetic basis of chronic mountain sickness in Andean highlanders. *American Journal of Human Genetics* 93(3):452–462.

Zhou, W., X. Wang, and M.G. Rosenfeld. 2009. Histone H2A ubiquitination in transcriptional regulation and DNA damage repair. *International Journal of Biochemistry & Cell Biology* 41(1):12–15.

Zhou, G.Q., Y. Zhai, X.J. Dong, et al. 2004. Haplotype structure and evidence for positive selection at the human IL13 locus. *Molecular Biology and Evolution* 21(1):29–35.

Zhu, T., B.T. Korber, A.J. Nahmias, et al. 1998. An African HIV-1 sequence from 1959 and implications for the origin of the epidemic. *Nature* 391(6667):594–597. doi:10.1038/35400

Zhu, T., H. Mo, N. Wang, et al. 1993. Genotypic and phenotypic characterization of HIV-1 patients with primary infection. *Science* 261(5125):1179–1171.

Zhuang, J., T. Droma, S. Sun, et al. 1993. Hypoxic ventilatory responsiveness in Tibetan compared with Han residents of 3,658 m. *Journal of Applied Physiology* 74(1):303–311.

Zlojutro, M., R. Rubicz, and M.H. Crawford. 2009. Mitochondrial DNA and Y-chromosome variation in five eastern Aleut communities: evidence for genetic substructure in the Aleut population. *Annals of Human Biology* 36:511–526.

Zuckerkandl, E., and L. Pauling. 1962. Molecular disease, evolution, and genic heterogeneity. In M. Kasha, and B. Pullman (eds.), *Horizons in Biochemistry*. Academic Press, New York, pp. 189–225.

Index

Italic page numbers refer to figures and **bold** page numbers refer to tables.

A Companion to Anthropological Genetics, First Edition. Edited by Dennis H. O'Rourke.

Printed and bound by CPI Group (UK) Ltd, Croydon, CR0 4YY
16/09/2026
14963461-0007